# STRUCTURED-POPULATION

# MODELS IN MARINE,

# TERRESTRIAL, AND

# FRESHWATER SYSTEMS

Population and Community Biology Series
Principal Editor
Michael B. Usher
*Chief Scientific Advisor and Director of Research and Advisory Services,*
*Scottish Natural Heritage, UK*

Editors

D.L. DeAngelis
*Department of Biology, University of Florida, USA*

B.F.J. Manly
*Director, Centre for Applications of Mathematics and Statistics,*
*Universtiy of Otago, New Zealand*

The study of both populations and communities is central to the science of ecology. This series of books explores many facets of population biology and the processes that determine the structure and dynamics of communities. Although individual authors are given freedom to develop their subjects in their own way, these books are scientifically rigorous and a quantitive approach to analysing population and community phenomena is often used.

Already published

1. **Population Dynamics of Infectious Diseases: Theory and Applications**
   R.M. Anderson (ed.) (1982) 368pp. Hb.

2. **Food Webs**
   S.L. Pimm (1982) 219pp. Hb/Pb.

3. **Predation**
   R.J. Taylor (1984) 166pp. Hb/Pb.

4. **The Statistics of Natural Selection**
   B.F.J. Manly (1985) 484pp. Hb/Pb.

5. **Multivariate Analysis of Ecological Communities**
   P. Digby and R. Kempton (1987) 206pp. Hb/Pb.

6. **Competition**
   P. Keddy (1989) 202pp. Hb/Pb.

7. **Stage-Structured Populations: Sampling, Analysis and Simulation**
   B.F.J. Manly (1990) 200pp. Hb.

8. **Habitat Structure: The Physical Arrangement of Objects in Space**
   S.S. Bell, E.D. McCoy and H.R. Mushinsky (1991) 452pp. Hb.

9. **Dynamics of Nutrient Cycling and Food Webs**
   D.L. DeAngelis (1992) 285pp. Pb.

10. **Analytical Population Dynamics**
    T. Royama (1992) 387pp. Hb.

11. **Plant Succession: Theory and Prediction**
    D.C. Glenn-Lewin, R.K. Peet and T.T Veblen (1992) 361pp. Hb.

12. **Risk Assessment in Conservation Biology**
    M.A. Burgman, S. Ferson and R. Akcakaya (1993) 324pp. Hb.

13. **Rarity**
    K.J. Gaston (1994) 224pp. Hb/Pb.

14. **Fire and Plants**
    W.J. Bond & B.W.van Wilgen (1995) 272pp. Hb.

15. **Biological Invasions**
    M. Williamson (1996) c. 256pp. Pb.

16. **Regulation and Stabilization: Paradigms in Population Ecology**
    P.J. den Boer & J.Reddingius (1996) ca. 400pp. Hb.

17. **Biology of Rarity**
    W. Kunin & K. Gaston (eds) (1996)

18. **Structured-Population Models in Marine, Terrestrial, and Freshwater Systems**
    S. Tuljapurkar & H. Caswell (eds) (1996)

# STRUCTURED-POPULATION MODELS IN MARINE, TERRESTRIAL, AND FRESHWATER SYSTEMS

Edited by

## SHRIPAD TULJAPURKAR

President, Mountain View Research, Los Altos, California

## HAL CASWELL

Biology Department, Woods Hole Oceanographic Institution, Woods Hole, Massachusetts

CHAPMAN & HALL

I(T)P International Thomson Publishing

New York • Albany • Bonn • Boston • Cincinnati • Detroit • London • Madrid • Melbourne
Mexico City • Pacific Grove • Paris • San Francisco • Singapore • Tokyo • Toronto • Washington

## JOIN US ON THE INTERNET
### WWW: http://www.thomson.com
### EMAIL: findit@kiosk.thomson.com

*thomson.com* is the on-line portal for the products, services and resources available from International Thomson Publishing (ITP). This Internet kiosk gives users immediate access to more than 34 ITP publishers and over 20,000 products. Through *thomson.com* Internet users can search catalogs, examine subject-specific resource centers and subscribe to electronic discussion lists.You can purchase ITP products from your local bookseller, or directly through *thomson.com*.

Visit Chapman & Hall's Internet Resource Center for information on our new publications, links to useful sites on the World Wide Web and an opportunity to join our e-mail mailing list. Point your browser to: or **http://www.chaphall.com** or **http://www.chaphall.com/chaphall/lifesce.html** for Life Sciences

A service of **ITP**

Dedicated to Shubha and Anjali, and Solange and Erin
Copyright © 1997 by Chapman & Hall

Printed in the United States of America

For more information, contact:

Chapman & Hall
115 Fifth Avenue
New York, NY 10003

Chapman & Hall
2-6 Boundary Row
London SE1 8HN
England

Thomas Nelson Australia
102 Dodds Street
South Melbourne, 3205
Victoria, Australia

Chapman & Hall GmbH
Postfach 100 263
D-69442 Weinheim
Germany

International Thomson Editores
Campos Eliseos 385, Piso 7
Col. Polanco
11560 Mexico D. F.
Mexico

International Thomson Publishing - Japan
Hirakawacho-cho Kyowa Building, 3F
1-2-1 Hirakawacho-cho
Chiyoda-ku, 102 Tokyo
Japan

International Thomson Publishing Asia
221 Henderson Road #05-10
Henderson Building
Singapore 0315

1 2 3 4 5 6 7 8 9 10 XXX 01 00 99 98 97

Structured-population models in marine, terrestrial, and freshwater
  systems / Shripad Tuljapurkar and Hal Caswell, editors.
        p.   cm.
    Includes bibliographical references and index.
    ISBN 0-412-07271-8
    (pbk. : alk. paper)
    1. Population biology—Mathematical models.   I. Tuljapurkar,
  Shripad, 1951 -  .   II. Caswell, Hal.
  QH352.S86   1996
  574.5'248'0151—dc20                                96-29407
                                                     CIP

**British Library Cataloguing-in-Publication Data available**

To order this or any other Chapman & Hall book, please contact **International Thomson Publishing,**
**7625 Empire Drive, Florence, KY 41042.** Phone: (606) 525-6600 or 1-800-842-3636.
Fax: (606) 525-7778. e-mail: order@chaphall.com.

For a complete listing of Chapman & Hall's titles, send your request to **Chapman & Hall, Dept. BC,**
**115 Fifth Avenue, New York, NY 10003.**

# Contents

# Preface

In the summer of 1993, twenty-six graduate and postdoctoral students and fourteen lecturers converged on Cornell University for a summer school devoted to structured-population models. This school was one of a series to address concepts cutting across the traditional boundaries separating terrestrial, marine, and freshwater ecology. Earlier schools resulted in the books *Patch Dynamics* (S. A. Levin, T. M. Powell & J. H. Steele, eds., Springer-Verlag, Berlin, 1993) and *Ecological Time Series* (T. M. Powell & J. H. Steele, eds., Chapman and Hall, New York, 1995); a book on food webs is in preparation. Models of population structure (differences among individuals due to age, size, developmental stage, spatial location, or genotype) have an important place in studies of all three kinds of ecosystem.

In choosing the participants and lecturers for the school, we selected for diversity—biologists who knew some mathematics and mathematicians who knew some biology, field biologists sobered by encounters with messy data and theoreticians intoxicated by the elegance of the underlying mathematics, people concerned with long-term evolutionary problems and people concerned with the acute crises of conservation biology. For four weeks, these perspectives swirled in discussions that started in the lecture hall and carried on into the sweltering Ithaca night. Diversity may or may not increase stability, but it surely makes things interesting.

This book tries to convey the breadth and some of the excitement of those lectures and discussions. The first part of the book presents the basic methods for structured-population modeling. The second part contains chapters that apply those methods to a variety of organisms, ecosystems, and questions. This part includes chapters based on work done by student groups during and after the school.

Although there are books describing a variety of structured-population models, there has been no comprehensive treatment of all the major types of models. One of our main goals here is to provide such a treatment, in the hope that it will make the entire spectrum of structured-population models more easily available to potential users. Those users include population ecologists, evolutionary biologists, human demographers, and managers concerned with resources, conservation of endangered species, and disease control. Our goal was a book that can usefully serve as both a reference for researchers and a useful text for advanced students.

We have attempted, with noticeable but incomplete success, to make this book readable and as nearly uniform in style as possible—in spite of the diversity of authors and material. This effort was guided by the meticulous editorial work of Jean Doble. Carl Boe provided most of the LaTeX expertise and enormous associated effort. They have our gratitude for the organization and clarity that we hope readers find in this book; all remaining errors may be blamed on us. Cheryl Nakashima did much of the first round of translation into TeX. Without the heroic efforts of this team we would never have transformed a recalcitrant pile of manuscripts into a book.

We are grateful for financial support from the Office of Naval Research (grant N00014-92-J-1527), the National Science Foundation (grant OCE-9302874), and the Center for Applied Mathematics at Cornell University. We thank John Guckenheimer for providing facilities at the Center for Applied Mathematics, and Colleen Martin and Dolores Pendell for invaluable logistic support at Cornell.

Finally, our thanks to the participants in the course. Without their enthusiasm and curiosity, we would have nothing to report.

S.T.
H.C.

PART I

# THEORY AND METHODS

CHAPTER 1

# Structured-Population Models: Many Methods, a Few Basic Concepts

## Hal Caswell, Roger M. Nisbet, André M. de Roos, and Shripad Tuljapurkar

This is a book about structured-population models. If there is a starting point for thinking about populations, it must be the individual. Individual organisms are born, grow, develop, mature, move, reproduce, and eventually die. The rates at which these processes occur determine whether the population increases or decreases, persists or becomes extinct, expands or contracts, fluctuates or remains stable. The environment affects the population through its effects on these individual processes. To model the dynamics of a population, and their response to the environment, requires a link between individual rates (collectively called the "vital rates") and population processes. That link is provided by structured-population models, the subject of this book.

> The vital rates (i.e., birth, growth, development, and mortality rates) on which demography depends describe the development of individuals through the life cycle. The response of these rates to the environment determines population dynamics in ecological time and the evolution of life histories in evolutionary time. (Caswell 1989, p. 1)
>
> One fundamental impediment to elucidating the concept of stress in ecology is that the simplest measurements to undertake are often on *individuals*, while our primary interest is likely to be in effects at the *population* level. ... [A] potentially valuable tool in the study of ecological stress will be *structured population models* which aim to predict the dynamics of a population, given a well-posed dynamic

specification of the response of individual members of that population
to external factors. (Nisbet et al. 1989, p. 80)

In population models the basic unit is the individual. Therefore
it is the task of the model builder to translate his/her knowledge
about mechanisms on the individual level into models for the change
in the number of such individuals. ... What is needed, therefore, is
a modeling methodology which in principle can accommodate any
necessary amount of biological detail and yet is sufficiently near to
the mainstream of applied mathematics that its tools can be brought
to bear. (Metz & Diekmann 1986, p. 3)

Developmental processes cause individuals to differ depending
on their age, size, or developmental stage. That old animals have
different reproductive rates than young ones, or that large plants
experience different mortality risks than small seedlings, is so obvious as to hardly need mention. However, models written in terms of
the total numbers or biomass of a population ignore these individual differences. What makes models "structured" is the inclusion
of such individual differences.

The information necessary to determine the response of an individual to its environment is the state of the individual, or *i*-state.
All the models discussed here assume that all individuals experience the same environment, so that all individuals with the same
*i*-state respond identically. Thus, the state of the population is
given by a distribution function showing the number of individuals
in each *i*-state category.

## 1 What Do Structured Models Look Like?

A structured model of a single population describes the distribution of individuals in the population among the possible categories
of important individual differences. If age differences are most important, then the population is described by an age distribution.
If size reveals more about an individual than does age, the population is better described by a size distribution. In some cases, it
may be necessary to track individuals by both age and size, or by
other descriptors. The important categories that are used to classify individuals in a structured model are called *i*-state variables.
Examples include size, age, and sex. Subsequent chapters provide
many other examples.

The model itself is a mathematical rule that specifies the change
over time of a distribution function, which specifies the numbers
of individuals in each of the stages considered. This rule may be

written in many mathematical forms (and one of the purposes of this book is to show you those forms), but it is always derived from a description of the movement of individuals through their life cycle.

In an important sense, structured population models are *individual-based models*, because they are based on a description of the changes in state of individual organisms. However, that term is now often reserved for a different kind of model: for simulation models that describe populations by keeping track of each individual in the population. The latter, called "configuration" models, keep track of the characteristics of every individual, whereas this book is concerned with "distribution" models, which keep track of the numbers of individuals in each *i*-state category (Caswell & John 1992).

In most situations, a model that specifies changes for every individual can also be written as a model for changes in a distribution function. In the literature of physics, the latter equations are often called "transport" equations or "master" equations (van Kampen 1981; Gardiner 1985; Grünbaum 1994).

There are three main mathematical forms used to describe the dynamics of the distribution function. These will be introduced in detail in Chapters 2–6, but to help provide an overview and vocabulary, we describe them briefly here, listing some of their advantages and disadvantages.

## *Models Using Matrices*

Matrix population models classify a population into discrete stages and project it in discrete time. The state of the population is given by a vector, $\mathbf{n}(t) = [n_1(t), \ldots, n_2(t)]^{\mathrm{T}}$. This vector is projected from time $t$ to $t+1$ by a population projection matrix $\mathbf{A}$:

$$\mathbf{n}(t+1) = \mathbf{A}\mathbf{n}(t). \tag{1}$$

The $ij$th entry of $\mathbf{A}$ tells how many individuals of stage $i$ appear at time $t+1$ per individual of stage $j$ at time $t$; these entries are determined by the stage-specific rates of survival, growth, development, reproduction, etc. Their interpretation depends on how the population has been divided into stages (i.e., on the choice of an *i*-state variable).

Matrix models may be linear and deterministic ($\mathbf{A}$ is a constant matrix), nonlinear ($\mathbf{A}$ changes in response to population density), or stochastic ($\mathbf{A}$ changes over time in a stochastic fashion).

*Advantages.* Matrix models are easy to construct, either from a graphic description of the life cycle or from data on individuals followed over time. They are easy to simulate, since equation (1) can be iterated by simple matrix multiplication. Stochastic matrix models are relatively easy to construct, since they avoid the technical difficulties of continuous-time stochastic processes (Nisbet & Gurney 1982). Matrix models are relatively easy to analyze, especially given modern computer software (MATLAB, Mathematica, MathCad).

*Disadvantages.* Matrix models require discrete stages. This is no problem if natural stages exist (e.g., eggs, larvae, pupae, adults), but it requires some work if they do not (e.g., size or age). All individuals within a stage are treated as identical; this may make the dynamics, especially in nonlinear models, sensitive to the choice of stages.

## Models Using Delay Differential Equations

Delay-differential-equation (DDE) models classify the population into discrete stages, while describing its dynamics in continuous time. They describe the population by a vector of stage abundances, $\mathbf{n}(t) = [n_1(t), \ldots, n_2(t)]$, and write an ordinary differential equation for each of the $n_i$. The equation incorporates time delays to account for the time required for individuals to move through the stages.

Consider a model with two stages. Individuals in stage $i$ die at a rate $\mu_i$. The duration of stage 1 is $\tau_1$, and the rate of production of offspring by stage 2 is $m_2$. Then we can write

$$\frac{dn_1}{dt} = -\mu_1 n_1(t) + m_2 n_2(t),$$

$$\frac{dn_2}{dt} = m_2 n_2(t - \tau_1) e^{-\mu_1 \tau_1} - \mu_2 n_2(t).$$

The time-lag term in the second equation expresses the number of juveniles becoming adults at time $t$ as the number of new juveniles born $\tau_1$ time units previously times the probability that they survive for the $\tau_1$ time units necessary to reach adulthood.

*Advantages.* Delay-differential-equation models lend themselves to parameterization in terms of instantaneous rates of mortality, reproduction, and development. They allow the developmental rate to be studied as a dynamically varying quantity. They seem particularly well suited for models of interacting species (see, e.g., Briggs 1993; Briggs et al. 1993).

*Disadvantages.* DDE models are difficult to simulate numerically, although this problem is alleviated somewhat by the existence of the package SOLVER, which implements an efficient numerical algorithm for the solution of such equations. They do not lend themselves to parameterization from discrete transition data. Stochastic versions can be constructed, but care must be taken in describing stochastic changes in continuous time (Nisbet & Gurney 1982).

*Models Using Partial Differential Equations*

Partial-differential-equation (PDE) models describe the population in terms of a continuous variable (e.g., age or size), in continuous time. The state of the population is given by a function, $n(a,t)$, which gives the density of individuals in stage $a$ at time $t$. The dynamics of $n(a,t)$ are given by two equations. Suppose for now that $a$ is age; the equations are

$$\frac{\partial n(a,t)}{\partial t} + \frac{\partial n(a,t)}{\partial a} = -\mu(a)n(a,t)\,,$$

$$n(0,t) = \int_0^\infty m(a)n(a,t)\,da\,,$$

where $\mu(a)$ and $m(a)$ are the mortality rate and reproductive rate as functions of age. The first of these equations describes changes in $n(a,t)$ as individuals age and die. The second describes the production of newborn individuals through reproduction by the extant population. Similar equations are derived for individuals classified by size or physiological stage (see Chapter 5, by de Roos).

*Advantages.* PDE models can include the most detailed information on the vital rates, because they do not lump individuals into discrete stages or divide time into discrete intervals. They permit coupling to the environment either through direct density dependence of the parameters or through dynamic feedback from resources. Like DDE models, they can take advantage of parametric forms for the vital rates.

*Disadvantages.* PDE models are the most difficult of all three types to study numerically, although the recent development of the "Escalator Boxcar Train" method (de Roos 1988; de Roos et al. 1990, 1992; see also Chapter 5) has made their study much easier. They are analytically difficult; much of the theory available for nonlinear matrix models or DDE models simply does not exist yet for nonlinear PDE models (e.g., de Roos et al. 1990). Although PDE models are natural for continuous $i$-state variables, they are less so for naturally discrete developmental stages, although they can be applied to such cases. Including either demographic or environmental stochasticity in these models is difficult at best. Finally, they do not lend themselves to situations where the vital rates are not given in parametric form.

## 2 Other Kinds of Structure

The focus in this book is mainly on structure in individual vital rates due to differences in age, size, developmental stage, and similar characteristics. Yet biologists are often concerned with structure at different levels or of different kinds. A heirarchy of structure exists in biology between levels of organization, for example, in descending order of complexity, from ecosystems, through communities, species, populations, and individuals, to cells.

Heirarchical structure is not a direct concern of this book, although some of the chapters do examine models that are structured in this way (e.g., by trophic levels, in Chapter 14, by Monger et al.; by species, in Chapter 21, by Hatfield & Chesson). In most cases, however, this book deals mainly with the significance of individual structure at the individual level, and the ways in which this structure affects biological phenomena.

A different kind of population structure is internal, with individuals classified by genotype and/or sex in addition to such things as age and/or size. Some of the work described here is concerned with the ways in which genetic and/or sexual differences interact with structural dynamics to determine biological patterns.

Finally, populations can be structured because of the physical environment. In marine populations (e.g., see Chapter 13, Hofmann), space interacts in important ways with individual behavior and vital rates. Although some of the material in this book considers spatial effects, the emphasis here is on the interplay between space and the dynamics of individual vital rates. For an introduction to the range of issues that spatially structured models seek to address, consult Levin (1974), Levin et al. (1993), Botsford et al. (1994), Hastings and Higgins (1994), and Shigesada et al. (1995).

## 3 Why Include Population Structure?

First, structure in population models has important consequences for dynamics. Because it takes time for an individual to move through the life cycle, population structure defines some of the important time scales for population dynamics. A population responds to environmental changes with time lags that reflect individual development. This gives structured populations a dynamic memory that can lead to oscillatory behavior (damped or otherwise), which does not appear in unstructured models (Jansen et al. 1990). When these oscillatory tendencies are stimulated by externally driven, environmental variability, the resulting dynamics can be profoundly different from those in unstructured models.

Second, because structured-population models are written in terms of the vital rates of individuals, their parameters often have a clear operational definition, and so are amenable to direct measurement, in contrast to parameters in typical unstructured models. The carrying capacity in the simple (and unstructured) logistic equation is notoriously difficult to measure, because it is not defined in terms that let it be measured from individuals. Since the parameters of structured models are defined in terms of individual processes, and environmental factors directly affect individuals, it is relatively easy to define the coupling of the parameters to the environment in structured models.

These features of structured models have already led to rich and sophisticated connections between theoretical models and experimental or observational data. These can be seen, for example, in delay-differential equation models of insects (Gurney et al. 1980; Crowley et al. 1987; Gordon et al. 1988), host-parasite systems (Gordon et al. 1991; Murdoch et al. 1992), or *Daphnia* (Gurney et al. 1990; McCauley et al. 1990). Or the partial-differential-equation models based on physiological processes in *Daphnia* (de Roos et al. 1992; Kooijman and Metz 1984). Or recent applications of matrix models to dissect the demography of plant and animal populations (Aberg 1992*a,b*; Kalisz & McPeek 1992; Silvertown et al. 1992; Brault & Caswell 1993; Horvitz & Schemske 1995). Or the work on *Tribolium* summarized by Costantino and Desharnais (1992), Costantino et al. (1995), and Dennis et al. (1995). Compared with similar analyses a decade ago, these studies clearly include more biologically significant mechanisms, more experimentally measurable parameters, more environmental variability, and more important nonlinear processes. Subsequent chapters in this book provide more examples.

Beyond its consequences for dynamics and implications for connection with data, population structure is important because evolution is fueled by biological diversity. If the differences among individuals include any genetic component, then individual differences lead to evolutionary dynamics in the population. Life-history evolution in stochastic environments is particularly sensitive to population structure (Orzack & Tuljapurkar 1989; Tuljapurkar 1990).

## 4 Analysis of Structured-Population Models

This book introduces and describes various mathematical methods for analyzing structured-population models. A reader unfamiliar with these methods will find it helpful to note that this diversity of methods is designed to answer a relatively small set of questions, and that while these questions can be expressed mathematically, they are, in essence, biological questions.

Population models (like models of other kinds) can be classified as linear or nonlinear, and as deterministic or stochastic. Stochastic models are those that contain a random component; their dynamics are not "totally" random (whatever that means), as some students seem to believe. Mathematically, the analysis of the linear deterministic model is easy, but the analysis of nonlinear stochastic models is hard. The other two combinations fall somewhere in between; there is a set of things that one can do relatively easily, and unless you ask questions that fall outside that set, much progress can be made.

|           | Deterministic | Stochastic |
|-----------|:-------------:|:----------:|
| Linear    | *easy*        | *less easy* |
| Nonlinear | *harder*      | *hard*      |

### Asymptotic Behavior

The biological question "what happens to this population if the mechanisms described here are allowed to operate for a long time?" corresponds to the mathematical question "what happens to this set of equations as $t \to \infty$?" This question concerns asymptotic behavior, an important feature of any model.

For linear models (deterministic and stochastic), the long-term dynamics are characterized by a rate of exponential growth and by a stable (in an appropriate sense) population structure. The convergence of the population, perhaps after some initial fluctuations, to this pattern of growth is referred to as *ergodicity*. Some of the most important theorems in population biology describe the conditions under which ergodicity does, or does not, hold.

In nonlinear deterministic models, the long-term dynamics are often described by *attractors*: sets of population states that are invariant, so that once the population enters them it does not leave. Attractors can be equilibria (the population remains at a constant size with fixed structure), cycles, quasicycles (a kind of almost-periodic motion composed of two or more incommensurate frequencies), or strange attractors.

Much less is known about the asymptotic dynamics of nonlinear stochastic systems, but ergodic results may still hold, and the long-term dynamics can be described by an invariant measure (an unchanging probability distribution) over an attracting set (e.g., a probability cloud around a point).

Asymptotic dynamics in all these flavors of model are often described by an appropriate kind of exponential growth. The exponential rate of this growth is measured by quantities known, in mathematical circles, as Lyapunov exponents. In the linear deterministic case, population growth rate is given by the dominant eigenvalue of a population-projection matrix; this is the dominant Lyapunov exponent. In the linear stochastic case, the long-term average growth rate converges to a quantity that is also a dominant Lyapunov exponent (although it is not a dominant eigenvalue of any matrix). In nonlinear models with an equilibrium point, the convergence of trajectories to that equilibrium (implying stability) is described by the dominant eigenvalue of a local stability matrix; this number is a Lyapunov exponent. If the equilibrium is an unstable point, the Lyapunov exponents describe the rate at which the population moves away. For nonlinear systems with more-complicated attractors, Lyapunov exponents describe the stability, and in chaotic cases the existence, of attractors.

If you watch a mathematical modeler at work, you usually see that one of the first things he or she does is to address asymptotic dynamics in one way or another. It is important to remember that there may be good biological reasons for declaring asymptotic dynamics uninteresting (e.g., the system is so frequently perturbed that it never has a chance to reach its long-term dynamics). In biological oceanography, for example, models of zooplankton populations are usually followed for only a few weeks or months after a specific initial condition (often that at the beginning of the growing season; see Chapter 13, by Hofmann). This may or may not be appropriate, but it is certainly customary.

## Transient Behavior

The analysis of zooplankton for a few months following the spring bloom is an example of transient analysis, which focuses on the short-term consequences of particular initial conditions. In the linear case, the asymptotic dynamics are characterized by exponential growth, but transient dynamics may involve oscillations. Transient dynamics can be characterized by the frequency of these oscillations and their rate of decay as the system approaches its long-term attractor. The rates of convergence to the asymptotic dynamics are also characterized by Lyapunov exponents.

In the nonlinear case, transient behavior is characterized by the rate of approach to attractors (deterministic) or invariant measures (stochastic).

Transient dynamics obviously depend on the initial conditions from which the population starts (after all, if the population starts out at its stable structure, or on an attractor, there are no transients at all). In some nonlinear systems there may be more than one attractor, and part of transient analysis is partitioning the state space into regions ("basins of attraction") within which trajectories converge to each attractor.

## Perturbation Analysis

Perturbation analysis approaches population modeling by asking what would happen if something in the model were to change. It may be that the researcher expects such a change to occur, but that is not necessary; it can be informative to make perturbation analyses describing the results of changes that are known to be biologically unrealistic.

In the linear case, perturbation analysis focuses on sensitivity analysis of the population growth rate, asking "how would the rate of growth change if the vital rates of some stage in the population were to change?" Mathematically, this is done by examining the sensitivity of the growth to changes in the parameters of the model.

In the nonlinear case, local stability analysis focuses on the sensitivity of asymptotic dynamics to changes in the initial conditions by looking at the effect of small perturbations from an equilibrium.

Bifurcation analysis of nonlinear models addresses the effects of parameter changes on the qualitative nature of the attractor (stable vs. unstable, fixed point vs. cycle, etc.). Bifurcations are said to occur when small quantitative parameter changes produce large qualitative changes in the dynamics. In one sense, bifurcation analysis is the next step beyond local stability analysis, in that it describes the fate of large perturbations from an equilibrium.

Most population-model papers, including the chapters in the rest of this book, contain some combination of asymptotic analysis, transient analysis, and perturbation analysis. The mathematical machinery may appear very different depending on whether the model is deterministic or stochastic, linear or nonlinear, matrix or PDE or DDE. The questions, however, are the same: what are the effects of these processes if allowed to operate for a long time, what are their short-term consequences from specific initial conditions, and how will these conclusions change if the parameters or the initial conditions are changed?

## 5  Models and Modeling: Some General Remarks

Anyone reading this book is probably convinced that mathematical models can play a useful role in biology, but it may be useful to summarize our view of the purposes of modeling. The overarching objective of modeling is always to help make sense of the world as we observe it, but the kind of model used in a specific case depends on our immediate goals. Possible goals include prediction, explanation, exploration, synthesis, estimation, and data analysis.

The structure, mathematical complexity, and biological complexity of any model depends on its intended use. There is a spectrum of complexity in the making of models, and models should be made, and judged, on the basis of what they can deliver (Nisbet & Gurney 1982). This is a more pragamatic view than that of Levins (1968), who classified models in terms of reality, precision, and generality.

The structured models in this book provide a systematic way of sorting through the biological components of a possible model, in relation to the scientific objectives one has in mind. The chapters of this book continually revisit the basic themes that are laid out in this chapter, focusing on one or another of a diverse set of goals. Perhaps the greatest advance that has come with using structured models is in the confrontation between models and laboratory or field data, as illustrated in Chapters 5 (de Roos), 6 (Cushing), 9 (Desharnais), and 13 (Hofmann). But these models also serve effectively in addressing "strategic" questions about aspects of evolution (Chapters 8, by Orzack; 10, by Kumm et al.;

TABLE 1. *Types of Models and Their Applications in the Chapters of This Book*

|       | TH  | EV  | LA  | EC       | CO      | DA  |
|-------|-----|-----|-----|----------|---------|-----|
| LDM   | 2   | 11  |     | 7,11,14  | 7,15,16 | 7   |
| LSM   | 3   | 8   |     | 12,18    | 15,16   | 18  |
| NDM   | 2,4 | 17  | 9   | 10,11,12 |         | 19  |
| NDD   | 6   | 20  | 5   | 10,20    |         |     |
| NDP   | 5   |     | 6   | 13,14    |         |     |
| NSD   | 21  |     |     | 21       |         |     |

NOTE.—The labels used are these: TH(eory); EV(olution); LA(boratory); EC(ology); CO(nservation); DA(ta analysis); L(inear) D(eterministic) M(atrix); L(inear) S(tochastic) M(atrix); N(onlinear) D(eterministic) M(atrix); N(onlinear) D(eterministic) D(ifferential) equation, ordinary or delayed; N(onlinear) D(eterministic) P(artial) differential equation; N(onlinear) S(tochastic) D(ifferential) equation, ordinary or delayed.

21, by Hatfield & Chesson), "applied" questions about population assessment (Chapters 15, by Nations & Boyce; 16, by Dixon et al.), and the important subject of model estimation (Chapter 19, by Wood).

The surest path toward the development of models that effectively serve biological understanding is to make, analyze, use, and test a variety of models. This book is dedicated to helping more biologists do this.

## 6 A Guide to the Rest of the Book

This book is divided into two parts. Part I, "Theory and Methods," contains expository chapters describing each of the main kinds of structured-population model. The goal of these chapters is to provide detailed introductions to these classes of model. Part II, "Applications," contains chapters that use structured-population models to investigate experimental, observational, or theoretical problems.

At least in principle, if you master the material in Part I, the application chapters should be a piece of cake. Perhaps cake with some unexpected crunchy bits in it. Table 1 summarizes the location of methods and applications in the book.

## Acknowledgments

We gratefully acknowledge support from National Science Foundation grant OCE-9302874 and Office of Naval Research grant URIP N00014-92-J-1527. This is Woods Hole Oceanographic Institution Contribution 9225.

## Literature Cited

Aberg, P. 1992*a*. A demographic study of two populations of the seaweed *Ascophyllum nodosum*. *Ecology* 73: 1473–1487.

———. 1992*b*. Size-based demography of the seaweed *Ascophyllum nodosum* in stochastic environments. *Ecology* 73: 1488–1501.

Botsford, L. W., C. L. Moloney, A. Hastings, J. L. Largier, T. M. Powell, K. Higgins, and J. F. Quinn. 1994. The influence of spatially and temporally varying oceanographic conditions on meroplanktonic population dynamics. *Deep-Sea Research II*: 107–145.

Brault, S., and H. Caswell. 1993. Pod-specific demography of killer whales (*Orcinus orca*) in British Columbia and Washington State. *Ecology* 74: 1444–1454.

Briggs, C. J. 1993. Competition among parasitoid species on a stage-structured host and its effect on host suppression. *American Naturalist* 141: 372–397.

Briggs, C. J., R. M. Nisbet, and W. W. Murdoch. 1993. Coexistence of competing parasitoid species on a host with a variable life cycle. *Theoretical Population Biology* 44: 341–373.

Caswell, H. 1989. *Matrix Population Models: Construction, Analysis, and Interpretation*. Sinauer, Sunderland, Mass.

Caswell, H., and A. M. John. 1992. From the individual to the population in demographic models. Pp. 36–61 *in* D. DeAngelis and L. Gross, eds., *Individual-Based Models and Approaches in Ecology*. Chapman & Hall, New York.

Costantino, R., and R. Desharnais. 1992. *Population Dynamics and the 'Tribolium' Model*. Springer-Verlag, New York.

Costantino, R. F., J. M. Cushing, B. Dennis, and R. A. Desharnais. 1995. Experimentally induced transitions in the dynamic behavior of insect populations. *Nature* 375: 227–230.

Crowley, P. H., R. M. Nisbet, W. S. C. Gurney, and J. H. Lawton. 1987. Population regulation in animals with complex life-histories: Formulation and analysis of a damselfly model. *Advances in Ecological Research* 17: 1–59.

Dennis, B., R. A. Desharnais, J. M. Cushing, and R. F. Costantino. 1995. Nonlinear demographic dynamics: Mathematical models, statistical methods, and biological experiments. *Ecological Monographs* 65: 261–281.

de Roos, A. M. 1988. Numerical methods for structured population models: The Escalator Boxcar Train. *Numerical Methods for Partial Differential Equations* 4: 173–195.

de Roos, A. M., J. A. J. Metz, E. Evers, and A. Leipoldt. 1990. A size dependent predator-prey interaction: Who pursues whom? *Journal of Mathematical Biology* 28: 609–643.

de Roos, A. M., O. Diekmann, and J. A. J. Metz. 1992. Studying the dynamics of structured population models: A versatile technique and its application to *Daphnia. American Naturalist* 139: 123–147.

Gardiner, C. W. 1985. *Handbook of Stochastic Methods.* 2nd ed. Springer-Verlag, Berlin.

Gordon, D. M., W. S. C. Gurney, R. M. Nisbet, and R. K. Stewart. 1988. A model of *Cadra cautella* growth and development. *Journal of Animal Ecology* 57: 645–658.

Gordon, D. M., R. M. Nisbet, A. de Roos, W. S. C. Gurney, and R. K. Stewart. 1991. Discrete generations in host-parasite models with contrasting life cycles. *Journal of Animal Ecology* 60: 295–308.

Grünbaum, D. 1994. Translating stochastic density-dependent individual behaviour with sensory constraints to an Eulerian model of animal swarming. *Journal of Mathematical Biology* 33: 139–161.

Gurney, W. S. C., S. P. Blythe, and R. M. Nisbet. 1980. Nicholson's blowflies revisited. *Nature* 287: 17–21.

Gurney, W. S. C., E. McCauley, R. M. Nisbet, and W. W. Murdoch. 1990. The physiological ecology of *Daphnia*: Formulation and tests of a dynamic model of growth and reproduction. *Ecology* 71: 716–732.

Hastings, A., and K. Higgins. 1994. Persistence of transients in spatially structured ecological models. *Science* 263: 1133–1136.

Horvitz, C. C., and D. W. Schemske. 1995. Spatiotemporal variation in demographic transitions of a tropical understory herb: Projection matrix analysis. *Ecological Monographs* 65: 155–192.

Jansen, V. A. A., R. M. Nisbet, and W. S. C. Gurney. 1990. Generation cycles in stage structured populations. *Bulletin of Mathematical Biology* 52: 375–396.

Kalisz, S., and M. A. McPeek. 1992. Demography of an age-structured annual: Resampled projection matrices, elasticity analyses, and seed bank effects. *Ecology* 73: 1082–1093.

Kooijman, S. A. L. M., and J. A. J. Metz. 1984. On the dynamics of chemically stressed populations: The deduction of population consequences from effects on individuals. *Ecotoxicology & Environmental Safety* 8: 254–274.

Levin, S. A. 1974. Dispersion and population interactions. *American Naturalist* 108: 207–228.

Levin, S. A., T. M. Powell, and J. H. Steele. 1993. *Patch Dynamics.* Springer-Verlag, New York.

Levins, R. 1968. *Evolution in Changing Environments.* Princeton University Press, Princeton, N.J.

McCauley, E., W. W. Murdoch, R. M. Nisbet, and W. S. C. Gurney. 1990. The physiological ecology of *Daphnia*: Development of a model of growth and reproduction. *Ecology* 71: 703–715.

Metz, J. A. J., and O. Diekmann. 1986. *The Dynamics of Physiologically Structured Populations.* Springer-Verlag, New York.

Murdoch, W. W., R. M. Nisbet, R. F. Luck, H. C. J. Godfray, and W. S. C. Gurney. 1992. *Journal of Animal Ecology* 61: 533–541.

Nisbet, R. M., and W. S. C. Gurney. 1982. *Modelling Fluctuating Populations.* Wiley, New York.

Nisbet, R. M., W. S. C. Gurney, W. W. Murdoch, and E. McCauley. 1989. Structured population models: A tool for linking effects at the individual and population level. *Biological Journal of the Linnean Society* 37: 79–99.

Orzack, S. H., and S. Tuljapurkar. 1989. Population dynamics in variable environments. VII. The demography and genetics of iteroparity. *American Naturalist* 133: 901–923.

Shigesada, N., K. Kawasaki, and Y. Takeda. 1995. Modeling stratified diffusion in biological invasions. *American Naturalist* 146: 229–251.

Silvertown, J., M. Franco, and K. McConway. 1992. A demographic interpretation of Grime's triangle. *Functional Ecology* 6: 130–136.

Tuljapurkar, S. D. 1990. *Population Dynamics in Variable Environments.* Springer-Verlag, New York.

van Kampen, N. G. 1981. *Stochastic Processes in Physics and Chemistry.* North-Holland, Amsterdam.

CHAPTER 2

# Matrix Methods for Population Analysis

## Hal Caswell

Matrix models for structured populations were introduced by P. H. Leslie in the 1940's (Leslie 1945, 1948). Although they are in some ways the simplest of the mathematical approaches to structured population modeling (see Chapter 1), their analysis requires computational power. For this reason, and because ecologists of the day viewed matrix algebra as an esoteric branch of advanced mathematics, they were largely neglected until the late 1960's, when they were rediscovered by ecologists (Lefkovitch 1965) and human demographers (Goodman 1967; Keyfitz 1967). In the 1970's, matrix models were adopted by plant ecologists, who discovered that they could easily handle the complexity of plant life cycles in which size or developmental stage was more important than chronological age in determining the fate of individuals (Sarukhán & Gadgil 1974; Hartshorn 1975; Werner & Caswell 1977).

This chapter introduces the construction and analysis of matrix population models. I will not try to be comprehensive; I have done that elsewhere in book form (Caswell 1989a) and twice in simplified form with a focus on particular taxa (Caswell 1986; McDonald & Caswell 1993). Instead, I try to convey the basics of matrix population models clearly and briefly. Wherever possible, I use different derivations than before (Caswell 1989a), so you may find some new ways to understand the source of some familiar results. I rarely cite my book (Caswell 1989a) (in spite of having done so three times in this paragraph); almost every topic presented here could be followed by the instruction, "see the book for more information." My focus here is on methods; I am sparing in my use of examples, because they can be found in many of the other chapters in this volume.

A note about notation. I use boldface symbols to denote vectors (lower case, as in $\mathbf{n}$) and matrices (uppercase, as in $\mathbf{A}$). Entries of vectors and matrices are lowercase letters with subscripts, so the $i$th entry of $\mathbf{n}$ is $n_i$, and the element in the $i$th row and $j$th column of $\mathbf{A}$ is $a_{ij}$. Sometimes I use parenthetical superscripts to label matrices or vectors. Thus, $\mathbf{A}_m$ or $\mathbf{A}^{(m)}$ might both be used to denote the $m$th in a series of matrices; the $ij$th element of this matrix is written $a_{ij}^{(m)}$. The transpose of the matrix $\mathbf{A}$ is $\mathbf{A}^{\mathrm{T}}$. If $x = a + bi$ is a complex number, the complex conjugate is denoted by $\bar{x} = a - bi$. The complex-conjugate transpose of $\mathbf{A}$ is $\mathbf{A}^*$. The scalar product of two vectors is $\langle \mathbf{x}, \mathbf{y} \rangle = \mathbf{y}^* \mathbf{x}$.

## 1 Formulating Matrix Models

A matrix population model operates in discrete time, projecting a population from $t$ to $t + 1$. The first step in formulating a matrix model is to define the time scale for the projection; this is called the *projection interval*. Models for the same population with different projection intervals may look quite different.

The second step is to choose a set of state variables for individuals (*i-state variables*); these provide the information necessary to determine the response of an individual to the environment, over a projection interval. Examples of $i$-state variables include age, size, developmental stage, and geographical location.

A matrix model uses discrete stages, so the third step is to define a set of discrete categories for each $i$-state variable. Some $i$-states are naturally discrete (e.g., instars), while others are naturally continuous and must be made discrete (e.g., size). Dividing continuous variables into discrete categories involves trade-offs. A model treats all individuals within a category as identical, so creating only a few large categories reduces the accuracy of the $i$-state dynamics. Creating many small categories, alternatively, leads to a large model and may make it hard to estimate parameter values because sample sizes in each category are small.

The stages describe the life cycle, or as much of it as we believe to be demographically important. The next step is to translate them into a model. The life-cycle graph is a useful tool for this translation.

## The Life-Cycle Graph

A life-cycle graph describes the transitions an individual can make, during a projection interval, among the $i$-state categories that define its life cycle. To construct the graph, first draw a numbered point (a "node" in graph-theory terminology) for each $i$-state category. If, for example, size is the $i$-state variable, then the life-cycle graph contains a node for each size class. If age and size are both $i$-state variables, then the life-cycle graph contains a node for each age-size category. Draw arrows, "directed arcs," between nodes to indicate where it is possible for an individual in one stage to contribute individuals to another stage over a single projection interval. The head of the arrow shows the direction in which individuals move. If individuals can contribute in both directions between two stages, draw two arrows, rather than an arrow with a head on both ends. Contributions from one stage to another can result from the movement of individuals from one stage to another (e.g., by growth or aging) or from production of new individuals (e.g., by birth).

With each arrow is associated a coefficient; the coefficient on the arrow from stage $j$ to stage $i$ is denoted $a_{ij}$ (the ordering of the subscripts is important; it corresponds to the arrangement of coefficients in the resulting matrix model). The coefficient $a_{ij}$ gives the number of stage $i$ individuals at $t+1$ per stage $j$ individual at time $t$.

So far, we have made no decisions about the nature of these coefficients; I return to this below.

Figure 1$a$ shows a life-cycle graph for an age-classified model with the age interval equal to the projection interval. Individuals in one age class can contribute to another only by surviving to the next older age class or by reproduction to the first age class. Figure 1$b$ shows the graph for a size-classified model in which individuals may grow to the next size class, remain in their own size class, and possibly reproduce new individuals in the first size class. Suppose some individuals in the first size class grow so rapidly that after one projection interval they are in the third size class. This would require the modification shown in Figure $c$, 1 where an arrow has been drawn from stage 1 to stage 3.

The interpretation of the coefficients depends on the identity of the stages and the processes involved in the transitions. In Figure 1$a$, the $P_i$ are age-specific survival probabilities and the $F_i$ are age-specific fertilities. In Figures 1$b$ and 1$c$, the $G_i$ are probabilities of surviving $and$ growing, the $P_i$ are probabilities of surviving

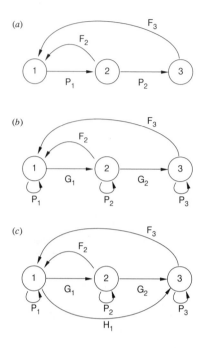

FIGURE 1. *Three life-cycle graphs.* (a) *An age-classifed model with three age classes; the $P_i$ are age-specific survival probabilities, and the $F_i$ are age-specific fertilities.* (b) *A size-classified model with three size classes; the $G_i$ are size-specific probabilities of survival and growth, the $P_i$ are size-specific probabilities of surviving and remaining in the same size class, and the $F_i$ are size-specific fertilities.* (c) *The same life-cycle graph as in* (b), *but with an additional transition ($H_1$) from size class 1 to size class 3.*

and not growing, and the $F_i$ are size-specific fertilities. In Figure 1c, the coefficient $H_1$ is the probability of surviving *and* growing enough to move from size class 1 to size class 3. Demographers use the term *vital rates* to refer collectively to the rates of survival, growth, reproduction, and any other important demographic processes.

## A Set of Difference Equations

The life-cycle graph corresponds directly to a model written as a set of difference equations. For the size-classified graph in Figure 1b, remembering the definitions of the coefficients, the set of equations describing the population is

$$n_1(t+1) = P_1 n_1(t) + F_2 n_2(t) + F_3 n_3(t) \,,$$
$$n_2(t+1) = G_1 n_1(t) + P_2 n_2(t) \,, \tag{1}$$
$$n_3(t+1) = G_2 n_2(t) + P_3 n_3(t) \,.$$

It is worth looking at these equations for a moment. The first states that the number of individuals in stage 1 at $t+1$ is the sum of those remaining in stage 1 from time $t$ and those contributed by reproduction from stages 2 and 3. The second equation states that the number in stage 2 at $t+1$ is the sum of those growing into stage 2 from stage 1 and those remaining in stage 2 from time $t$. The third equation says the same thing for stage 3.

The equations corresponding to Figure 1c are

$$n_1(t+1) = \qquad\quad P_1 n_1(t) + F_2 n_2(t) + F_3 n_3(t) \,,$$
$$n_2(t+1) = \qquad\qquad\quad G_1 n_1(t) + P_2 n_2(t) \,, \tag{2}$$
$$n_3(t+1) = \quad H_1 n_1(t) + G_2 n_2(t) + P_3 n_3(t) \,.$$

It would be possible to write down these equations directly, without using the life-cycle graph, if we were clear about the nature of the possible transitions, which in turn depends on the definition of the stages. But using the life-cycle graph makes it easier, and helps to guard against mistakes in defining the stages and transitions.

## The Matrix Model

The system of difference equations derived from the life-cycle graph can be written more simply in matrix form:

$$\mathbf{n}(t+1) = \mathbf{A}\mathbf{n}(t) \,, \tag{1}$$

where

$$\mathbf{n}(t) = \begin{bmatrix} n_1(t) \\ n_2(t) \\ \vdots \\ n_k(t) \end{bmatrix} \tag{2}$$

is a *stage-distribution vector* and $\mathbf{A}$ is a *population-projection matrix*. The elements of this matrix can be obtained from the system of difference equations or directly from the life-cycle graph: the $ij$th entry of $\mathbf{A}$ is the coefficient on the arrow from stage $j$ to stage $i$. The reason for the order of the subscripts is to guarantee this correspondence.

Applying this rule to the life-cycle graphs in Figure 1 yields

$$\mathbf{A}^{(a)} = \begin{bmatrix} 0 & F_2 & F_3 \\ P_1 & 0 & 0 \\ 0 & P_2 & 0 \end{bmatrix}, \tag{3}$$

$$\mathbf{A}^{(b)} = \begin{bmatrix} P_1 & F_2 & F_3 \\ G_1 & P_2 & 0 \\ 0 & G_2 & P_3 \end{bmatrix}, \tag{4}$$

$$\mathbf{A}^{(c)} = \begin{bmatrix} P_1 & F_2 & F_3 \\ G_1 & P_2 & 0 \\ H_1 & G_2 & P_3 \end{bmatrix}. \tag{5}$$

The age-classified model produces a special matrix, with positive entries only on the first row (fertilities) and the subdiagonal (survival probabilities). Such a matrix is often called a Leslie matrix, in recognition of the early papers of Leslie (1945, 1948).

I have said nothing about how the numerical values of the coefficients $a_{ij}$ are obtained. This obviously important question deserves its own chapter (see Chapter 19, by Wood, for one approach), but here I assume that the matrix is at hand and focus on how to analyze it.

## Types of Matrix Models

The coefficient $a_{ij}$ is the contribution of each individual in stage $j$ to the number of individuals in stage $i$ during one projection interval. What happens in the next projection interval? Depending on the answer to this question, matrix models fall into three classs, each with its own analytical approach.

*Linear, constant-coefficient models.* If the coefficients $a_{ij}$ are constants, the resulting model is linear and time-invariant:

$$\mathbf{n}(t+1) = \mathbf{A}\mathbf{n}(t). \tag{6}$$

This is the simplest case; it can be analyzed in great detail, and it is widely used. But in reality the vital rates are not constants, so the biological interpretation of these results requires great care.

*Nonlinear models.* If the $a_{ij}$ are not constant but depend on the current state of the population, the resulting model is nonlinear:

$$\mathbf{n}(t+1) = \mathbf{A}_n \mathbf{n}(t), \tag{7}$$

where $\mathbf{A}_n$ is the transition matrix evaluated at $\mathbf{n}$. The nonlinearity may result from density dependence (e.g., competition for resources), frequency dependence (e.g., competition for mates), or both.

*Time-varying models.* The coefficients may also change with time, independently of $\mathbf{n}(t)$, resulting in the model

$$\mathbf{n}(t+1) = \mathbf{A}_t \mathbf{n}(t). \tag{8}$$

Deterministic, periodic variation is often used to model seasonality or other kinds of environmental periodicity. Alternatively, the coefficients may vary stochastically, reflecting some random environmental process (see Chapter 3, by Tuljapurkar). Time-varying models may be either linear or nonlinear.

## Objectives of Analysis

The analysis of each of these types of model, although requiring different mathematical tools, addresses a set of similar questions. Imagine that you are in possession of a matrix population model. What you should do with it depends on the question you want to answer.

1. *Transient analyses* describe the short-term dynamics resulting from specific initial conditions.

2. *Asymptotic analyses* describe the long-term dynamics of the population.

   (a) *Population growth rate*: what is the asymptotic rate of population growth or decline?

   (b) *Population structure*: what are the relative abundances of the different stages in the life cycle?

   (c) *Ergodicity*: are the dynamics, including the growth rate and the population structure, asymptotically independent of initial conditions?

   (d) *Attractors* (mainly in density-dependent models): what are the qualitative properties of the asymptotic dynamics (fixed point, cycle, quasiperiodicity, chaos, etc.)?

3. *Perturbation analyses* examine the effects of changes in parameter values or initial conditions on the results of the models.
   Three questions are of particular importance:

   (a) *Sensitivity and elasticity analysis of population growth rate*:
       how does the growth rate respond to changes in vital rates?
   (b) *Stability analysis of equilibria*: if initial conditions are per-
       turbed slightly away from an equilibrium point, does the
       solution return to or depart from the neighborhood of the
       equilibrium?
   (c) *Bifurcation analysis*: what happens to the asymptotic be-
       havior of a nonlinear model as a parameter in the model is
       changed?

The methods used to address these questions depend on the nature
and sometimes on the details of the model, but any population-
modeling project that does not address short-term dynamics, long-
term dynamics, and the effects of perturbations on those dynamics
has left something out.

## 2 Analysis: The Linear Case

We begin with the linear time-invariant model (6), in which $\mathbf{A}$ is
a constant matrix. There are two justifications for spending time
on this model, in spite of the fact that the vital rates of any real
population are certainly not constant. The first is theoretical: understanding population dynamics in the simplest case is a fundamental step in understanding more-complicated cases. The second
is practical: when interpreted as a projection rather than a prediction (Keyfitz 1968; Caswell 1989a), the results of a linear model
provide a valuable characterization of the current environment by
calculating the purely hypothetical consequences of maintaining
that environment forever. Linear matrix population models are
frequently used in this way, as a form of demographic analysis
of vital-rate data, rather than as a prediction of future population
dynamics.

*Exponential Solutions and the Characteristic Equation*

One approach to equation (6) is to conjecture that, like other linear
equations, it has an exponential solution(s),

$$\mathbf{n}(t) = \lambda^t \mathbf{w} \qquad (9)$$

for some fixed vector $\mathbf{w}$. Substituting this into (6) gives

$$\lambda^{t+1}\mathbf{w} = \lambda^t \mathbf{A}\mathbf{w}.$$

A scalar $\lambda$ and a vector $\mathbf{w}$ that satisfy this relation are called an eigenvalue and eigenvector of $\mathbf{A}$, respectively. They must satisfy

$$(\mathbf{A} - \lambda\mathbf{I})\mathbf{w} = 0,$$

which has a nonzero solution for $\mathbf{w}$ only if the determinant of the matrix $\mathbf{A} - \lambda\mathbf{I}$ equals zero. This is called the characteristic equation:

$$\det(\mathbf{A} - \lambda\mathbf{I}) = 0. \qquad (10)$$

*The Spectral Decomposition of* $\mathbf{A}$

An alternative approach is to begin by solving equation (6), starting from a specified initial population $\mathbf{n}(t_0)$. By repeatedly applying (6), we see that $\mathbf{n}(t_0 + 1) = \mathbf{A}\mathbf{n}(t_0)$, $\mathbf{n}(t_0 + 2) = \mathbf{A}^2\mathbf{n}(t_0)$, and in general

$$\mathbf{n}(t_0 + t) = \mathbf{A}^t\mathbf{n}(t_0). \qquad (11)$$

Thus, to understand population dynamics over time we need only understand the behavior of $\mathbf{A}^t$.

One approach to the problem is via the *spectral decomposition* of $\mathbf{A}$, which makes it possible to evaluate any function of $\mathbf{A}$, including $\mathbf{A}^t$. First, note a few facts about the eigenvalues and eigenvectors of a matrix. The vectors $\mathbf{w}$ and $\mathbf{v}$ are right and left eigenvectors of $\mathbf{A}$ if there is a (possibly complex) scalar $\lambda$ such that

$$\mathbf{A}\mathbf{w} = \lambda\mathbf{w}, \qquad (12)$$
$$\mathbf{v}^*\mathbf{A} = \lambda\mathbf{v}^*, \qquad (13)$$

where the asterisk denotes the complex-conjugate transpose. A left eigenvector $\mathbf{v}$ of $\mathbf{A}$, corresponding to $\lambda$, is a right eigenvector of $\mathbf{A}^*$ corresponding to $\bar{\lambda}$; that is,

$$\mathbf{A}^*\mathbf{v} = \bar{\lambda}\mathbf{v}. \qquad (14)$$

The eigenvalues are found as the solutions of the characteristic equation (10).

If $\mathbf{A}$ is a $k \times k$ matrix, the characteristic equation is a polynomial of degree $k$ and has $k$ solutions $\lambda_i$, $i = 1, 2, \ldots, k$. The corresponding eigenvectors are $\mathbf{w}_i$ and $\mathbf{v}_i$, $i = 1, 2, \ldots, k$. I assume that these eigenvalues are all distinct, as seems to be true in practice for population-projection matrices. This assumption guarantees that the right eigenvectors and left eigenvectors, respectively, are linearly independent sets.

Let $\langle \mathbf{w}, \mathbf{v} \rangle = \mathbf{v}^* \mathbf{w}$ denote the scalar product of $\mathbf{w}$ and $\mathbf{v}$. The left and right eigenvectors can always be scaled so that $\langle \mathbf{v}_i, \mathbf{w}_i \rangle = 1$. In addition, the left and right eigenvectors corresponding to different eigenvalues are orthogonal, so that $\langle \mathbf{v}_i, \mathbf{w}_j \rangle = 0$ if $i \neq j$.

Any matrix $\mathbf{A}$ with distinct eigenvalues can be written in the form

$$\mathbf{A} = \lambda_1 \mathbf{Z}_1 + \cdots + \lambda_k \mathbf{Z}_k \, ,$$

where the matrices $\mathbf{Z}_i$, known as the *constituent matrices* of $\mathbf{A}$, are given by

$$\mathbf{Z}_i = \mathbf{w}_i \mathbf{v}_i^* \, . \tag{15}$$

That is, $\mathbf{Z}_i$ is a matrix whose columns are all proportional to $\mathbf{w}_i$ and whose rows are all proportional to $\mathbf{v}_i^*$.

The constituent matrices have two important properties. First,

$$
\begin{aligned}
\mathbf{Z}_i^2 &= \mathbf{w}_i \mathbf{v}_i^* \mathbf{w}_i \mathbf{v}_i^* \\
&= \mathbf{w}_i \langle \mathbf{w}_i, \mathbf{v}_i \rangle \mathbf{v}_i^* \\
&= \mathbf{Z}_i.
\end{aligned}
\tag{16}
$$

(Such matrices are called idempotent.) Second, multiplying two different constituent matrices yields a zero matrix:

$$
\begin{aligned}
\mathbf{Z}_i \mathbf{Z}_j &= \mathbf{w}_i \mathbf{v}_i^* \mathbf{w}_j \mathbf{v}_j^* \\
&= \mathbf{w}_i \langle \mathbf{w}_j, \mathbf{v}_i \rangle \mathbf{v}_j^* \\
&= \mathbf{0} \, .
\end{aligned}
\tag{17}
$$

These properties are useful because, together, they imply that

$$\mathbf{A}^2 = \left( \sum_i \lambda_i \mathbf{Z}_i \right) \left( \sum_j \lambda_j \mathbf{Z}_j \right) = \sum_i \lambda_i^2 \mathbf{Z}_i \, . \tag{18}$$

Multiplying repeatedly by $\mathbf{A}$, it is not hard to see that

$$\mathbf{A}^t = \sum_i \lambda^t \mathbf{Z}_i \, . \tag{19}$$

This result, together with equation (13), yields our desired expression for the dynamics of a population described by (8):

$$\mathbf{n}(t_0 + t) = \sum_i \lambda_i^t \mathbf{Z}_i \mathbf{n}(t_0) \, . \tag{20}$$

The only parts of the right-hand side of (20) that vary with time are the factors $\lambda_i^t$. The behavior of $\lambda_i^t$ depends on the sign of $\lambda_i$ and on whether $\lambda_i$ is real or complex. If $\lambda_i$ is real and positive, $\lambda_i^t$ grows or decays exponentially, depending on whether $\lambda_i$ is greater or less than one. If $\lambda_i$ is real and negative, $\lambda_i^t$ oscillates between positive and negative values, growing or decaying in magnitude depending on whether $|\lambda_i|$ is greater or less than one. If $\lambda_i$ is complex, $\lambda_i^t$ oscillates in a sinusoidal pattern, growing or decaying in magnitude depending on whether $|\lambda_i|$ is greater or less than one.

Since the dynamic properties of the population are determined by the eigenvalues of $\mathbf{A}$, it behooves us to see what we can say, a priori and in general, about these eigenvalues.

### Eigenvalues, Eigenvectors, and the Perron-Frobenius Theorem

We can safely assume that the elements of $\mathbf{A}$ are nonnegative. Negative elements in $\mathbf{A}$ imply the possibility of negative individuals, which I prefer not to deal with. Perhaps surprisingly, this simple assumption tells us almost everything we want to know about the eigenvalues and eigenvectors of $\mathbf{A}$, thanks to a mathematical result known as the Perron-Frobenius theorem. In order to state the theorem, we need two more properties of $\mathbf{A}$: *irreducibility* and *primitivity*.

A matrix $\mathbf{A}$ is irreducible if and only if its life-cycle graph is connected, that is, if there is a path, following the direction of the arrows, from every stage to every other stage. A matrix $\mathbf{A}$ is primitive if and only if there is some integer $k$ such that every element of $\mathbf{A}^k$ is strictly greater than zero. A more biologically revealing criterion is based on the life-cycle graph. Define a *loop* as a sequence of arrows, traversed in the direction of the arrows, that begins and ends at the same node, without passing through any node twice. The matrix $\mathbf{A}$ is primitive if and only if the greatest common divisor of the lengths of the loops in the life-cycle graph is one. Any primitive matrix is also irreducible. Most population-projection matrices encountered in practice are both irreducible and primitive.

What about matrices that are reducible or imprimitive (i.e., not primitive)? A reducible matrix has some stages that make no contribution to some other stages; the life-cycle graph breaks into two (or more) pieces with only one-way communication. The most common example is a life cycle with post-reproductive stages; from

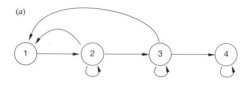

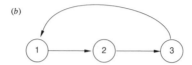

FIGURE 2. (a) A life-cycle graph corresponding to a reducible matrix. Stage 4 represents post-reproductive females; there is no pathway from this stage to any of the earlier stages. (b) A life-cycle graph corresponding to an imprimitive matrix. This is a semelparous age-classified model; individuals that survive to age class 3 reproduce and die.

such a stage there is no pathway back to the part of the life cycle that does reproduce. Figure 2*a* shows an example; a graph of this form appears in a stage-classified model for killer whales (Brault & Caswell 1993). An imprimitive life cycle has some underlying periodicity, so that the loops in the life-cycle graph are all multiples of some common loop length. Imprimitive matrices are sometimes called "cyclic" to reflect this fact. The most common example is a semelparous age-classified life cycle with a fixed age at reproduction (Fig. 2*b*). Only a single loop appears in such a life-cycle graph, with a length determined by the age at reproduction. Some kinds of seasonal models for annual organisms also produce imprimitive matrices, reflecting the periodicity imposed by the annual cycle of the seasons. The graphs in Figure 2 contain no coefficients because reducibility and primitivity depend on the form of the graph but not on the values of the coefficients.

The Perron-Frobenius theorem states that a nonnegative, irreducible, primitive matrix has three properties:

1. a simple (i.e., non-repeated) eigenvalue $\lambda_1$ that is real, positive, and strictly greater in magnitude than any of its other eigenvalues,

2. a right eigenvector $\mathbf{w}_1$ corresponding to $\lambda_1$, which is strictly positive (or can be made so by multiplying by a scalar) and is the only nonnegative right eigenvector, and

3. a left eigenvector $\mathbf{v}_1$ corresponding to $\lambda_1$, which is also strictly positive and is the only nonnegative left eigenvector.

The Perron-Frobenius theorem also describes the eigenvalues and eigenvectors of imprimitive and reducible matrices (see Caswell 1989$a$ and, of course, many matrix texts, e.g., Seneta 1981; Horn & Johnson 1985).

*Demographic Ergodicity*

The eigenvalue $\lambda_1$ (the dominant eigenvalue) plays a central role in the asymptotic analysis of linear matrix models. In equation (20), the growth of $\mathbf{n}(t_0 + t)$ is given by a sum of terms involving the eigenvalues of $\mathbf{A}$ raised to higher and higher powers. Intuitively, as $t$ gets large, $\lambda_1^t$ increases more quickly, or decreases more slowly, than $\lambda_i^t$ for $i \neq 1$. Asymptotically, we expect the growth of the population to be determined by $\lambda_1$, whereas all the eigenvalues contribute to short-term transient behavior. More precisely,

$$
\lim_{t \to \infty} \frac{\mathbf{n}(t_0 + t)}{\lambda_1^t} = \lim_{t \to \infty} \sum_i \left( \frac{\lambda_i}{\lambda_1} \right)^t \mathbf{Z}_i \mathbf{n}(t_0)
$$
$$
= \mathbf{Z}_1 \mathbf{n}(t_0) \qquad (21)
$$
$$
= \mathbf{w}_1 \mathbf{v}_1^* \mathbf{n}(t_0).
$$

This gives the following results on asymptotic dynamics, conditional on the primitivity of $\mathbf{A}$.

1. The population eventually grows geometrically at a rate given by $\lambda_1$ (the *population growth rate* or *rate of increase*).

2. Population structure eventually becomes proportional to $\mathbf{w}_1$ (the *stable stage distribution*).

3. The constant of proportionality relating population structure and $\mathbf{w}$ is a weighted sum of the initial numbers in each stage ($\mathbf{v}_1^* \mathbf{n}(t_0)$). The weights are the elements in $\mathbf{v}_1$; the vector $\mathbf{v}_1$ thus gives the relative contributions of the stages to eventual population size (*not* population growth rate) and is called the *reproductive-value vector*.

The population eventually converges to the stable stage distribution, growing at a rate given by the dominant eigenvalue, regardless of the initial conditions (except, of course, the special case of a zero initial population). The property of forgetting the past and growing at a rate determined by the vital rates rather than by initial conditions is called *ergodicity*.

Because $\lambda_1$, $\mathbf{w}_1$, and $\mathbf{v}_1$ are properties of the vital rates rather than initial conditions, they are widely used as demographic statistics. They can provide valuable insight into the vital rates and the environmental conditions that determine them, but they cannot predict actual population dynamics; everyone "knows" that populations cannot grow geometrically forever. These statistics must be interpreted as projections of what *would* happen if the vital rates were to remain constant, rather than as predictions of what *will* happen. They characterize the present environment, not the future of the population.

Similar ergodic results hold for stochastic models (see Chapter 3) and density-dependent models. In each case, the asymptotic properties provide demographic statistics that are determined by the vital rates rather than by the historical accidents of initial conditions. They can be used just as $\lambda_1$, $\mathbf{w}_1$, and $\mathbf{v}_1$ are used in the linear case.

## 3 Perturbation Analysis

Only rarely are we interested in one precisely specified model. We can usually imagine that the model might change in some way, and would like to know how such changes would affect the results of the analysis. Perturbation analyses address this problem. In density-independent models, perturbation analyses focus on the eigenvalues and eigenvectors, whereas in density-dependent models perturbation analyses focus on the stability and bifurcation of equilibria.

A perturbation analysis of the eigenvalues of a population-projection matrix can answer several questions.

1. What are the effects of potential changes in the vital rates, as might result from strategies designed to protect endangered species (by increasing $\lambda$) or control pest species (by reducing $\lambda$)?

2. Where should efforts to improve the estimates of the vital rates be focused in order to improve the accuracy of the estimate of $\lambda$? All else being equal, the biggest payoff comes from improving the estimates of the vital rates to which $\lambda$ is most sensitive, since errors in those estimates have the biggest effect.

3. Genetic variation produces individuals whose vital rates are perturbed from the overall population values; from these, natural selection chooses those perturbations whose carriers increase most rapidly. Which vital rates are under the greatest selective pressure?

4. Suppose that some environmental differences (either natural or the result of experimental manipulation) have produced differences in the vital rates, and hence in $\lambda$, among two or more populations. How much do each of the vital-rate differences contribute to these observed differences in $\lambda$?

Fortunately, it is easy to calculate the sensitivity of $\lambda$ to a change in any of the vital rates, once we know the eigenvectors. The next subsection presents these calculations. Formulas also exist for the sensitivities of the eigenvectors $\mathbf{w}$ and $\mathbf{v}$, for the sensitivities of $\lambda$ for periodic time-varying models (Caswell & Trevisan 1994), and for the sensitivities of the sensitivities themselves (Caswell 1996$b$).

## Sensitivity and Elasticity of Eigenvalues

The sensitivity of population growth rate to changes in the vital rates can be calculated as the derivative of $\lambda$ to changes in the matrix elements $a_{ij}$. Suppose that $\lambda$, $\mathbf{w}$, and $\mathbf{v}$ satisfy

$$\mathbf{A}\mathbf{w} = \lambda\mathbf{w}, \tag{22}$$

$$\mathbf{v}^*\mathbf{A} = \lambda\mathbf{v}^*, \tag{23}$$

$$\langle \mathbf{w}, \mathbf{v} \rangle = \mathbf{v}^*\mathbf{w} = 1. \tag{24}$$

Now consider a perturbed matrix $\mathbf{A} + d\mathbf{A}$, where $d\mathbf{A}$ is a matrix of small perturbations $da_{ij}$. The eigenvalues and eigenvectors of the new matrix satisfy

$$(\mathbf{A} + d\mathbf{A})(\mathbf{w} + d\mathbf{w}) = (\lambda + d\lambda)(\mathbf{w} + d\mathbf{w}). \tag{25}$$

Expanding the products and eliminating second-order terms yields

$$\mathbf{A}\mathbf{w} + \mathbf{A}(d\mathbf{w}) + (d\mathbf{A})\mathbf{w} = \lambda\mathbf{w} + \lambda(d\mathbf{w}) + (d\lambda)\mathbf{w}, \tag{26}$$

which simplifies to

$$\mathbf{A}(d\mathbf{w}) + (d\mathbf{A})\mathbf{w} = \lambda(d\mathbf{w}) + (d\lambda)\mathbf{w}. \tag{27}$$

Multiplying both sides by $\mathbf{v}^*$ yields

$$\mathbf{v}^*\mathbf{A}(d\mathbf{w}) + \mathbf{v}^*(d\mathbf{A})\mathbf{w} = \lambda\mathbf{v}^*(d\mathbf{w}) + (d\lambda)\mathbf{v}^*\mathbf{w}. \tag{28}$$

The first term on the left-hand side is the same as the first term on the right-hand side (because of eq. 23), and the last term on the right-hand side simplifies to $d\lambda$ (because of eq. 24), leaving

$$\mathbf{v}^* d\mathbf{A}\mathbf{w} = d\lambda. \tag{29}$$

If $d\mathbf{A}$ contains only a single nonzero element $da_{ij}$, a change in only the $ij$th element of $\mathbf{A}$, we obtain the fundamental sensitivity equation:

$$\frac{\partial \lambda}{\partial a_{ij}} = \bar{v}_i w_j. \tag{30}$$

(The bar over $v_i$ has been ignored in most presentations of this formula, including mine. It is irrelevant to the case of the dominant eigenvalue of a population-projection matrix, which always has real eigenvectors, but it must be included for calculations involving other eigenvalues.)

Equation (30) says that the sensitivity of $\lambda$ to changes in $a_{ij}$ is proportional to the product of the reproductive value of stage $i$ and the representation of stage $j$ in the stable stage distribution.

The sensitivity of $\lambda$ to changes in other parameters can be calculated using the chain rule: for some parameter $x$,

$$\frac{\partial \lambda}{\partial x} = \sum_{i,j} \frac{\partial \lambda}{\partial a_{ij}} \frac{\partial a_{ij}}{\partial x}. \tag{31}$$

The sensitivity of $\lambda$ gives the effect of a small additive change in one of the vital rates. The effect of a small *proportional* change in a vital rate is given by the elasticity of $\lambda$:

$$e_{ij} = \frac{a_{ij}}{\lambda} \frac{\partial \lambda}{\partial a_{ij}}. \tag{32}$$

In addition to giving the proportional change in $\lambda$ resulting from a proportional change in the $a_{ij}$, the elasticities also measure the contribution of the $a_{ij}$ to overall population growth rate. To be precise, $\sum_{i,j} e_{ij} = 1$ (for a simple proof, see Mesterton-Gibbons 1993), and $e_{ij}$ can be interpreted as the proportion of $\lambda$ contributed by $a_{ij}$.

Elasticities to other parameters can also be calculated:

$$\begin{aligned} e(x) &= \frac{x}{\lambda} \frac{\partial \lambda}{\partial x} \\ &= \frac{x}{\lambda} \sum_{i,j} \frac{\partial \lambda}{\partial a_{ij}} \frac{\partial a_{ij}}{\partial x}. \end{aligned} \tag{33}$$

The elasticities of $\lambda$ with respect to other parameters do not in general sum to one, and they cannot be interpreted as contributions to population growth rate.

*Sensitivity or Elasticity?*

Some authors seem to believe that sensitivities and elasticities are alternatives, that one is superior to the other, or that one or the other is biased in some way. This is not so; they provide accurate answers to different questions. The difference between them is comparable to the difference between plotting the same set of numbers on arithmetic (sensitivity) or logarithmic (elasticity) axes. Neither kind of graph is wrong, but one or the other may be better at revealing interesting patterns in the numbers. For more discussion, see Chapter 7, by Horvitz et al.

*Life-Table-Response Experiments and Comparative Demography*

Life-table-response experiments (LTRE's) are manipulative experiments or comparative observations in which the dependent variable is a complete set of vital rates (loosely speaking, a life table; Caswell 1989*b*). The different environmental conditions (the "treatments") cause changes in the vital rates, which in turn affect population dynamics. LTRE's are often summarized by using the rate of increase, $\lambda$, as a demographic statistic to integrate the treatment effects on survival and reproduction throughout the life cycle.

Knowing that a treatment produces a particular value of $\lambda$ leaves unresolved the question of how the manifold changes in the vital rates contribute to the effect on growth rate. After all, some vital rates can be changed a great deal without affecting $\lambda$ (e.g., the survival of a post-reproductive age class), whereas small changes in other vital rates produce large changes in $\lambda$. In addition, most environmental factors have differential effects on the different vital rates. A given treatment may affect survival, growth, and fertility differently, with different effects on those rates in different stages.

Treatment effects on $\lambda$ can be decomposed into contributions from the effects on each of the vital rates (Caswell 1989*b*, 1996*a*). This decomposition makes it possible to pinpoint where in the life history the treatment has its greatest impact. The decomposition uses a first-order linear approximation to the effect on $\lambda$. I outline the simplest case here: a set of $M$ treatments $T_m$, $m = 1, \ldots, M$, each of which produces its own matrix $\mathbf{A}_m$ and population growth rate $\lambda^{(m)}$. (I use parenthetical superscripts to denote treatments when subscripts distinguish matrix elements.)

Choose some condition as a reference treatment; this might be the mean of all the treatments, a control treatment, or any other condition of particular interest. The reference matrix is denoted $\mathbf{A}_r$, and treatment effects on $\lambda$ are measured relative to $\lambda^{(r)}$.

To first order, we can write

$$\lambda^{(m)} - \lambda^{(r)} \approx \sum_{ij} \left( a_{ij}^{(m)} - a_{ij}^{(r)} \right) \left. \frac{\partial \lambda}{\partial a_{ij}} \right|_{(\mathbf{A}_m + \mathbf{A}_r)/2} \tag{34}$$

for $m = 1, \ldots, M$. Each term in the summation gives the contribution of the effect of treatment $m$ on one of the vital rates to its overall effect on $\lambda$. That contribution is the product of the vital-rate effect and the sensitivity of $\lambda$ to changes in that vital rate. If either of these terms is small—if the treatment doesn't effect $a_{ij}$ or if $\lambda$ is insensitive to $a_{ij}$—then the contribution of effects on $a_{ij}$ to effects on $\lambda$ is small. The converse is also true.

The sensitivities in (34) must be calculated from some particular matrix; here they are calculated from a matrix "halfway between" the two matrices ($\mathbf{A}_m$ and $\mathbf{A}_r$) being compared. There is some theoretical justification for this (Caswell 1989*b*), and it works well in practice.

The reason for using $\lambda$ as a statistic to summarize the results of an LTRE is that it integrates the diverse and stage-specific effects of the treatments. The decomposition analysis complements this use; it pinpoints the source, within the life cycle, of the effects on $\lambda$. Experience with this kind of analysis shows that it is not safe to assume that the biggest changes in vital rates are responsible for the effects of a treatment on $\lambda$. Without some analysis like (34), half of the information contained in an LTRE is wasted.

Equation (34) describes a simple, one-way, fixed-effect experimental design. The approach has been extended to factorial designs, random designs, and regression designs (Caswell, in press). It can also be applied to statistics other than $\lambda$ (as long as a perturbation theory is available for the statistic) and to parameters other than matrix elements (Caswell 1989*c*, 1996*a*, in press).

*Prospective and Retrospective Analyses*

The preceding subsections outline two ways of using perturbation analysis. Sensitivity and elasticity calculations are *prospective* analyses; they predict the results of perturbations of the vital rates before they happen. Indeed, they even show the results of perturbations that are biologically impossible. They tell nothing about which vital rates are actually responsible for an observed change in $\lambda$. The LTRE decomposition analysis answers this kind of *retrospective* question. It does so by combining sensitivity analysis with information on the actual variance in the $a_{ij}$.

Don't confuse these two kinds of analysis, especially in ambiguous questions such as, "which of the vital rates is most important to population growth?" One way to answer this question is to find the rate with the biggest sensitivity (or elasticity); that rate is most important in the sense that if you were to change all the rates by the same amount (or same proportion), it would have the biggest impact. Another answer is based on an LTRE analysis; the most important vital rate is the one with the variation that makes the biggest contribution to the variability in $\lambda$. The two answers are usually different. Both are valid, but they answer different questions, the first prospective, the second retrospective. Chapter 7, by Horvitz et al., explores these issues further.

## 4 Density-Dependent Matrix Models

The models analyzed so far have been linear and time-invariant; the vital rates are independent of population density or temporal changes. Time-varying models are discussed in Chapter 3; here I consider the inclusion of density dependence, which makes the model nonlinear. Although the mathematical tools are different from those used in the linear case, the focus is still on asymptotic behavior and perturbation. Unlike linear models, density-dependent models do not grow exponentially. Instead, solutions tend to converge to limited subsets of the state space, called attractors. The attractors may be *fixed points* (also called *equilibria*), cycles, or more-complicated structures.

Two kinds of perturbation analysis are important. One asks the effect of perturbing initial conditions. This is of no interest in the linear case, since the ergodic theorem guarantees convergence to the stable population structure and asymptotic growth rate from any nonzero initial condition. In a density-dependent model, however, small perturbations of initial conditions can lead to very different dynamics, depending on the stability of a fixed point. The second kind of perturbation considers changes in the parameters of the model. Often, small changes in parameters leave the qualitative asymptotic behavior unchanged. But sometimes small parameter changes have big effects on asymptotic dynamics: stable trajectories become unstable, fixed points are created or destroyed, attractors change from fixed points to cycles, cycles give way to chaos, etc. These qualitative changes are called bifurcations, and finding them is a major pastime of people who study density-dependent models.

There are many ways of incorporating density dependence in matrix models (Caswell 1989a; see also Cushing 1988; Getz & Haight 1989; Silva & Hallam 1992; Logofet 1993; Dennis et al. 1995). "Density" can be defined as the abundance of a stage or set of stages, as a weighted combination of the abundances of a set of stages, or simply as the total number $N = \sum n_i$. Density can affect the vital rates at many points in the life cycle or at only one. Each of the vital rates may be a different function of the density, or the entire set of vital rates may be affected by density in the same way. In this section, I introduce three models that incorporate simple density effects on reproduction, on growth, and on survival, using them to demonstrate some dynamic consequences of density dependence.

*Basic Formulations*

Consider a two-stage model, with a matrix

$$\mathbf{A} = \left[ \begin{array}{cc} P_1 & F_2 \\ G_1 & P_2 \end{array} \right], \tag{35}$$

and suppose that the matrix entries are given by

$$\begin{array}{rcl} P_1 & = & \sigma_1(1 - \gamma_1), \\ G_1 & = & \sigma_1\gamma_1, \\ P_2 & = & \sigma_2, \end{array}$$

where $\sigma_1$ and $\sigma_2$ are survival probabilities, and $\gamma_1$ is the probability of growth from stage 1 to stage 2, approximated by $1/\tau_1$, where $\tau_1$ is the mean duration of stage 1.

This simple model contains the rates of reproduction, growth, and survival, each of which can be made density-dependent.

*Density-dependent reproduction.* Let fertility at zero density be given by $f_0$, and suppose that fertility declines exponentially with increases in density. Then,

$$F_2(N) = f_0 \exp(-bN), \tag{36}$$

where $b$ is a constant that measures the strength of density dependence. The resulting matrix is

$$\mathbf{A}^{(f)} = \left[ \begin{array}{cc} P_1 & f_0 e^{-bN} \\ G_1 & P_2 \end{array} \right]. \tag{37}$$

*Density-dependent survival.* Let $\sigma_{1,0}$ and $\sigma_{2,0}$ denote the survival probabilities of stages 1 and 2 at zero density. Assume that both juvenile and adult survival probabilities are affected in the same way by density, so that

$$\sigma_1(N) = \sigma_{1,0} \exp(-bN), \tag{38}$$

$$\sigma_2(N) = \sigma_{2,0} \exp(-bN). \tag{39}$$

The resulting matrix is

$$\begin{aligned} \mathbf{A}^{(s)} &= \left[ \begin{array}{cc} \sigma_{1,0}(1-\gamma_1)e^{-bN} & F_2 \\ \sigma_{1,0}\gamma_1 e^{-bN} & \sigma_{2,0}e^{-bN} \end{array} \right] \\ &= \left[ \begin{array}{cc} P_{1,0}e^{-b_1 N} & F_2 \\ G_{1,0}e^{-b_1 N} & P_{2,0}e^{-b_2 N} \end{array} \right]. \end{aligned} \tag{40}$$

*Density-dependent growth.* Suppose that the mean duration of the juvenile stage increases with density:

$$\tau(N) = \tau_0 \exp(bN);$$

then the growth probability is given by

$$\begin{aligned} \gamma(N) &= \frac{1}{\tau(N)} \\ &= \gamma_0 \exp(-bN). \end{aligned} \tag{41}$$

The resulting matrix is

$$\begin{aligned} \mathbf{A}^{(g)} &= \left[ \begin{array}{cc} \sigma_1(1-\gamma_0 e^{-bN}) & F_2 \\ \sigma_1 \gamma_0 e^{-bN} & P_2 \end{array} \right] \\ &= \left[ \begin{array}{cc} \sigma_1(1-\gamma_0 e^{-bN} & F_2 \\ G_{1,0}e^{-bN} & P_2 \end{array} \right]. \end{aligned} \tag{42}$$

The assumption of exponential density dependence ("overcompensation" in the language of fisheries biology) has important dynamic consequences. A strictly compensatory form (e.g., $1/(1+bN)$), or a depensatory form with some range of positive density dependence, may produce different dynamics (Caswell 1989a; Silva & Hallam 1992).

*Equilibria and Stability*

The equilibria of a density-dependent model are population vectors $\hat{\mathbf{n}}$ that satisfy

$$\hat{\mathbf{n}} = \mathbf{A}_{\hat{n}}\hat{\mathbf{n}}. \qquad (43)$$

Depending on the form of the density dependence, there may be more than one equilibrium. In the absence of immigration, $\hat{\mathbf{n}} = 0$ is always an equilibrium. With a little luck, equation (43) can be solved analytically for $\hat{\mathbf{n}}$. More often than not, however, the equilibria must be found numerically.

*Density-dependent reproduction.* The equilibrium $\hat{\mathbf{n}}$ is defined by

$$\begin{align}
\hat{n}_1 &= P_1\hat{n}_1 + f_0 e^{-b\hat{N}}\hat{n}_2, &(44)\\
\hat{n}_2 &= G_1\hat{n}_1 + P_2\hat{n}_2. &(45)
\end{align}$$

The second of these equations can be solved for $\hat{n}_1$:

$$\hat{n}_1 = \frac{1-P_2}{G_1}\hat{n}_2; \qquad (46)$$

substituting this in the first equation yields

$$\hat{N} = -\frac{1}{b}\log\left(\frac{(1-P_1)(1-P_2)}{f_0 G_1}\right). \qquad (47)$$

Combining these two results gives an expression for $\hat{n}_2$,

$$\hat{n}_2 = \frac{G_1\hat{N}}{(1+G_1-P_2)}; \qquad (48)$$

and $\hat{n}_1$ can be found as $\hat{n}_1 = \hat{N} - \hat{n}_2$.

*Density-dependent survival.* The equations defining the equilibrium are

$$\begin{align}
\hat{n}_1 &= P_{1,0}e^{-b\hat{N}}\hat{n}_1 + F_2\hat{n}_2, \\
\hat{n}_2 &= G_{1,0}e^{-b\hat{N}}\hat{n}_1 + P_{2,0}e^{-b\hat{N}}\hat{n}_2.
\end{align} \qquad (49)$$

The second of these equations can be solved for $\hat{n}_2$:

$$\hat{n}_2 = \frac{G_{1,0}e^{-b\hat{N}}}{(1-P_{2,0})e^{-b\hat{N}}}\hat{n}_1. \qquad (50)$$

When substituted into the first equation this eventually yields

$$P_{1,0} - P_{1,0}P_{2,0}e^{-b\hat{N}} + F_2 G_{1,0} + P_{2,0} = e^{b\hat{N}}. \qquad (51)$$

Multiplying both sides by $e^{b\hat{N}}$ gives a quadratic equation in the new variable $y = e^{b\hat{N}}$. Solving this equation for $\hat{y}$ and substituting leads to

$$\hat{N} = \log \hat{y} b \, ,$$

$$\hat{n}_1 = \log \hat{y} \left( b \left( 1 + \frac{G_{1,0}}{\hat{y} - P_{2,0}} \right) \right)^{-1} , \tag{52}$$

$$\hat{n}_2 = \hat{N} - \hat{n}_1 \, .$$

*Density-dependent growth.* The equations to be solved for the equilibrium are

$$\hat{n}_1 = \sigma_1 \left( 1 - \gamma_0 e^{-b\hat{N}} \right) \hat{n}_1 + F_2 \hat{n}_2 \, ,$$
$$\hat{n}_2 = \sigma_1 \gamma_0 e^{-b\hat{N}} \hat{n}_1 + P_2 \hat{n}_2 \, . \tag{53}$$

As before, the second equation can be solved for $\hat{n}_2$,

$$\hat{n}_2 = \frac{\sigma_1 \gamma_0 e^{-b\hat{N}}}{1 - P_2} \hat{n}_1 \, ; \tag{54}$$

substituting this relation into the first equation eventually leads to

$$\hat{N} = -\frac{1}{b} \log \left( \frac{(1 - \sigma_1)(1 - P_2)}{\sigma_1 \gamma_0 (F_2 + P_2 - 1)} \right) \tag{55}$$

None of these analytical solutions for $\hat{\mathbf{n}}$ is particularly informative at first glance. This is typical; only in exceptional circumstances are the formulas for the equilibria in a matrix population model simple enough to appear informative.

An equilibrium is said to be locally stable if small perturbations remain close to the equilibrium, and locally asymptotically stable if small perturbations eventually return to the equilibrium. The adjective "local" refers to the smallness of the perturbations; it may well happen that small perturbations return to an equilibrium, but large ones are attracted to another equilibrium, or to some other kind of attractor (a cycle, for example, or a strange attractor).

The local stability of a fixed point is determined by approximating the nonlinear density-dependent model by a linear model that is accurate for small perturbations. Begin by defining a vector $\mathbf{x}$ of deviations from the equilibrium $\hat{\mathbf{n}}$:

$$\mathbf{x}(t) = \mathbf{n}(t) - \hat{\mathbf{n}} \, . \tag{56}$$

The dynamics of $\mathbf{x}$, near the equilibrium, are given by

$$\mathbf{x}(t+1) = \mathbf{B}\mathbf{x}(t). \tag{57}$$

where the constant matrix $\mathbf{B}$, called the Jacobian matrix, is the linear approximation to the full nonlinear system near the equilibrium.

To illustrate, let us write a two-dimensional matrix model as

$$
\begin{aligned}
n_1(t+1) &= a_{11}n_1(t) + a_{12}n_2(t) = f_1(\mathbf{n}), \tag{58} \\
n_2(t+1) &= a_{21}n_1(t) + a_{22}n_2(t) = f_2(\mathbf{n}), \tag{59}
\end{aligned}
$$

with the understanding that $a_{ij} = a_{ij}(\mathbf{n})$.

The Jacobian matrix $\mathbf{B}$ is

$$
\mathbf{B} = \left[ \begin{array}{cc} \frac{\partial f_1}{\partial n_1} & \frac{\partial f_1}{\partial n_2} \\[2mm] \frac{\partial f_2}{\partial n_1} & \frac{\partial f_2}{\partial n_2} \end{array} \right]. \tag{60}
$$

Differentiating $f_1$ gives

$$
\frac{\partial f_1}{\partial n_1} = a_{11} + n_1 \frac{\partial a_{11}}{\partial n_1} + n_2 \frac{\partial a_{12}}{\partial n_1}, \tag{61}
$$

$$
\frac{\partial f_1}{\partial n_2} = a_{12} + n_1 \frac{\partial a_{11}}{\partial n_2} + n_2 \frac{\partial a_{12}}{\partial n_2}. \tag{62}
$$

The derivatives of $f_2$ have the same form. Thus, $\mathbf{B}$ is given by

$$
\begin{aligned}
\mathbf{B} &= \mathbf{A}_{\hat{n}} + \left[ \begin{array}{cc} n_1 \frac{\partial a_{11}}{\partial n_1} + n_2 \frac{\partial a_{12}}{\partial n_1} & n_1 \frac{\partial a_{11}}{\partial n_2} + n_2 \frac{\partial a_{12}}{\partial n_2} \\[2mm] n_1 \frac{\partial a_{21}}{\partial n_1} + n_2 \frac{\partial a_{22}}{\partial n_1} & n_1 \frac{\partial a_{21}}{\partial n_2} + n_2 \frac{\partial a_{22}}{\partial n_2} \end{array} \right] \\[3mm]
&= \mathbf{A}_{\hat{n}} + \left[ \begin{array}{cc} \frac{\partial \mathbf{A}}{\partial n_1} \hat{\mathbf{n}} & \frac{\partial \mathbf{A}}{\partial n_2} \hat{\mathbf{n}} \end{array} \right]. \tag{63}
\end{aligned}
$$

All of the derivatives are evaluated at the equilibrium.

Equation (63) makes the calculation of the Jacobian straightforward for matrix population models (it is due to Beddington 1974). Given the equilibrium vector $\hat{\mathbf{n}}$, calculate the derivatives of $\mathbf{A}$ with respect to each of the $n_i$ and multiply these matrices by $\hat{\mathbf{n}}$; put the resulting vectors as columns in a matrix, and add this to $\mathbf{A}$ evaluated at the equilibrium.

The equilibrium $\hat{\mathbf{n}}$ is asymptotically stable if the eigenvalues of $\mathbf{B}$ are all less than one in magnitude (or are "within the unit circle," referring to the circle with radius one in the complex plane). It is unstable if any of the eigenvalues are outside the unit circle. An eigenvalue falling exactly on the unit circle (i.e., with magnitude exactly equal to one) signals a transition from stability to instability, or vice versa, and is considered in the next section.

Remember that (63) describes the dynamics of perturbations, not the dynamics of numbers of individuals. Thus, the entries of $\mathbf{x}$ can be negative, and the matrix $\mathbf{B}$ often contains negative entries. The Perron-Frobenius theorem is of little use in evaluating the eigenvalues of $\mathbf{B}$, except in special cases (DeAngelis et al. 1980; Caswell 1989$a$, Example 9.3), and the eigenvalue of largest magnitude may be negative or complex.

The Jacobian matrices for our three density-dependent models can be written down easily. Note that all density effects depend on $N = n_1 + n_2$, so that

$$\frac{\partial a_{ij}}{\partial n_1} = \frac{\partial a_{ij}}{\partial n_2} = \frac{\partial a_{ij}}{\partial N}. \tag{64}$$

Thus, the matrix $\mathbf{B}$ reduces to

$$\mathbf{B} = \mathbf{A}_{\hat{n}} + \frac{\partial \mathbf{A}}{\partial N} \begin{bmatrix} \hat{n}_1 & \hat{n}_1 \\ \hat{n}_2 & \hat{n}_2 \end{bmatrix}. \tag{65}$$

The resulting Jacobian matrices for the three example models are as follows: for density-dependent reproduction,

$$\mathbf{B} = \begin{bmatrix} P_1 & f_0 e^{-b\hat{N}} \\ G_1 & P_2 \end{bmatrix} - b f_0 e^{-b\hat{N}} \begin{bmatrix} \hat{n}_2 & \hat{n}_2 \\ 0 & 0 \end{bmatrix}; \tag{66}$$

for density-dependent survival,

$$\mathbf{B} = \begin{bmatrix} P_{1,0} e^{-b\hat{N}} & F_2 \\ G_{1,0} e^{-b\hat{N}} & P_{2,0} e^{-b\hat{N}} \end{bmatrix}$$

$$- b e^{-b\hat{N}} \begin{bmatrix} P_{1,0}\hat{n}_1 & P_{1,0}\hat{n}_1 \\ G_{1,0}\hat{n}_1 + P_{2,0}\hat{n}_2 & G_{1,0}\hat{n}_1 + P_{2,0}\hat{n}_2 \end{bmatrix};$$

and for density-dependent growth,

$$\mathbf{B} = \begin{bmatrix} \sigma_1(1 - \gamma_0 e^{-b\hat{N}}) & F_2 \\ G_{1,0} e^{-b\hat{N}} & P_2 \end{bmatrix}$$

$$+ b G_{1,0} e^{-b\hat{N}} \begin{bmatrix} \hat{n}_1 & \hat{n}_1 \\ -\hat{n}_1 & -\hat{n}_1 \end{bmatrix}.$$

In each case, $\hat{N}$ and $\hat{\mathbf{n}}$ are evaluated using the appropriate equilibrium formulas. Note that I have not attempted to insert the formulas for the equilibria. In general, it is a lucky circumstance in which such expressions simplify usefully. More often, the best that can be done is to write the Jacobian in terms of the equilibria and to carry out further analyses numerically.

In studying a linear model, the response of the eigenvalues to changes in parameters is usually more interesting than their values for one specific parameter set. Similarly, in studying a nonlinear model it is usually more interesting to see how the stability properties change as parameters are varied than to characterize stability for one parameter set. A qualitative change in stability is called a bifurcation.

*Bifurcations*

Roger Tory Peterson published the first edition of *A Field Guide to the Birds* in 1934. At the time, the standard guide for the would-be birdwatcher was Chapman's (1932) *Handbook of Birds of Eastern North America*, first published in 1895. Chapman's book contained detailed dichotomous keys to the bird species, most of which would be useless without having the bird in one hand (and, most likely, a smoking shotgun in the other). Peterson's innovation was to sacrifice rigor but to focus on characters that would distinguish most of the species, most of the time, in the field.

My goal in this section is to provide a kind of field guide to bifurcations of equilibria of nonlinear matrix models. I describe the basic ideas of bifurcation theory, without assuming that my reader is equipped with the powerful tools of rigorous bifurcation theory, but assuming the ability to conduct numerical simulations. The question is, what to look for that will help to identify the bifurcations that occur.

Be forewarned that, just as there are situations where Peterson had to admit that a reliance on field marks was misleading or impossible, there are no doubt situations in which my suggestions here will be incorrect. (In a notorious case, Chapman distinguished the Alder and Least flycatchers of the genus *Empidonax* on the basis of whether the wing is longer or shorter than 2.60 inches and whether the lower mandible is flesh-colored or strongly tinged with brownish. Peterson, by contrast, said that "it is quite risky to attempt to tell them apart by mere variations in color" and relied on calls during the breeding season.) In such cases, you can make use of the many detailed treatments of bifurcation theory (two excellent recent examples are Wiggins 1990 and Hale & Koçak 1991; see also the review paper Whitley 1983) that provide the mathamatical analogue of the birdwatcher's shotgun. See Chapter 6, by Cushing, for an introduction to the application of this theory to matrix models.

*Types of bifurcations.*    An equilibrium point can bifurcate when it loses stability in response to a change in a parameter, that is, when the dominant eigenvalue $\lambda$ of the Jacobian matrix $\mathbf{B}$ in equation (63) is equal to one. This can happen in three ways:

$$\begin{aligned} \lambda &= 1, & &\text{a ``+1 bifurcation,''} \\ \lambda &= -1, & &\text{a ``-1 bifurcation,''} \\ \lambda &= a \pm bi, & &\text{a ``complex conjugate pair'' bifurcation.} \end{aligned} \qquad (67)$$

When an eigenvalue (or a complex conjugate pair) crosses the unit circle, the others must still be inside, or there would be no change in stability, since the equilibrium would already be unstable. (I am ignoring the rare cases in which more than one eigenvalue crosses the unit circle simultaneously.) Perhaps surprisingly, the bifurcation does not depend on what those other eigenvalues are doing, nor on how many of them there are. The mathematical expression of this fact is the Center Manifold Theorem (e.g., Wiggins 1990). Roughly, it says that the state space can be divided into two parts, one associated with the eigenvalue leaving the unit circle and the second associated with all the other eigenvalues. The dynamics on the second part remain stable, while the bifurcation can be studied on the first part. Thus all the basic behaviors of +1 and −1 bifurcations can be studied in one-dimensional models, while the Hopf bifurcation can be studied in two-dimensional models.

What different bifurcations are possible in response to a parameter change? First, +1 bifurcations are of three types.

1. At a *transcritical bifurcation*, two fixed points, one stable and the other unstable, collide and exchange stability. This occurs in all density-dependent matrix models as a bifurcation of the zero equilibrium. On one side of the bifurcation the zero equilibrium is stable. An unstable equilibrium exists, but it is negative and usually ignored in our thinking about populations. However, it does exist. On the other side of the bifurcation, the zero equilibrium is unstable, and there is a stable positive equilibrium. See Chapter 6, by Cushing, for a detailed description of this bifurcation.

2. On one side of the *saddle-node* bifurcation, there is no fixed point at all. At the bifurcation point, two fixed points appear, one stable and the other unstable. What you see in a saddle-node bifurcation depends on where you look. If you follow the stable fixed point, it suddenly disappears (as it collides with the

unstable fixed point of which you were blissfully unaware), and trajectories move off to some other attractor in the state space.

Saddle-node bifurcations occur in population models when there is an Allee effect (also known as depensatory density dependence) present, that is, when, at low densities, survival or fertility depends *positively* on density. Typically, such a population has three fixed points: zero, an intermediate critical density, and a high density. Zero and the high density are both stable fixed points, while the intermediate critical density is unstable. Populations below the critical density decay to extinction, and those above it increase to the high-density fixed point. As some parameter is varied, the high-density fixed point and the intermediate-density fixed point collide and annihilate each other; the population at that point crashes to extinction. Cushing (Chapter 6) shows an example.

3. On one side of a *pitchfork* bifurcation, there is one stable fixed point. On the other side, one unstable fixed point is surrounded by two new, stable fixed points. At the bifurcation, the two stable fixed points appear close to the now-unstable fixed point; a trajectory may approach either of the two stable fixed points, depending on initial conditions. The canonical examples of pitchfork bifurcations occur in one-dimensional cubic maps. These maps appear in some genetic models (May 1979), but pitchfork bifurcations rarely appear in density-dependent matrix models.

Second are $-1$ bifurcations, or *flip* bifurcations. At this type of bifurcation point, a stable fixed point becomes unstable, and a stable 2-cycle appears. Just beyond the bifurcation point, the amplitude of the 2-cycle is small, and it surrounds the unstable fixed point.

The third type of bifurcation develops from complex-conjugate pairs of eigenvalues and is also called a Hopf or Naimark-Sacker bifurcation. As a complex conjugate pair of eigenvalues leave the unit circle, the fixed point loses its stability and an *invariant circle* forms around the now-unstable fixed point: An invariant circle is a continuous closed curve (not necessarily an exact circle). Any point on this curve is mapped to another point on the curve, hence the term "invariant."

Trajectories on the invariant circle may be periodic, or they may be quasiperiodic, rotating around the circle without ever repeating themselves, depending on the location of the eigenvalues when they cross the unit circle at the bifurcation. In general, trajectories immediately following a Hopf bifurcation are quasiperiodic, containing two periods that are not integer multiples of each other (see example below).

There are two other possibilities. One is called strong resonance and occurs when the bifurcation point happens at

$$\lambda = e^{2\pi k i}, \qquad \text{where} \qquad k = 1, \tfrac{1}{2}, \tfrac{1}{3}, \tfrac{1}{4}. \qquad (68)$$

The second is called *weak resonance* and occurs when

$$\lambda = e^{2\pi i p/q}, \qquad (69)$$

where $p/q$ is a ratio of small integers but does not equal any of the first four roots of unity. In this case, the trajectory on the invariant circle has period $q$.

Strong resonance is complicated. Mathematicians say things like "the strong resonances ... exhibit rather different behavior, the details of which are not yet fully understood" (Whitley 1983), or "the dynamics of such maps ... can be exceedingly complicated and the answer is not yet completely known" (Hale & Koçak 1991). I cannot improve on that, but it appears that bifurcations with strong resonance often have trajectories with period $1/k$ lurking in their dynamics, often mixed up with quasiperiodic and, eventually, chaotic dynamics (for an example, see Guckenheimer et al. 1977).

*Examples: bifurcations of the zero equilibrium.* The three simple density-dependent models in this section provide examples of these bifurcations. Zero is an equilibrium of all three models; to examine its bifurcations, evaluate the Jacobian matrix at zero and track the magnitude of its eigenvalues as parameters in the matrix change. In the examples below, I show bifurcations resulting from changes in reproductive output $F_2$ (or $f_0$ in the model with density-dependent fertility). Note, however, that Cushing (1988; Chapter 6) has shown in general that the bifurcation parameter can be taken to be the net reproductive rate (the expected number of offspring produced by an individual over its lifetime); he has shown how to calculate this quantity for arbitrary stage-structured matrices.

The Jacobian matrix for the zero equilibrium is particularly simple, because the second term on the right-hand side of equation (63) is zero. Thus, the Jacobian is simply the projection matrix evaluated at zero, and it is the same in all three examples.

Figure 3 shows the magnitude of the dominant eigenvalue of $\mathbf{B}$ as a function of $F_2$; it exceeds one when $F_2 \approx 0.8$. The lower panel shows the location in the complex plane of the dominant

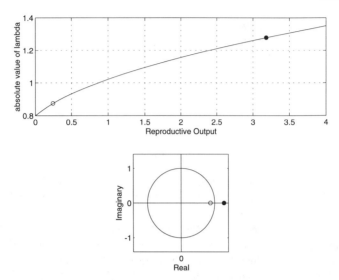

FIGURE 3. *The bifurcation of the zero equilibrium for the three two-stage examples.* Upper panel, *The magnitude of the dominant eigenvalue of the Jacobian matrix* **B** *as a function of the reproductive output* $F_2$. Lower panel. *The location, relative to the unit circle in the complex plane, of the eigenvalue just before* (open circle) *and after* (closed circle) *the bifurcation. Parameter values:* $\sigma_1 = 0.6$, $\sigma_2 = 0.8$, $\gamma_{1,0} = 0.2$, $b = 1$.

eigenvalue before and after the bifurcation; this is a $+1$ bifurcation since the eigenvalue leaves the unit circle at $+1$. Figure 4 shows the equilibrium (for total population size $N = n_1 + n_2$) as a function of $F_2$. The pattern is typical of a transcritical bifurcation. Note that all three models bifurcate from zero at the same value of $F_2$, because all three matrices are identical evaluated at $\hat{\mathbf{n}} = 0$, even though they differ in their subsequent behavior.

*Bifurcations of the positive equilibrium.* What happens to the stable positive equilibrium that appears after the zero equilibrium becomes unstable? Each of the three models shows a different bifurcation pattern. I show examples for each model, using $F_2$ (or $f_0$ in the model with density-dependent fertility) as the bifurcation parameter. There is no reason to expect that this exhausts the possibilities; other types of bifurcations may well be produced by varying other parameters.

For density-dependent reproduction, the bifurcation occurs when $f_0 \approx 95$. The eigenvalue leaves the unit circle at $-1$ (Fig. 5),

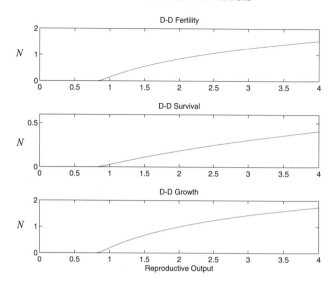

FIGURE 4. *Bifurcation plots for the three two-stage examples, showing the equilibrium value of $N = n_1 + n_2$ as a function of reproductive output ($F_2$; in the model with density-dependent fertility, $f_0$). Other parameter values as in Figure 3.*

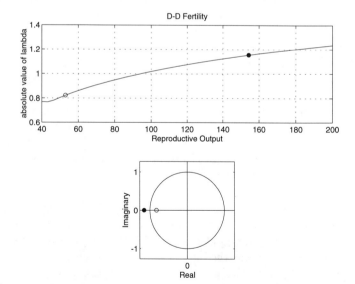

FIGURE 5. *The bifurcation of the positive equilibrium for the density-dependent fertility model as a function of reproductive output $f_0$. Interpretation and parameter values as in Figure 3.*

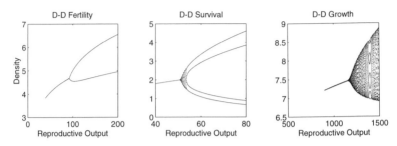

FIGURE 6. *Bifurcation plots for the three two-stage examples, showing the long-term behavior of total population size ($N = n_1 + n_2$) as a function of reproductive output ($F_2$; for the model with density-dependent fertility, $f_0$). Other parameters as in Figure 3.*

implying a flip bifurcation, as seen in Figure 6$a$. The fixed point is replaced by a 2-cycle.

For the model with density-dependent survival, the equilibrium becomes unstable when $F_2 \approx 50$. The eigenvalue leaves the unit circle very near (but not quite on) the imaginary axis (Fig. 7). This implies a Hopf bifurcation, but one very close to strong resonance. The bifurcation pattern is shown in Figure 6$b$. Just after the bifurcation point, trajectories are quasiperiodic on an invariant circle. As $F_2$ increases further, the invariant circle begins to look more like an invariant square, finally collapsing to a stable 4-cycle (Fig. 8). See Wikan and Mjolhus (1995) for an exploration of this period-4 behavior in a fully age-classified model with density-dependent survival.

For the model with density-dependent growth, the bifurcation occurs when $F_2 \approx 1200$. This time, the eigenvalues leave the unit circle as a complex conjugate pair (Fig. 9) that is not one of the first four roots of unity. The result is a Hopf bifurcation to quasiperiodic dynamics on an invariant circle (Fig. 10).

*Subcritical and supercritical bifurcations.* The flip, pitchfork, and Hopf bifurcations come in two varieties, the supercritical (which I have discussed so far) and the subcritical. For purposes of this discussion, suppose that the bifurcation happens when $f = f_c$, and that the fixed point is stable for $f < f_c$ and unstable for $f > f_c$. In the supercritical bifurcations, the unstable fixed point for $f > f_c$ is accompanied by a stable 2-cycle, two stable fixed points, or a stable invariant circle. In the subcritical bifurcations, the 2-cycle,

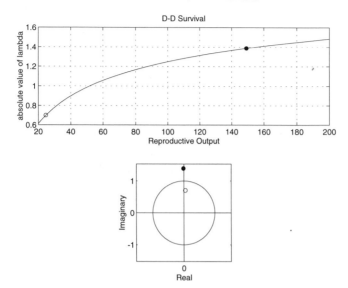

FIGURE 7. *The bifurcation of the positive equilibrium for the model with density-dependent survival as a function of reproductive output $F_2$. Interpretation and parameter values as in Figure 3.*

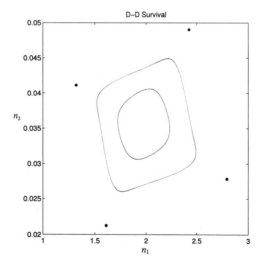

FIGURE 8. *Phase portraits for the asymptotic dynamics of the model with density-dependent survival. The inner invariant curve is for $F_2 = 52$; the outer curve, $F_2 = 53$; the 4-cycle, $F_2 = 55$. Other parameters as in Figure 3.*

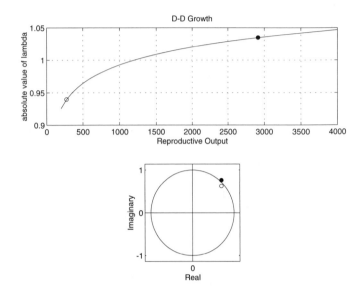

FIGURE 9. *The bifurcation of the positive equilibrium for the model with density-dependent growth as a function of reproductive output $F_2$. Interpretation and parameter values as in Figure 3.*

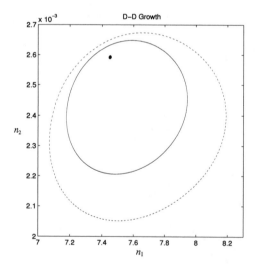

FIGURE 10. *Phase portraits for the asymptotic dynamics of the model with density-dependent growth. The inner equilibrium point is for $F_2 = 1150$; the invariant curves are for $F_2 = 1250$ and $F_2 = 1300$. Other parameters as in Figure 3.*

the two fixed points, or the invariant circle exist for $f < f_c$, but they are unstable. At the bifurcation point $f_c$, the original fixed point becomes unstable and the other structures disappear.

What happens after a subcritical bifurcation (i.e., for $f > f_c$) is hard to predict. Since the fixed point becomes unstable at $f_c$, trajectories do not remain close by, but no nearby stable fixed points, cycles, or invariant circles appear to attract trajectories. Therefore, solutions leave the vicinity of the original fixed point, and their dynamics depend on the location of other attractors in the system.

Subcritical bifurcations are not as common as supercritical ones, but they cannot be ignored. Neubert and Kot (1992) found that, in four discrete predator-prey models in which the prey alone exhibited a supercritical flip bifurcation, the system with both predator and prey exhibited a subcritical flip bifurcation, leading to extinction of the predator. Guckenheimer et al. (1977) and Levin (1981) found subcritical bifurcations in age-structured models with strongly resonant Hopf bifurcations, and Dennis et al. (1995) found a subcritical bifurcation in a stage-structured model fit to laboratory data on *Tribolium* populations.

Analytical methods exist for determining whether a bifurcation is supercritical or subcritical, on the basis of terms in the Taylor series expansion of the model near the fixed point. However, these methods require that the model first be reduced to one or two dimensions using the Center Manifold Theorem, which is not always easy.

*On beyond the first bifurcation.*   The positive fixed point of a matrix population model usually first loses stability through a supercritical flip or supercritical Hopf bifurcation and, then, is replaced by a 2-cycle or a quasiperiodic trajectory on an invariant circle. As the bifurcation parameter increases further, these attractors go through a series of bifurcations, often leading to chaos. The flip bifurcation is typically followed by a series of period-doublings. Successively higher periods are stable for smaller ranges of the bifurcation parameter; eventually, trajectories become chaotic. The resulting bifurcation diagram is similar to the familiar diagram for the discrete logistic equation.

Quasiperiodic solutions produced by a Hopf bifurcation usually go through a series of frequency locking events in which trajectories become periodic for a range of values of the bifurcation parameter.

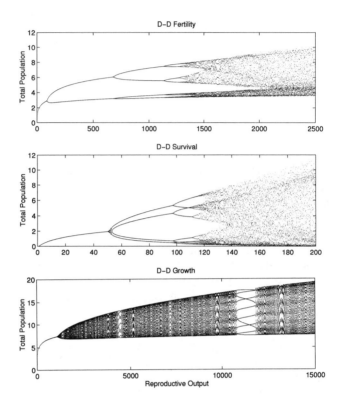

FIGURE 11. *Bifurcation plots beyond the first bifurcation for the three two-stage examples, showing the long-term behavior of total population size ($N = n_1 + n_2$) as a function of reproductive output ($F_2$ or, for the model with density-dependent fertility, $f_0$). Other parameters as in Figure 3.*

Frequency locking is followed by more quasiperiodicity. Eventually, the invariant circle may begin to break up, or become wrinkled or folded, leading to chaotic dynamics.

Both patterns can be found in our examples. Figure 11 shows the bifurcation diagrams for the three models as reproductive output is increased even further, and Figure 12 shows examples of the attractors in the chaotic regime. Note that although all three models eventually become chaotic, the structures of the resulting attractors are quite different.

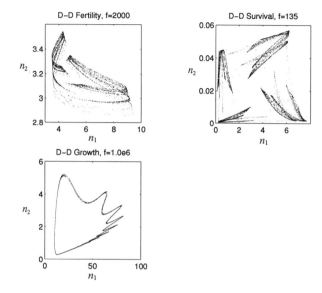

FIGURE 12. *Phase portraits for chaotic attractors for the three two-stage examples, plotting $n_2$ vs. $n_1$ for several thousand iterations, after transients have died out. For the growth plot, the vertical axis is $100,000 \times n_2$. Values of reproductive output ($F_2$ or, for the model with density-dependent fertility, $f_0$) are shown; other parameters as in Figure 3.*

## 5 Conclusion

This chapter introduces some of the most immediately useful methods for analyzing population dynamics with matrix models. In combination with Chapter 3 on stochastic models (Tuljapurkar), and Chapter 6 on density-dependent models (Cushing), it should provide the tools needed to begin the analysis of matrix models and to understand the relevant literature.

Remember that the objective of a matrix model (indeed, any structured-population model) is to describe the individual-level processes of development, growth, aging, survival, and reproduction to their population-level consequences. How much detail is required depends on the goals of the model. To characterize population response to a current environment, a linear, time-invariant model provides a powerful tool. To characterize the effects of population and environmental variability, the stochastic tools in Chapter 3 are necessary. To include feedback between the state of the population and the environmental conditions faced by the individuals, incorporate density-dependent nonlinearities and use the methods presented here and in Chapter 6. Regardless of the level of detail in the model, the analysis usually considers long-term dynamics, problems of ergodicity, and a perturbation analysis.

## Acknowledgments

Thanks to the students in the Summer School for their participation, and to Myriam Barbeau, Mark Hill, Mike Neubert, and Shripad Tuljapurkar for detailed comments on the manuscript. Financial support provided by National Science Foundation grants DEB-9211945 and OCE-9302874, and Office of Naval Research grant URIP N00014-92-J-1527. This is Woods Hole Oceanographic Institution Contribution 9224.

## Literature Cited

Beddington, J. 1974. Age distribution and the stability of simple discrete time population models. *Journal of Theoretical Biology* 47: 65–74.

Brault, S., and H. Caswell. 1993. Pod-specific demography of killer whales (*Orcinus orca*). *Ecology* 74: 1444–1454.

Caswell, H. 1986. Life cycle models for plants. *Lectures on Mathematics in the Life Sciences* 18: 171–233.

———. 1989a. *Matrix Population Models: Construction, Analysis, and Interpretation.* Sinauer, Sunderland, Mass.

———. 1989b. The analysis of life table response experiments. I. Decomposition of treatment effects on population growth rate. *Ecological Modelling* 46: 221–237.

———. 1989c. Life history strategies. Pp. 285–308 *in* J. M. Cherrett, ed., *Ecological Concepts.* Blackwell, Oxford, Engl.

———. 1996a. Demography meets ecotoxicology: Untangling the population level effects of toxic substances. Pp. 255–292 *in* M. C. Newman and C. H. Jagoe, eds., *Ecotoxicology: A Hierarchical Treatment.* Lewis Publishers, Boca Raton, Fla.

———. 1996b. Second derivatives of population growth rate: calculation and applications. *Ecology* 77: 870–879.

———. In press. Analysis of life table response experiments. II. Alternative parameterizations for size- and stage-structured models. *Ecological Modelling.*

Caswell, H. and M. C. Trevisan. 1994. The sensitivity analysis of periodic matrix models. *Ecology* 75: 1299–1303.

Chapman, F.M. 1932. *Handbook of Birds of Eastern North America.* 2nd, rev. ed. D. Appleton, New York.

Cushing, J. M. 1988. Nonlinear matrix models and population dynamics. *Natural Resource Modeling* 2: 539–580.

DeAngelis, D. L., L. J. Svoboda, S. W. Christensen, and D. S. Vaughan. 1980. Stability and return times of Leslie matrices with density-dependent survival: Applications to fish populations. *Ecological Modelling* 8: 149–163.

Dennis, B., R. A. Desharnais, J. M. Cushing, and R. F. Costantino. 1995. Nonlinear demographic dynamics: Mathematical models, statistical methods, and biological experiments. *Ecological Monographs* 65: 261–281.

Getz, W. M., and R. G. Haight. 1989. *Population Harvesting: Demographic Models of Fish, Forest, and Animal Resources*. Princeton University Press, Princeton, N.J.

Goodman, L. A. 1967. On the reconciliation of mathematical theories of population growth. *Journal of the Royal Statistical Society A* 130: 541–553.

Guckenheimer, J., G. Oster, and A. Ipaktchi. 1977. The dynamics of density dependent population models. *Journal of Mathematical Biology* 4: 101–147.

Hale, J., and H. Koçak. 1991. *Dynamics and Bifurcations*. Springer-Verlag, New York.

Hartshorn, G. S. 1975. A matrix model of tree population dynamics. Pp. 41–51 *in* F. B. Golley and E. Medina, eds., *Tropical Ecological Systems*. Springer-Verlag, New York.

Horn, R. A., and C. A. Johnson. 1985. *Matrix Analysis*. Cambridge University Press.

Keyfitz, N. 1967. Reconciliation of population models: Matrix, integral equation and partial fraction. *Journal of the Royal Statistical Society A* 130: 61–83.

———. 1968. *Introduction to the mathematics of population*. Addison-Wesley, Reading, Mass.

Lefkovitch, L. P. 1965. The study of population growth in organisms grouped by stages. *Biometrics* 21: 1–18.

Leslie, P. H. 1945. On the use of matrices in certain population mathematics. *Biometrika* 33: 183–212.

———. 1948. Some further notes on the use of matrices in population mathematics. *Biometrika* 35: 213–245.

Levin, S. A. 1981. Age-structure and stability in multiple-age spawning populations. Pp. 21–45 *in* T. L. Vincent and J. M. Skowronski, eds., *Renewable Resource Management*. Springer-Verlag, Heidelberg.

Logofet, D.O. 1993. *Matrices and Graphs: Stability Problems in Mathematical Ecology*. CRC Press, Boca Raton, Fla.

May, R. M. 1979. Bifurcations and dynamic complexity in ecological systems. *Annals of the New York Academy of Sciences* 316: 517–529.

McDonald, D. B., and H. Caswell. 1993. Matrix methods for avian demography. *Current Ornithology* 10: 139–185.

Mesterton-Gibbons, M. 1993. Why demographic elasticities sum to one: A postscript to de Kroon et al. *Ecology* 74: 2467–2468.

Neubert, M. G., and M. Kot. 1992. The subcritical collapse of predator populations in discrete-time predator-prey models. *Mathematical Biosciences* 110: 45–66.

Sarukhán, J., and M. Gadgil. 1974. Studies on plant demography: *Ranunculus repens* L., *R. bulbosus* L. and *R. acris* L. III. A mathematical model incorporating multiple modes of reproduction. *Journal of Ecology* 62: 921–936.

Seneta, E. 1981. *Non-Negative Matrices and Markov Chains.* 2nd ed. Springer-Verlag, New York.

Silva, J. A. L., and T. G. Hallam. 1992. Compensation and stability in nonlinear matrix models. *Mathematical Biosciences* 110: 67–101.

Werner, P. A., and H. Caswell. 1977. Population growth rates and age versus stage-distribution models for teasel (*Dipsacus sylvestris* Huds.). *Ecology* 58: 1103–1111.

Whitley, D. 1983. Discrete dynamical systems in dimensions one and two. *Bulletin of the London Mathematical Society* 15: 177–217.

Wiggins, S. 1990. *Introduction to Applied Nonlinear Dynamical Systems and Chaos.* Springer-Verlag, New York.

Wikan, A., and E. Mjolhus. 1995. Periodicity of 4 in age-structured population models with density-dependence. *Journal of Theoretical Biology* 173: 109–119.

CHAPTER 3

# Stochastic Matrix Models

## *Shripad Tuljapurkar*

This chapter, like Chapter 2, is about population models in which time and population structure are discrete, but here the models contain vital rates that vary randomly over time. Such random variation is ubiquitous and can strongly influence the dynamics and evolution of populations. I aim to present the main ideas and techniques that are used to study these influences. These methods are applied in Chapter 8 by Orzack, Chapter 15 by Nations and Boyce, and Chapter 16 by Dixon et al.

I begin with a brief review of commonly used models of random variables. In subsequent sections, the general theory of stochastic population models alternates with illustrations of the theory with specific models. Because the theory may seem abstract, the reader may find it helpful to set up computer programs for the specific models and run through the calculations described here. A convenient programming language is MATLAB, which was used to get the results given here. MATLAB, an interactive language built around matrix-vector analysis, is available for most computers from The Mathworks (for more information, consult their Internet site, http://www.mathworks.com). To help with such calculations, I include information on the numerical implementation of stochastic models. Some results are presented as exercises (with answers given in the last section of this chapter); you are urged to work through them. Rather than systematically citing original sources in the literature, I list instead a few key references for each of the main topics discussed. The theory presented here originated with Cohen's (1977) random-ergodic theorem; subsequent developments are described in Tuljapurkar (1990). The discussion here aims to get the reader started on using stochastic models, not on the theory per se.

I assume that you have read and understood Chapter 2. Here are a few words about notation. Vectors and matrices are in boldface type. Deterministic quantities are lower case letters, and stochastic ones are uppercase or Greek letters, except when they are not (this happens rarely). A superscript T indicates a transpose of a matrix or a vector. The components of a vector are a list: $(1, 2, 3)$ is a row vector, and $(1, 2, 3)^{\mathrm{T}}$ is a column vector. By itself, the word "vector" indicates a column vector. The special vector $\mathbf{e} = (1, 1, \ldots, 1)^{\mathrm{T}}$. The identity matrix is $\mathbf{I}$. A scalar product between vectors $\mathbf{q}$ and $\mathbf{r}$ is written $(\mathbf{q}, \mathbf{r})$. Vertical bars around a vector (as, $|\mathbf{q}|$) indicate the sum of all elements in the vector (i.e., $(\mathbf{e}, \mathbf{q})$); note that this is simply a sum, not a norm, because I do not take squares or absolute values of the elements. Probabilities are indicated by Pr, averages by E, variances by var, covariances by cov.

## 1  Models of Randomness

Consider a statement that might be made about a series of observations on a population: "the survival rate of juveniles varies randomly from year to year." To represent this mathematically, survival rate is taken to be a random (equivalently, stochastic) variable, $S(t)$, where $t$ is the year. Randomness means that only a statistical description is possible, which for a single year is the probability distribution of $S(t)$; and for a set of years, $t_1, t_2, \ldots$ is the joint probability distribution of $S(t_1), S(t_2), \ldots$. In population models, random variation is generated by a random (stochastic) process. Such processes come in many flavors, but three are commonly used.

The simplest process assumes that $S(t)$ has the same probability distribution for every $t$. There is no serial correlation, meaning that $S(t_1)$ and $S(t_2)$ are uncorrelated when $t_1 \neq t_2$, so this is called the IID (Independent Identical Distributions) model. When using this model, choose a probability distribution that respects the constraints on the variable; for example, if $S$ is a survival rate, don't use a normal distribution for it. Scientists, inordinately fond of the normal distribution, sometimes sneak it in by transforming a variable. Remember, in such cases, that a random variable $X$ does not have the same statistical properties as nonlinear functions of it (such as $\exp(-X)$). Numerical work requires samples of IID random variables with specified distributions, and programs like MATLAB have built-in procedures for this. Because such procedures use a deterministic algorithm to generate the desired numbers, the results are more accurately called pseudo-random numbers. Modern random-number algorithms are good enough that we may ignore the "pseudo" aspect. However, it is a good idea to browse through a discussion of this, for example, the nice brief account by Ripley (1987) or the masterful but long account by Knuth (1981). Numerical work also requires specific values for the statistics that fix the distribution of the random variable under study, for example, its mean and variance. When using a numerically generated sample of random numbers, compute the sample mean and variance to ensure that they agree with their specified values. If they don't, rescale the numbers in the sample, by adding a constant to correct the mean and multiplying by a constant to correct the variance.

The second common model adds an effect of history, by asserting that the state of the environment at a time $t$ influences the probability of observing future environmental states. In the simplest such case, the state at time $t + 1$ depends explicitly on only the immediately preceding state at the time $t$; this assumption defines a Markov process (also called a Markov chain when time is measured in discrete units, as it is here). Obviously, the state at time $t + 1$ depends implicitly on the state at earlier times, $t - 1, t - 2$, and so on; this is serial correlation. The mathematical analysis of Markov processes is easiest if the environmental state takes only discrete values, and most models assume this to be the case.

The archetypal example is the two-state Markov chain, in which the environmental state is a random variable $S(t)$, which can equal either $s_1$ or $s_2$ for each $t$; these values are the possible states of the random process. Since the chain is Markovian, the probability of being in a particular state at time $t$ depends on the state of the chain at time $t-1$. The model therefore is defined by the conditional transition probabilities (indicated by the vertical bar):

$$\Pr\{S(t) = s_i | S(t) = s_j\} = p_{ij}. \qquad (1)$$

Note that two alternative conventions are in use when defining these transition probabilities. In definition (1) the subscripts $ij$ on the probability are ordered to indicate transitions from $j$ to $i$. Some writers, however (e.g., Karlin & Taylor 1975), do the opposite. To keep this clear, it can help to employ an arrow; definition (1) could be written

$$p_{ij} = p_{i \leftarrow j}.$$

From any state $j$ at time $t$, the environment must end up in some state at time $t + 1$, and therefore

$$\sum_i p_{ij} = 1. \tag{2}$$

This means that each column of the matrix $\mathbf{p} = (p_{ij})$ must sum to one; such a matrix is called column stochastic. For two states, we need specify only $p_{11} = c$ and $p_{22} = d$, yielding the transition probability matrix

$$\mathbf{p} = \begin{bmatrix} c & 1-d \\ 1-c & d \end{bmatrix}. \tag{3}$$

Now define the probabilities

$$\Pr\{S(t) = s_i\} = q_i(t) \quad \text{for } i = 1, 2. \tag{4}$$

From equation (1) it follows that

$$q_i(t) = \sum_j p_{ij}\, q_j(t-1).$$

Using this equation for all states leads to the vector equation

$$\mathbf{q}(t) = \mathbf{p}\,\mathbf{q}(t-1), \tag{5}$$

where the column vector $\mathbf{q}(t) = (q_1(t), q_2(t))^{\mathrm{T}}$. Observe that equation (5) is a matrix recursion involving the nonnegative matrix $\mathbf{p}$ and may therefore be analyzed by the methods presented by Caswell in Chapter 2.

Here is a problem whose solution (no fair peeking, try it) is given in the last section of this chapter. Show that the eigenvalues of $\mathbf{p}$ are 1 and $\rho = (c + d - 1)$. Find the left and right eigenvectors of $\mathbf{p}$ corresponding to these eigenvalues. Determine the spectral decomposition of $\mathbf{p}$. You will find that the right eigenvector corresponding to the dominant eigenvalue (which equals 1) is

$$\boldsymbol{\pi}_1 = \frac{1}{2 - c - d} \begin{bmatrix} 1-c \\ 1-d \end{bmatrix}. \tag{6}$$

From the material in Chapter 2, it should be clear that $\mathbf{q}(t) \to \boldsymbol{\pi}_1$ as time $t$ increases. The elements of $\boldsymbol{\pi}_1$ are called the stationary (also, equilibrium) probabilities of states 1 and 2 of this Markov chain. The rate at which the stationary state is approached depends on the magnitude of the subdominant eigenvalue $\rho$ (why?). Continuing with the problem, use the spectral decomposition to show that the covariance between values of $S(t)$ at different times is

$$\operatorname{cov} S(t+n)S(t) = \rho^n \operatorname{var} S(t). \tag{7}$$

The model above illustrates key concepts that apply to many stochastic models. When $0 < c, d < 1$, the two-state Markov chain converges to a stationary state $\boldsymbol{\pi}_1$. If, initially, $\mathbf{q}(t) = \boldsymbol{\pi}_1$, the chain generates a stationary stochastic process, whose probability distributions do not change with time. This chain is ergodic, meaning that stationary expectations are equal to time averages. For example,

$$\mathrm{E}\, S(t) = \lim_{n\to\infty} \frac{1}{n} \sum_{i=1}^{n} S(i). \qquad (8)$$

In order for a stochastic process to be ergodic, the probability distribution of the random variable must converge to a stationary distribution, and the stationary statistics calculated as in equation (8) must be independent of the initial conditions (i.e., values) of the random process. Finally, note that the magnitude of the serial correlation $|\rho| < 1$. Hence, in view of equation (7), the state of the environment for times $(t + n)$ becomes independent of conditions at or before time $t$ when $n$ becomes large. The latter property is described by saying that the stochastic process is "mixing." This term is best appreciated, as the mathematician Paul Halmos said, by watching a drop of vermouth mix through cold gin. A process that is not mixing will have an arbitrarily long memory; think about situations where this could happen. All the models here assume that mixing occurs.

It is a good exercise to set up a simulation of the two-state chain. How is this done? The basic step: suppose that a random variable $X$ has probability $q_i$ of equaling integer $i$ in the set $(1, 2, \ldots, k)$. Make the probability distribution whose elements are $(q_1, q_1 + q_2, q_1 + q_2 + q_3, \ldots)$. Now use the computer to generate a random number $U$ that is uniformly distributed on the interval $(0, 1)$. If $U < q_1$, set $X$ equal to 1; if $q_1 \leq U < q_1 + q_2$, set $X$ equal to 2; and so on. To simulate $S(t)$ from the chain, choose an arbitrary initial vector $\mathbf{q}(0)$ in equation (1), and use the preceding algorithm to choose $S(0)$. Then use the conditional probabilities to choose $S(1)$; repeat as desired. It is instructive to plot the resulting sequences, check ergodic results such as equation (8), and explore the effects of changing $c$ and $d$ to make $\rho$ negative, zero, or positive.

Simulations of the two-state chain make concrete the notion of sample paths. Every simulation of a stochastic process is likely to yield a different sequence of values of $S(1), \ldots, S(t)$. Identify such a sequence of values by a label $\omega$. This means that ten simulations yield ten sequences, $\omega_1$, $\omega_2$, and so on through $\omega_{10}$. The stochastic model determines a probability for each possible sequence or, in other words, yields a probability distribution for the sequence label $\omega$. The sequence $S(t, \omega)$ is called the sample path of the process identified by the label $\omega$.

The third common kind of stochastic model is a time-series model, for example,

$$S(t) = c\, S(t-1) + d\, \epsilon(t), \tag{9}$$

or

$$S(t) = c_1\, S(t-1) + c_2\, S(t-2) + d\, \epsilon(t). \tag{10}$$

The $\epsilon(t)$ in these models is a random IID sequence of "shocks," also called "innovations." These are usually taken to be normally distributed, which would imply that $S(t)$ is normally distributed. (Hence, such a model may not be used for a survival rate without first making a nonlinear transformation.) The model (9) is a first-order autoregressive process, whereas (10) is a second-order autoregressive process. The standard assumption is that $\mathrm{E}[\epsilon(t)] = 0$ and that $\mathrm{var}\,[\epsilon(t)] = \sigma_\epsilon^2$ is a fixed variance for the shocks. In both models, the shock at time $t$ is independent of the values of $S(t-j)$ when $j > 0$. With these assumptions, show the following (solutions at the end of the chapter). The mean value in both processes is $\mathrm{E}S(t) = 0$. The variance in equation (9) is

$$\mathrm{var}\, S(t) = \frac{d^2\, \sigma_\epsilon^2}{1 - c^2}. \tag{11}$$

The serial autocovariance for equation (9) has the form of equation (7) but with $c$ replacing $\rho$. Equation (11) reveals the important fact that serial autocorrelation $c$ acts as a variance multiplier, in that increasing autocorrelation ($c \to \pm 1$) produces an increase in the variance of $S(t)$ even when the variance of the shocks is fixed. The considerable difference between large positive and large negative autocorrelations is best appreciated by simulating equation (9) on a computer. When $c > 0$, there tend to be runs in successive values of $S(t)$; if, say, $S(1)$ is large, then $S(2)$ tends to be large and so on. As time goes by, the effect of random shocks is to suddenly switch the pattern. In contrast, when $c < 0$, values of $S(t)$ tend to switch sign, and the series of values of $S(t)$ oscillates. In the process of equation (10), the serial covariance between $S(t)$ and $S(t + n)$ does not always die out geometrically with increasing $n$

(as it does in eq. 7). Instead (problem for the reader), its value can oscillate as $n$ increases, for suitable values of $c_1$ and $c_2$. Such behavior is relevant in modeling random fluctuations that have a noticeable cyclic component. In empirical work, time-series models are popular because most observations are in fact time series; in addition, good standard software is available to estimate such models. In theoretical analyses, time-series models result from the linearization of any nonlinear stochastic-population model.

Here are some good sources of further information. On stochastic processes, see Karlin and Taylor (1975) for an accessible mathematical discussion, Bartlett (1978) for a more biological but narrower account, and Gardiner (1985) for a useful if somewhat physics-oriented treatment. For time-series models, a compact account is by Chatfield (1989), and the classic treatment is by Box et al. (1994). A readable and rich treatment of Markov chains is given by Kemeny and Snell (1960). These sources contain more than beginning modelers need to know, but no single efficient and concise source is available.

## 2 Structure and Ergodicity

*Basic Equations*

The models here describe a population that has $k$ stages (age or size classes, for example). At time $t$ the population is described by a vector $\mathbf{N}(t)$ of numbers in the successive stages, the total population number is $M(t) = |\mathbf{N}(t)| = \text{sum}\big(\text{elements of } \mathbf{N}(t)\big)$, and the population structure is the vector of proportions-by-stage defined as

$$\mathbf{Y}(t) = \mathbf{N}(t)/|\mathbf{N}(t)|. \tag{12}$$

Changes in population are described by a projection matrix $\mathbf{X}(t)$, examples of which are given in Chapter 2 in cases when the vital rates are constant. The projection matrix need not be a Leslie matrix but can be more general, for example, a matrix describing changes in size structure. The basic dynamic equation is

$$\mathbf{N}(t+1) = \mathbf{X}(t+1)\,\mathbf{N}(t). \tag{13}$$

In this equation, the rates that apply to $\mathbf{N}(t)$ are assigned the time label $(t+1)$; some writers use $t$. Add components on both sides of equation (13) to find that

$$|\mathbf{N}(t+1)| = |\mathbf{X}(t+1)\,\mathbf{N}(t)|.$$

Divide equation (13) by the corresponding sides of the above equation, and use equation (12) to see that

$$\mathbf{Y}(t+1) = \frac{\mathbf{N}(t+1)}{|\mathbf{N}(t+1)|} = \frac{\mathbf{X}(t+1)\,\mathbf{N}(t)}{|\mathbf{X}(t+1)\,\mathbf{N}(t)|}.$$

Divide the numerator and denominator of the rightmost ratio by $|\mathbf{N}(t)|$, and again use equation (12) to find that

$$\mathbf{Y}(t+1) = \frac{\mathbf{X}(t+1)\,\mathbf{Y}(t)}{|\mathbf{X}(t+1)\,\mathbf{Y}(t)|}. \tag{14}$$

This important equation describes population structure and has two advantages over equation (13). In numerical work the elements of $\mathbf{N}(t)$ may become very small or very large, whereas the structure vector, $\mathbf{Y}(t)$, always has a sum of one. Further, equation (14) allows us to study population structure without worrying about the growth of the total population, $M(t)$.

The models here assume that the elements of $\mathbf{X}(t)$ are described by a stochastic process that is stationary, ergodic, and mixing. Define the time-independent average matrix by $\mathbf{a} = E\mathbf{X}(t)$, and deviations around this mean by:

$$\mathbf{H}(t) = \mathbf{X}(t) - \mathbf{a},$$
$$E\mathbf{H}(t) = 0.$$

The models here also assume that each matrix $\mathbf{X}(t)$ is nonnegative and that the average matrix, $\mathbf{a}$, has all the nice properties described in Chapter 2. The dominant eigenvalue of $\mathbf{a}$ is called $\lambda_0$; the corresponding left and right eigenvectors are called, respectively, $\mathbf{v}_0$ (this is the deterministic reproductive value) and $\mathbf{u}_0$ (this is the deterministic stable structure). I define these so that $|\mathbf{u}_0| = |\mathbf{v}_0| = 1$.

Equation (14) is easiest to analyze when the vital rates produce demographic weak ergodicity, which is the property that $\mathbf{Y}(t)$ becomes independent of its distant past. To make this idea clear, suppose that the random rates follow a sample path $\omega$, so that successive matrices in equation (14) form the sequence $\mathbf{X}(t, \omega)$. Starting with some initial population structure $\mathbf{y}_1$, generate the sequence of population structures:

$$\mathbf{Y}_1(1, \omega) = \frac{\mathbf{X}(1, \omega)\mathbf{y}_1}{|\mathbf{X}(1, \omega)\mathbf{y}_1|},$$

$$\mathbf{Y}_1(2, \omega) = \frac{\mathbf{X}(2, \omega)\mathbf{Y}_1(1, \omega)}{|\mathbf{X}(2, \omega)\mathbf{Y}_1(1, \omega)|},$$

and so on. Using the same matrices and the same procedure but a different initial population structure, $\mathbf{y}_2 \neq \mathbf{y}_1$, generate the sequence of population structures $\mathbf{Y}_2(1, \omega), \mathbf{Y}_2(2, \omega)$, and so on. Demographic weak ergodicity means that

$$|\mathbf{Y}_1(t, \omega) - \mathbf{Y}_2(t, \omega)| \to 0 \quad \text{as } t \to \infty.$$

When does a particular model for $\mathbf{X}(t)$ produce demographic weak ergodicity? A sufficient condition is that every possible product of the growth matrices eventually becomes a positive matrix (whose elements are all positive). In making models we often begin with an average matrix, $\mathbf{a}$, that satisfies strong demographic ergodicity (see Chapter 2) and then add random components to the elements of $\mathbf{a}$. If the resulting random matrices have positive or zero elements in the same places as $\mathbf{a}$, then the sufficient condition for weak ergodicity is satisfied (why?). Necessary and sufficient conditions for demographic weak ergodicity are not known; to explore this problem further, consult Seneta (1981).

*Example 1: Diapause*

To illustrate the above concepts, consider a model of a population in which some individuals can diapause (Tuljapurkar & Istock 1993). At time $t$ the population is divided into a diapause pool, $D(t)$, and a breeding pool, $F(t)$. A fraction $s$ of the current diapause pool survives to the next time period and breaks diapause to enter the breeding pool. With a per capita reproductive rate of $R(t)$, there are $E(t) = R(t)F(t)$ offspring produced at time $t$. Of these, a fraction $f$ forms the diapause pool in the next time period, and the remainder develop directly to enter the breeding pool in the next time interval. Thus, the breeding pool at time $(t + 1)$ is of size $F(t) = (1 - f)E(t) + sD(t)$. These assumptions are summarized in the matrix equation

$$\begin{bmatrix} E(t) \\ D(t) \end{bmatrix} = \begin{bmatrix} (1-f)R(t) & sR(t) \\ f & 0 \end{bmatrix} \begin{bmatrix} E(t-1) \\ D(t-1) \end{bmatrix}. \quad (15)$$

This model has the form of equation (13). I assume that $R(t) = 1/G(t)$, where $G(t)$ is an IID sequence with a gamma distribution. This distribution yields only positive reproductive rates (a sensible constraint) and allows a complete analytical development. Here I present only results from simulations of the model. Let $R_0 = \mathrm{E}R(t)$ and $c^2 = \mathrm{var}\, R(t)/R_0^2$.

Set $R_0 = 1.2, f = 0.2$, and $s = 0.75$; and consider two initial structures, $\mathbf{y}_1 = (0.5, 0.5)^{\mathrm{T}}$ and $\mathbf{y}_2 = (0.1, 0.9)^{\mathrm{T}}$. Problem: starting

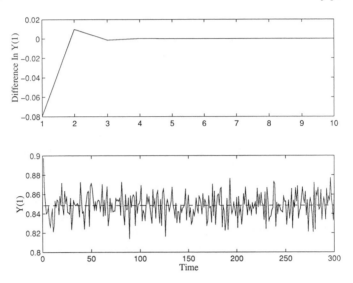

FIGURE 1. Upper panel, *Difference in projected proportions of individuals in stage 1 using the same random vital rates but different initial population structures.* Lower panel, *Trajectory of proportion of individuals in stage 1.* Dashes, *Stable proportion that would result if vital rates were fixed at their averages.*

with matrix (15), write the equation for the population structure, in the form of (14), and a computer program that will perform iterations of the equation (see "Solutions"). Using a random-number generator to produce 300 values of $R(t)$, compute the population structure over time; let $\mathbf{Y}_i(t)$ be the sequence that results from the initial structure $\mathbf{y}_i$, for $i = 1, 2$.

Demographic weak ergodicity is illustrated by results from such a simulation. The upper panel of Figure 1 displays the difference between the first components, $Y_1(1, t)$ and $Y_2(1, t)$, of the population structures. Although the initial structures are quite different, the two trajectories of population structure converge rapidly, and the difference in initial conditions is lost in about four iterations. The lower panel displays the continuing variability in the trajectory of $Y_1(1, t)$ driven by fluctuations in $R(t)$. Obviously, demographic weak ergodicity does not imply that the population's structure becomes fixed. If the model is iterated with $R(t)$ equal to $R_0$, the sequence $Y_1(1, t)$ rapidly converges to $u_0(1) = 0.85$, which is the deterministic stable proportion in population class 1, indicated by the dashed line in the lower panel of Figure 1. The stochastic proportion varies between about 0.82 and 0.88. In this simulation, $R(t)$ has a coefficient of variation of $c$ equal to 0.1, and this produces a coefficient of variation of about 0.015 in $Y(1,t)$. This response of order $c^2$ is typical in stochastic structured models.

Demographic theory also asserts that the statistical distribution of the structure becomes stationary. To check this, run several independent simulations over the same time interval, always starting with the same initial structure. Then plot histograms of $Y(1,t)$ for different values of $t$. As $t$ increases, these histograms start to resemble each other closely. Analytical calculations of the stationary distribution of $[Y(1,t)/Y(2,t)]$ are rare because the equation that defines the stationary distribution is usually difficult to solve (but see Tuljapurkar & Istock 1993).

*Growth Rates*

Consider now the growth of the total population $M(t)$. Returning to equation (13), add components on both sides of the equation to see that

$$M(t+1) = |\mathbf{N}(t+1)| = |\mathbf{X}(t+1)\,\mathbf{N}(t)|.$$

Therefore, using equation (12), the growth rate of the population between $t$ and $(t+1)$ is

$$\lambda_{t+1} = [M(t+1)/M(t)] = |\mathbf{X}(t+1)\,\mathbf{Y}(t)|. \qquad (16)$$

Since the structure vector and the matrix change with time, so will the growth rate $\lambda_t$.

Suppose now that the random environments follow a particular sample path $\omega$. The cumulative growth of population between time 0 and time $t$, assuming a starting population size of $m(0)$, is

$$\Lambda_t(\omega) = [M(t,\omega)/m(0)] = \lambda_t(\omega)\lambda_{t-1}(\omega)\ldots\lambda_1(\omega). \qquad (17)$$

Here, $\omega$ indicates changes along one particular sample path. The random process determines the probability distribution of $\omega$, that is, of the sample paths. A central theoretical result is that (if the random environmental process is ergodic and mixing, and there is demographic weak ergodicity) the logarithm of the cumulative growth rate increases at a characteristic long-term rate,

$$a = \lim_{t \to \infty} \frac{1}{t} \operatorname{E} \log \Lambda_t(\omega)$$
$$= \lim_{t \to \infty} \frac{1}{t} \log \Lambda_t(\omega). \tag{18}$$

The first of these equations means that $a$ equals the average (over the distribution of sample paths) of the long-run rate of change of the logarithm of the cumulative growth rate. The second means that $a$ equals the long-run rate of change of the logarithm of the cumulative growth rate for *any* sample path. In addition, the logarithm of $\Lambda_t(\omega)$ eventually becomes normally distributed, with mean $a\,t$ and variance $\sigma^2\,t$, where

$$\sigma^2 = \lim_{t \to \infty} \frac{1}{t} \operatorname{var}\left[\log \Lambda_t(\omega)\right]. \tag{19}$$

The expectation in (18) and the variance in (19) are computed with respect to the distribution of $\omega$. In numerical practice, we simulate a large number of sample paths with the same (large) length $t$, find $\Lambda_t(\omega)$ for each sample path, and then calculate the mean and variance from this set of numbers.

The number $a$ is called the long-run growth rate of the stochastic population, or sometimes simply the stochastic growth rate. Equations (18) and (19) are based on an important paper by Furstenberg and Kesten (1960) and on its demographic extension (Tuljapurkar & Orzack 1980). The number $a$ is also the leading Liapunov exponent of the random matrix products generated by the matrices $\mathbf{X}(t)$. Liapunov exponents are generalizations of eigenvalues, and they appear both in stochastic linear models such as (13) and in nonlinear models; Ott (1993) gave a relatively straightforward discussion of these exponents in nonlinear models.

In the second line of equation (18), note that

$$\log \Lambda_t(\omega) = \sum_{i=1}^{t} \log \lambda_t(\omega),$$

and therefore,

$$a = \lim_{t \to \infty} \frac{1}{t} \sum_{i=1}^{t} \log \lambda_t(\omega)$$
$$= \operatorname{E} \log \lambda_t(\omega). \tag{20}$$

The second line here uses ergodicity, because it equates the time average over a single sample path and the stationary average with respect to the probability distribution of sample paths. From the above equation and equation (16), we deduce that the stochastic growth rate $a$ is also equal to the stationary average

$$a = \mathrm{E}\log(\lambda_t) = \mathrm{E}\log(|\mathbf{X}(t)\,\mathbf{Y}(t-1)|), \qquad (21)$$

where the rightmost expectation is calculated with respect to the stationary probability distributions of both $\mathbf{X}(t)$ and $\mathbf{Y}(t-1)$. In the case of IID random rates, these variables have independent distributions, but with Markovian rates they have a joint stationary distribution. The calculation in (21) is difficult to carry out, because the distribution of population structure is difficult to compute analytically, or even numerically (for an example, see Cohen 1977).

In practice, $a$ is found by simulation via equation (20) or by using an approximation (Tuljapurkar 1990, Chapter 12) that is described below.

*Example 1, Continued*

Consider again the diapause model, equation (15). Figure 2 (*upper panel*) displays $\log \Lambda_t$, the logarithm of the cumulative growth rate, as a function of elapsed time $t$ since the start of the simulation. This plot is for a single sample path. In agreement with equation (18), this plot rapidly turns into a straight line of constant slope.

The lower panel of Figure 2 shows values of the slope of the logarithm of cumulative growth rate, that is, values of $(1/t)\log \Lambda_t$ for increasing $t$. Note the large oscillations at small $t$, followed by apparent convergence to a limit. The value when $t = 300$ is the estimated stochastic growth rate $a$, 0.1124, which is less than the deterministic $r$, 0.1139. The fluctuations in the estimated stochastic growth rate, even for large $t$, are expected because of the limiting normal distribution mentioned above. These fluctuations die out as $t$ becomes large.

To check the convergence in the distribution of growth rates, I ran 100 independent simulations of the model (these are 100 sample paths). Figure 3 (*upper panel*) is a quantile-quantile plot of the quantities $(1/250)\log \Lambda_{250}$ and $(1/300)\log \Lambda_{300}$, using the 100 values of each. The plot is linear, indicating that the two variables have essentially the same distribution. The lower panel shows a normal probability plot of $(1/300)\log \Lambda_{300}$. Again, this plot is linear, meaning that the variable is normally distributed, to a good approximation.

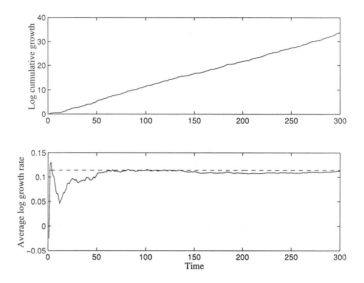

FIGURE 2. Upper panel, *The logarithm of cumulative growth rate increases linearly at long times.* Lower panel, *Estimates of stochastic growth rate, a, converge as the base period increases.*

## 3 Stochastic Growth Rate

*Analytical Approximation*

The stochastic growth rate, $a$, can be approximated by an analytical formula when the variability in vital rates is small, meaning that the coefficient of variation in the rates needs to be much less than one. I now present the approximation for the case of IID variability in vital rates; for a derivation and for correlated rates see (Tuljapurkar 1990, Chapter 12). Do not bother with the derivation if you just want to apply the result.

Begin with the average vital rates as contained in the matrix **a**. From Chapter 2, recall the definition of the elasticities $e_{ij}$ of this matrix with respect to its elements. Here, $r = \log(\lambda_0)$, the intrinsic rate of increase obtained with this average matrix. For each location $ij$ at which there is a nonzero entry in the random matrix $\mathbf{X}(t)$, identify the deviation $H_{ij}(t)$ of the element from its average value. For every pair of nonzero elements $(i, j), (k, \ell)$ in the random matrix, compute the scaled covariance,

$$\operatorname{cov}(ij, k\ell) = \operatorname{E}\left(\frac{H_{ij}}{a_{ij}}\right)\left(\frac{H_{k\ell}}{a_{k\ell}}\right). \tag{22}$$

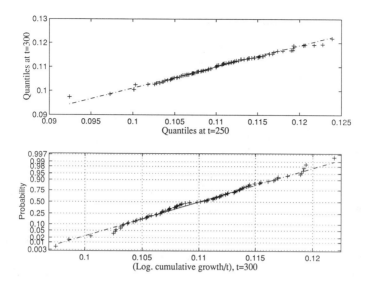

FIGURE 3. Upper panel, *Stationarity of the distribution of* $(1/t)\log\Lambda_t$, *illustrated by a quantile-quantile plot comparing* $t = 250$ *with* $t = 300$. Lower panel, *This stationary distribution is asymptotically normal.*

The approximation to $a$ is

$$a \simeq r - \frac{1}{2} \sum_{(ij),(k\ell)} e_{ij} e_{k\ell} \mathrm{cov}\,(ij, k\ell). \qquad (23)$$

The sum is taken over all pairs of nonzero elements in the matrix. A key property of $a$ revealed by this approximation is that $a < r$ when the random variation is IID. This is true even if some covariances between growth rates are negative, because the total variance term, the second term in equation (22), must always be greater than zero if there is any variation. In the presence of serial correlation, the inequality $a < r$ need not hold. The difference between $a$ and $r$ is at the heart of arguments for the importance of including random variability when one studies life-history evolution and extinction.

Note that equation (23) involves the scaled covariances in equation (22), in which the random elements are divided by their averages. Then, equation (23) can be written in terms of elasticities rather than sensitivities. Rewrite (22) and (23) in terms of unscaled covariances and sensitivities (see "Solutions") to get the $a$ expansion in the form that is used by Dixon et al. in Chapter 16.

*Example 2: Approximating a*

The best way to understand equation (23) is to use it. Consider the random projection matrix

$$
\mathbf{X}(t) = \begin{bmatrix}
0 & F_2(t) & F_3(t) & F_4(t) \\
P_1(t) & S_2(t) & 0 & 0 \\
0 & p_2 & s_3 & 0 \\
0 & 0 & p_3 & 0
\end{bmatrix}. \tag{24}
$$

This matrix describes a population with four stages. Random variation affects the fertilities in the first row of the matrix; the survival rate of stage 1, the (2,1) element; and the fraction of individuals who stay on in stage 2, the (2,2) element. All other elements of the matrix are constant over time. Suppose now that all fertilities respond to one random factor (such as food supply), while the survival rate and the other transition rate respond to an independent random factor (such as temperature). Express these assumptions by setting

$$
\begin{aligned}
F_i(t) &= f_i[1 + \epsilon(t)], \\
P_1(t) &= p_1[1 + \eta(t)], \\
S_2(t) &= s_2[1 + \eta(t)],
\end{aligned} \tag{25}
$$

where $\epsilon(t)$ and $\eta(t)$ are independent IID sequences. Set $\sigma_\epsilon$ and $\sigma_\eta$ to be the standard deviations of these variables (their means are taken to be zero).

The matrix elements in equations (25) are the only ones that enter into the covariance terms in equation (23) for the growth rate $a$. From equation (25) check that

$$
\begin{aligned}
H_{1i}/a_{1i} &= \epsilon(t) \quad \text{for } i = 2, 3, 4, \\
H_{2i}/a_{2i} &= \eta(t) \quad \text{for } i = 1, 2.
\end{aligned}
$$

The necessary covariances are therefore given by

$$
\begin{aligned}
\operatorname{cov}(1i, 1i) &= \sigma_\epsilon^2 \quad \text{for } i = 2, 3, 4, \\
\operatorname{cov}(1i, 1j) &= \sigma_\epsilon^2 \quad \text{for } i \neq j, \ i, j = 2, 3, 4, \\
\operatorname{cov}(2i, 2i) &= \sigma_\eta^2 \quad \text{for } i = 1, 2, \\
\operatorname{cov}(2i, 2j) &= \sigma_\eta^2 \quad \text{for } i \neq j, \ i, j = 1, 2.
\end{aligned}
$$

From equation (24), the average matrix is deduced to be

$$\mathbf{a} = \begin{bmatrix} 0 & f_2 & f_3 & f_4 \\ p_1 & s_2 & 0 & 0 \\ 0 & p_2 & s_3 & 0 \\ 0 & 0 & p_3 & 0 \end{bmatrix}. \tag{26}$$

The methods of Chapter 2 allow the computation of $r$ and elasticities $e_{ij}$ for this matrix.

Inserting the above information into equation (23), and collecting terms, verify that

$$a \simeq r - \frac{1}{2}\sigma_\epsilon^2 \left( e_{12} + e_{13} + e_{14} \right)^2 + \frac{1}{2}\sigma_\eta^2 \left( e_{21} + e_{22} \right)^2. \tag{27}$$

This approximation may be checked against a numerical simulation of equation (24). To generate the sequence $\epsilon(t)$, I create a sequence of random numbers uniformly distributed on $(0, 1)$. From this sequence, I subtract the sample mean (to ensure that the resulting sequence has a sample mean of zero). Next, I estimate the sample standard deviation, $s$, and then multiply each number by the ratio $(\sigma_\epsilon/s)$ (ensuring that the sample has a variance of $\sigma_\epsilon^2$). A similar procedure is used to generate the sequence $\eta(t)$. It is important to make the adjustments for sample means and variances when running simulations, because, for example, a nonzero sample mean amounts to a change in the average matrix of vital rates. Figure 4 displays the excellent agreement between the analytical equation (27) and estimated values of $a$ from simulations over 300 iterations of the model. The comparison covers a range of values of $\sigma_\epsilon$, with $\sigma_\eta$ fixed at 0.08. A more careful comparison could be made by computing and displaying the sample standard error of the simulation estimates.

## Properties of $a$

The stochastic growth rate has properties quite distinct from those of the deterministic $r$. The biological consequences are best illustrated by examples.

*Example 3.* Consider two Leslie matrices,

$$\mathbf{a}_1 = \begin{bmatrix} 0.2 & 3 \\ 0.25 & 0 \end{bmatrix}, \quad \mathbf{a}_2 = \begin{bmatrix} 0.9 & 0.1 \\ 1 & 0 \end{bmatrix}. \tag{28}$$

The logarithms of the dominant eigenvalues of these matrices are, respectively, $r_1 = -0.0286$ and $r_2 = 0$. Suppose that a pop-

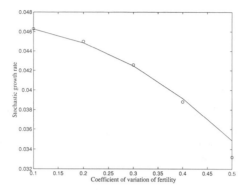

FIGURE 4. *Predicted stochastic growth rate from the analytical approxi-mation, compared with simulation estimates.*

ulation's growth is determined by $a_1$ or $a_2$, reflecting state 1 or state 2 of a random environment . Choose the environmental state randomly at each time, with $\Pr\{\text{state1}\} = \Pr\{\text{state2}\} = 0.5$. Numerical simulation of this stochastic growth yields an estimated stochastic growth rate, $a$, of 0.183, which makes it certain that $a > 0$, whatever the sampling error. This is a surprise, because if the environment is fixed in state 1, the population eventually declines, whereas if the environment is fixed in state 2, the population eventually becomes stationary. Thus, randomness produces a synergy, resulting in positive growth over time.

The logarithm of the geometric mean of the dominant eigenvalues of the matrices in equation (28) has been proposed (following Schaffer 1974) as an approximation to the growth rate, $a$. But here the logarithm of the geometric mean is $0.5(r_1 + r_2) = -0.014$, clearly a bad approximation to $a$. The approximation of equation (23), in contrast, yields $a \simeq 0.258$, substantially larger than the simulation estimate of 0.183. However, the latter at least predicts the sign of $a$ correctly, arguably the most important thing to get right. Equation (23) does poorly here because it underestimates the variation. The coefficient of variation of the 22 element of the random matrix here is 0.94, whereas equation (23) assumes that variation is small.

A final note about this deceptively simple example. Suppose that the matrices $\mathbf{a}_1$ and $\mathbf{a}_2$ occur alternately. Then the long-run growth rate is half of the logarithm of the dominant eigenvalue of the product $\mathbf{a}_1\mathbf{a}_2$, which turns out to be 0.578, larger than $a$. The reader will find it instructive to simulate a model in which the two matrices (28) are chosen randomly according to a two-state Markov chain. You should find that $a$ approaches the cyclic growth rate when the autocorrelation of the chain approaches $-1$.

*Example 1, again.* The difference between the biological implications of structured and unstructured models has been explored in the context of diapause (Tuljapurkar & Istock 1993). A related example, in a model of delayed flowering in biennial plants, is by Roerdink (1987, 1989). These papers are also interesting because they present explicit analytical expressions for the stochastic growth rate, $a$, of their models.

*Example 4.* The complexity of environmental pattern can potentially affect population dynamics in unusual ways. To model an environment that has several distinct phases, a two-state Markov chain may be generalized to a chain with several states. Markov chains with more than two states have the interesting property that sample paths generated by running the chain forward in time may have a different distribution than those generated by running the chain backward in time. Key (1987) has shown that this difference can lead to different Liapunov exponents for the forward and backward chains. He uses three matrices,

$$\mathbf{a}_1 = \begin{bmatrix} 1 & 1 \\ 0 & 0 \end{bmatrix}, \quad \mathbf{a}_2 = \begin{bmatrix} 0 & 1 \\ 1 & 0 \end{bmatrix}, \quad \mathbf{a}_3 = \begin{bmatrix} 1 & 0 \\ 0 & 0 \end{bmatrix}. \tag{29}$$

These matrices are chosen according to a three-state Markov chain whose transition probabilities are given (recall eq. 3) by the matrix

$$\mathbf{p} = \begin{bmatrix} 0 & 0 & \frac{1}{2} \\ 1 & 0 & 0 \\ 0 & 0 & \frac{1}{2} \end{bmatrix}. \tag{30}$$

Using the elements of this matrix, convince yourself that this chain generates sample paths of the type $1 \rightarrow 2 \rightarrow 3 \rightarrow 1 \rightarrow \ldots$ or $1 \rightarrow 2 \rightarrow 3 \rightarrow 3 \rightarrow \ldots$. Check that $\mathbf{a}_3\mathbf{a}_3 = \mathbf{a}_3$ and $\mathbf{a}_3\mathbf{a}_2\mathbf{a}_1 = 0$, so that a population whose growth matrices are chosen according to these sample paths will surely go to zero. In other words, for such a population, $a = -\infty$.

Now consider the chain of matrix (30) running backward in time. Start with state 1, ask which state could have preceded it, choose

one such preceding state, and repeat. Convince yourself that the
backward chain generates sample paths of the form $1 \to 3 \to 2 \to
1 \to \ldots$ or $1 \to 3 \to 2 \to 1 \to 3 \to \ldots$. Check that

$$\mathbf{a}_2\mathbf{a}_3\mathbf{a}_1 = \begin{bmatrix} 0 & 0 \\ 1 & 1 \end{bmatrix},$$

whereas $\mathbf{a}_1\mathbf{a}_2\mathbf{a}_3\mathbf{a}_1 = \mathbf{a}_1$, and $\mathbf{a}_3\mathbf{a}_1 = \mathbf{a}_1$. Combining this with the
sample-path information, show that a starting vector of the form

$$\begin{bmatrix} 1 \\ 0 \end{bmatrix}$$

is preserved by the action of any random sequence $\mathbf{X}(t)\mathbf{X}(t-1)$
$\ldots \mathbf{X}(1)$. Therefore, the sum of the elements of the vector remains
constant, implying that the logarithmic growth rate, $a$, is zero. In
other words, there is a certain decline of the "population" with the
forward chain but constancy of the "population" with the backward chain.

This example may be only a curiosity, because it is contrived and
does not use demographic projection matrices. Complicated environmental sequences do occur in nature, however, and they can be
irreversible in the sense of this example. The possible consequences
of irreversible environments on biological dynamics deserve further
exploration.

## 4 Other Aspects of Stochastic Dynamics

At least three additional aspects of stochastic dynamics are interesting: reproductive value, the rate of convergence to stability, and
finite-population effects.

Reproductive value measures the contribution to future generations made by individuals currently alive in different population
stages. Let $W(t,i)$ be the total number of descendants of an individual in state $i$ at time $t$. This individual produces $X_{ji}(t+1)$
individuals of type $j$ at time $(t+1)$, and thus

$$W(t,i) = \sum_j W(t+1,j)\, X_{ji}(t+1),$$

$$\mathbf{W}(t) = \mathbf{X}^{\mathrm{T}}(t+1)\,\mathbf{W}(t+1).$$

The second line rewrites the first line as a matrix recursion. Because the elements of $\mathbf{W}(t)$ change exponentially over time (since
they obey a linear matrix recursion), the stochastic reproductive-
value vector is best defined as a vector of proportions, $\mathbf{V}(t) =$

$\mathbf{W}(t)/|\mathbf{W}(t)|$. This definition is analogous to the definition of population structure in equation (12). In the same way in which we obtained equation (14), we deduce that changes in reproductive value are described by the equation

$$\mathbf{V}(t) = \frac{\mathbf{X}^{\mathrm{T}}(t+1)\mathbf{V}(t+1)}{|\mathbf{X}^{\mathrm{T}}(t+1)\mathbf{V}(t+1)|}. \tag{31}$$

In contrast to equation (14) for the population-structure vector, equation (31) runs backward in time. To understand the significance of this point, consider the difference between population structure and reproductive value.

Suppose that our model of random rates generates "forward-in-time" sequences of projection matrices $\mathbf{X}(1,\omega), \ldots, \mathbf{X}(t,\omega), \ldots$. Starting with $\mathbf{Y}(0,\omega) = \mathbf{y}(0)$, subsequent population structures are given by

$$\mathbf{Y}(1,\omega) = \frac{\mathbf{X}(1,\omega)\mathbf{y}(0)}{|\mathbf{X}(1,\omega)\mathbf{y}(0)|},$$

$$\mathbf{Y}(2,\omega) = \frac{\mathbf{X}(2,\omega)\mathbf{Y}(1,\omega)}{|\mathbf{X}(2,\omega)\mathbf{Y}(1,\omega)|},$$

and so forth. At long times $t$, demographic weak ergodicity implies that the population structure becomes independent of the initial structure $\mathbf{y}(0)$ and is determined by the "forward" matrix product $\mathbf{X}(t,\omega)\ldots\mathbf{X}(1,\omega)$.

In contrast, for every sample path, reproductive value $\mathbf{V}(t,\omega)$ is found by working backwards in time from a known reproductive value $\mathbf{V}(t_{\mathrm{v}},\omega) = \mathbf{v}(t_{\mathrm{v}})$ at some future time $t_{\mathrm{v}} > 0$. Thus,

$$\mathbf{V}(t_{\mathrm{v}} - 1, \omega) = \frac{\mathbf{X}^{\mathrm{T}}(t_{\mathrm{v}},\omega)\mathbf{v}(t_{\mathrm{v}})}{|\mathbf{X}^{\mathrm{T}}(t_{\mathrm{v}},\omega)\mathbf{v}(t_{\mathrm{v}})|},$$

$$\mathbf{V}(t_{\mathrm{v}} - 2, \omega) = \frac{\mathbf{X}^{\mathrm{T}}(t_{\mathrm{v}} - 1,\omega)\mathbf{V}(t_{\mathrm{v}} - 1,\omega)}{|\mathbf{X}^{\mathrm{T}}(t_{\mathrm{v}} - 1,\omega)\mathbf{V}(t_{\mathrm{v}} - 1,\omega)|},$$

and so forth. Here, too, demographic weak ergodicity applies: the distribution of $\mathbf{V}(0,\omega)$ is independent of the starting vector $\mathbf{v}(t_{\mathrm{v}})$ when $t_{\mathrm{v}}$ is large. The statistics of $\mathbf{V}(0,\omega)$ are determined by the "backward" matrix product $\mathbf{X}^{\mathrm{T}}(1,\omega)\mathbf{X}^{\mathrm{T}}(2,\omega)\ldots\mathbf{X}^{\mathrm{T}}(t_{\mathrm{v}},\omega)$. This difference between reproductive value and population structure relates to the discussion in Example 4, above. If we were to generate the projection matrices by a Markov chain, the probabilities of "forward" products are determined by the transition probabilities of the "forward" chain, whereas the probabilities of "back-

ward" products are determined by the transition probabilities of
the "backward" chain.

Stochastic reproductive value has been studied in an analysis
of the stochastic version of the stable equivalent population (Tuljapurkar & Lee, in press; for a deterministic analysis of this idea,
see Keyfitz 1985).

Next, consider convergence to stability. When there is a fixed,
ergodic projection matrix, two initially different population structures converge over time to the stable structure. In the stochastic
case, the analogous question concerns the rate at which two sequences started with different population structures converge over
time. Elsewhere, I have discussed ways of measuring convergence
rate (Tuljapurkar 1990, Chapter 4). Benton and Grant (1996) examined convergence rate in several examples, concluding that the
deterministic convergence rate is often a good estimate of the
stochastic convergence rate, especially when variability is small.
There is little other published work on convergence.

This chapter ignores randomness in population dynamics that is
produced by sampling finite populations. For instance, consider $n$
individuals in a particular population stage with a survival rate $P$.
If $P$ is a probability of survival and individuals die independently,
the number of survivors is binomially distributed ($n$ trials each
with a probability of success $P$). If $EP = p$ and the environmental
variance is var $P = \sigma^2$, the reader may show that the variance in
the number of survivors is

$$p(1 - p)n + \sigma^2(n^2 - n).$$

Here, the first term is sampling variance, whereas the second term
is environmental variance of the sort discussed here. As $n$ increases,
the environmental variance outweighs the sampling variance. Thus,
the models studied here are more accurate for large populations
than for small ones, and it may be necessary to include sampling
variance when analyzing small populations, such as populations
close to extinction.

## 5 Invasion and ESS

Demographic models are often used to study evolutionary changes
in genotypic and phenotypic composition. A typical approach assumes that a phenotype of interest (e.g., the age pattern of mortality) is influenced by two alleles at a single autosomal locus. Initially, one allele is assumed to be common, and a second, invading,

allele to be rare. The object is to determine conditions under which the rare allele increases in frequency. Such analysis identifies an E(volutionarily) S(table) S(tate) as an allelic phenotype that can invade populations containing other alleles but that cannot be invaded when it is common. The analysis of invasions relies on the method described below. For applications of this approach, and a valuable biological perspective, see Metz et al. (1992).

To analyze invasion by a rare allele into a randomly mating population, start with a population that contains one resident allele, $A$, so that all individuals have genotype $AA$. Next, suppose that an invading allele $B$ is introduced into the population at some low frequency, $p_0 \ll 1$. With random mating, the relative frequencies of $AA$, $AB$, and $BB$ genotypes are $(1 - p_0)^2$, $2p_0(1 - p_0) \simeq 2p_0$, and $p_0^2 \simeq 0$, respectively. The implication is that the dynamics of allele $B$ during an invasion are described by the dynamics of the $AB$ heterozygote.

Now suppose that the dynamics of a homogeneous population of one genotype are described by the (possibly) density-dependent equation

$$\mathbf{N}(t + 1) = \mathbf{X}[t + 1, \mathbf{N}(t)] \, \mathbf{N}(t). \tag{32}$$

A homogeneous population of $AA$ genotypes is described by the steady-state dynamics of

$$\mathbf{N}_{AA}(t + 1) = \mathbf{X}[t + 1, \mathbf{N}_{AA}(t)] \, \mathbf{N}_{AA}(t). \tag{33}$$

The fate of an invasion is determined by the dynamics of $AB$ heterozygotes,

$$\begin{aligned}\mathbf{N}_{AB}(t + 1) &= \mathbf{X}[t + 1, \mathbf{N}_{AA}(t), \mathbf{N}_{AB}(t)] \, \mathbf{N}_{AB}(t) \\ &\simeq \mathbf{X}[t + 1, \mathbf{N}_{AA}(t)] \, \mathbf{N}_{AB}(t).\end{aligned} \tag{34}$$

The second line of (34) sets the negligible numbers of the rare $AB$ genotype to zero. Equation (34) is in the form of equation (13), except that in the density-dependent case it contains $\mathbf{N}_{AA}(t)$, which is obtained as a random sequence from equation (33). Therefore, equation (34) can be analyzed to find a stochastic growth rate for the number of $AB$ heterozygotes; if this growth rate is positive, allele $B$ increases in frequency and invades the population.

## 6 Parting Words

This chapter should allow readers to construct and analyze stochastic structured models and to read papers that use them. As pointed out at the start of this chapter, several applications are to be found elsewhere in this book. A useful application of stochastic models that we have not discussed here is the development of methods for estimating the vital rates of structured populations from data. Such estimation methods (Manton & Stallard 1988) are desirable in ecology, although rarely used.

## 7 Solutions

### *Two-State Chain*

The eigenvalues of $\mathbf{p}$, which are the solutions to the characteristic equation, are one and $\rho = c + d - 1$. Using these values, solve the eigenvector equation (e.g., $\mathbf{p}\boldsymbol{\pi}_1 = (1)\boldsymbol{\pi}_1$) for the left eigenvectors,

$$\boldsymbol{\pi}_1 = \frac{1}{2 - c - d} \begin{bmatrix} 1 - c \\ 1 - d \end{bmatrix}, \; \boldsymbol{\pi}_2 = \frac{1}{2 - c - d} \begin{bmatrix} 1 - c \\ -(1 - d) \end{bmatrix},$$

Here, by definition, $|\boldsymbol{\pi}_1| = 1$. Next, the corresponding right eigenvectors are found to be $\mathbf{e}$ and $\boldsymbol{\psi} = (1, -1)^{\mathrm{T}}$. From equation (5), deduce that powers $\mathbf{p}^n$ of the matrix $\mathbf{p}$ describe the $n$-step transition probabilities

$$\Pr\{S(t+n) = s_i | S(t) = s_j\} = (\mathbf{p}^n)_{ij} = p_{ij}^{(n)} \quad \text{for } n > 0.$$

With all eigenvectors known, the spectral decomposition of Chapter 2 can be written as

$$\mathbf{p}^n = \boldsymbol{\pi}_1 \mathbf{e}^{\mathrm{T}} + \rho^n \boldsymbol{\pi}_2 \boldsymbol{\psi}^{\mathrm{T}}.$$

Here $\boldsymbol{\pi}_1 \mathbf{e}^{\mathrm{T}} + \boldsymbol{\pi}_2 \boldsymbol{\psi}^{\mathrm{T}} = \mathbf{I}$, since these are the component matrices in the spectral decomposition.

To examine the memory of process, suppose that the chain is in its equilibrium state at time $t$, with $\mathbf{q}(t) = \boldsymbol{\pi}_1$. Writing the components of $\boldsymbol{\pi}_1$ as $\pi_i, i = 1, 2$, the time-independent average of $S(t)$ is

$$\mathrm{E}\, S(t) = \pi_1 s_1 + \pi_2 s_2 = \; < S > .$$

Let the deviations from this average be

$$\boldsymbol{\delta} = \begin{bmatrix} s_1 - <S> \\ s_2 - <S> \end{bmatrix} = \begin{bmatrix} \delta_1 \\ \delta_2 \end{bmatrix}.$$

The variance can be written in a convenient form by defining the vector

$$\boldsymbol{\mu} = \begin{bmatrix} \pi_1 \delta_1 \\ \pi_2 \delta_2 \end{bmatrix}.$$

Using this vector, compute the variance

$$\begin{aligned}
\operatorname{var} S(t) &= \operatorname{E}\left[S_t - <S>\right]^2 \\
&= \sigma_S^2, \\
&= \sum_i \pi_i\, \delta_i^2, \\
&= \boldsymbol{\delta}^{\mathrm{T}} \boldsymbol{\mu}.
\end{aligned}$$

The first line is merely the definition. Line two defines the symbol $\sigma_S^2$ to emphasize the fact that we are calculating a stationary (time-independent) variance. Line three inserts the values that constitute the expectation in the first line; and line four rewrites the sum as a scalar product of vectors.

To find the covariance between $S(t)$ and $S(t+n)$ for any positive $n$, recall that if $S(t) = s_j$, then the probability that $S(t+n) = s_i$ is given by the $ij$ element of $\mathbf{p}^n$. Thus, $\pi_j\, p_{ij}^{(n)}$ is the joint probability of these events, yielding the first two lines below,

$$\begin{aligned}
\operatorname{cov} S(t+n)S(t) &= \operatorname{E}\left[S(t+n) - <S>\right]\left[S(t) - <S>\right], \\
&= \sum_{i,j} p_{ij}^{(n)} \delta_i \pi_j \delta_j \\
&= \boldsymbol{\delta}^{\mathrm{T}} \mathbf{p}^n \boldsymbol{\mu} \\
&= \boldsymbol{\delta}^{\mathrm{T}}\left[\boldsymbol{\pi}_1 \mathbf{e}^{\mathrm{T}} + \rho^n\, \boldsymbol{\pi}_2 \boldsymbol{\psi}^{\mathrm{T}}\right]\boldsymbol{\mu} \\
&= \rho^n \boldsymbol{\delta}^{\mathrm{T}} \boldsymbol{\pi}_2 \boldsymbol{\psi}^{\mathrm{T}} \boldsymbol{\mu} \\
&= \rho^n \boldsymbol{\delta}^{\mathrm{T}}\left[\mathbf{I} - \boldsymbol{\pi}_1 \mathbf{e}^{\mathrm{T}}\right]\boldsymbol{\mu} \\
&= \rho^n \boldsymbol{\delta}^{\mathrm{T}} \boldsymbol{\mu} \\
&= \rho^n \operatorname{var} S(t).
\end{aligned}$$

Between lines two and three above, use the definition of $\boldsymbol{\mu}$; between lines three and four, use the spectral decomposition. The definition of $<S>$ implies that $\boldsymbol{\delta}^{\mathrm{T}} \boldsymbol{\pi}_1 = 0$, which takes us from line four to line five. Between five and six, use the fact that the component matrices of the spectral decomposition add to the unit matrix. From six to seven, use the argument that got us from four to five.

*Autoregressive Processes*

For equation (9) and equation (10), average both sides term by term and assume that $\operatorname{E}S(t) = \operatorname{E}S(t-1) = <S>$, to find $<S> = 0$.

In equation (9), square both sides, assume stationarity, and find

$$\begin{aligned}
\mathrm{E}S(t)^2 &= \operatorname{var} S(t) \\
&= \sigma_S^2 \\
&= c^2 \operatorname{var} S(t-1) + 2cd\,\mathrm{E}[S(t-1)\epsilon(t)] + d^2 \operatorname{var} \epsilon(t) \\
&= c^2 \operatorname{var} S(t-1) + d^2\,\sigma_\epsilon^2 \\
&= d^2\,\sigma_\epsilon^2/(1-c^2).
\end{aligned}$$

Line two emphasizes the fact that $\operatorname{var} S(t)$ is time-independent (the variance is the same for each $t$). In line three, the middle term has value zero because the shock is independent of the environmental state, leading to line four. Setting line four equal to $\operatorname{var} S(t)$, and collecting terms, solve to find the expression in line five.

Stationarity means that the joint probability distribution of any set of environmental state—say $S(t), S(t+n), S(t-m)$ for any $n > 0$ and $m > 0$—depends not on the value of $t$ but only on the differences $n$ and $m$. Ergo, stationary covariances obey the rules

$$\mathrm{E}[S(t+n)S(t)] = \mathrm{E}[S(t-n)S(t)] = R(n).$$

To compute the autocovariance, multiply equation (9) by $S(t-m)$ on both sides and find the expected value of every term. On the right, $\mathrm{E}\,[S(t-n)\epsilon(t)] = 0$ if $n > 0$. Thus,

$$R(n) = c\,R(n-1) = c^n\,R(0) = c^n\,\sigma_S^2,$$

which is the same form as equation (7), but with $c$ replacing $\rho$.

The second-order process (10) is a bit more complicated. Again, let $R(n)$ be the stationary covariance between $S(t+n)$ and $S(t)$, and define the stationary autocorrelation $\rho(n)$ by

$$\rho(n) = R(n)/\sigma_S^2.$$

Note (stationarity!) that $R(n) = R(-n)$ and so $\rho(n) = \rho(-n)$. Multiply both sides of equation (10) by $S(t-n)$ and compute the expectation to find

$$R(n) = c_1\,R(n-1) + c_2\,R(n-2) \quad \text{when } n > 0.$$

Here, we have again used the independence of $\epsilon(t)$ and $S(t-n)$ when $n > 0$. Dividing through by the variance of $S(t)$ leads to

$$\rho(n) = c_1\,\rho(n-1) + c_2\,\rho(n-2) \quad \text{when } n > 0.$$

It is obvious that $\rho(0) = 1$. Setting $n = 1$ in the last displayed equation yields

$$\rho(1) = c_1/(1-c_2).$$

With $\rho(0)$ and $\rho(1)$ known, the equation for $\rho(n)$ is a linear difference equation with the solution (for $n > 0$)

$$\rho(n) = b_1\, g_1^n + b_2\, g_2^n,$$

where the $g_i$ are the roots of

$$g^2 - c_1 g - c_2 = 0$$

and where $b_1$ and $b_2$ are found by using the known values of $\rho(0)$
and $\rho(1)$. If the $g_i$ are complex, the values of $\rho(n)$ oscillate with
$n$, as stated in the text. To find the variance $\sigma_S^2 = R(0)$, multiply
equation (10) by $S(t)$, and take expectations to get

$$
\begin{aligned}
R(0) &= c_1\, R(-1) + c_2\, R(-2) + \mathrm{E}[S(t)\epsilon(t)], \\
&= c_1\, R(0)\rho(1) + c_2\, R(0)\rho(2) + d\,\sigma_\epsilon^2.
\end{aligned}
$$

The last term in line one is evaluated by multiplying equation (10)
by $\epsilon(t)$ on both sides and taking the expectation of all terms. We
already know all the $\rho(n)$, so we are done.

*Example 1*

In equation (15), let $\mathbf{Y}(t)$ be the structure vector for the population, and employ equation (14) with

$$\mathbf{X}(t) = \left[ \begin{array}{cc} (1-f)R(t) & sR(t) \\ f & 0 \end{array} \right].$$

The component equations for the components of $\mathbf{Y}(t)$ are therefore

$$Y(1,t) = \frac{(1-f)R(t)Y(1,t-1) + sR(t)Y(2,t-1)}{[(1-f)R(t)+f]Y(1,t-1) + sR(t)Y(2,t-1)},$$

$$Y(2,t) = \frac{fY(1,t-1)}{[(1-f)R(t)+f]Y(1,t-1) + sR(t)Y(2,t-1)}.$$

Instead of using this pair of equations, it is faster to set

$$U(t) = \frac{Y(2,t)}{Y(1,t)},$$

divide the second equation of the pair by the first, and get

$$U(t) = \frac{f}{(1-f)R(t) + sR(t)U(t-1)}.$$

This is a one-dimensional, scalar, difference equation.

*Expansion of a*

In place of equation (22), define pairwise covariances of the random parts of the matrix by

$$\sigma_{ij,k\ell} = \text{cov}\, H_{ij} H_{k\ell}.$$

Let the sensitivities of the average projection matrix be $s_{ij}$. Then the expansion for $a$ becomes

$$a \simeq r - \frac{1}{2\lambda_0^2} \sum_{(ij),(k\ell)} \sigma_{ij,k\ell}\, s_{ij}\, s_{k\ell}.$$

## Acknowledgments

I thank Hal Caswell, Steve Orzack, and Jean Doble for suggestions, some of which I ignored at my (and your) peril. National Science Foundation grant DEB-9420153 and National Institutes of Health grant HD 16640 provided support, as did the Morrison Institute for Population and Resource Studies at Stanford University, courtesy of Marc Feldman and Jean Doble. I am grateful to Anjali Tuljapurkar for suffering through part of the Cornell summer when these lectures were given, and for her enthusiastic company on the streams.

## Literature Cited

Bartlett, M. S. 1978. *An Introduction to Stochastic Processes.* 3d ed. Cambridge University Press.

Benton, T. G., and A. Grant. 1996. How to keep fit in the real world: Elasticity analyses and selection pressures on life histories in a variable environment. *American Naturalist* 147: 115–139.

Box, G. E. P., G. M. Jenkins, and G. C. Reinsel. 1994. *Time Series Analysis: Forecasting and Control.* 3rd ed. Prentice Hall, Englewood Cliffs, N.J.

Chatfield, C. 1989. *The Analysis of Time Series.* Chapman & Hall, London.

Cohen, J. E. 1977. Ergodicity of age structure in populations with Markovian vital rates. II. General states. *Advances in Applied Probability* 9: 18-37.

Furstenberg, H., and H. Kesten. 1960. Products of random matrices. *Annals of Mathematical Statistics* 31: 457–469.

Gardiner, C. W. 1985. *Handbook of Stochastic Methods for Physics, Chemistry, and the Natural Sciences.* Springer-Verlag, Berlin.

Karlin, S., and H. M. Taylor. 1975. *A First Course in Stochastic Processes.* 2nd ed. Academic Press, New York.

Kemeny, J. G., and J. L. Snell. 1960. *Finite Markov Chains.* Van Nostrand, Princeton, N.J.

Key, E. S. 1987. Computable examples of the maximal Lyapunov exponent. *Probability Theory & Related Fields* 75: 97–107.

Keyfitz, N. 1985. *Applied Mathematical Demography.* 2nd ed. Springer-Verlag, New York.

Knuth, D. E. 1981. *The Art of Computer Programming. Volume 2: Seminumerical Algorithms.* Addison-Wesley, Reading, Mass.

Manton, K. G., and E. Stallard. 1988. *Chronic Disease Modelling : Measurement and Evaluation of the Risks of Chronic Disease Processes.* Oxford University Press, New York.

Metz, J. A. J., S. A. H. Geritz, and R. M. Nisbet. 1992. How should we define fitness for general ecological scenarios? *Trends in Ecology & Evolution* 7: 198–202.

Ott, E. 1993. *Chaos in Dynamical Systems.* Cambridge University Press.

Ripley, B. D. 1987. *Stochastic Simulation.* Wiley, New York.

Roerdink, J. B. T. M. 1987. The biennial life strategy in a random environment. *Journal of Mathematical Biology* 26: 199–215.

———. 1989. The biennial life strategy in a random environment. Supplement. *Journal of Mathematical Biology* 27: 309–320.

Scaffer, W. M. 1974. Optimal reproductive efforts in fluctuating environments. *American Naturalist* 108: 783–790.

Seneta, E. 1981. *Non-Negative Matrices and Markov Chains.* Springer-Verlag, New York.

Tuljapurkar, S. 1990. *Population Dynamics in Variable Environments.* Springer-Verlag, New York.

Tuljapurkar, S., and C. Istock. 1993. Environmental uncertainty and variable diapause. *Theoretical Population Biology* 43: 251–280.

Tuljapurkar, S., and R. Lee. In press. Demographic uncertainty and the stable equivalent population. *Mathematical and Computer Modelling.*

Tuljapurkar, S., and S. H. Orzack. 1980. Population dynamics in variable environments. I: Long-run growth rates and extinction. *Theoretical Population Biology* 18: 314–342.

CHAPTER 4

# Delay-Differential Equations for Structured Populations

*Roger M. Nisbet*

This chapter develops an approach to continuous-time models that emphasizes the stage structure of a population. The key motivating idea is that, in many situations, biological differences among individuals within a stage may be unimportant in comparison with interstage differences. For example, the rate of egg production by adult insects may depend on their age, but much insight into population dynamics may be obtained from models that recognize only the distinction between nonreproductive individuals (juveniles) and reproductive individuals (adults). Likewise, an organism's susceptibility to parasitism may depend on its size but with a range of sizes at which the organism is not vulnerable. It is reasonable to explore the qualitative effects of the invulnerable class before investigating the detailed, quantitative consequences of the size structure.

The most immediate way in which stage structure influences population dynamics is by introducing time delays in the action of regulatory mechanisms, and delay-differential equations (DDE's) constitute a natural context within which to figure out the effects of the time delays without compromising the rigorous, individual-based philosophy advocated in Chapter 5, by de Roos. For example, in the model of giant kelp, shading by the largest plants reduces the rate of recruitment of the youngest stages. Some finite time is required before these young plants grow sufficiently to themselves contribute to the shading. In the host-parasitoid models, the finite developmental times of both species play a role in determining stability.

The mathematics of nonlinear DDE's is hard, but it has advanced rapidly in recent years. Recent texts for the mathematically battle-hardened include those by Gyori and Ladas (1991), Kuang (1993), and Diekmann et al. (1995). The reader interested in analyzing the characteristic equations that arise in the study of linear delay-differential equations will find MacDonald's (1989) monograph particularly valuable. My hope in writing this chapter, however, is to make it possible to glimpse some of the concepts and appreciate the explanatory power for ecology of delay equations through a less rigorous approach involving a series of examples of increasing complexity. Section 1 starts with a representation of the dynamics of a two-stage population in a constant environment. Section 2 then presents a model with density dependence acting only through the effect of adult density on juvenile recruitment. This type of model is already mathematically rich, and it has been used to model organisms as diverse as blowflies (Gurney et al. 1980), whales (May 1980), and giant kelp (Nisbet & Bence 1989). Section 3 introduces formalism, developed by Gurney et al. (1983), for models in which mortality in pre-adult stages depends on the environment. This formalism has proved particularly powerful in the study of insect host-parasitoid interactions, and Section 4 introduces a simple model of this type due to Murdoch et al. (1987). Recognizing that in nature, time delays vary in duration in response to population density and to changes in the environment, Section 5 outlines an approach to modeling time-dependent delays (Nisbet & Gurney 1983), which has been used in modeling populations of aquatic insects (Crowley et al. 1987) and zooplankton (Nisbet et al. 1989; McCauley et al., in press). Section 6 comments on the possibilities for more-complex models and on the limitations of the formalism.

The mathematical level here is deliberately not uniform. Sections 1 and 2 include a fair quantity of algebraic detail, in the belief that statements like "It can be shown that ..." invite any number of unprintable responses unless the reader has seen some of the details in at least one situation. The analysis of any but the simplest DDE's, however, quickly leads to heavy algebra, which can obscure the overview, and I make increasing use of references to the primary literature as the chapter progresses.

## 1   A Two-Stage Model of Population Growth in a Constant Environment

The simplest, continuous-time stage-structured model of a closed system recognizes only two stages, juveniles and adults. The juvenile-stage duration is $\tau$, and the per capita death rate is $\mu_J$. Adult per capita death rate is $\mu_A$, and fecundity (offspring produced per adult per unit of time) is $\beta$. The following notation is used throughout this chapter:

$N_A(t)$,  adult population at time $t$,

$R_A(t)$,  total adult recruitment rate (i.e., new adults per unit of time) at time $t$,

$D_A(t)$,  total adult death rate at time $t$,

$N_J(t)$,  juvenile population at time $t$,

$R_J(t)$,  total juvenile recruitment rate (i.e., new offspring per unit of time) at time $t$,

$M_J(t)$,  total maturation rate from the juvenile stage at time $t$, and

$D_J(t)$,  total juvenile death rate at time $t$.

Note the distinction between total and per capita rates. If, for example, a bottle contains 200 adult waterfleas, each releasing 2 neonates per day, $\beta = 2$ juveniles day$^{-1}$ adult$^{-1}$ and $R_J(t) = 400$ juveniles day$^{-1}$. As in all continuous-time models, per capita rates are defined on an instantaneous basis but must be expressed in terms of practical units (e.g., days, years). Given a per capita death rate $\mu$, the fraction of a population that survives $T$ time units is $\exp(-\mu T)$. Thus, if 50 of the 200 original waterfleas present at the start of one day die during that day, the per capita death rate is 0.288 day$^{-1}$ (i.e., $-\ln(0.75)$) and not 0.25 day$^{-1}$.

Derivation of equations describing the population dynamics of the juvenile and adult populations proceeds in two steps. First, each population is described by a population balance equation; these equations express in mathematical language the requirement that, for any stage, the rate of change of the population size must equal the difference between total recruitment and loss rates for that stage. Thus, in our example, the adult balance equation is

$$dN_A(t)/dt = R_A(t) - D_A(t), \qquad (1)$$

because all change in adult population is due to recruitment or death. The juvenile balance equation has an additional loss term since changes in a juvenile population occur through recruitment (i.e., birth), death, and maturation to the adult stage. Thus,

$$dN_J(t)/dt = R_J(t) - M_J(t) - D_A(t). \qquad (2)$$

The final component of the population balance recognizes that the rate of maturation from the juvenile population is equal to the rate of recruitment to the adult population, giving

$$R_A(t) = M_J(t). \tag{3}$$

The second step in formulating the model equations is setting out the model functions that relate rate processes (recruitment, death, maturation) to stage populations. The assumptions in the opening paragraph of this section give

$$D_A(t) = \mu_A N_A(t), \quad D_J(t) = \mu_J N_J(t), \quad R_J(t) = \beta N_A(t). \tag{4}$$

Furthermore, the rate of maturation from the juvenile stage at time $t$ is equal to the rate of recruitment to that stage at the earlier time $t - \tau$ multiplied by the proportion of recruits that survive, as juveniles, over the time interval $t - \tau$ to $t$. Thus,

$$M_J(t) = R_J(t - \tau) \exp(-\mu_J \tau). \tag{5}$$

Equations (1)–(5) describe the dynamics of the system, although not in the form of a standard set of ordinary differential equations. With a little manipulation involving equations (4) and (5), the differential equation for the adults can be written as

$$dN_A/dt = Q N_A(t - \tau) - \mu_A N(t), \text{where } Q = \beta \exp(-\mu_J \tau). \tag{6}$$

The parameter $Q$ can be thought of as the per capita rate of production of adults for the next generation, taking account of both fecundity and juvenile survival. Equation (6) is a delay-differential equation (DDE), so-called because the rate of change of $N_A$ at any time $t$ depends not only on its current value but also on its value at an earlier time, $t - \tau$. Equation (6) is also of the first order (no higher-order derivatives) and linear (the derivative and the right-hand side contain no nonlinear terms).

Linear, delay-differential equations share many properties with the more familiar linear, ordinary, differential equations, but there are many important differences. Probably the most important is that a solution is not specified uniquely by the initial value of the dependent variable; we need in addition to specify values over a time interval equal in length to the delay, for example, the interval $-\tau \le t < 0$, together with the value when $t = 0$. I call this information the initial history. Arguably, the simplest method of constructing a numerical solution to a first-order DDE is to consider successive intervals of duration $\tau$. Armed with the initial history, we know the values of $N_A(t - \tau)$ when $0 \le t < \tau$; thus, we can integrate the DDE up to the time at which $t = \tau$ using any one from the multiplicity of algorithms on offer for solving ordinary differential equations. But $0 \leq t < \tau$, $N_A(t)$ is equal to $N_A(t - \tau)$ when $\tau \leq t < 2\tau$, so we can integrate the DDE up to $t = 2\tau$; and so on. With simple DDE's (e.g., eq. 6), this method is easily implemented using computer algebra in a system such as MATHEMATICA or MAPLE. With more-complex equations, or with systems of DDE's with more than one time delay, there is canned software, for example, the SOLVER package (Gurney & Tobia 1995).

Sometimes an initial history is unavailable or even meaningless. For example, a laboratory population may be set up at some particular time (which we call $t = 0$). In such circumstances, it is natural to regard the population dynamics as being determined by the DDE together with the initial juvenile-age distribution (i.e., the number of individuals per age interval of age $a$ at time 0), denoted by $m(0, a)$, $0 \leq a < \tau$, following the notation used by de Roos (Chapter 5). The adult recruitment rate is then given by

$$R_A(t) = m(0, \tau - t) \exp\left[-\mu_J(\tau - t)\right] \quad \text{for} \quad t \leq \tau. \qquad (7)$$

Notice here that $\tau - t$ is the initial age of a juvenile maturing at time $t \leq \tau$. To obtain $N_A(t)$ when $0 \leq t < \tau$, solve

$$dN_A/dt = R_A(t) - \mu_A N_A(t), \qquad (8)$$

with $R_A(t)$ evaluated using equation (7). When $t \geq \tau$, the integration of the DDE proceeds as in the preceding paragraph. This procedure is particularly simple if the initial juvenile population is zero. Then $m(0, a) = 0$, and there is no adult recruitment earlier than when $t = \tau$.

This model is a particular example of a continuous-time, linear, age-structured model. From the general theory developed in Chapter 5 by de Roos, we expect the population to eventually grow or decline exponentially at some long-run growth rate $\lambda$, achieving a stationary stage structure (i.e., a constant ratio of juveniles to adults). To derive the long-run growth rate, assume that $N_A(t) \propto \exp(\lambda t)$; then,

$$N_A(t - \tau) \propto \exp\left[\lambda(t - \tau)\right] = \exp(-\lambda \tau) \exp(\lambda t). \qquad (9)$$

Substituting into equation (6) shows that $\lambda$ must be the real root of the characteristic equation

$$\lambda + \mu_A = Q \exp(-\lambda t). \qquad (10)$$

A simple graphical argument (Fig. 1) shows that the long-run growth rate is positive if $Q > \mu_A$ (i.e., $\beta \exp[-\mu_J \tau] > \mu_A$) and negative if this inequality is reversed. This is consistent with intuition

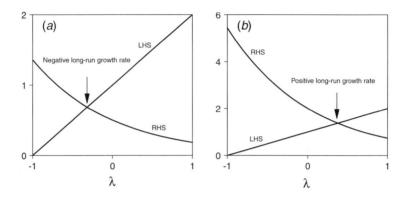

FIGURE 1. *The real root of the characteristic equation: left-hand side and right-hand sides of equation (10) against $\lambda$ for $\mu_A = 1$ and (a) $Q = 0.5$, (b) $Q = 2.0$. The long-run growth rate is the value of $\lambda$ at the intersection point.*

if we note that the mean lifetime offspring production per adult is $\beta/\mu_A$, since $\mu_A^{-1}$ is the mean adult lifetime. When juvenile survival is taken into account, the mean lifetime contribution to the next generation of adults is $\beta\exp(-\mu_J\tau)/\mu_A$; the condition for a positive long-run growth rate is that this contribution be greater than one.

The characteristic equation (10) may be derived as a special case of the general age-structured model in de Roos' chapter (Chapter 5) by evaluating the terms in his characteristic equation (eq. 27), taking his general $i$-state variable $x$ to be age. His birth function $b(E_c, x)$ is $\beta$ if $x \geq \tau$ and zero otherwise. His growth function $g(E_c, x) = 1$. Because there is no upper limit to age in this model, $x_m = \infty$. The survival function $U(E_c, x) = \exp\left[-\mu_J\tau - \mu_A(x - \tau)\right]$ when $x \geq \tau$. De Roos' equation (27) then becomes

$$1 = \beta e^{-\mu_J\tau} \int_\tau^\infty e^{-\lambda x} e^{-\mu_A(x-\tau)} dx$$

$$= Q e^{-\lambda\tau} \int_0^\infty e^{-\lambda(x-\tau)} e^{-\mu_A(x-\tau)} dx \qquad (11)$$

$$= Q e^{-\lambda\tau}/(\lambda + \mu_A),$$

which is equivalent to equation (10), above.

For a population growing at the long-run growth rate, the stationary stage distribution is most easily obtained by substituting

(a)

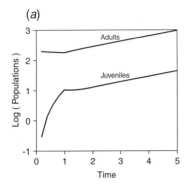

(b)

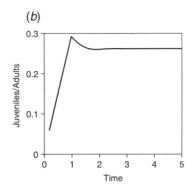

FIGURE 2. *Illustration of the approach to long-run growth rate and a stable age distribution in a numerical solution of equations (1)–(5) with parameters $\beta = 0.3$, $\mu_J = 0.1$, $\mu_A = 0.05$, $\tau = 1$ and initial conditions $N_A = 10$ and $N_J = 0$ at $t = 0$, $N_A = N_J = 0$ when $0 \leq t < \tau$. (a) Logarithm of juvenile and adult populations; (b) ratio of juveniles to adults.*

from equations (3)–(5) into equation (2), with the result that

$$N_J(t)/N_A(t) = \beta \left\{1 - \exp\left[-(\lambda + \mu_J)\tau\right]\right\}/(\lambda + \mu_J). \qquad (12)$$

In the special situation of a stationary population, $\lambda = 0$, and a little algebra using equations (10) and (12) leads to a simple result:

$$N_J(t)/N_A(t) = (\beta - \mu_A)/\mu_J, \qquad (13)$$

which is consistent with the requirement that in a stationary population, total recruitment rate $(\beta N_A)$ must balance total death rate $(\mu_J N_J + \mu_A N_A)$. Figure 2 shows an example of the approach to a stationary age distribution and a stable long-run growth rate.

## 2  A Two-Stage Model with Density Dependence in the Adult Stage

As noted in earlier chapters, individual birth and death rates vary in response to the environment. One component of the environment that may be of particular importance is the density (i.e., number per unit area or volume) of individuals in a particular stage, and it is well known that populations with density-dependent vital rates may be regulated and, indeed, may even approach some stable equilibrium. For an ecologist's perspective on the significance of density dependence, see Murdoch (1994).

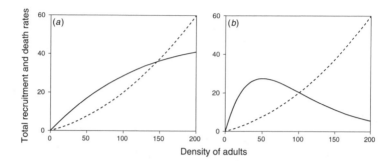

FIGURE 3. *Plausible forms of adult density dependence of total adult recruitment rate* (solid line) *and total adult death rate* (dashed line)*:* (a) *no overcompensation in recruitment;* (b) *overcompensation in recruitment. The curves intersect at the adult equilibrium population. The slopes of the tangents to the two curves at the equilibrium define the quantities* $\alpha$ *and* $\gamma$ *used in the local stability analysis.*

The simplest situation to model includes adult fecundity and mortality dependent on adult population density, while retaining the assumption of constant juvenile mortality. The population dynamics can still be described by a single DDE, but the equation is nonlinear. In this section I investigate the stability of an equilibrium state of such a nonlinear DDE, showing how time-delayed negative feedback can lead to instability and persisting population oscillations.

*Formalism*

The new model retains all the assumptions of the preceding section except that adult fecundity and death rates are now functions of the density of adults; that is, $\beta(t) = \beta\,[N_A(t)]$ and $\mu_A(t) = \mu_A\,[N_A(t)]$. This implies that the juvenile recruitment rate $R_J(t)$ is a function of $N_A(t)$ and, hence, that the adult recruitment rate $R_A(t)$ is a function of $N_A(t - \tau)$. In the absence of cooperative interactions (e.g., an Allee effect influencing recruitment), we expect fecundity to decrease with adult density; thus, the dependence of $R_A$ on adult density can take one of two forms, illustrated in Figure 3. Likewise, we expect adult per capita mortality to increase (or at least not decrease) with adult density; then, $D_A(t)$ increases at least linearly with $N_A(t)$. With these assumptions, the adult population dynamics are fully described by the DDE

$$dN_A(t)/dt = R_A \left[ N_A(t - \tau) \right] - D_A \left[ N_A(t) \right]. \qquad (14)$$

At equilibrium, recruitment and loss rates are perfectly balanced for some time interval of a duration of at least $\tau$. Thus, at any equilibrium, $N^*$, of equation (14),

$$R_A(N^*) = D_A(N^*). \qquad (15)$$

The equilibrium is locally stable if all small populations eventually decay to zero. To analyze local stability, we investigate the fate of small perturbations $n_A(t)$, defined by

$$n_A(t) = N_A(t) - N^* \quad \text{with} \quad |n_A(t)| \ll N^*. \qquad (16)$$

The right-hand side of equation (14) is approximated by the leading terms of a Taylor expansion, as in the standard local stability analysis of models using ordinary differential equations (see, e.g., Nisbet & Gurney 1982, Chapters 2 and 4), with the result that

$$dn_A(t)/dt = \alpha n_A(t - \tau) - \gamma n_A(t), \qquad (17)$$

where

$$\alpha = (dR_A/dN_A)_{N_A=N^*}, \quad \gamma = (dD_A/dN_A)_{N_A=N^*}. \qquad (18)$$

The coefficients $\alpha$ and $\gamma$ can be interpreted as the slopes at equilibrium of the adult recruitment curve and the death curves, respectively (see Fig. 3).

Equation (17) is similar to the linear DDE (eq. 6) that arose in the analysis of density-independent population growth. However, as is evident from Figure 3, there is one key difference: the coefficient $\alpha$, unlike the coefficient $Q$, may be positive or negative, depending on whether there is overcompensation in recruitment. This opens new possibilities for the dynamics. As before, we assume a solution proportional to $\exp(\lambda t)$ and obtain the characteristic equation

$$\lambda + \gamma = \alpha \exp(-\lambda \tau). \qquad (19)$$

If $\alpha$ is positive, the analysis above holds: this equation has one real root, which is positive if $\alpha < \gamma$ and negative if this inequality is reversed. When the real root is negative (Fig. 3a), the perturbation eventually decays to zero, and the equilibrium is stable. Thus, unless there is overcompensation in recruitment, the equilibrium in this model is locally stable.

Now suppose that $\alpha$ is negative—that is, there is overcompensation in recruitment—and the equilibrium is to the right of the "hump" in the recruitment curve (Fig. 3b). The equilibrium is certainly stable if $\tau = 0$, and it is natural to ask whether increasing the value of $\tau$ can ever destabilize the system? The answer is "yes" if (and only if) $|\alpha| > \gamma$, that is, if the rate of fall in recruitment is larger than the rate of increase in the death rate. To prove this, first note that equation (19) may have complex roots of the form $\lambda = \xi + i\omega$. Linear combinations of solutions proportional to $\exp(\lambda t)$ then represent oscillations, which grow in amplitude if $\xi > 0$ and decay toward zero if $\xi < 0$. The period of the oscillations is $2\pi/\omega$, and $\omega$ is commonly called the angular frequency (in radians per unit of time). The stability when $\tau = 0$ suggests that for small values of the delay $\tau$, all roots of equation (19) have negative real parts; that is, $\xi < 0$. As $\tau$ increases, the roots of equation (19) trace paths in the complex plane, and it is possible that a conjugate pair of roots may cross the imaginary axis; this would represent a transition from stability to instability of the equilibrium. Suppose that $\lambda = i\omega$ ($\omega > 0$) is a root of equation (19). Then, using the identity $\exp(-ix) = \cos x - i \sin x$,

$$i\omega + \gamma = \alpha \exp(-i\omega\tau) = \alpha \cos \omega\tau - i\alpha \sin \omega\tau. \qquad (20)$$

Equating real and imaginary parts yields

$$\cos \omega\tau = \gamma/\alpha \quad \text{and} \quad \sin \omega\tau = -\omega/\alpha. \qquad (21)$$

Since a cosine must take values in the range $[-1, 1]$, the first of these equations yields a contradiction unless $|\alpha| > \gamma$, thereby proving the condition given above. To find the critical value of the delay at which instability occurs, find the angular frequency by squaring and adding the two equations (21), recalling that $\cos^2 + \sin^2 = 1$, and obtaining

$$\omega = (\alpha^2 - \gamma^2)^{1/2}. \qquad (22)$$

From equation (21), the delay at the point of marginal stability is given by

$$\tau = \cos^{-1}(\gamma/\alpha)/\omega = \cos^{-1}(\gamma/\alpha)/(\alpha^2 - \gamma^2)^{1/2}, \qquad (23)$$

where $\cos^{-1}$ means "the angle, expressed in radians, whose cosine is." An immediate implication of this formula is that the critical delay is short if the slope of the recruitment curve at the equilibrium point is steep, that is, if density dependence is strong at this density. See the kelp model, below.

Finally, equation (21) tells us that when $\alpha$ is negative, $\cos \omega \tau < 0$ and $\sin \omega \tau > 0$. This implies that $\omega \tau$ must lie in the range $(\pi/2, \pi)$. Since the period of the cycles is $2\pi/\omega$, the cycle period must lie in the range $(2\tau, 4\tau)$. The lower limit is of particular interest because it tells us that population cycles caused by the action of delayed negative feedback must have a period of at least twice the developmental delay. This can be a powerful diagnostic property when interpreting ecological data.

The analysis to this point tells only what happens to a system poised exactly at the point of marginal stability. Often, however, an oscillatory instability indicates the occurrence of a stable limit cycle, a periodic solution of the DDE whose form is independent of the initial history. Mathematical proof of the existence of limit cycles in DDE's is a formidable task (see, e.g., Gyori & Ladas 1991), so it is common to revert at this point to numerical solutions of the equations. In the remainder of this section, some properties of periodic solutions are illustrated by two models, the blowfly model of Gurney et al. (1980) and a model of giant kelp due to Nisbet and Bence (1989).

*Example: Blowfly Model*

Gurney et al. (1980; see also Nisbet & Gurney 1982, Chapter 8) constructed a model of a laboratory blowfly population, one of the classic experimental demonstrations of population cycles (Nicholson 1954, 1957). The adult recruitment and death functions were

$$R_A(N_A) = q N_A \exp(-N_A/N_0), \quad D(N_A) = \mu_A N_A, \qquad (24)$$

where $q$ and $N_0$ are constants. For this model, the equilibrium population is

$$N^* = N_0 \ln(q/\mu_A). \qquad (25)$$

The coefficients $\alpha$ and $\gamma$ are evaluated from equation (18). The resulting expressions involve $N_A^*$; after substituting from equation (25) and using a few lines of algebra, they can be reduced to the form

$$\alpha = \mu_A[1 - \ln(q/\mu_A)], \quad \gamma = \mu_A. \qquad (26)$$

Substitution in equation (23) now gives the critical value for the delay at the onset of instability.

Figure 4 shows a series of numerical solutions of the full nonlinear

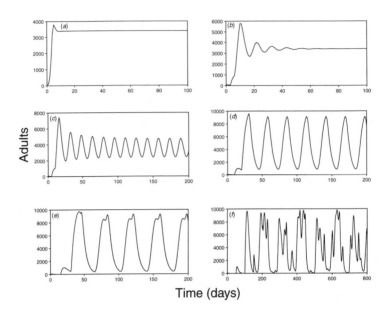

FIGURE 4. *Effect on population dynamics of increasing time delay. Adult population plotted against time for the blowfly model with $q = 6$, $N_0 = 1,000$, $\mu_A = 0.2$, and (a) $\tau = 1$ (stable equilibrium; no significant oscillations), (b) $\tau = 3$ (stable equilibrium; damped oscillations), (c) $\tau = 5$ (unstable equilibrium; small-amplitude limit cycles), (d) $\tau = 10$ (large-amplitude limit cycles), (e) $\tau = 15$ (limit cycles with complex structure), (f) $\tau = 50$ (chaos).*

DDE, with parameters taking values in the range appropriate for the blowfly populations. With these parameters, the critical value of the delay for instability is 4.6 days. The numerical solutions are consistent with this calculation: when $\tau = 1$ or 3, the equilibrium is stable; when $\tau = 5$, there is a small-amplitude limit cycle. Predicting the nonlinear dynamics with larger values of $\tau$ is beyond the scope of the linear analysis presented here, but Figure 4 illustrates some possibilities. With a further increase in $\tau$, the limit-cycle amplitude initially increases ($\tau = 10$) and then develops a double peak ($\tau = 15$), which, interestingly, is also present in the experimental data. Finally, with a long delay, chaotic solutions are found. More detail on the possibilities for complex dynamics is given by Blythe et al. (1982).

*Example: Kelp Model*

The analysis of the characteristic equation, and the numerical study of the blowfly model, establishes the possibility of sustained oscillations (limit cycles) caused by delayed negative feedback on population vital rates. The period of the cycles exceeds twice the developmental delay. In one instructive model, the limit cycle is approached in a finite time and has a form that can be determined analytically. The model, a particular case of equation (14), was first studied by an der Heiden and Mackay (1982) and has since been used as the starting point for a study of the dynamics of California giant kelp (Nisbet & Bence 1989).

The model is an idealization of a kelp forest and describes the dynamics of "adult" plants. Spores are assumed to be available at all times and to originate beyond boundaries of the system. Successful recruitment of young plants requires light levels at the ocean bed to exceed some critical threshold value; however, light is attenuated by shading from adult plants, and recruitment is suppressed if adult plant density exceeds some critical level $N_0$. Adult plants are assumed to experience density-independent mortality. Thus,

$$R_A(t) = \begin{cases} R_0 & \text{if } N_A(t - \tau) \leq N_0 \\ 0 & \text{otherwise} \end{cases} \quad \text{and} \quad D_A(t) = \mu_A N_A. \quad (27)$$

These forms are sketched in Figure 5*a*, from which it is clear that if $\mu_A$ is sufficiently large, the equilibrium is stable (since $\alpha$ in eq. 17 is zero). However, if $\mu_A$ is smaller, then $\alpha = -\infty$, and the equilibrium is unstable for any nonzero time delay. The form of the limit cycle is shown in Figure 5*b* and can be understood by assuming an initial history in which $N(t) < N_0$ for all $t \leq 0$. The population grows initially until it reaches $N_0$, at which time juvenile recruitment ceases. Adult population growth continues for an additional $\tau$ units of time, at which time adult recruitment ceases. The adult population then declines exponentially, juvenile recruitment resumes when the adult population passes $N_0$, and adult recruitment starts $\tau$ time units later. The period of the cycle, clearly, must exceed $2\tau$.

The adult population dynamics on the upswing obey the simple, linear, ordinary differential equation

$$dN_A/dt = R_0 - \mu_A N_A, \quad (28)$$

which can be solved explicitly. On the downswing, the adult pop-

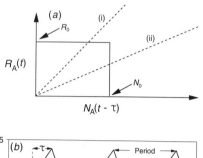

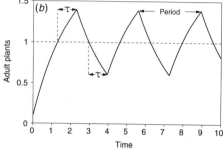

FIGURE 5. (a) *Assumed form for the density dependence of adult recruit-ment rate in the kelp model, and two possible curves for total adult death rate:* (i) *with the equilibrium stable and overdamped;* (ii) *with no limit cycles.* (b) *A typical limit cycle in the kelp model. Parameter values: $R_0 = 1$, $N_0 = 1$, $\mu_A = 0.5$. Segments of duration $\tau$ illustrate the mechanism that ensures that the cycle period must exceed $2\tau$. The exact value of the cycle period is 3.327 in agreement with equation (30).*

ulation declines exponentially. With some algebraic labor (or by consulting Nisbet & Bence 1989, Appendix), it can be shown that the highest population, $N_H$, the lowest population, $N_L$, and the cycle period, $P$, are given by

$$N_H = N_0 \exp(-\mu_A \tau) + R_0[1 - \exp(-\mu_A \tau)]/\mu_A \,,$$
$$N_L = N_0 \exp(-\mu_A \tau) \,, \tag{29}$$

$$P = 2\tau + \mu_A^{-1} \Big\{ \ln(N_H/N_0) \\ + \ln\left[(R_0 - \mu_A N_L)/(R_0 - \mu_A N_0)\right] \Big\}. \tag{30}$$

Equation (30) confirms the lower limit of $2\tau$ for the cycle period discussed earlier.

## 3 Toward More-General Stage-Structured Models: A Two-Stage Model with Environment-Dependent Juvenile Mortality

The models in Section 2 took the particularly simple form of a single DDE because the environment was assumed to influence rate processes of the adult population. The formalism becomes more complex if the environment influences mortality in preadult stages. For example, any explicit or implicit time dependence of the juvenile death rate in our two-stage model invalidates equation (5). With certain restrictions, it is still possible to use DDE's to describe population dynamics when juvenile mortality rate changes in response to the environment. The model in this section is a particular case of a family of models proposed by Gurney et al. (1983). The essential assumption leading to a DDE formalism is that a single factor ("physiological age" or "developmental index") determines the transition from one life stage to its successor. Here, this single factor is chronological age, and all life stages have a fixed duration. The methodology has been extended in a number of papers (e.g., Nisbet & Gurney 1983, 1986; Blythe et al. 1984; Gurney et al. 1986) and is still being modified for new biological contexts (see, e.g., Shea et al., in press).

*Formalism*

As in Section 1, a population consists of juveniles and adults, with juveniles maturing to the adult stage after a time delay $\tau$. All notation is the same except that the juvenile per capita death rate, $\mu_J(t)$, can now vary with time. The balance equations (1)–(3) remain valid, and the only change in the model functions specified in equation (4) is that the total juvenile death rate is now given by

$$D_J(t) = \mu_J(t)N_J(t). \tag{31}$$

The key difference is that equation (5), which relates maturation rate out of the juvenile stage at time $t$ to recruitment rate at the earlier time $t - \tau$, is no longer valid. To derive the appropriate relationship, consider a cohort of $x_0$ juveniles recruited at time $t - \tau$ and maturing at time $t$. The size of this cohort, denoted by $x(t)$, decreases in accordance with the ordinary differential equation

$$dx(t')/dt' = -\mu_J(t')x(t')$$
$$\text{for } t - \tau \le t' \le t \text{ and } x(t - \tau) = x_0, \tag{32}$$

where the "dummy" variable $t'$ represents a time between $t - \tau$ and $t$. Equation (32) can be solved to give the number maturing at time $t$:

$$x(t) = x_0 \exp\left(-\int_{t-\tau}^{t} \mu_J(t')dt'\right). \tag{33}$$

Now define $S_J(t)$ as the juvenile through-stage survival for individuals maturing at time $t$, that is, the proportion of a cohort recruited at time $t - \tau$ that survives to mature to the adult stage at time $t$. Equation (33) implies that

$$S_J(t) = \exp\left(-\int_{t-\tau}^{t} \mu_J(t')dt'\right), \tag{34}$$

and the equation replacing equation (5) is

$$M_J(t) = R_J(t - \tau)S_J(t). \tag{35}$$

Similar reasoning can address the problem of the maturation of members of the initial juvenile population, and the analogue of equation (7) is

$$M_J(t) = R_A(t)$$
$$= m(0, \tau - t)\exp\left(-\int_{0}^{t} \mu_J(t')dt'\right) \quad \text{for } t \le \tau. \tag{36}$$

This complication can be ignored if the initial population has no juveniles.

Completing the formulation of the model in terms of a set of DDE's requires some trick to avoid the need to evaluate the integrals on the right-hand side of equations (34) and (36). This is most easily achieved by defining a new variable,

$$\Phi(t) \equiv \int_{0}^{t} \mu_J(t')dt'. \tag{37}$$

Since differentiation and integration are inverse operations, this new variable obeys the ordinary differential equation

$$d\Phi(t)/dt = \mu_J(t) \quad \text{with initial condition } \Phi(0) = 0. \tag{38}$$

The juvenile through-stage survival (whose value is needed only

TABLE 1. *General Two-Stage Model with Time-Dependent Juvenile Mortality*

| | | |
|---|---|---|
| Populations | $N_J(t)$ | number of juveniles at time $t$ |
| | $N_A(t)$ | number of adults at time $t$ |
| | $m(0, a)$ | initial juvenile age distribution |
| Properties of individuals | $\tau$ | juvenile-stage duration |
| | $\beta$ | adult fecundity |
| | $\mu_A$ | adult per capita death rate |
| | $\mu_J(t)$ | juvenile per capita death rate |

Vital processes

$$R_J(t) = \beta N_A(t) \quad \text{juvenile recruitment rate at time } t$$

$$R_A(t) = \begin{cases} R_J(t - \tau)\exp[\Phi(t - \tau) - \Phi(t)] & \text{if } t \geq \tau \\ m(0, \tau - t)\exp[-\Phi(t)] & \text{if } t < \tau \end{cases}$$

adult recruitment rate

Balance equations

$$dN_J(t)/dt = R_J(t) - R_A(t) - \mu_J(t)N_J(t) \quad \text{juveniles}$$
$$dN_A(t)/dt = R_A(t) - \mu_A N_A(t) \quad \text{adults}$$

Additional equation

$$d\Phi(t)/dt = \mu_J(t)$$

Initial conditions

$$N_J(0) = \int_0^\tau m(0, a)da$$
$$N_A(0) \quad \text{arbitrary}$$
$$\Phi(0) = 0$$

NOTE.—To implement the model, three differential equations (the two balance equations and the "additional equation") must be solved simultaneously with the initial conditions as specified. All quantities required to evaluate the right-hand sides of the differential equations are defined earlier in the table.

when $t > \tau$) is then given by

$$S_J(t) = \exp\left[\Phi(t - \tau) - \Phi(t)\right]. \tag{39}$$

This completes the recipe for formulating a single-species model. The equations that need to be solved, together with initial conditions, are summarized in Table 1.

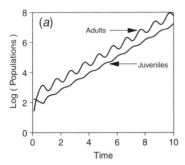
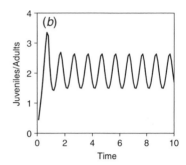

FIGURE 6. *Population dynamics and fluctuations in the age distribution in a numerical solution of the equations in Table 1 with parameters $\beta = 5$, $\mu_A = 0.5$, $\tau = 0.78$. The juvenile death rate varies sinusoidally with period $T$ and amplitude $b$; thus, $\mu_J(t) = \mu_{J0}(1 + b\cos[2\pi t/T])$, with $\mu_{J0} = 1.5$, $T = 1.0$, $b = 1.0$. Initial conditions: $N_A = 10$ and $N_J = 0$ at $t = 0$; $N_A = N_J = 0$ for $0 \le t < \tau$. The graphs show the variation with time of* (a) *the logarithm of juvenile and adult populations and* (b) *the ratio of juveniles to adults.*

## Example: Model with Density-Independent, Time-Dependent Juvenile Mortality

Figure 6 shows the long-term dynamics of a model with all vital rates constant except the juvenile death rate, which is assumed to vary sinusoidally with time (cf. Fig. 2). The logarithm of the stage populations still has a linear upward trend, but cyclic fluctuations are superimposed. Similarly, the instantaneous growth rate settles down to a pattern of cyclic fluctuations around a stable mean. This is consistent with the expected behavior of a linear model in a variable environment discussed by Tuljapurkar (Chapter 3).

## Example: Model with Density-Dependent Juvenile Mortality

Gurney et al. (1983; see also Gurney & Nisbet 1985) proposed a particularly simple model of a laboratory moth population regulated by competition among larvae for food. In the language of our present model, larvae (juveniles) have an instantaneous death rate proportional to the number of larvae present:

$$\mu_J(t) = \alpha N_J(t)/[1 - \eta N_J(t)]. \tag{40}$$

It is left as an exercise to the reader to show, using the equations in Table 1, that the population has an equilibrium at which

$$N_J^* = \ln(\beta/\mu_A)/[\alpha\tau + \eta\ln(\beta/\mu_A)],$$
$$N_A^* = \mu_J(N_J^*)N_J^*/(\beta - \mu_A). \tag{41}$$

Local stability analysis of this model involves some heavy algebra (for details, see Gurney & Nisbet 1985; Jones et al. 1988). The procedure is to derive the elements of a matrix $\mathbf{A}(\lambda)$ such that the characteristic equation takes the form $\det[\mathbf{A}(\lambda)] = 0$. All values of $\lambda$ satisfying this equation must have negative real parts for the local stability of an equilibrium. The matrix elements involve some or all of the model parameters and the equilibrium values of the state variables.

At a point of marginal stability, a root crosses the imaginary axis, and "boundaries" in parameter space can be calculated by reasoning resembling that in Section 2 (especially eqs. 20–23). Set $\lambda = i\omega$ and denote by $\{p_i\}$ the set of model parameters. On the stability boundary, both the real and imaginary parts of $\det[\mathbf{A}(i\omega)]$ equal zero. As with the present model, explicit expressions describe the equilibria in terms of the model parameters. Then, for $K$ parameters, there are two equations and $K + 1$ unknowns (all the parameters plus $\omega$). Specify the values of all but two of the parameters, yielding two equations in three unknowns; this defines a curve or set of curves that can be derived by any of a large number of numerical root-finding routines (e.g., those in MATHEMATICA). The program CONTOUR in the SOLVER package (Gurney & Tobia 1995) has a graphics overlay that facilitates these calculations. The main practical difficulty derives from the fact that these nonlinear equations typically have an infinite number of roots, and the method above locates a curve for which any pair of roots crosses the imaginary axis. For stability calculations, we are only interested in the situation where the root with the largest real part crosses the axis. The easiest way to resolve this problem is to perform some set of numerical experiments with the model and find some set of parameters close to a point of marginal stability. These parameters can then be used as the initial guess in the stability calculation.

The outcome of the local stability analysis for this model is that the primary determinant of stability is the ratio of the time delay to the mean adult lifetime, that is, the product $\mu_A\tau$. With long adult lifetimes, the equilibrium is stable; with short adult lifetimes, the

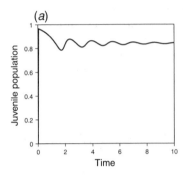
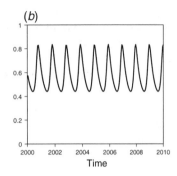

FIGURE 7. (a) *Stable equilibrium and* (b) *single-generation cycles in the model with density-dependent juvenile mortality given by equation (40). Parameter values:* $\beta = 300$, $\tau = 1.0$, $\alpha = 1.0$, $\eta = 1.0$, *and* (a) $\mu_A = 1.5$, (b) $\mu_A = 50.0$.

population cycles. Examples are shown in Figure 7. However, these cycles are *not* of the "delayed-feedback" type encountered in the models with juvenile recruitment regulated by adult density. The period of each cycle is close to $\tau + \mu_A^{-1}$, the mean lifetime of an individual, and it is natural to interpret them as single-generation cycles. This interpretation is confirmed by detailed calculation of the generation structure (Jansen et al. 1990).

## 4  Host-Parasitoid Dynamics

The richest set of applications to date of the formalism in Section 3 has concerned the dynamics of insect host-parasitoid systems (see, e.g., Murdoch et al. 1987, 1992, in press; Godfray & Hassell 1989; Godfray & Waage 1991; Gordon et al. 1991; Briggs 1993; Briggs et al. 1993; Murdoch & Briggs, in press; Shea et al., in press). The basic model for these applications, due to Murdoch et al. (1987), was motivated by studies of California red scale, a citrus pest whose population has been controlled at low levels by the parasitoid *Aphytis*. This basic model recognizes two stages for the host (red scale): juveniles, which are susceptible to attack by the parasitoid; and adults, which are not vulnerable to parasitism. Parasitoids lay eggs in juvenile hosts, and a host that is attacked by a parasitoid is assumed to die instantly, yielding recruits to the juvenile parasitoid population. The egg takes some finite time to mature to become an adult parasitoid.

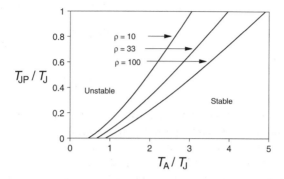

FIGURE 8. *Red scale stability boundaries for three values of the dimen-sionless ratio* $\rho = \beta/\mu_A$. *Other parameter values:* $\sigma_{JP} = 1$; $\mu_P T_J = 8$.

This system thus involves two interacting populations, with dy-namics corresponding to the situations covered in Sections 2 and 3, respectively. The "environment" variable that affects the *para-sitoid* dynamics is the density of juvenile hosts; the rate process affected is juvenile (parasitoid) recruitment. Thus, the parasitoids can be described by a single DDE, as with the models of Section 2. The "environment" variable for the *hosts* is the density of adult parasitoids, and the rate process affected is juvenile (host) mor-tality, the formalism of Section 3 applies, and three equations are necessary to describe the dynamics. The resulting four-equation model is detailed in Table 2.

The calculation of equilibria and the numerical investigation of local stability proceed as in previous sections. Here, however, *two* developmental delays may influence stability. Figure 8 shows how these delays affect stability; interestingly, some insight from the single-species models carries over to this model of interacting pop-ulations. First, the parasitoid time delay is destabilizing, as would be expected from the single-species models of Section 2. Second, the equilibrium is stable if the mean adult-host lifetime is long in comparison with the juvenile-stage duration, as would be expected from the model in Section 3. However, the single-species models are not perfect predictors of the dynamics, since the population cycles observed in the regions of parameter space where the equilibrium is unstable are not of the single-generation type but have periods and properties more like those of simple (unstructured) Lotka-Volterra predator-prey models.

TABLE 2. *The Stage-Structured Host-Parasitoid Model*

| | | |
|---|---|---|
| Populations | $J(t)$ | number of juvenile hosts at time $t$ |
| | $A(t)$ | number of (invulnerable) adult hosts at time $t$ |
| | $P(t)$ | number of adult parasitoids at time $t$ |
| | $m(0, a)$ | initial age distribution for juvenile hosts |
| Properties of individuals | $T_J$ | juvenile-host duration |
| | $T_{JP}$ | juvenile-parasitoid duration |
| | $\beta$ | adult-host fecundity |
| | $\alpha$ | parasitoid attack rate |
| | $\mu_A$ | adult-host per capita death rate |
| | $\mu_P$ | adult-parasitoid per capita death rate |
| | $\sigma_{JP}$ | through-stage survival for juvenile parasitoids |

Vital processes

$$R_J(t) = \beta N_A(t) \qquad \text{juvenile-host recruitment rate at time } t$$

$$\Pi(t) = \alpha J(t) P(t) \qquad \text{total parasitism rate at time } t$$

$$\Phi(t) = \mu_J t + \alpha \int_0^t P(x)dx \qquad \text{cumulative juvenile-host mortality}$$

$$R_A(t) = \begin{cases} R_J(t - T_J) \exp\left[\Phi(t - T_J) - \Phi(t)\right] & \text{if } t \geq T_J \\ m(0, T_J - t) \exp\left[-\Phi(t)\right] & \text{if } t < T_J \end{cases}$$

adult-host recruitment rate at time $t$

$$R_P(t) = \sigma_{JP}\Pi(t - T_{JP}) \qquad \text{adult-parasitoid recruitment rate at time } t$$

Balance equations

$$dJ(t)/dt = R_J(t) - R_A(t) - \mu_J(t) - \Pi(t) \qquad \text{juvenile hosts}$$

$$dA(t)/dt = R_A(t) - \mu_A A(t) \qquad \text{adults}$$

$$dP(t)/dt = R_P(t) - \mu_P P(t) \qquad \text{adult parasitoids}$$

Additional relationship

$$d\Phi(t)/dt = \mu_J + \alpha P(t)$$

## 5  Dynamically Varying Time Delays

The models discussed thus far allow the environment to influence rate processes but assume that the values of the time delays remain fixed. This is unrealistic in many contexts; for example, the duration of many insect stages is a function of temperature. The formalism in Section 3 can be extended to cover this situation provided that for any given stage an individual can be assigned a "developmental index" (DI), denoted by $q$, which describes that individual's progression through the stage. As the environment changes, so does the rate of increase (denoted by $g(t)$) of the DI; if individuals mature to the next stage when the DI attains some critical value, fluctuations in the environment lead to fluctuations in the stage duration. For example, experimental data might suggest that a fish or a particular species attains reproductive maturity when it reaches a particular length. In this situation, $q$ represents length, and $g(t)$ represents the growth rate. The developmental index need not have a simple physical interpretation; for example, larval development in a moth is well described by a developmental index that is a linear combination of age and weight (Gordon et al. 1988).

I again illustrate the formalism by considering a two-stage model. The model is an elaboration of that in Section 3, with the additional feature that the juvenile developmental rate may change in response to the environment. I retain all notation from previous models but assume, in addition, that juvenile developmental rate $g_J(t)$ varies with time in response to the environment. There is no loss of generality in assuming that maturation to the adult stage occurs when the developmental index equals one. Thus, if $\tau(t)$ denotes the time that is spent in the juvenile stage by an individual maturing at time $t$, then

$$\int_{t-\tau(t)}^{t} g_J(t')dt' = 1. \tag{42}$$

A new variable $\Theta(t)$, analogous to $\Phi$ in Section 3, is defined as follows:

$$\Theta(t) = \int_0^t g_J(t')dt', \tag{43}$$

in terms of which, equation (42) implies that

$$\Theta(t) - \Theta\left[t - \tau(t)\right] = 1. \tag{44}$$

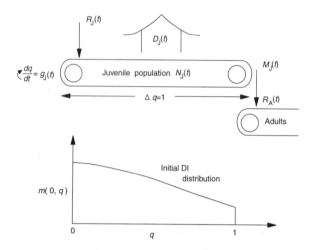

FIGURE 9. *Conveyor-belt analogy of the model in Section 5 (after Auslander et al. 1974). See the text for explanation.*

As with the model in Section 3, a new equation is needed to relate the rates of recruitment to, and maturation from, the juvenile stage. The result (derived in Nisbet & Gurney 1983) is that when $t > \tau(t)$ (or, equivalently, $\Theta(t) > 1$),

$$M_J(t) = g_J(t)\left\{R_J\left[t - \tau(t)\right]/g_J\left[t - \tau(t)\right]\right\}S_J(t). \qquad (45)$$

This expression is most readily understood with the aid of a "conveyor-belt" analogy due originally to G. F. Oster and shown in Figure 9. The developmental-index distribution, $m(t, q)$ at time $t$, is analogous to the density of grains of sand on the belt, and the juvenile developmental rate, $g_J(t)$, is analogous to the speed of advance, which can vary with time in response to the environment. New sand is supplied at a rate $R_J(t)$ to the left-hand end, and sand is constantly leaking off the belt (analogous to mortality). Any sand that survives to reach the right-hand end falls onto the start of a second belt representing the adult stage. In this visualization, $M_J(t)$ is given by the product of the density of sand grains at the right-hand end and the speed of the belt. The density at the right-hand end at time $t$ is in turn equal to the density at the left-hand end at the time of recruitment (given in braces in eq. 45) multiplied by the through-stage survival.

Similar reasoning allows an evaluation of the maturation rate from the juvenile stage for times $t < \tau(t)$, that is, the maturation of survivors of the initial juvenile population (see Fig. 9, *lower panel*). An individual maturing at time $t$ has a DI equal to $1 - \Theta(t)$ at time $t = 0$. The probability of surviving from time zero to time $t$ is $\exp[-\Phi(t)]$. Thus, for $t < \tau(t)$ (or, equivalently, $\Theta(t) < 1$),

$$M_J(t) = g_J(t) m \Big[ 0, 1 - \Theta(t) \Big] \exp \Big[ - \Phi(t) \Big]. \tag{46}$$

As with the model in Section 3, this complication can be ignored if the initial population has no juveniles.

This completes the specification of the model. However, the delay $\tau(t)$ is defined in an inconvenient manner (eq. 44) that requires numerical interpolation in many applications. To avoid this, note that by making use of the rules for differentiating an integral and rearranging terms, it can be shown from equation (42) that the delay itself now obeys a DDE, namely,

$$d\tau(t)/dt = 1 - g_J(t)/g_J \Big[ t - \tau(t) \Big]. \tag{46}$$

The complete recipe for implementing the model is summarized in Table 3.

# 6  More-Complex Models

Ecological modelers have to face the dilemma of how much detail to include in models, recognizing that added realism frequently carries two distinct costs: added model complexity and reduced biological generality. The models discussed in this chapter are sufficiently simple that they can be used to develop intuition about the influence of simple life-history features on population dynamics. However, it can be argued that a particularly powerful route to ecological insight is the development of clusters of models that include both simple models, as aids to understanding, and more-complex models, as the basis for experimental tests on particular systems. I conclude this chapter with some comments on the possibilities for using DDE's in more-complex simulations.

First, there is nothing special about two-stage models. The formalism is applicable to an organism with any number of recognizably distinct developmental stages. The primary restriction is that all individuals within a stage must respond in an identical way to their environment. Even this restriction may be relaxed under some circumstances (see, e.g., Blythe et al. 1984; Gurney et al. 1986), but one restriction seems unavoidable: the instantaneous death rate for individuals within a stage must be independent of the developmental index. Although I cannot prove this assertion,

TABLE 3. *General Two-Stage Model with Time Dependence in Both Juvenile Mortality and Developmental Rates*

| | | |
|---|---|---|
| Populations | $N_J(t)$ | number of juveniles at time $t$ |
| | $N_A(t)$ | number of adults at time $t$ |
| | $m(0,q)$ | initial juvenile developmental-index distribution |
| | $\tau(t)$ | juvenile-stage duration for individual maturing at time $t$ |
| Properties of individuals | $\beta$ | adult fecundity |
| | $\mu_A$ | adult per capita death rate |
| | $\mu_J(t)$ | juvenile per capita death rate |
| | $g_J(t)$ | juvenile developmental rate |

Vital processes

$$R_J(t) = \beta N_A(t) \quad \text{juvenile recruitment rate at time } t$$

$$R_A(t) = \begin{cases} g_J(t)\dfrac{R_J[t-\tau(t)]}{g_J[t-\tau(t)]}S_J(t) & \text{if } t \geq \tau(t) \\ g_J(t)m[0, 1-q(t)]\exp[-\Phi(t)] & \text{if } t < \tau(t) \end{cases}$$

adult recruitment rate

Balance equations

$$dN_J(t)/dt = R_J(t) - R_A(t) - \mu_J(t)N_J(t) \quad \text{juveniles}$$
$$dN_A(t)/dt = R_A(t) - \mu_A N_A(t) \quad \text{adults}$$

Additional equations

$$d\Phi(t)/dt = \mu_J(t)$$
$$d\Theta/dt = g_J(t)$$
$$\Theta(t) - \Theta[t-\tau(t)] = 1$$
$$\rightarrow \frac{d\tau}{dt} = \begin{cases} 1 - g_J(t)/g_J[t-\tau(t)] & \text{if } \Theta(t) > 1 \\ 1 & \text{otherwise} \end{cases}$$

Initial conditions

$$N_J(0) = \int_0^\tau m(0,a)da; \quad N_A(0) \text{ arbitrary}$$
$$\Phi(0) = 0$$
$$\tau(0) = 0$$

NOTE.—To implement the model, five differential equations (two balance equations and the "additional equations") are solved simultaneously, with the initial conditions as specified.

I know of no exceptions, and it is consistent with related, more-abstract studies of "linear-chain trickery" by Metz and Diekmann (1991).

Although there is no serious difficulty in formulating more-complex models, they may exhibit unexpected dynamics. Simple models often, though not invariably, prove the key to qualitative insight, with more-complex models merely adding quantitative detail. For example, in contrast with the models presented in this chapter, several studies of models with *two* preadult stages exhibit subcritical Hopf bifurcations (see, e.g., Hastings 1987; Murdoch et al. 1992), the practical implication of which is that the long-term dynamics may depend on initial conditions as well as on model parameters.

Many mathematical challenges remain in extending the formalism to incorporate biological information not directly related to the developmental index. Shea et al. (in press) have developed a host-parasitoid model for certain parasitoids (including *Aphytis*) that typically eat smaller hosts as a source of protein for developing eggs and, after some time interval, lay the eggs in larger hosts. Whether the parasitoid feeds on a host or oviposits on it also depends on the parasitoid's current egg load; parasitoids with few eggs are more likely to host feed. Thus, one component of the "structure" of the parasitoid population (i.e., one *i*-state variable) must be egg load. A serious complication in the structure of the model by Shea et al. is that a parasitoid may lay one or more eggs while maturing a host meal. This problem is one example of a more general challenge: the incorporation of simple representations of energy acquisition and use into the DDE framework. However, it is important to recognize that the range of applicability of DDE's is limited. To be of value, any resolution of the remaining problems concerning model formulation must be simple and/or intuitive, since the more general approach to continuous-time modeling in the next chapter (by de Roos) provides a rigorous, if mathematically demanding, framework for energetic (and other) models.

### Acknowledgments

My understanding of much of the material in this chapter derives from collaborative work with Bill Gurney. The enthusiasm of Bill Murdoch and Ed McCauley for ecological applications of the DDE models stimulated much of the work described in the later sections. I thank Steve Blythe, Odo Diekmann, Hans Metz, Norman MacDonald, André de Roos, and many others for expert guidance through mathematical mine fields. Recent work on host-parasitoid models with Bill Murdoch, Cheryl Briggs, and Katriona Shea has provided encouragement for continuing development of the formalism. I thank Cheryl Briggs, Bill Murdoch, and Will Wilson for comments on an earlier draft. The research was supported by grants from the National Science Foundation (DEB-9319301) and the Environmental Protection Agency (R819433-01-0).

## Literature Cited

an der Heiden, U., and M. C. Mackay. 1982. Dynamics of destruction and renewal. *Journal of Mathematical Biology* 16: 75–101.

Auslander, D. M., G. F. Oster, and C. B. Huffaker. 1974. Dynamics of interacting populations. *Journal of the Franklin Institute* 297: 345-376.

Blythe, S. P., R. M. Nisbet, and W. S. C. Gurney. 1982. Instability and complex dynamic behavior in population models with long time delays. *Theoretical Population Biology* 22: 147–176.

———. 1984. The dynamics of population models with distributed maturation periods. *Theoretical Population Biology* 25: 289–311.

Briggs, C. J. 1993. Competition among parasitoid species on an age-structured host, and its effect on host suppression. *American Naturalist* 141: 372–397.

Briggs, C. J., R. M. Nisbet, and W. W. Murdoch. 1993. Coexistence of competing parasitoid species on a host with a variable life cycle. *Theoretical Population Biology* 44: 341–373.

Crowley, P. H., R. M. Nisbet, W. S. C. Gurney, and J. H. Lawton. 1987. Population regulation in animals with complex life histories: Formulation and analysis of a damselfly model. *Advances in Ecological Research* 17: 1–59.

Diekmann, O., S. A. Van Gils, S. M. V. Lunel, and H. O. Walther. 1995. *Delay Equations: Functional-, Complex-, and Nonlinear Analysis.* Springer-Verlag, New York.

Godfray, H. C. J., and M. P. Hassell. 1989. Discrete and continuous insect populations in tropical environments. *Journal of Animal Ecology* 58: 153–174.

Godfray, H. C. J., and J. K. Waage. 1991. Predictive modelling in biological control: The mango mealy bug (*Rastrococcus invadens*) and its parasitoids. *Journal of Applied Ecology* 28: 434–453.

Gordon, D. M., W. S. C. Gurney, R. M. Nisbet, and R. K. Stewart. 1988. A model of *Cadra cautella* larval growth and development. *Journal of Animal Ecology* 57: 645–658.

Gordon, D. M., R. M. Nisbet, A. M. de Roos, W. S. C. Gurney, and R. K. Stewart. 1991. Discrete generations in host-parasitoid models with contrasting life cycles. *Journal of Animal Ecology* 60: 295–308.

Gurney, W. S. C., and R. M. Nisbet. 1985. Fluctuation periodicity, generation separation and expression of larval competition. *Theoretical Population Biology* 28: 150–180.

Gurney, W. S. C., and S. Tobia. 1995. SOLVER—A program template for initial value problems expressible as sets of coupled ordinary or delay differential equations. STAMS, University of Strathclyde, Glasgow.

Gurney, W. S. C., S. P. Blythe, and R. M. Nisbet. 1980. Nicholson's blowflies revisited. *Nature* 287: 17–21.

Gurney, W. S. C., R. M. Nisbet, and J. H. Lawton. 1983. The systematic formulation of tractable single-species population models incorporating age structure. *Journal of Animal Ecology* 52: 479–495.

Gurney, W. S. C., R. M. Nisbet, and S. P. Blythe. 1986. The systematic formulation of models of stage structured populations. Pp. 474–494 *in* J. A. J. Metz and O. Diekmann, eds., *Lecture Notes in Biomathematics: The Dynamics of Physiologically Structured Populations*. Springer-Verlag, Berlin.

Gyori, I., and G. Ladas. 1991. *Oscillation Theory of Delay Differential Equations: With Applications*. Oxford University Press.

Hastings, A. 1987. Cycles in cannibalistic egg-larval interactions. *Journal of Mathematical Biology* 24: 651–666.

Jansen, V. A. A., R. M. Nisbet, and W. S. C. Gurney. 1990. Generation cycles in stage structured populations. *Bulletin of Mathematical Biology* 52: 375–396.

Jones, A. E., R. M. Nisbet, W. S. C. Gurney, and S. P. Blythe. 1988. Period to delay ratio near stability boundaries for systems with delayed feedback. *Journal of Mathematical Analysis and Applications* 135: 354–368.

Kuang, Y. 1993. *Delay Differential Equations: With Applications in Population Dynamics*. Academic Press, Boston.

MacDonald, N. 1989. *Biological Delay Systems: Linear Stability Theory*. Cambridge University Press, New York.

May, R. M. 1980. Mathematical models in whaling and fisheries management. Pp. 1–63 *in* G. F. Oster, ed., *Lectures on Mathematics in the Life Sciences* 13. American Mathematical Society, Providence, R.I.

McCauley, E. D., R. M. Nisbet, A. M. de Roos, W. W. Murdoch, and W. S. C. Gurney. In press. Structured population models of herbivorous zooplankton. *Ecology*.

Metz, J. A. J., and O. Diekmann. 1991. Exact finite dimensional representations of models for physiologically structured populations. I: The abstract foundations of linear chain trickery. Pp. 269–289 *in* J. A. Goldstein, F. Kappel, and W. Schappacher, eds., *Differential Equations with Applications in Biology, Physics, and Engineering*. Marcel Dekker, New York.

Murdoch, W. W. 1994. Population regulation in theory and practice. *Ecology* 75: 271–287.

Murdoch, W. W., and C. J. Briggs. In press. Theory for biological control: Recent developments. *Ecology.*

Murdoch, W. W., R. M. Nisbet, S. P. Blythe, W. S. C. Gurney, and J. D. Reeve. 1987. An invulnerable age class and stability in delay-differential parasitoid-host models. *American Naturalist* 129: 263–282.

Murdoch, W. W., R. M. Nisbet, R. F. Luck, H. C. J. Godfray, and W. S. C. Gurney. 1992. Size-selective sex-allocation and host feeding in a parasitoid-host model. *Journal of Animal Ecology* 61: 533-541.

Murdoch, W. W., C. J. Briggs, and R. M. Nisbet. In press. Competitive displacement and biological control in parasitoids: A model. *American Naturalist.*

Nicholson, A. J. 1954. An outline of the dynamics of animal populations. *Australian Journal of Ecology* 2: 9–65.

———. 1957. The self-adjustment of populations to change. *Cold Spring Harbor Symposia* 22: 153–173.

Nisbet, R. M., and J. R. Bence. 1989. Alternative dynamic regimes for canopy-forming kelp: A variant on density-vague population regulation. *American Naturalist* 134: 377–408.

Nisbet, R. M., and W. S. C. Gurney. 1982. *Modelling Fluctuating Populations.* Wiley, Chichester, Engl.

———. 1983. The systematic formulation of population models for insects with dynamically varying instar duration. *Theoretical Population Biology* 23: 114–135.

———. 1986. The formulation of age-structure models. Pp. 95–115 *in* T. G. Hallam and S. A. Levin, eds., *Mathematical Ecology.* Springer-Verlag, Berlin.

Nisbet, R. M., W. S. C. Gurney, W. W. Murdoch, and E. McCauley. 1989. Structured population models: A tool for linking effects at individual and population levels. *Biological Journal of the Linnean Society* 37: 79–99.

Shea, K., R. M. Nisbet, W. W. Murdoch, and J. S. Yoo. In press. The effect of egg limitation in insect host-parasitoid population models. *Journal of Animal Ecology.*

CHAPTER 5

# A Gentle Introduction to Physiologically Structured Population Models

## André M. de Roos

This chapter is intended as a gentle introduction to models of physiologically structured populations, as developed by Metz and Diekmann (1986) (hereafter, PSP models; the origin of the adjective "physiologically structured" is described in Section 2 after some necessary definitions). You will need a basic knowledge of mathematical modeling but no familiarity with PSP models or the ensuing partial differential equations. My aim is to equip you with enough skills to be able to use moderately complex PSP models for specific biological applications. Hence, I emphasize the formulation of the models, the biological interpretation of the equations, and the tools for studying the models. The mathematical background of the modeling framework and the justification of the equations are discussed only when necessary for a better understanding of the biological aspects of the models.

Like other "individual-based" models, PSP models describe the dynamics of a population in terms of the behavior of its constituent individuals. This approach is related to the central principle of population dynamics, that the dynamics of a population are nothing more (and nothing less) than the cumulative result of the actions of its individuals. As a consequence, PSP models can be considered a bookkeeping system, tracking the actions of single individuals and how these actions influence themselves and change their environment. For example, individuals of a consumer species usually compete for a resource. The amount of food they are able to collect determines not only their growth and the number of their offspring but also the reduction of resource levels. Consequently, it determines their future performance and that of their conspecifics. Together, these events determine the dynamics of the entire population, and PSP models specify these dynamics by summarizing the individual events.

The construction of a PSP model takes place at two different levels: first, and primarily, at the level of the individual; and subsequently, at the population level. In addition, it is crucial to distinguish the environment that the individuals live in as a component of the system. In the consumer example, the resource level defines the environment. The consumption, and consequent reduction, of the resource is the feedback loop by which the behavior of an individual influences its own future. These three components—the individual, its environment, and the population—are pivotal to any PSP model.

Exactly which models are included in the class of PSP models and which are not is, in my opinion, largely a matter of taste. The assumptions that are made while modeling a population, based on the behavior of its individuals, determine the appropriate model framework (Metz & de Roos 1992). These assumptions also determine the relation between PSP models and other "individual-based" models (Metz & de Roos 1992; see also the introductory chapter in this volume). Without discussing the basic assumptions, I limit the scope of this chapter by imposing the following restrictions.

1. Spatial distribution is completely neglected. All individuals experience the same, spatially homogeneous environment. More technically, individuals are assumed to mix quickly and randomly, such that the "mass-action law" applies, and the interactions can be completely described in terms of the densities (de Roos et al. 1991; Metz & de Roos 1992).

2. Only large populations are considered. In principle, the population can consist of infinitely many types of individuals, but *of each type* a sufficiently large number is assumed to be present that demographic stochasticity plays no role. Assuming this "law of large numbers" ensures that an inherently stochastic process, such as the death of individuals, can be approximated by a deterministic description.

3. The growth and development of individuals is assumed to be deterministic. This implies that two identical individuals develop in exactly the same way. As a consequence, identical individuals always remain identical.

From these restrictions it is clear that the PSP models considered here are deterministic approximations of essentially stochastic processes. They constitute only part of the entire class of PSP models (for other situations, see the more in-depth discussions in Metz & Diekmann 1986; Metz & de Roos 1992; Diekmann et al. 1994; Diekmann & Metz 1995). It should also be noted that the theory of PSP models is still actively evolving. According to recent theory, PSP models should be formulated as integral equations (or abstract renewal equations), as opposed to the partial differential equations (PDE's) used up to now. This "cumulative" formulation in terms of integral equations is claimed to be more natural, more elegant, and technically superior (Diekmann et al. 1994; Diekmann & Metz 1995). I do not discuss these recent developments but stick to the PDE formalism to stay in line with the tradition of writing population-dynamic models as differential equations. For the practical purposes of this introduction, this choice is irrelevant. (As an aside, note that although the technique for numerical study of PSP models was developed on the basis of the PDE formalism, it is actually formulated in terms of integral quantities and, hence, is close in spirit to the new "cumulative" formulation of the models.)

The traditional sequence for discussing the theory of structured populations begins with age-structured models, followed by size-structured models, and eventually leads to models with more than one structuring variable. This is a roughly chronological introduction, since age-structured PDE models were introduced by McKendrick (1926) and VonFoerster (1959), size-structured population models date back to the work of Bell and Anderson (1967), and Sinko and Streifer (1967) discussed a structured-population model that distinguishes both age and size. I take a different approach and discuss the formulation and analysis of a size-structured model for a specific biological situation. The resulting model is, however, representative of the broader class of PSP models. In addition to (and mixed in between) the formulation and analysis of this example, I discuss the general theory of PSP models, insofar as it is important for applications to other specific biological examples. The formulation and analysis of the example model in Sections 1, 3, 6, 8, and 9 constitutes a closed, independent story. The remaining sections discuss the theory in a more general context.

As this chapter is intended to be a *gentle* introduction, I refrain from giving an overview of the literature on PSP models, since this would encompass a large number of mathematical, theoretical, and biological publications. More about the theory can be found in the book by Metz and Diekmann (1986) and other, more recent, publications ( Metz & de Roos 1992; Diekmann et al. 1994; Diekmann & Metz 1995).

## 1 Modeling Individual *Daphnia*

The biological system that I use here to illustrate the formulation and analysis of a PSP model is the waterflea *Daphnia pulex*, which feeds on the alga *Chlamydomonas rheinhardii*. This system has frequently been the focus of structured-population modeling (see, e.g., Nisbet et al. 1989) because of the large amount of biological information available at both the individual and population levels. I frequently refer to the *Daphnia* and algae as "consumers" and "food," respectively, and sometimes as "predator" and "prey." I focus on the *Daphnia* population, paying little attention to the algae (they are considered the "environment" of the *Daphnia* population; see also Section 2).

*Daphnia* exploit their algal food source without apparent interference between the *Daphnia* individuals themselves. There is good evidence (Kooijman & Metz 1984; Kooijman 1986; McCauley et al. 1990$a$) that the amount of food consumed depends nonlinearly on the ambient food density, following a Holling Type-II functional response. Moreover, *Daphnia* conversion efficiency and individual mortality are approximately constant. In natural environments, the algae grow approximately logistically in the absence of consumers. Hence, the simplest, *un*structured-population model for this interaction is the Rosenzweig-MacArthur model:

$$\frac{dF}{dt} = \alpha F \left( 1 - \frac{F}{K} \right) - I_{\max} \frac{F}{F_{\mathrm{h}} + F} C , \qquad (1a)$$

$$\frac{dC}{dt} = \epsilon I_{\max} \frac{F}{F_{\mathrm{h}} + F} C - \mu C , \qquad (1b)$$

in which $F$ and $C$ represent the density of algae and *Daphnia*, respectively; $\alpha$ and $K$ denote the maximum growth rate and the carrying capacity of the algae, respectively; $I_{\max}$ and $F_{\mathrm{h}}$ are the maximum feeding rate and the half-saturation food level of the *Daphnia* functional response; and $\epsilon$ and $\mu$ are the conversion efficiency and per capita mortality rate of *Daphnia*, respectively.

The basic model (1) has been successfully used to model some aspects of the *Daphnia*-algal interaction (W. W. Murdoch, pers. comm.). However, it assumes that all *Daphnia* individuals are identical or can be represented by some "average" type, since it only keeps track of $C$, the total number of individuals present. A large amount of biological information shows that the size of an individual *Daphnia* greatly influences its behavior: larger individuals have higher food consumption, basal metabolism, and reproduction rate. To account for these influences of individual size on behavior, it is necessary to replace the variable $C$ in equation (1a) by something more sophisticated that accounts for the influence of individual size on feeding rate. We also must replace equation (1b), which describes the dynamics of $C$, with an equation that describes the dynamics of both the number of *Daphnia* and their sizes. Incorporating these changes is the goal of this chapter, but note that the basic layout of the more complicated, size-structured model is analogous to that of model (1): it consists of an equation describing the dynamics of the mathematical object representing the number and size composition of the *Daphnia* population and an equation describing the dynamics of the algae, in which some statistic (not just the total number) of the *Daphnia* population plays a role.

To incorporate size-dependent feeding and reproduction into the model, these dependences must be described in more detail. Many models describe the feeding, reproduction, and growth of *Daphnia* as a function of size and food density. The model presented here was first proposed by Kooijman and Metz (1984) and models these relations using simple assumptions about the gathering, assimilation, and allocation of energy. Although simple, the model incorporates features common to most models of individual *Daphnia* behavior (Paloheimo et al. 1982; Kooijman 1986; Gurney et al. 1990):

1. the feeding rate of individual *Daphnia* strongly increases with individual size and is an increasing but decelerating function of food density,

2. individual *Daphnia* mature on reaching a fixed size, and

3. ultimate size and growth rate increase with food availability.

In contrast to the original Kooijman-Metz model, the model used here assumes, for ease of presentation, that *Daphnia* individuals can shrink. Biologically this assumption is hard to defend, but it does not influence the properties and stability of the population equilibrium and has only minor effects on the dynamics studied at the end of the chapter (see de Roos et al. 1990). This is only a summary of the model; for a complete derivation and experimental underpinning, see Kooijman & Metz (1984) and Metz et al. (1988).

The length of an individual *Daphnia* is denoted by $\ell$, and it is assumed that individuals of different length are isomorphic, such that their surface area and their volume scale as $\ell^2$ and $\ell^3$, respectively. At birth the individual length of *Daphnia* tends to vary relatively little, and hence, individuals are assumed to be born with a fixed length indicated by $\ell_b$. Individual length also determines, to a large extent, the distinction between juveniles and adults; maturation occurs at a more or less fixed length, denoted by $\ell_j$.

Assume that the feeding of *Daphnia* is a surface-related process; then, the isomorphy assumption guarantees that the maximum ingestion rate, $I_{max}$, scales with the squared length:

$$I_{max} = \nu \ell^2 ,$$

where $\nu$ is a proportionality constant that can be interpreted as the maximum ingestion rate per unit of surface area. Together with the Type-II functional response already incorporated in the unstructured model (1), the allometric relation for $I_{max}$ implies that the food ingestion rate $I(F, \ell)$ of an individual *Daphnia* with length $\ell$ at food density $F$ is

$$I(F, \ell) = \nu \frac{F}{F_h + F} \ell^2. \tag{2}$$

In the Kooijman-Metz model, the conversion efficiency of ingested food into assimilated food is assumed constant, as are the energy required to produce a single offspring and the energy needed per unit of biomass increase via somatic growth. The pivotal assumption in the model is that a *fixed proportion* of the assimilate is used for development and reproduction, while the remaining part is used to cover maintenance and, if possible, somatic growth. From this assumption, it follows that the fecundity of adult *Daphnia* is directly proportional to food ingestion and can hence be described by the function

$$b(F, \ell) = \begin{cases} r_m \dfrac{F}{F_h + F} \ell^2 & \text{if } \ell > \ell_j , \\ 0 & \text{if } \ell \le \ell_j , \end{cases} \tag{3}$$

where $b(F, \ell)$ denotes the rate at which adult *Daphnia* produce offspring. The parameter $r_m$ is a proportionality factor indicating the maximum reproduction rate per unit of surface area. Juvenile individuals ($\ell < \ell_j$) are assumed to channel the same proportion of their energy into reproduction and development as adults do, but to use it first to develop their reproductive organs.

The remaining amount of assimilated energy is spent on basal metabolism and growth. Metabolic requirements are assumed to be proportional to individual volume ($\ell^3$) and take precedence over growth. Using these assumptions, together with a bit of formula manipulation (Kooijman & Metz 1984; de Roos et al. 1990), leads to the expression for the rate of growth in length, $g(F, \ell)$:

$$\frac{d\ell}{dt} = g(F, \ell) = \gamma \left( \ell_m \frac{F}{F_h + F} - \ell \right). \qquad (4)$$

Here, $\gamma$ is again a proportionality constant, while $\ell_m$ can be interpreted as the maximum length a *Daphnia* reaches under conditions of abundant food. Note that equation (4) allows $g(F, \ell)$ to become negative if $F$ is small, which implies that somatic tissue is used to cover maintenance requirements with a concurrent decrease in length. This reflects the assumption that individuals can shrink under low food conditions, mentioned above. If the food density is constant, such that $F = F_c$, equation (4) predicts that an individual grows according to the von Bertalanffy growth curve (von Bertalanffy 1957), approaching asymptotically the ultimate size at that food level, $\ell_m F_c / (F_h + F_c)$ (see Fig. 1):

$$\frac{d\ell}{da} = \gamma \left( \ell_m \frac{F_c}{F_h + F_c} - \ell \right)$$

$$\Rightarrow \quad \ell(a) = \ell_m \frac{F_c}{F_h + F_c} - \left( \ell_m \frac{F_c}{F_h + F_c} - \ell_b \right) e^{-\gamma a}. \qquad (5)$$

In subsequent sections, the function $h(F)$ often denotes the Type-II functional response:

$$h(F) = \frac{F}{F_h + F},$$

and the symbol $\ell_\infty$ indicates the maximum attainable size at a particular food density:

$$\ell_\infty = \ell_m h(F) = \ell_m \frac{F}{F_h + F}.$$

Mortality is modeled as in the unstructured model (1): individuals are assumed to have a constant probability of death per unit of time. Hence, the mortality rate $d(F, \ell)$ has a simple form:

$$d(F, \ell) = \mu. \qquad (6)$$

All necessary parts of the individual *Daphnia* behavior have now been specified. The algal population is assumed to follow the same

Individual size

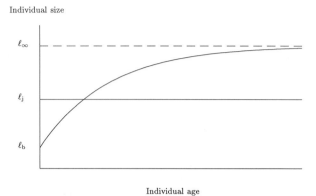

Individual age

FIGURE 1. *The individual length as a function of age for a* Daphnia *living in a constant food density, $F = F_c$. The resulting length-age relation is the von Bertalanffy growth curve illustrated here. $\ell_\infty$ indicates the ultimate size reached at the current constant food density and equals $\ell_m F_c / (F_h + F_c)$.*

logistic growth function used in the unstructured model (1). Equations from this section that appear in later sections are summarized in Table 1. The appropriate parameter values will be given when the model is studied numerically.

## 2 Modeling the Individual and Its Environment

Section 1 describes the model at the individual level for the *Daphnia* example. This section discusses modeling individual behavior in the context of the general theory of PSP models.

"Individual behavior" refers to processes like the reproduction and death of individuals and to processes indirectly important for reproduction and death (e.g., feeding behavior). From the viewpoint of population dynamics, the birth and death of individuals are the only processes of interest. However, the probability of giving birth or of dying may not be the same for each individual. The realization that differences between individuals, like body size in *Daphnia*, influence reproduction and mortality is the motivation for structured-population models. First, "differences between individuals" must be defined.

TABLE 1. Daphnia *Behavior and Algal Growth Dynamics*

| Growth rate | $g(F,\ell)$ | $=$ | $\gamma\left(\ell_{\mathrm{m}}\dfrac{F}{F_{\mathrm{h}}+F}-\ell\right)$ |
|---|---|---|---|
| Reproduction rate | $b(F,\ell)$ | $=$ | $\begin{cases}r_{\mathrm{m}}\dfrac{F}{F_{\mathrm{h}}+F}\ell^2 & \text{if }\ell_{\mathrm{j}}<\ell\\0 & \text{if }\ell\le\ell_{\mathrm{j}}\end{cases}$ |
| Mortality rate | $d(F,\ell)$ | $=$ | $\mu$ |
| Feeding rate | $I(F,\ell)$ | $=$ | $\nu\dfrac{F}{F_{\mathrm{h}}+F}\ell^2$ |
| Autonomous algal dynamics | $R(F)$ | $=$ | $\alpha F\left(1-\dfrac{F}{K}\right)$ |

NOTE.—*Daphnia* growth, reproduction, and mortality and autonomous algal growth dynamics as functions of the individual length $\ell$ and the density of algae $F$, for the size-structured example model.

## The Individual and Its State

Some differences in individual behavior can be related to specific characteristics of the individuals that are compared: for example, larger *Daphnia* individuals have higher reproductive rates. In addition, individuals may exhibit purely random variability in behavior, unrelated to apparent differences in individual characteristics. In the introduction I indicate that I am not considering differences in behavior between otherwise identical individuals, that is, purely random variability in behavior. Therefore, the focus here is on differences in individual behavior that are related to apparent individual traits.

From a biological point of view, every individual is unique. From a modeling point of view, we have to characterize that uniqueness and, since differences between individuals are of interest, to distinguish individuals from each other. This leads to the concept of the individual state, or *i*-state. Formally, the *i*-state can be defined as follows.

The individual state is a collection of individual, usually physiological, properties

1. that at any time completely determine, possibly together with the present state of its environment, the individual's probability of dying or giving birth and its influence on the environment (e.g., its contribution to the overall population dynamics) and

2. whose future values are completely determined by their present values plus the intervening environmental history, as encountered by the individual.

For *Daphnia*, the size of an individual has a major influence on its reproduction, feeding, growth, and development, as is the case for many animal species. Hence, size is an appropriate choice for the *i*-state. When discussing the general theory, $x$ denotes the *i*-state. Age, mass, and energy reserves are other physiological variables that have often been used as *i*-states.

From the definition it is clear that the *i*-state can consist of more than a single variable, in which case $x$ is a vector of physiological quantities. The elements of this vector are referred to as *individual state variables* or *i-state variables*. A very early paper on PSP models by Sinko and Streifer (1967) characterized the individuals of a *Daphnia* (!) population by their age and their size. The combination of age and size is often used, if the *i*-state is higher dimensional at all.

The *individual state space*, or *i-state space*, is the entire collection of values that the *i*-state can possibly attain. For *Daphnia*, the *i*-state space is the entire range of lengths that can possibly be observed in the species, which according to Section 1 equals all sizes in the range from $\ell_b$ to $\ell_m$. The *i*-state space is often indicated by $\Omega$.

Now we see what the term "physiological structure" refers to, in models of physiologically structured populations. Given the choice of the *i*-state, the physiological structure of a population refers to the number and type of individuals that constitute it, where the individuals are identified and distinguished from each other by their (physiological) *i*-state.

## The Environment and Its State

The environment encountered by an individual determines, to a large extent, its behavior in terms of reproduction, death, and so on. This environment consists of a multitude of biotic and abiotic factors. Analogous to the characterization of the individual by its *i*-state, it is also necessary to identify its environment by the *envi-*

*ronmental state*, or *E*-state. Formally, the environmental state is a collection of biotic and abiotic factors that characterize the environment in which an individual lives and determines the influence of that environment on the individual's behavior.

As for the *i*-state, the *E*-state is a collection of environmental characteristics selected by the modeler as being the determining factors influencing the individual behavior. The *E*-state can include more than a single aspect of the environment, but for practical purposes it should be restricted. For *Daphnia* it is known that temperature influences the behavior of individuals, because it changes the energy requirements for basal metabolism. However, the main point of interest in the example is the interaction of the individuals with their food source. It is logical to adopt food density as the important environmental characteristic and to assume that all other environmental factors, like temperature, are constant (simply because we do not want to study their influence at this point).

The *E*-state may be either a single quantity or a vector of *environmental state variables* (cf. the individual state variables); it is denoted by *E*. In the *Daphnia* example, where the food density (i.e., the biomass concentration of the alga *Chlamydomonas rheinhardii*) constitutes the *E*-state, the symbol $F$ is used, as in the equations in Section 1.

The environment that an individual experiences plays a fundamental role in population dynamics (and in other fields like evolution!). Its conceptual importance cannot be underestimated. Since random differences in individual behavior have been excluded, the present and future behavior of an individual would be determined if the time course of the relevant environmental characteristics were known. The simplest case is a constant environment, but it suffices that the *E*-state be given at every point in time. In such a "known" environment, the reproduction, survival, and development of an individual is fixed, since all these processes are determined (see the preceding subsection) by the combination of the *i*-state and *E*-state. When the environmental dynamics are given, the individuals of a biological population are therefore fully independent.

For a given individual, the rest of the population is part of its environment. Because we assume large numbers of individuals, this remaining part of the population can be equated with the population as a whole (the contribution of a single individual is negligible). Therefore, both the interactions between individuals and the interaction between an individual and itself some time later act via the common environment of all individuals in the population. The environment constitutes the route of all feedback loops; this is a fundamental aspect of physiologically structured population models.

From this discussion we infer that there are different categories of environmental variables.

*External forcing factors.* This term indicates characteristics of the environment that are externally imposed upon the population. Neither a single individual nor the population as a whole can influence these characteristics. Examples are temperature, irradiance, and the size of a predator population, if the latter is not determined by the abundance of the population of interest. The important distinction between this category and the next is that these factors can at most *modulate* the dynamics (e.g., growth rate) of a population; they cannot *regulate* population size within certain bounds. External forcing factors are always explicit functions of time.

*Feedback mechanisms.* These are characteristics of the environment that can regulate a population and hence constitute some form of feedback. This class of environmental characteristics includes, for example, total population size and food availability. These factors are the vehicles of density-dependent interactions and belong basically to two classes.

The first class of mechanisms is *feedback functions*. If individual behavior is directly influenced by the total number of individuals around (or any other measure of population size, like total biomass), we speak of "direct density dependence." This would be the case, for example, if predators interfere with each other while searching for prey. "Contest competition" is also a form of direct density dependence. The essence of feedback functions is that some statistic of the population has a direct (usually negative) influence on individual behavior and can consequently regulate the population density. As with the external forcing functions, direct density-dependent factors are explicit functions of time, because at any time the determining quantity is a statistic of the population, the state of which is known at that time. Because they are explicitly specified, these quantities are called feedback functions.

The second class of feedback mechanisms is *feedback loops*, which include factors that are themselves dynamically changing and that are only partly influenced by actions of individuals. Food availability is a classic example. This type of environmental variable can also regulate the population, but it exerts a more indirect density-

dependent influence. In the *Daphnia*-algal example, the algae have
(logistic) growth dynamics of their own, but they are also reduced
by *Daphnia* feeding. The reduced algal density influences the behavior of individual *Daphnia*. This form of density dependence is,
hence, a two-step process. These factors are called feedback loops
in the spirit of their indirect action. Feedback loops require a dynamic description, generally in terms of an ordinary differential
equation (ODE) that incorporates both the autonomous dynamics
of the environmental variable (that is, the dynamics in the absence
of the structured population; cf. the logistic algal growth in the
*Daphnia*-algal example) and the influence of the structured population on the value of this environmental variable (cf. the feeding
of the *Daphnia* individuals). The required ODE turns out to resemble the differential equation (1a) for the algal density in the
simple example model, as is seen below.

### The Individual Behavior

Once the $i$-state and $E$-state are selected as the characteristics of
an individual and its environment that completely determine its
behavior, the form of this dependence has to be specified. The behavior of individuals consists essentially of four processes: development, which refers to the changes in $i$-state over time; mortality;
reproduction; and the influence of the individual on its environment, such as the influence of feeding on the current food density.
All these behavioral processes are specified in terms of instantaneous rates because the dynamics of the population are described
on a continuous-time basis in terms of differential equations. The
discussions to follow focus on the formal definition of these rates
and their relation to observable, biological data.

*Development.*   The $i$-state changes over time in a fully deterministic fashion, since two individuals with identical $i$-states are assumed to be identical forever and the $i$-state and $E$-state together
determine the development of the individual. Formally, the individual development rate, $g(E, x)$, is the rate of change of the individual state variable $x$ at the current time, as a function of the
variable itself and the current environmental state $E$.

Consider an individual, born with an individual state equal to
$x_b$ at $t = 0$. The change in $i$-state is described by the ODE

$$\frac{dx}{dt} = g(E, x). \tag{7}$$

This equation only fixes the current developmental rate of the individual. Because the environmental variable $E$ can change over time, as a result of external factors or influences of the population itself, this ODE determines the entire development of the individual only if the time course of the $E$-state is given. In the *Daphnia* example, food density and length determine the current individual growth rate, but without information on the entire future of the food density the future sizes of the individuals are also unknown.

To relate individual behavior to experimental data, it is useful to consider individuals that are all born with a fixed size $x_b$ and live after birth in a constant environment; that is, $E(t)$ is equal to some constant $E_c$ for all $t$. Such a scenario corresponds to a study in which a cohort of identical individuals is followed over their entire life under constant conditions. Under such conditions we can usually derive expressions for observable quantities such as the relation between the age and the $i$-state of the individuals in the cohort. These expressions involve the instantaneous rates in the model and can therefore be used to determine these rates experimentally. Hence, the scenario of individuals living in a constant environment $E_c$ recurs frequently in this and the following sections.

If individuals are born with $i$-state $x_b$ and live in a constant environment $E_c$, their $i$-state at every age is given by the formal solution of the ODE (7). At age $a$ such individuals have an $i$-state $X(E_c, a)$:

$$X(E_c, a) \;=\; x_b \;+\; \int_0^a g\big(E_c, X(E_c, \zeta)\big)\, d\zeta\,. \tag{8}$$

Little or no analytical investigation is possible if the integral in (8) has no closed form. I therefore assume that this integral can be solved analytically. Consequently, the $i$-state of individuals at age $a$ in the "constant-environment scenario" is explicitly given by a function $X(E_c, a)$, which can be referred to as the ($i$-state)-age relation.

In the *Daphnia* example, the specific choice of $g(F, \ell)$ allows for an explicit solution of the length-age relation when the food density is constant. The resulting curve is known as the von Bertalanffy growth curve, the general form of which is illustrated in Figure 1.

*Mortality.* Individual mortality is governed by the instantaneous mortality rate, $d(E, x)$, defined as the rate at which the probability of survival from birth to $i$-state $x$ decreases with time. This definition implies that the product $d(E, x)\,\Delta t$ equals the probability that an individual with $i$-state $x$, living in an environment $E$, dies within a small time interval $\Delta t$. To relate this rate to observable data, it is again useful to consider an individual that has been born with $i$-state $x_b$ and lives in a constant environment $E_c$. The survival probability for such an individual can be characterized by the survival function $S(E_c, x)$, defined as the probability that an individual survives and reaches an $i$-state $x$, while living in a constant environment $E_c$.

The individual is born with $i$-state $x_b$, and therefore $S(E_c, x_b) = 1$. According to the definition of the instantaneous mortality rate, the dynamics of $S(E_c, x)$ are governed by the ODE

$$\frac{dS(E_c, x)}{dt} = -d(E_c, x)\, S(E_c, x),$$

where $S(E_c, x_b) = 1$. The right-hand side of this ODE depends implicitly on time because the $i$-state $x$ changes over time, following the relation (8). For the changes of $S(E_c, x)$ with the $i$-state $x$,

$$\frac{dS(E_c, x)}{dx} = \frac{dS(E_c, x)}{dt}\frac{dt}{dx} = -\frac{d(E_c, x)}{g(E_c, x)}\, S(E_c, x)$$

$$\Rightarrow \quad S(E_c, x) = \exp\left(-\int_{x_b}^{x}\frac{d(E_c, \xi)}{g(E_c, \xi)}\, d\xi\right), \qquad (9)$$

which explicitly relates the survival function to the instantaneous growth and mortality rates.

The probability that an individual dies with $i$-state $x$ is given by the probability density function $H(E_c, x)$:

$$H(E_c, x) = -\frac{d\,S(E_c, x)}{dx}$$

$$= \frac{d(E_c, x)}{g(E_c, x)}\, S(E_c, x) \qquad (10)$$

$$= \frac{d(E_c, x)}{g(E_c, x)}\exp\left(-\int_{x_b}^{x}\frac{d(E_c, \xi)}{g(E_c, \xi)}\, d\xi\right).$$

If $g(E_c, x)$ has been determined from observations on the ($i$-state)-age relation in the constant environment $E_c$, each of these equations could be used to determine the instantaneous death rate

$d(E_\mathrm{c}, x)$. Most readily, it is determined from the decrease in the log-transformed survival function with increasing $i$-state $x$, which equals the quotient of the individual death and growth rates at $x$:

$$- \frac{d}{dx} \ln\bigl(S(E_\mathrm{c}, x)\bigr) \;=\; - \frac{1}{S(E_\mathrm{c}, x)} \frac{dS(E_\mathrm{c}, x)}{dt} \frac{dt}{dx} \;=\; \frac{d(E_\mathrm{c}, x)}{g(E_\mathrm{c}, x)} \,.$$

When chronological age is used as the $i$-state, the function $g$ always has the value 1, but the relations derived in this section still apply. For example, the survival function becomes

$$S(E_\mathrm{c}, a) \;=\; \exp\left( - \int_0^a d(E_\mathrm{c}, \zeta)\, d\zeta \right) ,$$

where the variable $a$ is now used instead of $x$ to identify age as the $i$-state. This relation seems to contain an integral over a rate $d(E_\mathrm{c}, a)$ with a dimension (time$^{-1}$). However, the integral must be a dimensionless quantity to appear in the exponential function, which it is not if the integrand (i.e., the quantity to be integrated) carries the dimension (time$^{-1}$). In reality, the function $d(E_\mathrm{c}, a)$ is divided by a rate function that is always equal to 1, which makes the dimension of the integrand equal to (age$^{-1}$) and the integral indeed dimensionless. This obscure occurrence of the unit developmental rate in age should be stressed, since it is encountered in later sections as well.

Formal expressions of the expected value of the $i$-state and age at death are given in Appendix A.

*Reproduction.*   Reproduction is modeled by specifying the instantaneous birth or reproduction rate, $b(E, x)$, defined as the rate of production of offspring by an individual of $i$-state $x$ experiencing an environmental state $E$. The quantity $b(E, x)\, \Delta t$ gives the number of offspring that an individual with an $i$-state equal to $x$, living in an environment with $E$-state $E$, produces within a small time interval $\Delta t$.

In a constant environment, $E_\mathrm{c}$, the integral

$$\int_0^a b\bigl(E_\mathrm{c}, X(E_\mathrm{c}, \zeta)\bigr)\, d\zeta$$

gives the cumulative number of offspring an individual living in these conditions has produced by the time it reaches age $a$, given that it did not die. In cohort studies, the cumulative number of offspring can hence be used to determine $b(E, x)$. By a change of variables in the equation above, this cumulative number of offspring can be expressed in terms of $x$:

$$\int_0^a b\big(E_c, X(E_c, \zeta)\big) \, d\zeta = \int_{x_b}^{X(E_c, a)} b(E_c, x) \frac{da}{dx} \, dx$$

$$= \int_{x_b}^{X(E_c, a)} \frac{b(E_c, x)}{g(E_c, x)} \, dx \, .$$

It follows that the cumulative number of offspring produced by a surviving individual living under constant environmental conditions $E_c$ between its birth with $i$-state $x_b$ and the time it reaches the $i$-state $x$ is equal to

$$\int_{x_b}^x \frac{b(E_c, \xi)}{g(E_c, \xi)} \, d\xi \, . \tag{11}$$

If $g(E_c, x)$ has been determined from experimental observations on the ($i$-state)-age relation in the constant environment, $E_c$, the instantaneous reproduction rate, $b(E_c, x)$, is most readily determined from the increase in the cumulative number of offspring with $x$. This increase equals the ratio of the individual reproduction and growth rates at $x$:

$$\frac{d}{dx} \int_{x_b}^x \frac{b(E_c, \xi)}{g(E_c, \xi)} \, d\xi = \frac{b(E_c, x)}{g(E_c, x)} \, .$$

Experimental observations often involve the first reproduction event of an individual. The $i$-state at which reproduction first occurs can be calculated using the same approach used for survival. The probability that an individual has not yet reproduced by the time it has reached an $i$-state $x$, given that it survives, equals

$$\exp\left(-\int_{x_b}^x \frac{b(E_c, \xi)}{g(E_c, \xi)} \, d\xi\right) \, .$$

The probability that it first reproduces when it has an $i$-state $x$, again conditional on survival, is

$$\frac{b(E_c, x)}{g(E_c, x)} \exp\left( -\int_{x_b}^{x} \frac{b(E_c, \xi)}{g(E_c, \xi)} \, d\xi \right).$$

These expressions are completely analogous to the survival function and the probability distribution of the state at death, respectively, in the discussion on the mortality rate. Their derivation proceeds along the same lines and is therefore left as an exercise.

Appendix A gives formal expressions for the expected $i$-state and age of the individual when it first reproduces.

*Influence on the environment.* The last component of individual behavior to be modeled is its influence on the environment. As described in the preceding subsection, the characteristics of the environment that can be influenced by an individual can be categorized into feedback functions and feedback loops.

Feedback functions are statistics, such as population size, the number of juveniles or adults, and total biomass, that exert a direct density-dependent influence. Deriving expressions for such population statistics (see Section 4) requires introducing the population state. In general, feedback functions involve a weighting function, $\phi(x)$, which indicates the contribution of an individual of $i$-state $x$ to the feedback function. For the total population size, all individuals are weighted equally and $\phi(x) \equiv 1$. To determine the number of juvenile (adult) individuals, set $\phi(x) \equiv 1$ for all $i$-states that represent juveniles (adults) and $\phi(x) \equiv 0$ otherwise. If individual mass is taken as the $i$-state $x$ and the total population biomass is the statistic of interest, $\phi(x) = x$, because individuals contribute to the population biomass in accordance with their current $i$-state. In general, the weighting function $\phi$ depends only on the $i$-state $x$. In special cases, it can also depend on the current $E$-state or the $i$-state of other individuals. For example, to model cannibalism of older or larger individuals on younger or smaller individuals, a weighting function $\phi(x', x)$ is needed that depends on both the $i$-state $x'$ of the cannibal and the $i$-state $x$ of its victim (van den Bosch et al. 1988).

Feedback loops involve environmental variables, the dynamics of which are specified by an ODE. For each environmental variable $E$, a function $I(E, x)$ must be formulated that specifies the rate at which an individual with $i$-state $x$ decreases or increases $E$. A classic example is the resource density experienced by an individual.

The function $I(E, x)$ in this case models the resource consumption rate (see the *Daphnia* example).

### The Dynamics of the Environment without Individuals

Environmental forcing functions (see above) are explicit functions of time that modulate individual behavior. Their functional form must be specified. Annual temperature fluctuations could, for example, be modeled as a sinusoidal function of time.

To model an environmental feedback loop, the dynamics of the environmental variable must be described in the absence of the structured population. The logistic growth rate of the algal population in the *Daphnia* example is typical.

### Some Remarks on Notation

Discussing the *Daphnia* example in parallel with the general theory of PSP models raises some problems of notation. In general, I use $x$ and $E$ to denote the $i$-state and $E$-state, respectively, as long as no specific physiological or environmental traits have been attached to these two states. In the *Daphnia* example, individual length and the food density appear as $i$-state and $E$-state, respectively, and therefore, I use $\ell$ and $F$ in discussions of this particular model. I use $g$, $d$, $b$, and $I$ to denote the functions that model development, mortality, reproduction, and environmental interactions of the individuals. These functions can therefore appear with different arguments, for example, $g(E, x)$ as opposed to $g(F, \ell)$, depending on the context of the discussion.

### Recipe for Model Construction

1. Choose the set of physiological traits that determine individual behavior as *i-state variables*, denoted by $x$.

2. Choose the set of variables that characterize an individual's environment and influence its behavior as *E-state variables*, denoted by $E$.

3. Describe the individual behavior as a function of the $i$-state and the $E$-state, including developmental rate, $g(E, x)$; mortality rate, $d(E, x)$; reproduction rate, $b(E, x)$; and individual influences on its environment, the weighting functions $\phi(x)$ for the direct density-dependent factors and the rate functions $I(E, x)$ for indirect density-dependent factors.

4. Specify the "explicitly given" environmental forcing functions as a function of time, $E_{\exp}(t)$.

5. Describe the autonomous dynamics of the "dynamically varying" environmental variables, $R(E_{\mathrm{dyn}})$.

## 3 The Size-Structured *Daphnia* Population Model

Section 1 contains all the ingredients necessary for a size-structured model for *Daphnia* feeding on the alga *Chlamydomonas rhein-hardii*. From this point on, no further assumptions need be made; everything boils down to proper bookkeeping.

In the unstructured model (1), the total number of *Daphnia*, denoted by $C$, represents the biological population in the model. An analogous representation of the biological population is needed for the size-structured model. In addition to the total number of individuals, it must also contain information about their sizes. Given that individual size lies in the continuous interval $[\ell_{\mathrm{b}}, \ell_{\mathrm{m}})$ and the assumption that the number of individuals is large, an appropriate mathematical representation of the population is a density function. This density function, denoted by $n(t, \ell)$, gives the distribution of individuals as a function of $\ell$ at time $t$. It is important to understand that only the integral of $n(t, \ell)$ can be interpreted as a number of individuals; that is,

$$\int_{\ell_{\mathrm{b}}}^{\ell_{\mathrm{m}}} n(t, \ell)\, d\ell$$

equals the total population size at time $t$. In general, the integral

$$\int_{\ell_1}^{\ell_2} n(t, \ell)\, d\ell$$

gives the total number of *Daphnia* with length $\ell$ between $\ell_1$ and $\ell_2$ (see Fig. 2).

The interpretation of $n(t, \ell)$ itself (as opposed to its integral) is problematic. One can loosely think of the quantity $n(t, \ell)\,\Delta\ell$ as the number of individuals with a length in a (small) interval $[\ell, \ell + \Delta\ell)$. Since these $n(t, \ell)\,\Delta\ell$ individuals all have a feeding rate equal to $I(F, \ell)$ (Table 1), their total feeding rate equals the product $I(F, \ell)\, n(t, \ell)\,\Delta\ell$. Taking the limit $\Delta\ell \to 0$ and integrating

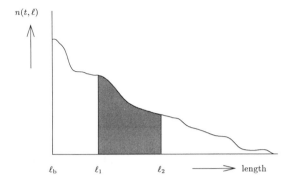

FIGURE 2. *The density function* $n(t, \ell)$ *constitutes the mathematical representation of the* Daphnia *population in the structured model. The hatched surface under the curve between* $\ell = \ell_1$ *and* $\ell = \ell_2$ *equals the integral of* $n(t, \ell)$ *over this length interval and represents the number of* Daphnia *with a length between these two bounds.*

the product function gives the total feeding rate of all individuals with a length between $\ell_1$ and $\ell_2$:

$$\int_{\ell_1}^{\ell_2} I(F, \ell)\, n(t, \ell)\, d\ell.$$

The ODE (1a) for algal density in the unstructured model (1) makes it clear that in the size-structured model the dynamics of algal density must also equal the balance between the growth of the algae and the feeding rate of the total *Daphnia* population. This total feeding rate is given by the integral of $I(F, \ell)\, n(t, \ell)$, given above with $\ell_1 = \ell_b$ and $\ell_2 = \ell_m$. Therefore, in the size-structured *Daphnia* model, the dynamics of algal density are described by

$$\frac{dF}{dt} = \alpha F \left(1 - \frac{F}{K}\right) - \int_{\ell_b}^{\ell_m} I(F, \ell)\, n(t, \ell)\, d\ell, \qquad (12)$$

which resembles the ODE (1a) in the unstructured model.

Once the state of the *Daphnia* population has been represented by the density function $n(t, \ell)$, a partial differential equation (PDE) governing its dynamics can be derived by considering a small size interval of width $\Delta \ell$ around a specific length $\ell = \ell'$ (see Fig. 3).

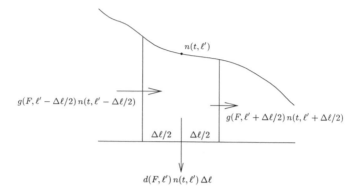

FIGURE 3. *The derivation of the partial differential equation for* $n(t, \ell)$.

Within a small time interval $\Delta t$ (small enough that $g(F, \ell') \Delta t \ll \Delta \ell$), the number of individuals in this length interval changes because (1) smaller individuals enter the interval by growing, (2) individuals at the upper end of the interval grow out of it, and (3) individuals disappear through death.

The flux across the lower bound ($\ell = \ell' - \Delta \ell/2$) is the rate at which individuals enter the interval as a result of growth. This flux is proportional to the rate at which a single individual grows across the boundary and to the number of individuals doing so, that is, to the product of the growth rate and the density at this value of $\ell$: $g(F, \ell' - \Delta \ell/2) \, n(t, \ell' - \Delta \ell/2)$. Analogously, the rate at which individuals leave the interval at the high end equals the product of the growth rate and the local density at $\ell = \ell' + \Delta \ell/2$: $g(F, \ell' + \Delta \ell/2) \, n(t, \ell' + \Delta \ell/2)$. Finally, the rate at which individuals disappear because of death equals the product of the death rate and the number of individuals with a length in the small interval of size $\Delta \ell$: $d(F, \ell') \, n(t, \ell') \, \Delta \ell$. Note that this last expression uses the "law of large numbers": in a small time interval, $\Delta t$, every individual in the interval around $\ell = \ell'$ has a probability of dying equal to $d(F, \ell') \, \Delta t$. The law of large numbers allows us to replace the *number* of individuals dying during the small time interval, which clearly is a stochastic variable, with its *expected* value!

Multiplying all rates by the considered time interval $\Delta t$, and adding the various contributions (see Fig. 3), yields a balance equation for the difference between the number of individuals,

$n(t + \Delta t, \ell')\,\Delta\ell$, at time $t + \Delta t$ and the number of individuals, $n(t, \ell')\,\Delta\ell$, at time $t$:

$$
\begin{aligned}
n(t + \Delta t, \ell')\,&\Delta\ell - n(t, \ell')\,\Delta\ell \\
&= -g(F, \ell' + \Delta\ell/2)n(t, \ell' + \Delta\ell/2)\Delta t \\
&\quad + g(F, \ell' - \Delta\ell/2)n(t, \ell' - \Delta\ell/2)\Delta t \\
&\quad - d(F, \ell')\,n(t, \ell')\,\Delta\ell\,\Delta t \,.
\end{aligned}
$$

It is instructive to check the dimensions in this equation. Note that the dimensions of $g(F, \ell)$, $n(t, \ell)$, and $d(F, \ell)$ are (length $\times$ time$^{-1}$), (number $\times$ length$^{-1}$), and (time$^{-1}$), respectively. Next, dividing both sides by the product $\Delta t\,\Delta\ell$ leads to

$$
\begin{aligned}
\frac{n(t + \Delta t, \ell') - n(t, \ell')}{\Delta t} \\
= -\frac{g(F, \ell' + \Delta\ell/2)\,n(t, \ell' + \Delta\ell/2)}{\Delta\ell} \\
-\frac{g(F, \ell' - \Delta\ell/2)\,n(t, \ell' - \Delta\ell/2)}{\Delta\ell} \\
- d(F, \ell')\,n(t, \ell') \,.
\end{aligned}
$$

Taking the limits $\Delta t \to 0$ and $\Delta\ell \to 0$ in such a way that $g(F, \ell')\,\Delta t \ll \Delta\ell$, the left-hand side of this equation can be recognized as the partial derivative of $n(t, \ell)$ at $\ell = \ell'$ with respect to time $t$, while the first term on the right-hand side equals the partial derivative of the product $g(F, \ell)\,n(t, \ell)$ at $\ell = \ell'$ with respect to length $\ell$. After dropping the primes, this limiting procedure results in the following partial differential equation for the density function $n(t, \ell)$:

$$
\frac{\partial n(t, \ell)}{\partial t} = -\frac{\partial g(F, \ell)\,n(t, \ell)}{\partial \ell} - d(F, \ell)\,n(t, \ell) \,. \tag{13}
$$

This PDE describes only the dynamics of the density function $n(t, \ell)$ in the interior of the interval $[\ell_{\mathrm{b}}, \ell_{\mathrm{m}})$ and incorporates the influence of individual growth and death in the two terms on the right-hand side. It must be supplemented with a boundary condition that specifies the density at the lower end ($\ell = \ell_{\mathrm{b}}$) of the interval, at every time $t$, to complete the model. Since individuals are born at length $\ell_{\mathrm{b}}$, it is clear that this boundary condition should involve reproduction. The form of the boundary condition can be derived by integrating the left and right sides of the PDE (eq. 13) from $\ell = \ell_b$ to $\ell = \ell_m$, as follows:

$$\int_{\ell_b}^{\ell_m} \frac{\partial n(t,\ell)}{\partial t}\, d\ell = -\int_{\ell_b}^{\ell_m} \frac{\partial g(F,\ell)\, n(t,\ell)}{\partial \ell}\, d\ell - \int_{\ell_b}^{\ell_m} d(F,\ell)\, n(t,\ell)\, d\ell$$

$$\Rightarrow \frac{d}{dt}\int_{\ell_b}^{\ell_m} n(t,\ell)\, d\ell = g(F,\ell_b)\, n(t,\ell_b) - g(F,\ell_m)\, n(t,\ell_m)$$

$$-\int_{\ell_b}^{\ell_m} d(F,\ell)\, n(t,\ell)\, d\ell\,.$$

The last equation has a clear biological interpretation. The left-hand side is the rate of change in the total number of individuals in the population. These changes should equal the balance between births and deaths. The last term on the right-hand side is the total mortality rate in the population (cf. the total population feeding rate in the ODE, eq. 12). Since $\ell_m$ is an asymptotic length that no individual can reach, $g(F,\ell_m)\, n(t,\ell_m) \equiv 0$. This leads to the conclusion that $g(F,\ell_b)\, n(t,\ell_b)$ should equal the total population birthrate. Analogous to the total population feeding rate in the ODE (eq. 12) and the total population death rate in the derivation above, this total population birthrate equals the integral of the product of the individual birthrate, $b(F,\ell)$, and the population density, $n(t,\ell)$, over the reachable size interval $[\ell_b, \ell_m)$. The boundary condition with which to supplement the PDE (eq. 13), therefore, has the form

$$g(F,\ell_b)\, n(t,\ell_b) = \int_{\ell_b}^{\ell_m} b(F,\ell)\, n(t,\ell)\, d\ell\,. \qquad (14)$$

It should be noted that the product $g(F,\ell_b)\, n(t,\ell_b)$ is natural, since it makes both sides of the condition a rate, with dimension (number $\times$time$^{-1}$).

This completes the size-structured model for *Daphnia*: starting from an initial state of both the *Daphnia* and the algal populations, the equations derived here determine the future dynamics of both. For the algal population, the initial state consists of an initial algal density. The initial state of the *Daphnia* population is more complicated; it must specify the initial size distribution as

TABLE 2. *Population-Level Equations of the Size-Structured Model for the Interaction between* Daphnia *and Algae*

---

| *Daphnia* dynamics | $\dfrac{\partial n(t,\ell)}{\partial t} + \dfrac{\partial g(F,\ell)\,n(t,\ell)}{\partial \ell} = -d(F,\ell)\,n(t,\ell)$ |
|---|---|

$$g(F,\ell_{\mathrm b})\,n(t,\ell_{\mathrm b}) \;=\; \int_{\ell_{\mathrm b}}^{\ell_{\mathrm m}} b(F,\ell)\,n(t,\ell)\,d\ell$$

| Algal dynamics | $\dfrac{dF}{dt} = \alpha F\left(1 - \dfrac{F}{K}\right) - \displaystyle\int_{\ell_{\mathrm b}}^{\ell_{\mathrm m}} I(F,\ell)\,n(t,\ell)\,d\ell$ |
|---|---|
| Initial conditions | $n(0,\ell) = \Psi(\ell)$ <br> $F(0) = F_0$ |

---

NOTE.—The functions $g(F,\ell)$, $d(F,\ell)$, $b(F,\ell)$, and $I(F,\ell)$, modeling the individual behavior, are specified in Table 1.

well as initial numbers. Thus, the density function $n(t,\ell)$ at $t = 0$ is specified as a function of $\ell$:

$$n(0,\ell) \;=\; \Psi(\ell)\,,$$

$$F(0) \;=\; F_0\,.$$

Here, $\Psi(\ell)$ denotes the density function characterizing the initial *Daphnia* population and $F_0$ denotes the initial algal density. The full set of equations for the size-structured *Daphnia* population model is summarized in Table 2, and Figure 4 shows a schematic illustration of the actions that these equations specify.

## 4 The Model at the Population Level

### The Population State and Its Statistics

Just as the $i$-state and $E$-state characterize the individual and its environment, the population state, or $p$-state, uniquely determines the size and composition of the structured population. The $p$-state is a mathematical object that represents the biological population of interest in the model. In unstructured models, in which individuals are not distinguished from each other, the total number of individuals fully characterizes the population and constitutes an

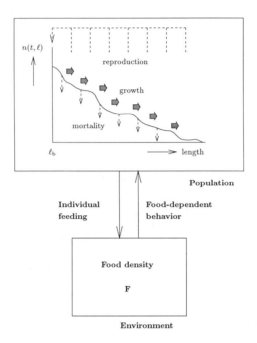

FIGURE 4. *Schematic representation of the* Daphnia *population model, specified in Table 2. Growth and mortality influences on $n(t, \ell)$ are described by the partial differential equation, the reproduction influence by the boundary condition in Table 2. Individual behavior, mainly reproduction and growth, depends on food density. Food dynamics are described by an ordinary differential equation incorporating the feeding influence of individual* Daphnia.

appropriate $p$-state. In models of populations with distinct year classes, or cohorts, such as fish populations with sharply pulsed reproduction, the population can be represented by a vector in which each element indicates the number of individuals in a particular cohort. In physiologically structured models, the state of the individuals can generally take its value from a continuous individual state space. As a consequence, the $p$-state in these models can be represented by a distribution or density function over the individual state space $\Omega$. (The choice of a density function as the $p$-state is connected to the PDE formalism. The recently developed "cumulative formulation" leads to different $p$-states.)

Note that the integral of the density function is interpreted as a number of individuals. Integration over the entire individual state space, $\Omega$, gives the total number of individuals in the population. Integration over part of $\Omega$ gives the number of individuals with an $i$-state in the part considered. More generally, any population statistic $\Phi(t)$ of biological interest can be written as a weighted integral of the population density function over the individual state space, $\Omega$:

$$\Phi(t) \;=\; \int_{\Omega} \phi(x)\, n(t, x)\, dx\,. \tag{15}$$

In this equation $\phi(x)$ is an arbitrary weighting function defined on $\Omega$. The total population size is obtained by assuming that $\phi(x) = 1$ for all $x$. The number of juvenile and adult individuals is obtained by assuming that $\phi(x) = 1$ for values of the $i$-state that pertain to juvenile and adult individuals, respectively, and assuming that $\phi(x) = 0$ otherwise. If $x$ refers to the weight of an individual, setting $\phi(x) = x$ gives the total population biomass at time $t$. These are just a few examples to show that population statistics of interest are weighted integrals of the density function $n(t, x)$ over the individual state space, $\Omega$.

Section 2 discusses how an individual can influence components of its environment, like the ambient food density, that in turn exert an influence on the individual behavior via a feedback loop. In the *Daphnia* model, the total population feeding rate appears in the ODE for food density as an integral of the density function $n(t, \ell)$ weighted by the individual feeding rate (see eq. 12). The behavior of an individual can also be influenced directly by the population itself (cf. feedback functions in Section 2). The acting variable in direct density dependence is generally also a weighted integral of $n(t, x)$ over the state space $\Omega$. Instead of appearing in an ODE that describes the dynamics of an environmental variable, it occurs as an independent variable in the individual growth, birth, or mortality rate.

### Other Individual State Variables

Although the size-structured model of Section 3 is formulated with specific choices for the functions $g(F, \ell)$, $d(F, \ell)$, $b(F, \ell)$, and $I(F, \ell)$ (Table 1), its form is representative of any structured-population model in which (1) both the $i$-state and the $E$-state are one-dimensional (i.e., consist of a single variable), (2) all individuals are born with a fixed identical state at birth, and (3) the $E$-state acts as a feedback loop for the structured population. To obtain a general formulation of such a structured-population model, the variables $\ell$ and $F$ in the population equations of Table 2 can be replaced everywhere with the unspecified variables $x$ and $E$. The general formulation is couched in terms of the functions $g(E, x)$, $d(E, x)$, $b(E, x)$, and $I(E, x)$, which model the individual behavior (see Section 2), and the density function $n(t, x)$ characterizing the state of the structured population. The bounds of the integrals that occur in these formulas (cf. Table 2) must be chosen so that the integration is over the appropriate range of reachable $i$-state values, usually the entire $i$-state space $\Omega$.

Models in which the chronological age $a$ of the individuals constitutes the $i$-state form a special class of these one-dimensional structured-population models. Age-structured models involve only the functions $d(E, a)$, $b(E, a)$, and $I(E, a)$, since the growth rate $g$ in age equals one. An age-structured model is therefore comparable to any other structured model, except that the function $g$ is replaced everywhere by the constant 1. This special form of $g$ (see also Section 2) disguises the exact interpretation of various terms in the equations, since rates and densities are no longer easily distinguished. For example, the boundary condition in an age-structured model is

$$n(t, 0) \;=\; \int\limits_0^\infty b(E, a)\, n(t, a)\, da$$

(cf. eq. 14). Notwithstanding its form, the left-hand side of this equation still has to be interpreted as a rate, since it is essentially the product of the rate of aging (one) and the density at age zero.

## More Environmental State Variables

Changing the interpretation of the environmental state variable, or characterizing the environment by more than a single variable, involves a relatively straightforward generalization of the model discussed in Section 3. This is in sharp contrast to adding more individual state variables, which is discussed in Section 10.

The $E$-state can be a vector of environmental variables, consisting of a mixture of forcing factors, feedback functions, and feedback loops (see Section 2). Each environmental forcing factor must be specified as an explicit function of time. Every feedback function has to be specified as a weighted integral of $n(t, x)$ (cf. eq. 15), using a weighting function $\phi(x)$. For every feedback loop, an ODE has to be added to the model to describe both the autonomous dynamics of the environmental variable and the influence of the structured population on its current value. Last but not least, the dependence of individual behavior on each environmental variable must be modeled. A higher-dimensional $E$-state makes the model more complicated but not qualitatively different.

## 5  Constant Environments: Linear, Density-Independent Models

In this section it is assumed that the environmental state is characterized by a vector of variables, $E_c$, that remain constant over time. This assumption eliminates density-dependent interactions between individuals because such interactions, by definition, act via the changing state of the environment. (The assumption of a constant environment and the absence of density-dependent interactions also exclude feedback functions among the environmental state variables; see Section 2.) In the absence of density dependence, the population either grows or declines exponentially, depending on the number of offspring produced, on the average, by a single individual during its life. The models discussed in this section are, therefore, linear and density-independent. Note that this discussion is important for the analysis of equilibria in density-dependent models as well, since the environment at equilibrium is also constant in time.

A *Daphnia* population in the laboratory, with constant ambient food density, is an example of a system considered here. Evidently, exponential population growth or decline means that such an experimental system cannot exist for a long time, but structured-population models allow us to compute, for example, the population growth rate as a function of the life-history characteristics of the individuals. For ease of presentation, the $i$-state is assumed to comprise a single variable, although the discussion also applies to higher-dimensional cases.

Since the environment is constant, the only equations of interest

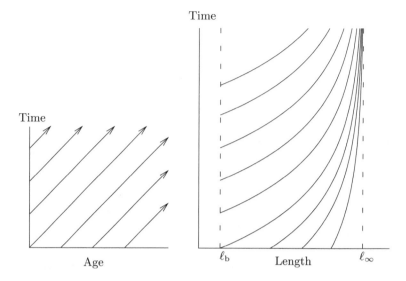

FIGURE 5. *Characteristics of the partial differential equation for an age-structured population* (left) *and for the size-structured,* Daphnia *model* (right). *For Daphnia, food density* $F_c$ *is assumed constant over time, such that the ultimate length* $\ell_\infty$ *equals* $\ell_m F_c/(F_h + F_c)$. *The characteristics can most easily be interpreted as the trajectories or paths through the (i-state)-time plane, followed by the individuals during their lifetime. The characteristics originating on the x-axis pertain to members of the initial population; those originating on the y-axis* (left) *or on the vertical line* $\ell = \ell_b$ (right) *to individuals born later.*

are those relating to the structured population:

$$\frac{\partial n(t,x)}{\partial t} = -\frac{\partial g(E_c,x)\,n(t,x)}{\partial x} - d(E_c,x)\,n(t,x)\,, \qquad (16a)$$

$$g(E_c,x_b)\,n(t,x_b) = \int_{x_b}^{x_m} b(E_c,x)\,n(t,x)\,dx\,, \qquad (16b)$$

$$n(0,x) = \Psi(x)\,. \qquad (16c)$$

Here, the $i$-state space $\Omega$ is the interval $[x_b, x_m)$, and individuals are born with $i$-state $x_b$. As usual, $n(t,x)$ is the density function characterizing the population state, and $g(E_c,x)$, $d(E_c,x)$, and $b(E_c,x)$ model development, mortality, and reproduction functions of the individual state $x$ in the (constant) environment, $E_c$.

First, the concept of characteristics has to be introduced. *Characteristics* are curves in the space, spanned by the *i*-state coordinate and time—here, the plane $\Omega \times \mathbb{R}^+$—along which the changes in the density function $n$ are described by an ordinary differential equation, if the derivative used in this ODE is the total (material) derivative of $n$:

$$\frac{Dn}{Dt} = \frac{\partial n(t,x)}{\partial t} + g(E_{\mathrm{c}}, x) \frac{\partial n(t,x)}{\partial x}.$$

In the general theory of partial differential equations, the characteristics are sometimes curves in the three-dimensional space spanned by the *i*-state $x$, time $t$, and density $n$. The characteristics, as defined here, are the projections of these latter curves on the two-dimensional plane spanned by the *i*-state and time. These two meanings of the term characteristics are used interchangeably. Most important, the characteristics, as defined here, can be interpreted biologically as the trajectories that individuals follow through the $(x,t)$-plane, if they do not die. Figure 5 shows the characteristics for an age-structured model and the size-structured model used in the *Daphnia* example. The ODE for the changes in the density $n$ along the characteristics turns out to be linear and is explicitly solvable. The phrase integration along characteristics refers to obtaining the solution of the PDE (16a) by deriving and integrating this linear ODE for the density $n$ along the characteristics. To explain the method of integration along characteristics, a bit of rather complicated notation is required.

Let the variable $\theta$ be proportional to time, with the moment in time when $\theta = 0$ not yet fixed. Mathematically, the characteristics (in the sense used here) can be defined as all those curves in the $(x,t)$-plane, for which

$$\frac{dt}{d\theta} = 1, \qquad (17\mathrm{a})$$

$$\frac{dx}{d\theta} = g(E_{\mathrm{c}}, x). \qquad (17\mathrm{b})$$

This definition implies that

$$\frac{dx}{dt} = g(E_{\mathrm{c}}, x), \qquad (18)$$

which is just the ODE describing individual development. The characteristics constitute an infinite collection of curves in the

$(x, t)$-plane with an identical form but different origins: if an individual is born at time $t_b$ with $i$-state $x_b$, its development through the $(x, t)$-plane follows the characteristic that originates in $(x_b, t_b)$. Alternatively, if an individual is already present when $t = 0$, with an $i$-state $x_0$, it follows the characteristic through the $(x, t)$-plane, which originates in $(x_0, 0)$ (see Fig. 5). Therefore, the set of characteristics comprises all the ($i$-state)-time relations of individuals that are born at different times $t_b$ with $i$-state $x_b$ or that are already present at time $0$ with $i$-state $x_0$. Let $\theta = 0$ correspond to a particular time and $i$-state, denoted by $t_{\theta=0}$ and $x_{\theta=0}$, respectively, and referred to as the *origin* of the characteristic. The origin is $(x_b, t_b)$ if the characteristic pertains to individuals born at time $t_b > 0$ with $i$-state $x_b$; it is $(x_0, 0)$ if the characteristic pertains to a member of the initial population. The variable $\theta$ indicates the position along the characteristic. For the characteristics followed by the members of the initial population, $\theta$ is identical to time because here $\theta = 0$ corresponds to $t = 0$. Analogously, for the characteristics followed by the individuals born after $t = 0$, $\theta$ is identical to the age of the individuals (see Fig. 5).

To derive the ODE for $n(t, x)$ along a particular characteristic, its derivative with respect to $\theta$ is calculated by considering $t$ and $x$ explicitly as functions of $\theta$:

$$
\begin{aligned}
\frac{dn\big(t(\theta), x(\theta)\big)}{d\theta} &= \frac{\partial n\big(t(\theta), x(\theta)\big)}{\partial t} \frac{dt}{d\theta} + \frac{\partial n\big(t(\theta), x(\theta)\big)}{\partial x} \frac{dx}{d\theta} \\[2mm]
&= \frac{\partial n\big(t(\theta), x(\theta)\big)}{\partial t} + g\big(E_c, x(\theta)\big) \frac{\partial n\big(t(\theta), x(\theta)\big)}{\partial x} \\[2mm]
&= -\frac{\partial g\big(E_c, x(\theta)\big)}{\partial x} n\big(t(\theta), x(\theta)\big) \\[2mm]
&\qquad - d\big(E_c, x(\theta)\big) n\big(t(\theta), x(\theta)\big).
\end{aligned}
$$

In the first two steps of the derivation, the chain rule of differentiation and the equations (17) for the characteristics are used to resolve the derivative $dn/d\theta$. The notations $t(\theta)$ and $x(\theta)$ indicate that both time and $i$-state, respectively, are considered explicit functions of $\theta$. In the last step, the PDE (16a) is used to eliminate the partial derivatives of $n$. The following, linear ODE results for the value of $n\big(t(\theta), x(\theta)\big)$ along a particular characteristic parameterized by $\theta$:

$$\frac{dn\big(t(\theta), x(\theta)\big)}{d\theta} = -\left[d\big(E_c, x(\theta)\big) + \frac{\partial g\big(E_c, x(\theta)\big)}{\partial x}\right] n\big(t(\theta), x(\theta)\big).$$

This ODE can be solved for the density $n\big(t(\theta), x(\theta)\big)$ when $\theta = \theta_1$:

$$n\big(t(\theta_1), x(\theta_1)\big) = n(t_{\theta=0}, x_{\theta=0})$$

$$\times \exp\left(-\int_0^{\theta_1} \left[d\big(E_c, x(\theta)\big) + \frac{\partial g\big(E_c, x(\theta)\big)}{\partial x}\right] d\theta\right). \quad (19)$$

Equation (19) shows that the density, $n$, at a specific point $\big(x(\theta_1), t(\theta_1)\big)$ is a function of its value at the origin $(x_{\theta=0}, t_{\theta=0})$ and the intervening values of $x(\theta)$ along the characteristic. The density changes through mortality and the change in developmental rate. The appearance of $d\big(E_c, x(\theta)\big)$ in (19) indicates that the density decreases along the characteristic because of individual deaths. In addition, the density can increase or decrease because characteristics converge or diverge, respectively, in the $(x, t)$-plane. This explains the occurrence of $\partial g\big(E_c, x(\theta)\big)/\partial x$ in (19). Where characteristics converge, individuals with smaller $x$ values have higher developmental rates, and hence, $\partial g\big(E_c, x(\theta)\big)/\partial x$ is negative. This results in an increase in the density along the characteristic (see eq. 19). Intuitively, the increase can be understood if we consider all individuals with $i$-state between $x_1$ and $x_2$ at time $t$. If no mortality takes place, the number of these individuals remains constant, which fixes the integral over the density function at a constant value. However, if the two characteristics passing through $x_1$ and $x_2$ at time $t$ converge, the range of $i$-state values, over which these individuals are distributed, gets smaller with time. Together with the constant integral, this decrease in range implies that the value of the density function has to increase, on the average (see also Fig. 5). For diverging characteristics, an analogous reasoning holds.

The use of $t(\theta_1)$ and $x(\theta_1)$ in equation (19) implies that the coordinates of the characteristics in the $(x, t)$-plane can be given explicitly as functions of the parameter $\theta$. This depends on the existence of a closed-form solution for the ODE's (17). Since $\theta$ is proportional to time, ODE (17a) implies that

$$T(\theta_1, t_{\theta=0}) = \theta_1 + t_{\theta=0},$$

where $T(\theta_1, t_{\theta=0})$ is the time coordinate of the point parameterized by $\theta = \theta_1$ along the characteristic with its origin at coordinates $(x_{\theta=0}, t_{\theta=0})$. For what follows, assume that an explicit solution for the ODE (17b) is known. The function $X(E_c, \theta_1, x_{\theta=0})$ indicates the $i$-state coordinate of the point along the characteristic with its origin at $(x_{\theta=0}, t_{\theta=0})$. The function $X(E_c, \theta_1, x_{\theta=0})$ depends on the form of the developmental rate $g(E_c, x)$ and cannot be given in general. Moreover, since development usually depends on the (constant) state of the environment, $E_c$ also appears as an argument in the function $X(E_c, \theta_1, x_{\theta=0})$.

In addition to the functions $T(\theta_1, t_{\theta=0})$ and $X(E_c, \theta_1, x_{\theta=0})$ for the coordinates along a specific characteristic, the derivation that follows also requires an explicit expression for the inverse of $X(E_c, \theta_1, x_{\theta=0})$ with respect to its second argument. This inverse is indicated by $\tau(E_c, x, x_{\theta=0})$ and can be interpreted as the time it takes to develop from an initial $i$-state $x_{\theta=0}$ to a final $i$-state $x$. Therefore,

$$\tau\big(E_c, X(E_c, \theta_1, x_{\theta=0}), x_{\theta=0}\big) \;=\; \theta_1$$

and

$$X\big(E_c, \tau(E_c, x_1, x_{\theta=0}), x_{\theta=0}\big) \;=\; x_1 \,.$$

To illustrate the rather complicated notation, the expressions for these functions will be derived for the *Daphnia* example model. The food density is assumed equal to the constant $F_c$. As discussed in Section 1, the growth in length $\ell$ under constant food conditions is

$$\frac{d\ell}{dt} \;=\; g(F_c, \ell) \;=\; \gamma\big(\ell_m\, h(F_c) \;-\; \ell\big),$$

in which the functional response is defined as

$$h(F) \;=\; \frac{F}{F_h + F}\,.$$

The ODE for $\ell$ can be solved explicitly. If a characteristic is considered with its origin located at the length and time coordinates $(\ell_{\theta=0}, t_{\theta=0})$, the analogue of the function $X(E_c, \theta, x_{\theta=0})$ is a function $L(F_c, \theta, \ell_{\theta=0})$, given by

$$L(F_c, \theta, \ell_{\theta=0}) \;=\; \ell_m\, h(F_c) \;-\; \big(\ell_m\, h(F_c) - \ell_{\theta=0}\big)\, e^{-\gamma\theta}$$

(cf. also the length-age relation in eq. 5). Figure 5 illustrates that all individuals born after $t = 0$ follow characteristics to the left and above the characteristic originating at $(\ell_b, 0)$. Hence, $\ell_{\theta=0}$ is taken to be equal to $\ell_b$ for these individuals and $t_{\theta=0}$ is identified with their time of birth. The members of the initial population all follow characteristics to the right and below the characteristic that originates at $(\ell_b, 0)$. As natural choices for the origins of these characteristics, let $t_{\theta=0}$ equal zero and $\ell_{\theta=0}$ be the length of these individuals at $t = 0$. The function $L(F_c, \theta, \ell_{\theta=0})$ specifies the length of individuals after growing for a period of $\theta$ time units at a constant food density $F_c$, given that they started out with initial length $\ell_{\theta=0}$. The analogue of the function $\tau(E_c, x, x_{\theta=0})$ is a function $A(F_c, \ell, \ell_{\theta=0})$, defined as

$$A(F_c, \ell, \ell_{\theta=0}) = \frac{1}{\gamma} \ln \left( \frac{\ell_m h(F_c) - \ell_{\theta=0}}{\ell_m h(F_c) - \ell} \right).$$

This equality is obtained by solving the relation

$$L(F_c, \theta, \ell_{\theta=0}) = \ell$$

for $\theta$. $A(F_c, \ell, \ell_{\theta=0})$ can be interpreted as the time it takes an individual at a constant food density, $F_c$, to grow from length $\ell_{\theta=0}$ to length $\ell$.

The functions $X(E_c, \theta, x_{\theta=0})$ and $\tau(E_c, x, x_{\theta=0})$ can be used to obtain a formal solution for the linear model (16). Equation (19) can be rewritten as

$$n\big(t(\theta_1), x(\theta_1)\big)$$

$$= n(t_{\theta=0}, x_{\theta=0}) \exp \left( -\int_0^{\theta_1} \left[ d(E_c, x(\theta)) + \frac{\partial g(E_c, x(\theta))}{\partial x} \right] d\theta \right)$$

$$= n(t_{\theta=0}, x_{\theta=0}) \exp \left( -\int_{x_{\theta=0}}^{x(\theta_1)} \left[ d(E_c, x) + \frac{\partial g(E_c, x)}{\partial x} \right] \frac{d\theta}{dx} dx \right)$$

$$= n(t_{\theta=0}, x_{\theta=0}) \exp \left( -\int_{x_{\theta=0}}^{x(\theta_1)} \frac{d(E_c, x) + \partial g(E_c, x)/\partial x}{g(E_c, x)} dx \right)$$

$$= n(t_{\theta=0}, x_{\theta=0}) \frac{g(E_c, x_{\theta=0})}{g\big(E_c, x(\theta_1)\big)} \exp \left( -\int_{x_{\theta=0}}^{x(\theta_1)} \frac{d(E_c, x)}{g(E_c, x)} dx \right).$$

$$(20)$$

For the members of the initial population, the density $n(t_{\theta=0}, x_{\theta=0})$ can be replaced by the initial density function $\Psi(x)$ from (16c) because $(x_{\theta=0}, t_{\theta=0})$ is $(x, 0)$. Substitution into equation (20) yields the following equation for the dynamics of this part of the population:

$$n\big(t, X(E_{\mathrm{c}}, t, x)\big) = \frac{g(E_{\mathrm{c}}, x)\Psi(x)}{g\big(E_{\mathrm{c}}, X(E_{\mathrm{c}}, t, x)\big)} \exp\left(-\int_{x}^{X(E_{\mathrm{c}}, t, x)} \frac{d(E_{\mathrm{c}}, \xi)}{g(E_{\mathrm{c}}, \xi)}\, d\xi\right).$$

$$(21)$$

By definition, members of the initial population with $i$-state $x$ at $t = 0$ have an $i$-state equal to $X(E_{\mathrm{c}}, t, x)$ at time $t$.

For individuals born after $t = 0$, the origin of the characteristic $(x_{\theta=0}, t_{\theta=0})$ is $(x_{\mathrm{b}}, t_{\mathrm{b}})$, where $t_{\mathrm{b}}$ refers to the time of the birth of these individuals. The dynamics of the part of the population born after $t = 0$ can therefore be described by

$$n(t, x) =$$

$$\frac{g(E_{\mathrm{c}}, x_{\mathrm{b}})n\big(t - \tau(E_{\mathrm{c}}, x, x_{\mathrm{b}}), x_{\mathrm{b}}\big)}{g(E_{\mathrm{c}}, x)} \exp\left(-\int_{x_{\mathrm{b}}}^{x} \frac{d(E_{\mathrm{c}}, \xi)}{g(E_{\mathrm{c}}, \xi)}\, d\xi\right).$$

$$(22)$$

Here, the time of birth $t_{\mathrm{b}}$ of individuals that have reached $i$-state $x$ at time $t$ has been replaced by

$$t_{\mathrm{b}} = t - \tau(E_{\mathrm{c}}, x, x_{\mathrm{b}}),$$

following the definition of the function $\tau(E_{\mathrm{c}}, x, x_{\mathrm{b}})$. Since $t_{\mathrm{b}} > 0$, it is clear that equation (22) holds only as long as $t > \tau(E_{\mathrm{c}}, x, x_{\mathrm{b}})$.

Together, equations (21) and (22) specify the entire solution of the linear PDE (16a) along its characteristics: equation (21) relates to the initial population, and equation (22) relates to individuals born after $t = 0$. In any realistic situation with a nonzero death rate, the initial population dies out, in the meantime making a diminishing contribution to the long-term dynamics. The initial population does play a role in short-term dynamics, and models exist in which its influence does not become negligible (see, e.g., Metz & Diekmann 1986, Chapter 2). In general, however, for the long-term behavior of the linear model (16), the dynamics of interest are described by equation (22).

If $B(E_{\mathrm{c}}, t)$ is defined as the total population birthrate for a population living in the constant environment $E_{\mathrm{c}}$, equation (22) can also be written as

$$n(t,x) = \frac{B\big(E_c, t - \tau(E_c, x, x_b)\big)}{g(E_c, x)} \exp\left(-\int\limits_{x_b}^{x} \frac{d(E_c, \xi)}{g(E_c, \xi)}\, d\xi\right).$$

Substitution into the boundary condition (16b) leads to the following integral equation:

$$B(E_c, t) =$$
$$\left[\int\limits_{x_b}^{x_m} B\big(E_c, t - \tau(E_c, x, x_b)\big) \frac{b(E_c, x)}{g(E_c, x)} \right. \tag{23}$$
$$\left. \times \exp\left(-\int\limits_{x_b}^{x} \frac{d(E_c, \xi)}{g(E_c, \xi)}\, d\xi\right) dx\right].$$

Provided that $\tau(E_c, x, x_b)$ (the time it takes an individual to develop from $x_b$ to $x$ when living in a constant environment, $E_c$) is a known function of its three arguments, this renewal equation relates the total population birthrate at time $t$ to its history.

Equation (23) shows that the linear model does not allow a stable-equilibrium solution. An equilibrium would imply that the birthrate $B$ is constant over time. Substitution of a constant birthrate $\tilde{B}(E_c)$ into (23) leads to an expression involving only parameters; no dynamic variables appear. Hence, the constant birthrate occurs for only very specific parameter combinations, such that every individual, on the average, produces exactly a single offspring during its lifetime. For almost all parameter combinations, individuals produce more or fewer offspring. Because an equilibrium is not feasible and the model does not incorporate any density dependence, the population ultimately grows or declines exponentially. Substituting the exponential trial solution,

$$B(E_c, t) = B_0\, e^{\lambda t},$$

into the renewal equation (23) leads to the following characteristic equation:

$$\Pi(E_c, \lambda) = 1, \tag{24}$$

in which

$$\Pi(E_c, \lambda) = \int\limits_{x_b}^{x_m} e^{-\lambda \tau(E_c, x, x_b)} \frac{b(E_c, x)}{g(E_c, x)} \exp\left(-\int\limits_{x_b}^{x} \frac{d(E_c, \xi)}{g(E_c, \xi)}\, d\xi\right) dx.$$

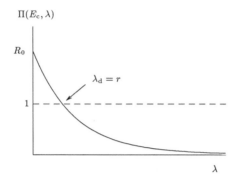

FIGURE 6. *The general form of the characteristic equation (25) for a structured-population model.*

The last exponential function appearing in the integrand is the survival function $S(E_c, x)$, given in equation (9). The characteristic equation can therefore also be written as

$$\Pi(E_c, \lambda) = \int_{x_b}^{x_m} e^{-\lambda \tau(E_c, x, x_b)} \frac{b(E_c, x)}{g(E_c, x)} S(E_c, x) \, dx = 1. \quad (25)$$

The integral expression for $\Pi(E_c, \lambda)$ when $\lambda = 0$ resembles equation (11), which gives the number of offspring produced by a surviving individual in a constant environment, $E_c$, between its birth with $i$-state $x_b$ and the time it reaches $i$-state $x$. In $\Pi(E_c, 0)$, however, the integration is carried out over $[x_b, x_m)$, that is, over the entire life history of the individual, and the reproductive contribution at every $i$-state $x$ is weighted with the probability $S(E_c, x)$ of surviving to that $i$-state. This quantity, $\Pi(E_c, 0)$, is often referred to as the net reproduction $R_0$:

$$R_0 = \Pi(E_c, 0) = \int_{x_b}^{x_m} \frac{b(E_c, x)}{g(E_c, x)} S(E_c, x) \, dx. \quad (26)$$

By definition, $R_0$ is the expected number of (female) offspring produced by one female individual throughout her life when density-dependent factors can be neglected, for example, when population densities are infinitesimally low.

Net reproduction, $R_0$, is sometimes also referred to as the expected lifetime reproduction of an individual. The integral in (24) can also be evaluated when $\lambda \neq 0$. The resulting quantity, $\Pi(E_c, \lambda)$, is a strictly decreasing function of $\lambda$ with a positive second derivative. Therefore, there exists exactly one simple real root to the characteristic equation (24) (see Fig. 6). This single real root is referred to as the dominant eigenvalue $\lambda_d$ of the linear model (16). It is the ultimate, exponential growth (or decline) rate of the population and is often called the intrinsic population growth rate. In the biological literature, this quantity is often indicated by $r$. The characteristic equation (24) is important because it relates the intrinsic population growth rate to the entire life history of the individuals and to the constant environmental conditions $E_c$. Figure 6 also makes clear that a positive (or negative) population growth rate corresponds to $R_0$'s being larger (or smaller) than unity:

$$R_0 > 1 \quad \Longleftrightarrow \quad r > 0, \qquad R_0 < 1 \quad \Longleftrightarrow \quad r < 0,$$

which is intuitively clear given the interpretation of $R_0$.

Although the population eventually grows or declines exponentially, and the density function $n(t, x)$ for every value $x$ does so as well, the form of the density function ultimately becomes constant over time:

$$n(t, x) \sim e^{-\lambda \tau (E_c, x, x_b)} \, \frac{b(E_c, x)}{g(E_c, x)} \, S(E_c, x)$$

$$= e^{-\lambda \tau (E_c, x, x_b)} \, \frac{b(E_c, x)}{g(E_c, x)} \, \exp \left( - \int_{x_b}^{x} \frac{d(E_c, \xi)}{g(E_c, \xi)} \, d\xi \right) \, dx. \tag{27}$$

This means that the composition of the population, in terms of the proportions of individuals with an $i$-state in various classes, becomes constant. The constant form of the density function is referred to as the *stable state distribution*.

For realistic choices of the functions $g$, $d$, and $b$, the characteristic equation (24) cannot be solved explicitly. For simple choices of the functions, the integral might have a closed-form solution, but even then the resulting equation is transcendental. Hence, the value of $\lambda_d$ has to be obtained numerically. In general, there are infinitely many roots (eigenvalues) of the characteristic equation (24), although only the dominant one is real. The others are complex and form a collection of complex conjugate pairs. The real parts of these complex pairs are, in all practical situations, smaller than the value of the single real eigenvalue, accounting for the adjective dominant. The remaining eigenvalues determine the rate at which the population attains the stable state distribution (27).

These equations can be simplified for age-structured populations, for which $x_b = 0$, $x_m = \infty$, $g(E_c, x) \equiv 1$, and most important, $\tau(E_c, x, x_b) = a$. With these substitutions and the replacement of the variable $x$ by $a$ to indicate age, the renewal equation (23) becomes

$$B(E_c, t) = \int\limits_0^\infty B(E_c, t - a) \, b(E_c, a) \, \exp\left(-\int\limits_0^a d(E_c, \zeta) \, d\zeta\right) \, da$$

$$= \int\limits_0^\infty B(E_c, t - a) \, b(E_c, a) \, S(E_c, a) \, da \, .$$

This equation is known as Lotka's integral equation for the dynamics of an age-structured population (Sharpe & Lotka 1911; Roughgarden 1979).

The characteristic equation (24) for the age-structured model becomes

$$\int\limits_0^\infty e^{-\lambda a} \, b(E_c, a) \, \exp\left(-\int\limits_0^a d(E_c, \zeta) \, d\zeta\right) \, da = 1$$

$$\Rightarrow \quad \int\limits_0^\infty e^{-\lambda a} \, b(E_c, a) \, S(E_c, a) \, da = 1 \, .$$

This is Euler's equation (Roughgarden 1979) for the computation of the intrinsic population growth rate of an age-structured population, in which the product of the natality function, $b(E_c, a)$, the survival function, $S(E_c, a)$, and an exponential weighting function $\exp(-\lambda a)$ is integrated over the entire individual life history. Therefore, the renewal equation (23) and the characteristic equation (24) can be interpreted as the analogues of Lotka's integral equation and Euler's equation, respectively, for an arbitrary $i$-state variable $x$.

## 6 The Equilibrium of the *Daphnia* Model

The equilibria of the *Daphnia* model can in principle be studied using the model formulation in Table 2. It turns out to be more conve-

nient, however, to reformulate the model as an age-dependent problem. Because of the positive mortality rate, the initial population
plays an exponentially diminishing role. The long-term dynamics,
including the approach to equilibrium, are determined by the individuals born after time zero. The deterministic growth in length,
described by the ODE (4), ensures that individuals follow a unique
growth trajectory, depending only on their initial length and the
food density. Since all individuals start at the same initial length $\ell_b$
and all experience the same environmental food density, individuals born at the same time remain identical, while they always differ
from individuals born at different times. In other words, after the
initial population becomes negligible, the length of an individual
is uniquely related to its chronological age. Of course, this length-
age relation can change over time owing to changing food density;
but at any moment in time, individual length and age form a one-
to-one relation. Thus, the *Daphnia* population can be completely
characterized in terms of a density function $m(t, a)$, representing
the age distribution of the population, together with a length-age
relation $L(t, a)$, which specifies at every time $t$ the unique relation between individual length and age. The possible age range is
$[0, \infty)$.

Like the integral of the population size distribution $n(t, \ell)$, the
integral

$$\int_{a_1}^{a_2} m(t, a)\, da$$

represents a total number of *Daphnia*, but now of individuals at
time $t$ with an age between $a_1$ and $a_2$. Since the function $L(t, a)$
relates age to length at time $t$, the following equality holds:

$$\int_{a_1}^{a_2} m(t, a)\, da = \int_{L(t,a_1)}^{L(t,a_2)} n(t, \ell)\, d\ell.$$

(The total number of *Daphnia* with an age between $a_1$ and $a_2$
must equal the total number of *Daphnia* with a length between
the corresponding bounds $L(t, a_1)$ and $L(t, a_2)$.) The right-hand

side of this equation can be rewritten by a change in variables as

$$\int_{L(t,a_1)}^{L(t,a_2)} n(t,\ell)\,d\ell = \int_{a_1}^{a_2} n\big(t, L(t,a)\big) \frac{d\ell}{dt}\frac{dt}{da}\,da$$

$$= \int_{a_1}^{a_2} n\big(t, L(t,a)\big)\, g\big(F, L(t,a)\big)\,da\,,$$

which shows that $m(t,a)$ and $n(t,\ell)$ are formally related to each other by

$$m(t,a) \equiv n\big(t, L(t,a)\big)\, g\big(F, L(t,a)\big)\,.$$

This equivalence can be used to express any (weighted) integral of $n(t,\ell)$ as an integral of $m(t,a)$. For example, the population birthrate can be rewritten as

$$\int_{\ell_b}^{\ell_m} b(F,\ell)\, n(t,\ell)\,d\ell = \int_0^\infty b\big(F, L(t,a)\big)\, n\big(t, L(t,a)\big) \frac{d\ell}{dt}\frac{dt}{da}\,da$$

$$= \int_0^\infty b\big(F, L(t,a)\big)\, n\big(t, L(t,a)\big)\, g\big(F, L(t,a)\big)\,da$$

$$= \int_0^\infty b\big(F, L(t,a)\big)\, m(t,a)\,da\,.$$

The PDE describing the dynamics of $m(t,a)$ can be derived using the same bookkeeping arguments as were used to derive the PDE for $n(t,\ell)$ (see Fig. 3). This leads to

$$\frac{\partial m(t,a)}{\partial t} = -\frac{\partial m(t,a)}{\partial a} - d\big(F, L(t,a)\big)\, m(t,a)\,. \qquad (28)$$

In this equation the death rate, $d\big(F, L(t,a)\big)$, indirectly depends on the individual age via its dependence on the individual length. In the *Daphnia* model this does not make a difference, because $d\big(F, L(t,a)\big) \equiv \mu$, which is independent of length and hence age. In the example model, the relevant PDE therefore equals

$$\frac{\partial m(t,a)}{\partial t} + \frac{\partial m(t,a)}{\partial a} = -\mu\, m(t,a)\,. \qquad (29)$$

The boundary condition for the PDE (28) can be derived by integrating both sides of the equation over the reachable age range

$[0, \infty)$ (cf. the derivation of eq. 14):

$$\int_0^\infty \frac{\partial m(t,a)}{\partial t}\, da \;=\; -\int_0^\infty \frac{\partial m(t,a)}{\partial a}\, da \;-\; \int_0^\infty d\big(F, L(t,a)\big)\, m(t,a)\, da$$

$$\Rightarrow \qquad \frac{d}{dt}\int_0^\infty m(t,a)\, da$$

$$= \; m(t,0) \;-\; m(t,\infty) \;-\; \int_0^\infty d\big(F, L(t,a)\big)\, m(t,a)\, da\,.$$

As in the derivation of the boundary condition (14), this leads to the conclusion that $m(t,0)$ should equal the population birthrate:

$$m(t,0) \;=\; \int_0^\infty b\big(F, L(t,a)\big)\, m(t,a)\, da\,. \qquad (30)$$

Note that the developmental rate $g$ does not disappear from the left-hand side but is simply equal to one.

In the *Daphnia* model, the birthrate, $b(t,\ell)$, equals zero when $\ell < \ell_j$. The age at which individuals reach this juvenile-to-adult threshold is indicated by the function $A_j(t)$, which equals the current duration of the juvenile stage. Because of the time dependence of the length-age relation, the value of $A_j(t)$ varies in time. The definition of $A_j(t)$ implies that it fulfills the condition

$$L\big(t, A_j(t)\big) \;=\; \ell_j\,. \qquad (31)$$

The juvenile period, $A_j(t)$, is only implicitly defined by (31), but an explicit expression is possible if the entire length-age relation is known. By substitution of the functional form for $b(F, \ell)$, given in Table 1, the population birthrate can be rewritten using the function $A_j(t)$, such that the boundary condition (30) becomes

$$m(t,0) \;=\; \int_{A_j(t)}^\infty r_m\, h(F)\,\big(L(t,a)\big)^2 m(t,a)\, da\,; \qquad (32)$$

$h(F)$ is used as before to denote the *Daphnia* functional response $F/(F_h + F)$.

The population feeding rate can also be rewritten using the density function $m(t,a)$ and the length-age relation, $L(t,a)$, so that the

ODE (12) describing the dynamics of the food density becomes

$$\frac{dF}{dt} = \alpha F \left(1 - \frac{F}{K}\right) - \int_0^\infty I\big(F, L(t,a)\big)\, m(t,a)\, da. \qquad (33)$$

After substituting the expression for $I\big(F, L(t,a)\big)$ (see Table 1), this ODE simplifies to

$$\frac{dF}{dt} = \alpha F \left(1 - \frac{F}{K}\right) - \int_0^\infty \nu\, h(F)\, \big(L(t,a)\big)^2 m(t,a)\, da. \qquad (34)$$

The equation for the dynamics of the length-age relation, $L(t,a)$, has to be derived separately. Clearly,

$$L(t,0) = \ell_{\mathrm{b}}, \qquad (35)$$

since at birth all individuals have length $\ell_{\mathrm{b}}$, whatever their time of birth. Consider the change in length in a small time interval $\Delta t$. An individual with age $a$ has length $L(t,a)$ at time $t$, and length $L(t+\Delta t, a+\Delta t)$ at the end of the interval $\Delta t$. The difference between these lengths is the growth increment during the interval, which equals $g\big(F, L(t,a)\big)\, \Delta t$. Equating the two yields an expression for the growth in length during the time interval $\Delta t$:

$$\frac{L(t+\Delta t, a+\Delta t) - L(t,a)}{\Delta t} = g\big(F, L(t,a)\big).$$

Expanding the term $L(t+\Delta t, a+\Delta t)$ to the first order and taking the limit as $\Delta t \to 0$ leads to the following PDE for the dynamics of the length-age relation, $L(t,a)$:

$$\frac{\partial L(t,a)}{\partial t} + \frac{\partial L(t,a)}{\partial a} = g\big(F, L(t,a)\big). \qquad (36)$$

Substituting the expression for $g(F, \ell)$ from Table 1 leads to the following result:

$$\frac{\partial L(t,a)}{\partial t} + \frac{\partial L(t,a)}{\partial a} = \gamma \big(\ell_{\mathrm{m}}\, h(F) - L(t,a)\big). \qquad (37)$$

The initial states of the *Daphnia* population and the algal food density now have to be specified by means of an initial condition for $m(0,a)$, an initial condition for $L(0,a)$, and the initial value for the food density $F(0)$:

$$m(0,a) = \Psi(a), \quad L(0,a) = \Lambda(a), \quad F(0) = F_0.$$

TABLE 3. *Equations of the* Daphnia *Population Model after Reformulation as an Age-Structured Problem*

| | |
|---|---|
| *Daphnia* dynamics | $\dfrac{\partial m(t,a)}{\partial t} + \dfrac{\partial m(t,a)}{\partial a} = -\mu\, m(t,a)$ |
| | $m(t,0) = \displaystyle\int_{A_j(t)}^{\infty} r_m\, h(F)\,\big(L(t,a)\big)^2 m(t,a)\, da$ |
| | $\dfrac{\partial L(t,a)}{\partial t} + \dfrac{\partial L(t,a)}{\partial a} = \gamma\big(\ell_m\, h(F) - L(t,a)\big)$ |
| | $L(t,0) = \ell_b$ |
| | $L\big(t, A_j(t)\big) = \ell_j$ |
| Algal dynamics | $\dfrac{dF}{dt} = \alpha F\left(1 - \dfrac{F}{K}\right) - \displaystyle\int_0^{\infty} \nu\, h(F)\,\big(L(t,a)\big)^2 m(t,a)\, da$ |
| Initial conditions | $m(0,a) = \Psi(a)$ |
| | $L(0,a) = \Lambda(a)$ |
| | $F(0) = F_0$ |

NOTE.—The function $h(F) = F/(F_h + F)$ refers to the *Daphnia* functional response. The expressions for the individual growth, death, reproduction, and feeding rates are already incorporated.

Here, $\Psi(a)$ and $\Lambda(a)$ are age-dependent distributions and $F_0$ is a scalar quantity. Since the reformulated, age-dependent model is used only for the computation and analysis of equilibria, these initial conditions are of lesser importance but are given for completeness. Table 3 lists the relevant equations.

This reformulation of a size-structured model into an age-structured model, involving a density function for the number of individuals and a function describing the unique relation between individual age and size, was first introduced by Murphy (1983). In her honor it has also been called the "Murphy trick" (Metz & Diekmann 1986); it makes the model more amenable to further analysis, such as the computation of equilibria and analysis of their stability. The entire set of equations derived in this section by applying the Murphy trick is summarized in Table 3. Once again, the long-term dynamics of the models summarized in Tables 2 and 3 are identical after the influence of the initial population has become negligibly small.

If an equilibrium of the *Daphnia* model exists, all dynamic variables should approach a constant value over time, including the environmental food density. The equilibrium food density is indicated by $\tilde{F}$. A constant food density leads to an equilibrium length-age relation, denoted by $\tilde{L}(\tilde{F}, a)$, which satisfies the PDE (37) and the accompanying boundary condition (35):

$$\frac{\partial \tilde{L}}{\partial a} = \gamma \left( \ell_{\mathrm{m}} h(\tilde{F}) - \tilde{L} \right), \qquad \tilde{L}(\tilde{F}, 0) = \ell_{\mathrm{b}}. \tag{38}$$

This equation can be solved explicitly for the equilibrium length-age relation:

$$\tilde{L}(\tilde{F}, a) = \ell_{\mathrm{m}} h(\tilde{F}) - \left( \ell_{\mathrm{m}} h(\tilde{F}) - \ell_{\mathrm{b}} \right) e^{-\gamma a}. \tag{39}$$

The equilibrium age at which individuals reach the juvenile length $\ell_{\mathrm{j}}$, and hence mature, is indicated by the function $\tilde{A}_{\mathrm{j}}(\tilde{F})$, which must fulfill its definition (31):

$$\tilde{L}\left(\tilde{F}, \tilde{A}_{\mathrm{j}}(\tilde{F})\right) = \ell_{\mathrm{m}} h(\tilde{F}) - \left( \ell_{\mathrm{m}} h(\tilde{F}) - \ell_{\mathrm{b}} \right) e^{-\gamma \tilde{A}_{\mathrm{j}}(\tilde{F})} = \ell_{\mathrm{j}},$$

which allows an explicit solution:

$$\tilde{A}_{\mathrm{j}}(\tilde{F}) = \frac{1}{\gamma} \ln \left( \frac{\ell_{\mathrm{m}} h(\tilde{F}) - \ell_{\mathrm{b}}}{\ell_{\mathrm{m}} h(\tilde{F}) - \ell_{\mathrm{j}}} \right). \tag{40}$$

Because of the occurrence of $h(\tilde{F})$ in this relation, $\tilde{F}$ is used as an explicit-function argument of $\tilde{A}_{\mathrm{j}}$.

The density function $m(t, a)$, representing the age distribution of the *Daphnia* population, is the only dynamic quantity remaining. When the system approaches an equilibrium, this function also becomes time- (but not age-) independent. The constant age distribution of the *Daphnia* population in equilibrium is denoted by $\tilde{m}(a)$. Because $\tilde{m}(a)$ has to fulfill the PDE (29), it can be obtained from

$$\frac{\partial \tilde{m}}{\partial a} = -\mu \tilde{m}, \tag{41}$$

which implies that

$$\tilde{m}(a) = \tilde{m}(0) e^{-\mu a}. \tag{42}$$

The equilibrium age distribution of the *Daphnia* is an exponentially decreasing function of age with a decay parameter equal to $\mu$. (The assumption that the death rate, $\mu$, is independent of length and food is drastic, but it makes the equilibrium calculations and the analysis of its stability much simpler; see Section 8.)

The equilibrium length-age relation (39), age distribution (42), and duration of juvenile period in equilibrium (40) can be substituted for $L(t, a)$, $m(t, a)$, and $A_j(t)$, respectively, in the boundary condition (29):

$$\tilde{m}(0) \;=\; \int_{\tilde{A}_j(\tilde{F})}^{\infty} r_m \, h(\tilde{F}) \left(\tilde{L}(\tilde{F}, a)\right)^2 \tilde{m}(a) \, da$$

$$=\; \int_{\tilde{A}_j(\tilde{F})}^{\infty} r_m \, h(\tilde{F}) \left(\tilde{L}(\tilde{F}, a)\right)^2 \tilde{m}(0) \, e^{-\mu a} \, da \, .$$

Both sides of this equation can be divided by $\tilde{m}(0)$ to yield the first condition for the equilibrium of the *Daphnia* model:

$$\int_{\tilde{A}_j(\tilde{F})}^{\infty} r_m \, h(\tilde{F}) \left(\tilde{L}(\tilde{F}, a)\right)^2 e^{-\mu a} \, da \;=\; 1 \, . \tag{43}$$

This equilibrium condition is derived using only the equations for the dynamics of the *Daphnia* population; the ODE (34) describing the food dynamics has not been used at all.

Apart from the growth, mortality, and reproduction parameters, the food density $\tilde{F}$ is the only unknown quantity in equation (43), which thus determines its equilibrium value. On the left-hand side of (43), the term $\exp(-\mu a)$ can be recognized as the survival function of an individual *Daphnia* (the probability that it survives to age $a$). The quantity $r_m \, h(\tilde{F}) \left(\tilde{L}(\tilde{F}, a)\right)^2$ is the reproduction rate at age $a$ (see also Table 1). The product of the survival function and the reproduction rate at age $a$ is integrated over the interval $[\tilde{A}_j(\tilde{F}), \infty)$, which represents the part of the life span during which an individual *Daphnia* is adult and reproduces. Thus, the left-hand side of the equilibrium condition (43) equals the expected number of offspring an individual *Daphnia* produces during its entire life when living at a constant food density $\tilde{F}$. At equilibrium, an individual should just replace itself; hence, the expected number of offspring equals one.

The interpretation of the equilibrium condition (43) indicates an important aspect of the *Daphnia*-algal system: in purely exploitative systems, in which individuals compete only for a common resource, the equilibrium resource density is fully set by the life-history characteristics of the consumer, in such a way that, on the average, every individual consumer just replaces itself. Population statistics of the consumer play no role in determining the equilibrium resource level. Only the equations for the *Daphnia* dynamics are necessary to obtain a condition for the equilibrium food density, $\tilde{F}$. This food density in turn completely determines the composition of the consumer population: the length-age relation (39) is fixed by the food density, and the age distribution (42) is fixed up to a multiplicative factor $\tilde{m}(0)$. The latter is also true when the death rate differs among individuals, since it can depend only on individual length and the food density, not on statistics of the *Daphnia* population. Equation (34) describing the food dynamics determines only the absolute size of the *Daphnia* population in equilibrium; the length structure is fixed by the food density imposed by the *Daphnia* themselves.

The expected number of offspring of a *Daphnia* individual, in an environment where the food density is equal to its carrying capacity $K$, can be calculated by substituting $\tilde{F} = K$ in the left-hand side of equation (43):

$$\int_{\tilde{A}_j(K)}^{\infty} r_m\, h(K) \left(\tilde{L}(K,a)\right)^2 e^{-\mu a}\, da\,. \tag{44}$$

If this expected number of offspring is larger than one, a *Daphnia* population can at least grow and establish itself in a "virgin" food environment. In addition, this guarantees the existence of at least one solution for $\tilde{F} \in [0, K]$ in equation (43), since its left-hand side equals zero when $\tilde{F} = 0$ ($h(0) = 0$). Because the functional response $h(F)$ is a monotonically increasing function, the equilibrium length-age relation (39) is an increasing function of $\tilde{F}$ as well. From equation (40) one can deduce that the age at maturation decreases monotonically with the equilibrium food density, $\tilde{F}$. These monotonicity arguments imply that the left-hand side of the equilibrium condition (43) increases with $\tilde{F}$ and that the equilibrium food density is unique.

The ODE (34), describing the food dynamics, governs the absolute size of the equilibrium *Daphnia* population. Both sides of this

ODE equal zero at equilibrium. Substitution of the equilibrium length-age relation (39) and equilibrium age distribution (42) for $L(t, a)$ and $m(t, a)$, respectively, in the right-hand side of the equation yields

$$\alpha \tilde{F} \left( 1 - \frac{\tilde{F}}{K} \right) - \int_0^\infty \nu \, h(\tilde{F}) \left( \tilde{L}(\tilde{F}, a) \right)^2 \tilde{m}(0) \, e^{-\mu a} \, da \; = \; 0 \,. \quad (45)$$

Because $\tilde{F}$ is determined by equation (43), the only unknown remaining is $\tilde{m}(0)$, the value of the equilibrium age distribution of the *Daphnia* population at age zero. The boundary condition (32) equates this value to the population birthrate of *Daphnia*. Let $\tilde{B}$ refer to the population birthrate of *Daphnia* at equilibrium. Substituting $\tilde{m}(0) = \tilde{B}$ in the equation above results in the second equilibrium condition for the *Daphnia* model:

$$\tilde{B} = \frac{\alpha \tilde{F} \left( 1 - \dfrac{\tilde{F}}{K} \right)}{\displaystyle\int_0^\infty \nu \, h(\tilde{F}) \left( \tilde{L}(\tilde{F}, a) \right)^2 e^{-\mu a} \, da} \,. \quad (46)$$

The equilibrium population birthrate is an explicit function of the equilibrium food density, which is determined by equation (43). Given $\tilde{B}$, the rest of the age distribution is calculated from equation (42).

The pair of scalar values $(\tilde{F}, \tilde{B})$ determined by the conditions (43) and (46) constitutes the equilibrium state of the *Daphnia* model. The entire composition of the population can be calculated from these two values using the relations derived above. The integrals in (43) and (46) can be solved explicitly, given the particular form of the equilibrium length-age relation (39) and the simple, exponential survival function $\exp(-\mu a)$ of the *Daphnia* model. However, the relation resulting from solving the integral in (43) is a nonlinear, implicit equality from which the value of $\tilde{F}$ can be determined only numerically. It is generally the case that the equilibrium equations can be solved only with numerical techniques, even for relatively simple models like the one considered here. Many numerical root-finding routines are available in popular computer software packages. Solving for the equilibrium also requires biologically realistic values for the parameters, which is sometimes the hardest problem of all.

When both the $i$-state and the $E$-state are one-dimensional, that is, consist of a single variable, the equations for the structured population completely determine the value of the $E$-state at equilibrium. This is true whether the $E$-state variable constitutes a feedback loop, as in the *Daphnia* model considered here, or a direct density-dependent feedback function. Because the individual behavior depends only on the $i$-state and $E$-state (see Section 2), this constant $E$-state fixes the relative composition of the structured population in equilibrium, that is, the age or size distribution of the population. If the $E$-state variable constitutes a feedback loop, the ODE describing its dynamics determines the absolute size of the structured population in equilibrium. In case of a feedback function, this absolute size is set by the equilibrium value of the $E$-state itself. For example, if the total population size directly influences individual behavior, its value at equilibrium is determined by the equations (PDE and boundary condition) describing the dynamics of the structured population. But since the total population size equals the integral of the density function, its equilibrium value immediately fixes the multiplication constant in the equilibrium density function of the population.

Even if the $E$-state is a vector of variables, the equilibrium of the structured-population model is fully specified in terms of a set of scalar values for all environmental state variables and a scalar value for the total population birthrate. Given these values, the definitions of the different state concepts (Section 2) ensure that the entire composition of the structured population can be calculated. That the $E$-state together with the total population birthrate completely determines the equilibrium of the structured population is intimately linked to the single $i$-state with which individuals can be born. Generalizations of this last point are difficult and are discussed briefly in Section 10.

## 7 Numerical Exploration of Dynamics

The dynamics of a structured-population model can be studied by integration of the governing set of equations (PDE, boundary condition, and accompanying ODE's) over time, starting from a specified initial state. Many methods have been developed for the numerical solution of such ordinary and partial differential equations. Ready-made recipes have been presented by, for example, Press et al. (1988). The combination of a PDE with a boundary condition that involves an integral of the solution $n(t, \ell)$ itself (eq. 14) is, however, rather unusual. A numerical method has therefore been developed specifically for physiologically structured population models, which is called the "Escalator Boxcar Train" (EBT; for its full mathematical description, see de Roos 1988). Its central idea is inspired by the biological origin of the equations, making it powerful and flexible enough to be applied to any kind of structured-population model (de Roos 1988).

Because its central idea is so close to the biological interpretation of the equations, the EBT method can also be derived on the basis of heuristic arguments. The method can be derived by generalizing an age-structured Leslie-matrix model to incorporate (1) size instead of age as the $i$-state variable, (2) competition for a dynamically varying food supply, and (3) a reproduction process that is continuous in time and hence leads to a continuous size distribution (de Roos et al. 1992). As a complementary approach, I derive the EBT method here from a mathematical perspective, that is, by considering the density function $n(t, \ell)$ as the object of study. Study of the more intuitive derivation (de Roos et al. 1992) is recommended for a more complete understanding, before applying the EBT method.

The central idea of the EBT method is to group individuals into cohorts, or classes. These cohorts are simply collections of individuals that are roughly similar, so that each cohort can be characterized realistically by the number of individuals it contains and their average length. By assumption, cohorts of individuals stay together as isolated groups, and individuals can leave a cohort only through death. The fate of each cohort is followed over time, and new cohorts arise only through reproduction. Biologically, the idea of cohorts is well known, except that in this case the members of a cohort are roughly, not fully, identical and are not all born at exactly the same time.

I begin by subdividing the range $[\ell_b, \ell_m)$ into small intervals, each spanning a certain range of individual lengths, such that individuals within a single interval can be classified as roughly similar. Assume, therefore, that a set of points $\ell_i$ has been chosen such that the entire range $[\ell_b, \ell_m)$ is covered by the intervals

$$\Omega_i = [\ell_i, \ell_{i+1}), \quad i = 1, \ldots, N.$$

The assumption implies that $\ell_1 = \ell_b$ and $\ell_{N+1} = \ell_m$. The selection of the intervals $\Omega_i$ is not described here; it will become clear that the method itself generates the subdivision of $[\ell_b, \ell_m)$ automatically. The total number of individuals with a length in the interval $\Omega_i$ is given by

$$\lambda_i(t) \;=\; \int\limits_{\ell_i}^{\ell_{i+1}} n(t, \ell)\, d\ell \tag{47}$$

(cf. Section 3); and the average length of these individuals is

$$\mu_i(t) \;=\; \frac{1}{\lambda_i(t)} \int\limits_{\ell_i}^{\ell_{i+1}} \ell\, n(t, \ell)\, d\ell \,. \tag{48}$$

The cohort with index $i$ is assumed to be fully characterized by its total number of individuals, $\lambda_i(t)$, and their average length, $\mu_i(t)$. Biologically, the population then is a collection of cohorts. Mathematically, the density function $n(t, \ell)$ is approximated by a set of delta functions (measures) of size $\lambda_i(t)$ at length $\ell = \mu_i(t)$.

Integrals of the type that occur in the boundary condition (14) and the ODE (12) can be written as a sum of integrals over the intervals $\Omega_i$:

$$\int\limits_{\ell_b}^{\ell_m} \phi(\ell)\, n(t, \ell)\, d\ell \;=\; \sum_{i=1}^{N} \int\limits_{\ell_i}^{\ell_{i+1}} \phi(\ell)\, n(t, \ell)\, d\ell \,.$$

In this expression the function $\phi(\ell)$ is used to denote a general weighting function for $n(t, \ell)$. Using a Taylor expansion of the function $\phi(\ell)$ around the value $\ell = \mu_i(t)$, the integral over the interval $\Omega_i$ can be approximated by

$$\int\limits_{\ell_i}^{\ell_{i+1}} \phi(\ell)\, n(t, \ell)\, d\ell = \int\limits_{\ell_i}^{\ell_{i+1}} \phi(\mu_i)\, n(t, \ell)\, d\ell + \int\limits_{\ell_i}^{\ell_{i+1}} \phi'(\mu_i)\, (\ell - \mu_i)\, n(t, \ell)\, d\ell$$

$$+ \frac{1}{2} \int\limits_{\ell_i}^{\ell_{i+1}} \phi''(\mu_i)\, (\ell - \mu_i)^2\, n(t, \ell)\, d\ell \;+\; \dots .$$

The definition (48) of $\mu_i(t)$ implies that the term involving $\phi'(\mu_i)$ vanishes in this expression. Therefore, the integral over the entire range $[\ell_b, \ell_m)$ can be approximated up to second-order precision by the following sum:

$$\int_{\ell_b}^{\ell_m} \phi(\ell)\, n(t,\ell)\, d\ell \;=\; \sum_{i=1}^{N} \phi\big(\mu_i(t)\big)\, \lambda_i(t)\,. \tag{49}$$

Hence, all population statistics of interest (those occurring in the boundary condition of eq. 14, the food-dynamics ODE of eq. 12, and the total population size or biomass) can be computed as a weighted sum over the cohort statistics.

For the dynamics of the quantities $\lambda_i(t)$ and $\mu_i(t)$, the EBT method assumes that, barring death, cohorts of individuals stay together indefinitely. Individuals never switch to other cohorts. This implies that the boundary values $\ell_i$ change over time following the ODE (4) describing the individual growth in length:

$$\frac{d\ell_i}{dt} \;=\; g(F,\ell_i)\,. \tag{50}$$

As long as all boundary values $\ell_i$ develop according to this ODE, individuals never cross the boundary and enter another cohort. The intervals $\Omega_i$ are moving through the length range $[\ell_b, \ell_m)$ and expanding (or shrinking), if the growth rate increases (or decreases) with $\ell$.

The dynamics of $\lambda_i(t)$ are described by the following ODE:

$$\frac{d\lambda_i}{dt} = \frac{d}{dt}\int_{\ell_i}^{\ell_{i+1}} n(t,\ell)\, d\ell$$

$$= \int_{\ell_i}^{\ell_{i+1}} \frac{\partial n(t,\ell)}{\partial t}\, d\ell \;+\; n(t,\ell_{i+1})\, \frac{d\ell_{i+1}}{dt} \;-\; n(t,\ell_i)\, \frac{d\ell_i}{dt}$$

$$= \int_{\ell_i}^{\ell_{i+1}} \frac{\partial n(t,\ell)}{\partial t}\, d\ell \;+\; n(t,\ell_{i+1})\, g(F,\ell_{i+1}) \;-\; n(t,\ell_i)\, g(F,\ell_i)$$

$$= \int_{\ell_i}^{\ell_{i+1}} \frac{\partial n(t,\ell)}{\partial t}\, d\ell \;+\; \int_{\ell_i}^{\ell_{i+1}} \frac{\partial g(F,\ell)\, n(t,\ell)}{\partial \ell}\, d\ell$$

$$= -\int_{\ell_i}^{\ell_{i+1}} d(F,\ell)\, n(t,\ell)\, d\ell\,.$$

The first step in this derivation uses Leibnitz' rule for the derivative of an integral when the integration bounds are not constant, and the last step exploits the PDE (13) to replace the time and length derivatives of $n(t,\ell)$. A more complicated derivation, proceeding along the same lines, yields the ODE for the dynamics of $\mu_i(t)$:

$$
\begin{aligned}
\frac{d\mu_i}{dt} &= \frac{d}{dt}\frac{1}{\lambda_i(t)}\int_{\ell_i}^{\ell_{i+1}} \ell\, n(t,\ell)\, d\ell \\
&= \frac{1}{\lambda_i(t)}\int_{\ell_i}^{\ell_{i+1}} \ell\frac{\partial n(t,\ell)}{\partial t} d\ell + \ell_{i+1}\, n(t,\ell_{i+1})\frac{d\ell_{i+1}}{dt} - \ell_i\, n(t,\ell_i)\frac{d\ell_i}{dt} \\
&\quad - \frac{1}{(\lambda_i(t))^2}\int_{\ell_i}^{\ell_{i+1}} \ell n(t,\ell)d\ell\frac{d\lambda_i}{dt} \\
&= \frac{1}{\lambda_i(t)}\int_{\ell_i}^{\ell_{i+1}} \ell\frac{\partial n(t,\ell)}{\partial t} d\ell + \ell_{i+1}n(t,\ell_{i+1})g(F,\ell_{i+1}) \\
&\quad - \ell_i n(t,\ell_i)g(F,\ell_i) - \frac{1}{\lambda_i(t)}\mu_i(t)\frac{d\lambda_i}{dt} \\
&= \frac{1}{\lambda_i(t)}\int_{\ell_i}^{\ell_{i+1}} \ell\frac{\partial n(t,\ell)}{\partial t} d\ell + \int_{\ell_i}^{\ell_{i+1}} \frac{\partial \ell n(t,\ell)g(F,\ell)}{\partial \ell} d\ell \\
&\quad - \frac{1}{\lambda_i(t)}\mu_i(t)\frac{d\lambda_i}{dt} \\
&= \frac{1}{\lambda_i(t)}\int_{\ell_i}^{\ell_{i+1}} g(F,\ell)n(t,\ell)\, d\ell - \frac{1}{\lambda_i(t)}\int_{\ell_i}^{\ell_{i+1}} (\ell-\mu_i(t))d(F,\ell)\, n(t,\ell)\, d\ell.
\end{aligned}
$$

The ODE's for $\lambda_i(t)$ and $\mu_i(t)$ do not form a solvable system, because they involve weighted integrals over the density function $n(t,\ell)$. To obtain a closed, solvable system, the functions $d(F,\ell)$ and $g(F,\ell)$ are approximated, as in the derivation of equation (49), by their Taylor expansion around $\ell=\mu_i(t)$. Higher-order terms involving squares and higher powers of the difference $\ell-\mu_i(t)$ are neglected. This leads to the following set of ODE's describing the approximate dynamics of $\lambda_i(t)$ and $\mu_i(t)$:

$$\begin{cases} \dfrac{d\lambda_i}{dt} & = & -d(F, \mu_i)\,\lambda_i \\[4mm] \dfrac{d\mu_i}{dt} & = & g(F, \mu_i) \end{cases} \qquad i = 1, \ldots, N. \qquad (51)$$

These equations show that the dynamics of the total number and average length of a cohort of individuals with lengths distributed over a small interval $\Omega_i$ can be approximated by the dynamics of a similar group of individuals, but now all having a length exactly equal to this average value. Of course, the approximation is better, the smaller the difference $\ell - \mu_i(t)$ and hence the interval $\Omega_i$.

The ODE's (51) describe the dynamics of cohorts that are already present in the population, but they do not account for reproduction. The cohorts that are already present are called internal cohorts. Since the value of the lowest interval bound, $\ell_1(t)$, changes over time following the ODE (50), all newborn individuals have a length within a widening interval $[\ell_b, \ell_1(t))$. This cohort is the boundary cohort. The boundary cohort is characterized by the total number of individuals it contains:

$$\lambda_0(t) = \int_{\ell_b}^{\ell_1} n(t, \ell)\, d\ell. \qquad (52)$$

Since the number of individuals within the interval is initially zero, their average length is initially not defined. In other words, equation (48) cannot be used to calculate the average length of the boundary cohort because the denominator might equal zero. A slightly different quantity is therefore employed to capture information on how the distribution of individuals within the interval $[\ell_b, \ell_1(t))$ changes over time:

$$\pi_0(t) = \int_{\ell_b}^{\ell_1} (\ell - \ell_b)\, n(t, \ell)\, d\ell. \qquad (53)$$

The choice of the weighting factor $(\ell - \ell_b)$ in this integral makes the resulting equations slightly simpler, since newborn individuals have $\ell = \ell_b$ and hence do not contribute to this statistic.

Formal differentiation of $\lambda_0(t)$ with respect to time yields (after some manipulation) the following ODE:

$$\frac{d\lambda_0}{dt} = -\int_{\ell_b}^{\ell_1} d(F,\ell)\,n(t,\ell)\,d\ell + g(F,\ell_b)\,n(t,\ell_b)\,.$$

The derivation of this ODE runs completely analogous to the derivation presented above for the ODE describing the dynamics of $\lambda_i(t)$. The special feature of the ODE for $\lambda_0(t)$ is the appearance of the term $g(F,\ell_b)\,n(t,\ell_b)$ in the right-hand side. This term occurs because the lower bound of the interval $[\ell_b, \ell_1(t))$ does not change with time as do all other bounds $\ell_i(t)$ (see eq. 50). This term is, however, the left-hand side of the boundary condition (14). It can therefore be replaced by the total population birthrate:

$$\frac{d\lambda_0}{dt} = -\int_{\ell_b}^{\ell_1} d(F,\ell)\,n(t,\ell)\,d\ell + \int_{\ell_b}^{\ell_m} b(F,\ell)\,n(t,\ell)\,d\ell\,.$$

Because the average length of newborn individuals in the boundary cohort is not defined as a characterizing statistic, the function $d(F,\ell)$ is instead expanded around the value $\ell = \ell_b$. Terms involving squared or higher powers of the difference $\ell - \ell_b$ are again neglected. Moreover, the integral representing the population reproduction rate is replaced by the appropriate summation (see eq. 49), involving the cohort measures $\lambda_i(t)$ and $\mu_i(t)$. These substitutions lead to the following ODE approximating the dynamics of $\lambda_0(t)$:

$$\frac{d\lambda_0}{dt} = -d(F,\ell_b)\,\lambda_0 - \frac{\partial}{\partial\ell}d(F,\ell_b)\,\pi_0 + \sum_{i=1}^{N} b(F,\mu_i)\,\lambda_i\,.$$

Although the summation in the last term is essentially over all cohorts, the contribution of those cohorts for which $\mu_i(t) \leq \ell_j$ is of course zero (see Table 1).

Formal differentiation of $\pi_0(t)$ with respect to time yields the ODE

$$\frac{d\pi_0}{dt} = \int_{\ell_b}^{\ell_1} g(F,\ell)\,n(t,\ell)\,d\ell - \int_{\ell_b}^{\ell_1} (\ell - \ell_b)\,d(F,\ell)\,n(t,\ell)\,d\ell\,.$$

Again, the functions $d(F,\ell)$ and $g(F,\ell)$ are replaced by their

Taylor expansion around the boundary value $\ell = \ell_b$, neglecting squared or higher powers of $\ell - \ell_b$. This yields an ODE for the approximate dynamics of $\pi_0(t)$:

$$\frac{d\pi_0}{dt} = g(F, \ell_b)\,\lambda_0 + \frac{\partial}{\partial\ell}g(F, \ell_b)\,\pi_0 - d(F, \ell_b)\,\pi_0\,.$$

The dynamics of the boundary cohort can therefore be approximated by the following set of ODE's:

$$\begin{cases} \dfrac{d\lambda_0}{dt} = -d(F, \ell_b)\,\lambda_0 - \dfrac{\partial}{\partial\ell}d(F, \ell_b)\,\pi_0 \\[2mm] \qquad\qquad + \displaystyle\sum_{i=1}^{N} b(F, \mu_i)\,\lambda_i\,, \\[4mm] \dfrac{d\pi_0}{dt} = g(F, \ell_b)\,\lambda_0 + \dfrac{\partial}{\partial\ell}g(F, \ell_b)\,\pi_0 - d(F, \ell_b)\,\pi_0\,. \end{cases} \qquad (54)$$

The dynamics described by the ODE's (51) and (54) for the $N$ internal cohorts and the single boundary cohort approximate the dynamics described by the PDE (13) and its boundary condition (14). They can be studied using any integration method to numerically solve systems of ordinary differential equations (e.g., the Runge-Kutta method). The boundary cohort cannot, however, be continued indefinitely, because the interval $[\ell_b, \ell_1(t))$ would become large and the approximation would break down. Therefore, a renumbering operation at regular time intervals $\Delta t$ is a crucial part of the EBT method. This renumbering operation transforms the current boundary cohort into an internal cohort and initializes a new, empty boundary cohort. At the same time, all internal cohorts are renumbered to make room for the new internal cohort and to keep them in order. Internal cohorts that have become negligible can be discarded.

If a renumbering procedure has just taken place when $t = t^*$, the population consists of internal cohorts characterized by the quantities $\lambda_i(t^*)$ and $\mu_i(t^*)$ $(i = 1, \ldots, N)$ and an empty boundary cohort for which $\lambda_0(t^*) = 0$ and $\pi_0(t^*) = 0$. Between $t^*$ and $t^* + \Delta t$, the ODE's (51) and (54) are solved numerically. At $t^* + \Delta t$, the following transformations are applied ($t^* + \Delta t^+$ is used in these equations to denote the value of the variables after the transformation):

$$
\left\{
\begin{array}{rcl}
\lambda_i(t^* + \Delta t^+) & = & \lambda_{i-1}(t^* + \Delta t) \\[4pt]
\mu_i(t^* + \Delta t^+) & = & \mu_{i-1}(t^* + \Delta t) \\[4pt]
\lambda_1(t^* + \Delta t^+) & = & \lambda_0(t^* + \Delta t) \\[4pt]
\mu_1(t^* + \Delta t^+) & = & \ell_{\mathrm{b}} + \dfrac{\pi_0(t^* + \Delta t)}{\lambda_0(t^* + \Delta t)} \\[10pt]
\lambda_0(t^* + \Delta t^+) & = & 0 \\[4pt]
\pi_0(t^* + \Delta t^+) & = & 0
\end{array}
\right.
\qquad i = 2, \ldots .
$$

$$(55)$$

The first two equations simply renumber the internal cohorts. The equation for $\mu_1(t^* + \Delta t^+)$ converts the quantity $\pi_0$ into the average length of the individuals in the new cohort with index one. The last two equations reset the values for the boundary cohort so that it is empty. Transforming the boundary cohort and starting a new, empty one increases the number of ODE's to solve by two; but the number of ODE's can decrease because some of the internal cohorts are discarded. The number of quantities with which the structured population is characterized, and hence the number of ODE's to solve, can therefore change in the renumbering procedure.

In the long run, the time interval $\Delta t$ between renumbering operations determines the intervals $\Omega_i(t)$. The choice of $\Delta t$ determines the initial width of the interval $[\ell_{\mathrm{b}}, \ell_1(t))$ at the moment that the boundary cohort is transformed into the internal cohort with index one. Because the bounds subsequently change according to the ODE (50), this initial width also determines the length interval spanned by the cohort at a later time. The initial population has only to be subdivided, one way or another, into a collection of cohorts; after that, the choice of $\Delta t$ completely determines this subdivision. Smaller values of $\Delta t$ imply that the dynamics of the PDE (13) are better approximated at the expense of a larger number of cohorts to keep track of and more ODE's to solve simultaneously.

The ODE (12) describing the dynamics of the food density must be solved simultaneously with the ODE's (51) and (54). Following equation (49), the integral occurring in (12) is replaced by the appropriate sum of the cohort statistics:

$$
\begin{cases}
\dfrac{dF}{dt} = \alpha F \left(1 - \dfrac{F}{K}\right) - \displaystyle\sum_{i=1}^{N} I(F, \mu_i)\, \lambda_i & \text{if } \lambda_0(t) = 0, \\[4ex]
\dfrac{dF}{dt} = \alpha F \left(1 - \dfrac{F}{K}\right) - \displaystyle\sum_{i=1}^{N} I(F, \mu_i)\, \lambda_i & \\[3ex]
\qquad\qquad - I\left(F, \ell_{\mathrm{b}} + \dfrac{\pi_0}{\lambda_0}\right) \lambda_0 & \text{if } \lambda_0(t) > 0.
\end{cases}
\tag{56}
$$

The second equation shows that the boundary cohort must be taken into account if $\lambda_0(t) > 0$ (note that this complication does not occur in the approximation of the total population birthrate in eq. 54, since the newborn individuals cannot yet reproduce). As for the internal cohorts, the function $I(F, \ell)$ must be evaluated at the average length of the individuals in the boundary cohort, which equals $\ell = \ell_{\mathrm{b}} + \pi_0(t)/\lambda_0(t)$.

This completes the formulation of the EBT method. For the *Daphnia* model, the resulting set of equations is summarized in Table 4. These equations lead to a consistent approximation of the original PDE and boundary condition (de Roos & Metz 1991). When the numerical solutions obtained with the EBT method can be compared with analytical results, the method has proved accurate (see, e.g., de Roos et al. 1990). The working of the EBT method is not immediately straightforward. In addition to the explanation given here, it is useful to study a more heuristic derivation (de Roos et al. 1992). A computer program that forms a template for implementing an arbitrary structured-population model can be obtained free of charge (unfortunately also free of support).

## 8 Stability Analysis of the *Daphnia* Equilibrium

In addition to numerical explorations of dynamics, studies of unstructured-population models usually involve some kind of bifurcation analysis of attractors (e.g., equilibria, limit cycles). Often this analysis is restricted to finding the boundary that separates parameter values for which the equilibrium is attractive from those for which it is repelling. Ideally, when studying a structured-population model, one would also like a complete inventory of the changes in dynamics that occur with changes in model parame-

TABLE 4. *"Escalator Boxcar Train" Formulation of the* Daphnia *Population Model*

Continuous-time dynamics for the boundary cohort, during cohort cycle interval

$$\frac{d\lambda_0}{dt} = -d(F, \ell_{\rm b})\,\lambda_0 - \frac{\partial}{\partial \ell}d(F, \ell_{\rm b})\,\pi_0$$

$$+ \sum_{i=1}^{N} b(F, \mu_i)\,\lambda_i$$

$$\frac{d\pi_0}{dt} = g(F, \ell_{\rm b})\,\lambda_0 + \frac{\partial}{\partial \ell}g(F, \ell_{\rm b})\,\pi_0$$

$$-d(F, \ell_{\rm b})\,\pi_0$$

Continuous-time dynamics for all other cohorts, during cohort cycle interval $(i = 1, \ldots, N)$

$$\frac{d\lambda_i}{dt} = -d(F, \mu_i)\,\lambda_i$$

$$\frac{d\mu_i}{dt} = g(F, \mu_i)$$

Transformations and new initial values for the boundary cohort at the end of the cohort cycle

$$\lambda_1(t^* + \Delta t^+) = \lambda_0(t^* + \Delta t)$$

$$\mu_1(t^* + \Delta t^+) = \ell_{\rm b} + \frac{\pi_0(t^* + \Delta t)}{\lambda_0(t^* + \Delta t)}$$

$$\lambda_0(t^* + \Delta t^+) = 0$$

$$\pi_0(t^* + \Delta t^+) = 0$$

Renumbering equations for all other cohorts at the end of the cohort cycle $(i = 2, \ldots)$

$$\lambda_i(t^* + \Delta t^+) = \lambda_{i-1}(t^* + \Delta t)$$

$$\mu_i(t^* + \Delta t^+) = \mu_{i-1}(t^* + \Delta t)$$

Dynamics of the food density in the environment

$$\frac{dF}{dt} = \alpha F\left(1 - \frac{F}{K}\right) - \sum_{i=1}^{N} I(F, \mu_i)\,\lambda_i \ {}^*$$

NOTE.—Formulation is based on the description of individual behavior given in Table 1 and the population-level dynamics specified in Table 2. $t^* + \Delta t$ indicates the time at which the cohort cycle ends; $t^* + \Delta t^+$, values just after the transformation and renumbering operation.

${}^*$ Include the boundary cohort in the summation if $\lambda_0 \neq 0$; use $\mu_0 = \ell_{\rm b} + \pi_0/\lambda_0$.

ters. Equilibria can disappear or become biologically irrelevant in certain parameter regions, a stable equilibrium can become unstable, or a limit cycle can arise that ultimately leads to persistent fluctuations in densities. Structured-population models can be expected to exhibit the same richness of dynamics found in unstructured models. In contrast to the theory for models formulated in terms of ODE's, the techniques and general theory for a complete bifurcation analysis of structured-population models is not readily available.

The *Daphnia* model, however, is simple enough to make a linear stability analysis feasible. The outcome indicates whether, and for which parameter values, the *Daphnia* population and the food density return to their equilibrium states (computed in Section 6) following a small perturbation away from those states. The issue to resolve is whether small perturbations of the equilibrium grow or decrease with time and, hence, whether the equilibrium is locally stable. In addition to numerical simulations, this analysis is most frequently applied in studies of structured-population models (see, e.g., de Roos et al. 1990; van den Bosch & de Roos, in press).

To explain the procedure for a structured-population model, it is useful to recapitulate the local stability analysis of the basic model (1). The first step is to linearize the equations, assuming that the system is close to the equilibrium. The resulting, linear equations can be solved explicitly by searching for exponential solutions. In all derivations below, I use exponential trial solutions right from the start to keep the presentation as short as possible.

Any internal equilibrium of the basic model (1) is a pair of nonzero values $(\tilde{F}, \tilde{C})$, for which the right-hand sides of the ODE's vanish. The stability of such an equilibrium is determined by assuming that the prey and predator populations are close, but not equal, to their equilibrium densities. The small deviations between actual and equilibrium densities are assumed to grow or decrease exponentially over time:

$$
\begin{aligned}
F(t) &= \tilde{F} + \Delta_{\mathrm{F}}\, e^{\lambda t}, \\
C(t) &= \tilde{C} + \Delta_{\mathrm{C}}\, e^{\lambda t},
\end{aligned}
\tag{57}
$$

where $\Delta_{\mathrm{F}}$ and $\Delta_{\mathrm{C}}$ are the initial perturbations of the prey and predators. These expressions for $F(t)$ and $C(t)$ are substituted into the ODE's (1) to obtain a set of equations describing the fate of the deviations $\Delta_F\,e^{\lambda t}$ and $\Delta_C\,e^{\lambda t}$:

$$\frac{d\left(\Delta_F\,e^{\lambda t}\right)}{dt} = \alpha\left(\tilde{F}+\Delta_F\,e^{\lambda t}\right)\left(1-\frac{\tilde{F}+\Delta_F\,e^{\lambda t}}{K}\right)$$

$$-\,I_{max}\,\frac{\tilde{F}+\Delta_F\,e^{\lambda t}}{F_h+\tilde{F}+\Delta_F\,e^{\lambda t}}\left(\tilde{C}+\Delta_C\,e^{\lambda t}\right),$$

$$\frac{d\left(\Delta_C\,e^{\lambda t}\right)}{dt} = \epsilon I_{max}\,\frac{\tilde{F}+\Delta_F\,e^{\lambda t}}{F_h+\tilde{F}+\Delta_F\,e^{\lambda t}}\left(\tilde{C}+\Delta_C\,e^{\lambda t}\right)$$

$$-\,\mu\left(\tilde{C}+\Delta_C\,e^{\lambda t}\right).$$

The left-hand sides of these equations involve only the time derivatives of $\Delta_F\,e^{\lambda t}$ and $\Delta_C\,e^{\lambda t}$ because $d\tilde{F}/dt = d\tilde{C}/dt = 0$. Since the deviations $\Delta_F\,e^{\lambda t}$ and $\Delta_C\,e^{\lambda t}$ are assumed to be small, the terms on the right-hand sides can be replaced by their first-order Taylor expansion around the equilibrium state $(\tilde{F},\tilde{C})$. All second- and higher-order terms involving $\Delta_F\,e^{\lambda t}$ and $\Delta_C\,e^{\lambda t}$ are hence neglected. Together with an evaluation of the time derivatives, this yields the following set of algebraic equations:

$$\lambda\Delta_F\,e^{\lambda t} \approx \alpha\left(1-2\frac{\tilde{F}}{K}\right)\Delta_F\,e^{\lambda t}$$

$$-\,I_{max}\frac{F_h}{(F_h+\tilde{F})^2}\tilde{C}\Delta_F\,e^{\lambda t} - I_{max}\frac{\tilde{F}}{F_h+\tilde{F}}\Delta_C\,e^{\lambda t},$$

$$\lambda\Delta_C\,e^{\lambda t} \approx \epsilon I_{max}\frac{F_h}{(F_h+\tilde{F})^2}\tilde{C}\Delta_F\,e^{\lambda t}$$

$$+\,\epsilon I_{max}\frac{\tilde{F}}{F_h+\tilde{F}}\Delta_C\,e^{\lambda t} - \mu\Delta_C\,e^{\lambda t}.$$

The last two terms in the second equation cancel, because $d\tilde{C}/dt = 0$ at equilibrium. Dividing both sides by $e^{\lambda t}$, the equations can also be expressed in matrix form as

$$\mathbf{J}(\lambda)\left[\begin{array}{c}\Delta_F \\ \Delta_C\end{array}\right] = 0, \tag{58}$$

in which

$$\mathbf{J}(\lambda) = \begin{bmatrix} \alpha\left(1-2\dfrac{\tilde{F}}{K}\right) - I_{\max}\dfrac{F_{\mathrm{h}}}{(F_{\mathrm{h}}+\tilde{F})^2}\,\tilde{C} - \lambda & -I_{\max}\dfrac{\tilde{F}}{F_{\mathrm{h}}+\tilde{F}} \\[3mm] \epsilon I_{\max}\dfrac{F_{\mathrm{h}}}{(F_{\mathrm{h}}+\tilde{F})^2}\,\tilde{C} & -\lambda \end{bmatrix}.$$

If both $\Delta_{\mathrm{F}}$ and $\Delta_{\mathrm{C}}$ are not zero, equation (58) holds only if

$$\det \mathbf{J}(\lambda) = 0\,. \tag{59}$$

Given initial perturbations $\Delta_{\mathrm{F}}$ and $\Delta_{\mathrm{C}}$ and the linear equations determining their time evolution, the change over time of these perturbations is, in general, described by a sum of contributions that are all of the form $\exp(\lambda t)$. The value of $\lambda$ in each contribution corresponds to a root of the equation (59), and the number of these roots determines the number of different contributions. It follows that as long as the real part of all the roots $\lambda$ is negative, the initial perturbations decline and the internal equilibrium is locally stable. The roots $\lambda$ of equation (59) are the eigenvalues of the basic model in the neighborhood of its equilibrium state $(\tilde{F}, \tilde{C})$. Equation (59) is called the characteristic equation of the model.

The characteristic equation (59) is a second-order polynomial in $\lambda$ with two, possibly complex roots $\lambda_1$ and $\lambda_2$. It can be shown that if a complex number $\sigma + i\omega$ is a root of equation (59), its complex conjugate, $\sigma - i\omega$, is also a root. The most interesting case occurs when the equilibrium loses its stability because such a pair of complex conjugates crosses the imaginary axis from the left to the right half of the complex plane (implying that $\sigma$ turns from negative to positive). Such a loss of stability may lead to a limit cycle, in which the prey and predator exhibit persistent periodic oscillations, although this is not universally true. At the particular parameter values at which this loss of stability occurs, the characteristic equation (59) has a pair of purely imaginary roots, $\lambda = \pm i\omega$ ($\sigma = 0$). If the *Daphnia* mortality rate $\mu$ is the parameter of interest, the value of $\mu$ where equilibrium stability changes can be found by solving for $\mu$ and $\omega$ the equation

$$\det \mathbf{J}(i\omega) = 0\,. \tag{60}$$

Equation (60) depends indirectly on $\mu$ because the equilibrium state $(\tilde{F}, \tilde{C})$ is a function of $\mu$. This condition is actually two equations, one for the real part and one for the imaginary part of $\det \mathbf{J}(i\omega)$. For the basic model (1), it can be shown that equation (60) permits simultaneous solutions of $\mu$ and $\omega$ only if the trace of the matrix $\mathbf{J}(\lambda)$, when $\lambda = 0$, is positive for the particular value of $\mu$. The matrix $\mathbf{J}(0)$ is the Jacobian matrix of the basic model (1), evaluated at the equilibrium state $(\tilde{F}, \tilde{C})$. A negative trace of the Jacobian matrix is one of the two Routh-Hurwitz criteria for an equilibrium of a set of two ODE's to be stable. The other Routh-Hurwitz criterion—for stability of such an equilibrium, the determinant of the Jacobian matrix should be positive—is always fulfilled for this model.

The linear stability analysis of the structured *Daphnia* model follows the same procedure. The states of the population and the food density are assumed to be close, but not identical, to the equilibrium state computed in Section 6. The deviations from the equilibrium are assumed to grow or decrease exponentially. A set of linear equations is derived for the time evolution of these deviations, in which only first-order terms are included. This allows the derivation of a characteristic equation analogous to (59) for the eigenvalues of the model in the neighborhood of the equilibrium state. The equation is solved for the value of a single parameter at which there exists a pair of purely imaginary eigenvalues, $\lambda = \pm i\omega$. This parameter value corresponds to a point on the stability boundary of the model. At this boundary the internal equilibrium loses its stability, and limit cycles can arise.

Although the conditions (43) and (46) show that the equilibrium state of the *Daphnia* model is entirely determined by the food density $\tilde{F}$ and the population birthrate $\tilde{B}$, it is convenient to start the stability analysis of the equilibrium from a characterization in terms of the equilibrium food density, $\tilde{F}$, the equilibrium age distribution, $\tilde{m}(a)$, the equilibrium length-age relation, $\tilde{L}(\tilde{F}, a)$, and the duration of the juvenile period in equilibrium, $\tilde{A}_j(\tilde{F})$. To simplify the notation, neglect the explicit food dependence in these last two quantities and simply write $\tilde{L}(a)$ and $\tilde{A}_j$ instead of $\tilde{L}(\tilde{F}, a)$ and $\tilde{A}_j(\tilde{F})$, respectively.

When the population and the food density are close to their equilibrium states and the deviations from the equilibrium are assumed to grow or decrease exponentially, the actual state of the system is

$$
\begin{array}{rcll}
F(t) & = & \tilde{F} + \Delta_{\mathrm{F}}\, e^{\lambda t}, & \\[4pt]
m(t,a) & = & \tilde{m}(a) + \Delta_{\mathrm{m}}(a)\, e^{\lambda t}, & \Delta_{\mathrm{m}}(0) = \Delta_{\mathrm{B}}, \\[4pt]
L(t,a) & = & \tilde{L}(a) + \Delta_{\mathrm{L}}(a)\, e^{\lambda t}, & \Delta_{\mathrm{L}}(0) = 0, \\[4pt]
A_{\mathrm{j}}(t) & = & \tilde{A}_{\mathrm{j}} + \Delta_{\mathrm{A}}\, e^{\lambda t}. &
\end{array}
\tag{61}
$$

For the basic model, it was sufficient to assume that prey and predator densities are close to their equilibrium values. Equation (61) incorporates deviations between the current and equilibrium values of both the population age distribution and the length-age relation at every age $a$. Then, $\Delta_{\mathrm{B}}$, which can be interpreted as the initial perturbation in the total population birthrate, is used as shorthand for the deviation in the population age distribution at age 0, $\Delta_{\mathrm{m}}(0)$. The relation $\Delta_{\mathrm{L}}(0) = 0$ follows from the assumptions that all individuals are born with the identical length at birth, $\ell_{\mathrm{b}}$, and that hence $L(t,0) = \ell_{\mathrm{b}}$ under all circumstances.

Appendix B shows in detail how substitution of the relations (61) into the model equations of Table 3 yields a set of equations for the time evolution of the deviations $\Delta_{\mathrm{F}}$, $\Delta_{\mathrm{m}}(a)$, $\Delta_{\mathrm{L}}(a)$, and $\Delta_{\mathrm{A}}$. Using Taylor expansion of the functions describing the individual behavior (Table 1) and neglecting all higher-order terms, this eventually leads to a system of equations for the eigenvalues $\lambda$ in the neighborhood of the equilibrium state:

$$
\mathbf{U}(\lambda)
\begin{bmatrix} \Delta_{\mathrm{B}} \\[4pt] \Delta_{\mathrm{F}} \end{bmatrix}
=
\begin{bmatrix} U_{11}(\lambda) & U_{12}(\lambda) \\[4pt] U_{21}(\lambda) & U_{22}(\lambda) \end{bmatrix}
\begin{bmatrix} \Delta_{\mathrm{B}} \\[4pt] \Delta_{\mathrm{F}} \end{bmatrix}
= 0.
\tag{62}
$$

Like equation (58), this equation is linear in the perturbation of the total population birthrate $\Delta_{\mathrm{B}}$ and the food density $\Delta_{\mathrm{F}}$. As in the unstructured model, the equation implies that the eigenvalues of the model are the roots of the following characteristic equation

$$
\det \mathbf{U}(\lambda) = 0.
\tag{63}
$$

In contrast to the matrix $\mathbf{J}$, however, the elements of $\mathbf{U}$ are nonlinear functions of $\lambda$. In Appendix B the following expressions are derived for the elements of $\mathbf{U}$:

$$U_{11}(\lambda) = \int_{\tilde{A}_j}^{\infty} r_{\mathrm{m}}\, h(\tilde{F})\left(\tilde{L}(a)\right)^2 e^{-(\mu+\lambda)a}\, da \; - \; 1,$$

$$U_{12}(\lambda) = \int_{\tilde{A}_j}^{\infty} r_{\mathrm{m}}\, h(\tilde{F})\left(\tilde{L}(a)\right)^2 \tilde{B}\, e^{-\mu a}\, da\, \frac{h'(\tilde{F})}{h(\tilde{F})}$$

$$+ \left\{ \frac{r_{\mathrm{m}}\ell_{\mathrm{m}}\left(h(\tilde{F})\right)^2 \ell_j{}^2}{\ell_{\mathrm{m}} h(\tilde{F}) - \ell_j}\, \tilde{B}\, e^{-\mu\tilde{A}_j}\, \frac{1 - e^{-(\lambda+\gamma)\tilde{A}_j}}{\lambda + \gamma} \right.$$

$$\left. + 2\gamma r_{\mathrm{m}}\ell_{\mathrm{m}}\left(h(\tilde{F})\right)^2 \int_{\tilde{A}_j}^{\infty} \tilde{L}(a)\tilde{B} e^{-\mu a}\, \frac{1 - e^{-(\lambda+\gamma)a}}{\lambda + \gamma}\, da \right\} \frac{h'(\tilde{F})}{h(\tilde{F})},$$

$$U_{21}(\lambda) = -\int_{0}^{\infty} \nu\, h(\tilde{F})\left(\tilde{L}(a)\right)^2 e^{-(\mu+\lambda)a}\, da,$$

$$U_{22}(\lambda) = R'(\tilde{F}) \; - \; \lambda \; - \; \int_{0}^{\infty} \nu\, h(\tilde{F})\left(\tilde{L}(a)\right)^2 \tilde{B}\, e^{-\mu a}\, da\, \frac{h'(\tilde{F})}{h(\tilde{F})}$$

$$+ \left\{ 2\gamma\nu\ell_{\mathrm{m}}\left(h(\tilde{F})\right)^2 \int_{0}^{\infty} \tilde{L}(a)\tilde{B} e^{-\mu a}\, \frac{1 - e^{-(\lambda+\gamma)a}}{\lambda + \gamma}\, da \right\} \frac{h'(\tilde{F})}{h(\tilde{F})}.$$

In these expressions, $R'(\tilde{F})$ and $h'(\tilde{F})$ denote the derivative with respect to food density of the algal growth rate $R(F)$ and the *Daphnia* functional response $h(F)$ evaluated at the equilibrium $\tilde{F}$. In general, the characteristic equation (63) has infinitely many roots that all correspond to eigenvalues of the model in the neighborhood of the equilibrium state. Because the equilibrium is unstable as soon as a single eigenvalue has a positive real part, the (pair of) eigenvalue(s) with the largest real part is crucial for stability. This is often referred to as the dominant (pair of) eigenvalue(s). In practice, this dominant (pair of) eigenvalue(s) is the only one that can be readily localized in numerical explorations. Hence, other roots can for practical purposes be ignored.

If the natural mortality rate of *Daphnia*, $\mu$, is our parameter of interest, the particular value of $\mu$ at which the equilibrium loses its stability and limit cycles may arise can be found by solving for $\mu$

and $\omega$ the equation

$$\det \mathbf{U}(i\omega) \;=\; 0\,. \tag{64}$$

Again, this is actually a set of two equations, one for the real part and one for the complex part of $\det \mathbf{U}$. Such equations can generally be solved with appropriate numerical root-finding procedures (e.g., the Newton-Raphson method; Press et al. 1922). Moreover, because the elements of the matrix $\mathbf{U}$ are functions of the food density $\tilde{F}$ and population birthrate $\tilde{B}$, the equilibrium conditions (43) and (46) must be solved simultaneously with (64). The entire set of equations may be so complicated that finding a solution is problematic.

The value for $\mu$ satisfying equation (64) is a point on the stability boundary of the model, as in the unstructured model. Such a stability boundary can be constructed in a plane of two parameters by varying one of the two in a stepwise manner over a range of values, while at every step solving for the value of the other parameter at which the equilibrium loses its stability. This procedure is illustrated in the next section. The nature of the dynamics once the equilibrium is unstable can be verified only by using numerical explorations close to the stability boundary.

## 9  Some Results and Implications for *Daphnia*

This section includes a discussion of some conclusions from numerical investigations of the equilibrium, stability, and dynamics of the *Daphnia* model. Many of these results are part of an extensive investigation of a slightly more complicated model (de Roos et al. 1990), which assumes that individual *Daphnia* have a maximum life span and cannot shrink during times of food shortage. The life-span assumption results in negligible, quantitative differences in the model predictions. The assumption that individuals cannot shrink implies that energy destined to be spent on reproduction must be rechanneled to cover basic metabolic needs during bouts of starvation, when the allocation rule of the model discussed here would predict individual weight loss. The equilibrium and its stability are completely unaffected by the nonshrinking assumption, because in equilibrium individuals never shrink. The only differences in the model predictions concern the oscillatory dynamics when the equilibrium is unstable. If individuals can shrink, the oscillations are far more pronounced, although the remaining char-

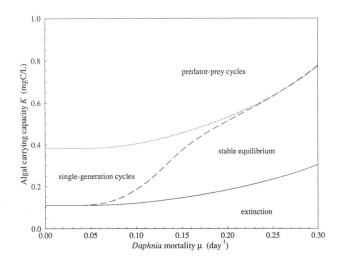

FIGURE 7. *Stability diagram of the size-structured model for the dynamics of a* Daphnia pulex *population feeding on* Chlamydomonas rheinhardii. *Parameters:* $\ell_{\mathrm{b}} = 0.6$, $\ell_{\mathrm{j}} = 1.4$, $\ell_{\mathrm{m}} = 3.5$, $F_{\mathrm{h}} = 0.164$, $\gamma = 0.11$, $r_{\mathrm{m}} = 1.0$, $\nu = 0.007$, $\alpha = 0.5$. Solid line, *Existence boundary;* dashed line, *stability boundary;* dotted line, *prey-escape boundary, approximately separating the parameter region with classical predator-prey oscillations from the region with single-generation oscillations (following de Roos et al. 1990).*

acteristics of the cycles in population structure and density are the same (de Roos et al. 1990).

### Existence and Stability of the Internal Equilibrium

Figure 7 shows the stability diagram as a function of the background *Daphnia* mortality rate $\mu$ and the carrying capacity $K$ of the algae. Three different regions can be distinguished in this diagram: (1) a region with parameter combinations $(\mu, K)$ in which the *Daphnia* population becomes extinct, (2) a region with a stable equilibrium, and (3) a region in which the *Daphnia* and algal populations cycle indefinitely. The existence boundary, separating the first and second regions, has been determined by numerically solving for the carrying capacity, $K$, at which the expected number of offspring produced by a single *Daphnia* individual equals unity (see eq. 44). This specific value of $K$ was determined for $\mu = 0.01, 0.02, 0.03, \ldots, 0.30$. Comparing expression (44) with the equilibrium condition (43), from which the equilibrium food density is determined, it can be concluded that this minimum value of $K$ for the existence of an equilibrium exactly equals the food density imposed by the *Daphnia* population in a stable equilibrium state. This equivalence of the minimum resource density for existence and the imposed equilibrium level is a general characteristic of purely exploitative systems.

The boundary between the stable equilibrium and the region of population oscillations has been determined by numerically solving for the value of the carrying capacity that satisfies both the bifurcation condition (64) and the equilibrium conditions (43) and (46). The particular $K$ value was computed for $\mu = 0.01, 0.02, 0.03,$ ..., $0.30$. Numerical studies of the dynamics were subsequently conducted to determine the type of dynamics arising when the equilibrium is unstable (de Roos et al. 1990), that is, for parameter combinations above the stability boundary. On the basis of these simulations, the region of population oscillations can be subdivided into two parts. In one, the fluctuations are similar to the oscillations found in the unstructured model (1). This type of oscillation is usually referred to as predator-prey, paradox-of-enrichment, or prey-escape cycles (Rosenzweig 1971; de Roos et al.1990). In the second part, roughly located at low values for both $\mu$ and $K$, the fluctuations are due to size-dependent growth and reproduction, and are referred to as single-generation or time-lag cycles (Gurney & Nisbet 1985).

An approximate boundary between the paradox-of-enrichment cycles and single-generation cycles can be determined by comparing the stability criteria for the structured *Daphnia* model and its unstructured analogue (1). This prey-escape boundary, marking the transition to the classical predator-prey cycles, is likely to occur in all models in which the individuals of an exploiter population with a saturating functional response compete with each other only indirectly for a (logistically) growing resource (de Roos et al. 1990). The transition is related to the ability of the resource to escape the control of the consumer population (for an approximate criterion for this prey-escape boundary, see de Roos et al. 1990). The loss of control is due to the saturating functional response. In the *Daphnia* model, as $K$ increases, the oscillations gradually change from single-generation to predator-prey cycles without a catastrophic transition. The prey-escape boundary is, therefore, not as sharp as the boundary that separates the stable and unstable regions. Nonetheless, it indicates the extent of the parameter region for which single-generation and predator-prey cycles can be expected to occur, a prediction that is confirmed by numerical simulations.

The main feature of Figure 7 is the large area of the parameter space in which classical predator-prey cycles are predicted by both the structured and unstructured *Daphnia* models. Eventually, these cycles always occur because the carrying capacity of the algae increases. As shown by McCauley and Murdoch (1990), this prediction by the model is at odds with experimental observations on natural and semi-natural populations of *Daphnia*, where prey-escape cycles are hard to find. The mechanism behind the prey-escape cycles suggests that any model of a purely exploitative consumer feeding on a logistically growing resource with a saturating functional response would disagree with the results of McCauley and Murdoch (1990) (de Roos et al. 1990). The discrepancy between the model prediction of predator-prey cycles and the apparent lack of such cycles in experimental and natural populations was addressed by Murdoch (pers. comm.), who discusses various hypotheses that could explain the discrepancy.

*Population Oscillations Induced by Size Structure*

Figure 8 shows a typical example of the single-generation oscillations that are induced by the size structure and size-dependent behavior of the *Daphnia* population. These results are obtained by numerically integrating the ODE's of Table 4. Both the *Daphnia* and algal populations fluctuate with a period of roughly 26 days. The maximum *Daphnia* density is approximately twice the minimum, whereas for the algae, the ratio of maximum to minimum density is less (Fig. 8, *top panel*). At any time, the *Daphnia* population contains one dominant cohort (*middle panel*). Both these features agree reasonably well with cycles observed in natural and laboratory *Daphnia* populations (Murdoch & McCauley 1985; McCauley & Murdoch 1987). This suggests that the aspects of individual behavior included in the *Daphnia* model play an important role in the dynamics of real-world *Daphnia* populations.

Figure 8 also shows results that shed doubt on this suggestion, however. The bottom panel shows that the single-generation oscillations are primarily due to fluctuations in the length of the juvenile period, which varies by a factor of about two. It appears that the number of adults at any time determines whether a new population birth pulse is generated. In contrast, adult fecundity fluctuates less and never reaches low values. This seems at odds with the experi-

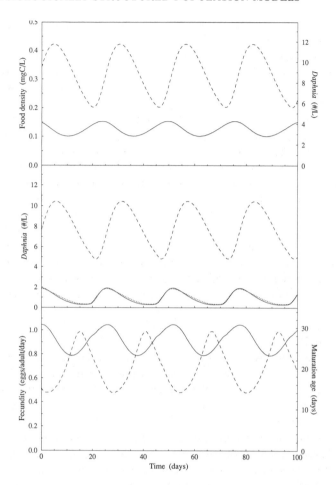

FIGURE 8. *Single-generation dynamics in the* Daphnia *model. Parameters:* Daphnia *mortality rate,* $\mu = 0.11$; *algal carrying capacity,* $K = 0.25$; *all others as in Figure 7. Top, Algal biomass* (solid) *and total number of* Daphnia (dashed) *per liter. Middle, Number of neonate* (dotted), *juvenile* (dashed), *and adult* (solid) Daphnia *per liter. Bottom, Length of the juvenile period* (dashed) *and adult fecundity* (solid).

mental observations presented by McCauley and Murdoch (1987). Although these authors concluded that real-world *Daphnia* populations display mainly delay-driven cycles, with retarded juvenile growth and suppressed adult fecundity as important causal mechanisms, it is actually the latter that seems to be the driving force behind the oscillations. Whenever fecundity is determined, it turns out to fluctuate dramatically (McCauley & Murdoch 1987, Figs. 7, 9). Although juvenile development is indeed slower than under high-food conditions, McCauley and Murdoch (1987) showed no evidence that this delay drives the cycles. Moreover, large fluctuations in adult fecundity agree with the observation in natural *Daphnia* populations that the surviving adults of a dominant cohort, at the end of their life, initiate a new dominant cohort (McCauley & Murdoch 1987).

Figure 9 shows more clearly why single-generation cycles are generated by the model. This figure displays the size structure of the *Daphnia* population during one entire oscillation of the simulation presented in Figure 8. A considerable cohort of individuals is stopped in their development just before reaching maturity. Within a short time (about four days), all mature, leading to a new birth pulse. As adults, they continue to reproduce while mortality decreases their number. The first of their offspring do not develop quickly enough to reach maturity before food conditions deteriorate and development slows. Hence, the front of the new cohort is again stopped shortly before reaching maturity. The individuals that are born later from the same adult cohort can catch up with the front, because smaller individuals are assumed to be more efficient. This leads to a "piling up" of individuals just before the size at maturation, $\ell_j$, which starts a new cycle after some time. It is mainly this piling-up mechanism (i.e., the variable delay of the juvenile period) that drives the single-generation cycles, as opposed to the length of the juvenile delay itself (de Roos et al. 1990). As an obvious consequence, it is not the last, surviving adults of a dominant cohort that start a new cycle (requiring a substantial fluctuation in adult fecundity) but the bulk of the dominant cohort shortly after they mature.

The predictions of the *Daphnia* model hence seem to agree superficially with observations of natural and laboratory populations of *Daphnia*, but they differ dramatically on closer inspection. As far as experimental data are available, the cycles in real-world *Daphnia* populations seem to be driven primarily by the changing fecundity of the adult *Daphnia*, whereas the model predicts dynamics that are driven primarily by the changing number of adults in the population. These changes in number come about by the large fluctuations in the length of the juvenile period. How the large fluctuations in adult fecundity arise in experimental populations, without inducing large-amplitude predator-prey cycles, is an unresolved question.

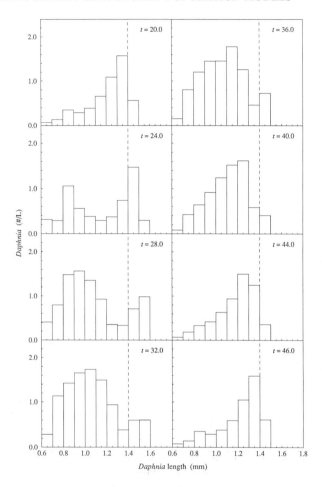

FIGURE 9. *Changes in the size distribution of* Daphnia *during one complete cycle of the simulation presented in Figure 8*. Dashed line, *Length* $\ell_j$, *at which individuals mature*.

## 10 General Perspective

This chapter has dealt with the formulation of models for physiologically structured populations and the techniques that are most commonly applied to analyze them. These techniques include equilibrium computations, stability analysis, and numerical simulations. Only relatively simple models could be dealt with in this chapter; the general class of PSP models is much broader and allows for the investigation of more-complicated scenarios. This last section puts the theory presented here in a more general perspective. Possible extensions below use the model formulation that is obtained after the application of the Murphy trick (Section 6).

Although the general discussions about the model at the individual and population levels (Sections 2, 4, and 5) are phrased in terms of an $i$-state and $E$-state of arbitrary dimension, they are illustrated with a model in which only a single variable for the $i$-state and $E$-state appear. Sections 2 and 4 show in some detail how to extend the $E$-state to any arbitrary, finite number of variables, but they do not touch on more $i$-state variables.

As it turns out, increasing the number of $i$-state variables is only moderately complicated, as long as all newborn individuals have the same state at birth. This can be illustrated by incorporating direct age dependence into the *Daphnia* model (Table 3). The functions describing individual growth, reproduction, mortality, and feeding now depend on three quantities: the current food density, the individual length, and the individual age. This dependence can be expressed by writing $g(F, L(t,a), a)$, $b(F, L(t,a), a)$, $d(F, L(t,a), a)$, and $I(F, L(t,a), a)$ to represent the model of the individual behavior. Then, all functions are seen to depend on age both directly and indirectly through the relation between age and length, $L(t,a)$. The direct age dependence of the mortality rate $d$ does not invalidate the PDE (28) for the *Daphnia* age distribution $m(t,a)$. A similar argument holds for the individual reproduction, feeding, and growth rates that occur in the equations (30), (33), and (36), respectively. All age-independent rates can hence be straightforwardly replaced by age-dependent functions. The basic set of PDE's, ODE's, and boundary conditions does not change.

Because of the additional age dependence, equilibrium computations and stability analysis are probably more difficult, but not qualitatively different. Only the sets of ODE's that are used to approximate the model for numerical simulations must be extended. Equation (51) must be supplemented with an equation for the mean age of the individuals in a cohort. The appropriate equation is identical to the ODE for $d\mu_i/dt$, except that the growth rate $g$ is replaced by the unit developmental rate in age. Similarly, the system of ODE's (54) has to be extended with an equation that is identical to the ODE for $d\pi_0/dt$ except for the substitution of $g$ by unity (and consequently all derivatives of $g$ by zero). The equation describes the development of the quantity that characterizes the age distribution of the individuals within the boundary cohort. (For a version of the *Daphnia* model that incorporates such additional age dependence, see de Roos et al. 1992.)

Extending the $i$-state with a physiological characteristic other than age is possible, at the expense of adding another PDE. For example, assume that the strict relation between individual length and weight assumed in the *Daphnia* model does not hold, but that the weight of an individual with a specific length can vary. Furthermore, assume that individuals are born with a fixed weight $w_{\mathrm{b}}$. A weight-age relation $W(t, a)$ can now be introduced to keep track of the weight of the individuals as a function of their age. The dynamics of $W(t, a)$ can be described by a PDE of the form

$$\frac{\partial W(t, a)}{\partial t} + \frac{\partial W(t, a)}{\partial a} = z\big(F, L(t, a), W(t, a)\big),$$

where $W(t, 0) = w_{\mathrm{b}}$ (cf. eq. 36; boundary condition 35). In this equation the function $z\big(F, L(t, a), W(t, a)\big)$ denotes the growth rate in weight, as a function of food density, individual length, and individual weight. These three variables now influence all individual-level functions. In addition to the function $z\big(F, L(t, a), W(t, a)\big)$, the model is specified in terms of the functions $g\big(F, L(t, a), W(t, a)\big)$, $b\big(F, L(t, a), W(t, a)\big)$, $d\big(F, L(t, a), W(t, a)\big)$, and $I\big(F, L(t, a), W(t, a)\big)$, modeling growth in individual length, reproduction, mortality, and feeding, respectively. As with the introduction of age dependence, adding weight as an $i$-state variable also requires extending the systems of ODE's (51) and (54) with equations that are analogous to the ODE's for $d\mu_i/dt$ and $d\pi_0/dt$. These equations must describe the change in mean weight of the individuals in an internal cohort and a boundary cohort, respectively, and hence involve the function $z\big(F, L(t, a), W(t, a)\big)$ instead of $g\big(F, L(t, a), W(t, a)\big)$. (For an example of the analysis of a PSP model of this sort, see van den Bosch & de Roos, in press.)

The crucial aspect of all these models is that the identical state at birth, together with deterministic growth and developmental rates, ensures an eventual one-to-one relation between the $i$-state and individual age. The members of the initial population can obviously have $i$-states that form a cloud of points in the three-dimensional $i$-state space defined by age, length, and weight. After their death, however, the $i$-states of all individuals at any time make up a one-dimensional curve through this three-dimensional $i$-state space, originating in the point that represents the state at birth of an individual. This holds even when the state at birth of the individuals changes over time! Essentially, more-complicated PSP models therefore involve either a larger number of possible states at birth or non-deterministic growth and development in the $i$-state.

If individuals can be born with one of a finite collection of $i$-states at birth, a PSP model can still be formulated (and possibly analyzed) using the framework discussed here. All individuals are identified with the state with which they are born, and the entire population can be subdivided on that basis into a finite number of subpopulations. The model would be formally analogous to a model with more than a single structured population, although the subpopulations may be coupled by the reproduction process when the state at birth of the offspring is not the same as the parent's state at birth.

Turning from the PDE framework to the "cumulative" formulation (Diekmann et al. 1994; Diekmann & Metz 1995), it may even be possible to allow for an infinite range of $i$-states at birth, or for non-deterministic growth and development in the $i$-state, which is one of the hardest restrictions to relax in the current framework. Although the model formulation might be theoretically possible, whether and what kind of analysis is possible for such a model remains absolutely unclear. I also know of no study that analyzed any specific, more complicated example. This reemphasizes the fact that the field of modeling physiologically structured populations is still developing, with respect both to the general theory and to the detailed studies of examples. Readers interested in more-complicated models and their analysis may find some solace in the book by Metz and Diekmann (1986) on which most of the theory presented here is based, but more likely they will have to wait until the theoretical state of the art has been developed further.

## Acknowledgments

I thank Odo Diekmann and Hans Metz for introducing me to physiologically structured population models. In addition, they played and continue to play crucial roles in developing the theory explained in this chapter. Over the years I have benefited substantially from discussions and collaborations with, among others, Frank van den Bosch, Roger Nisbet, Ed McCauley, Bill Gurney, Bill Murdoch, Lennart Persson, and Maurice Sabelis.

## Appendix A

*Some Formal Calculations Relating to Individual Death and Reproduction*

Consider a cohort of individuals living in a constant environment $E_c$, and assume that the individual state space $\Omega$ consists of the interval $[x_b, x_m)$, in which $x_b$ is the $i$-state at birth and $x_m$ is the asymptotic $i$-state that long-living individuals approach (formally, when their age, $a$, approaches infinity). Integration of the distribution of the state at death (10) from $x_b$ to $x_m$ yields

$$\int_{x_b}^{x_m} H(E_c, x)\, dx = \int_{x_b}^{x_m} \frac{d(E_c, x)}{g(E_c, x)} \exp\left( -\int_{x_b}^{x} \frac{d(E_c, \xi)}{g(E_c, \xi)}\, d\xi \right) dx$$

$$= \left. -\exp\left( -\int_{x_b}^{x} \frac{d(E_c, \xi)}{g(E_c, \xi)}\, d\xi \right) \right|_{x=x_b}^{x=x_m}$$

$$= S(E_c, x_b) - S(E_c, x_m).$$

$S(E_c, x_b)$ equals 1 by assumption. The value of $S(E_c, x_m)$ necessarily equals 0 because no individual lives long enough (infinitely long!) to attain the asymptotic $i$-state $x_m$. Essentially, this poses a condition on the instantaneous mortality rate, $d(E_c, x)$, which should ensure that $S(E_c, x_m) = 0$, that is, that indeed no individual lives forever. From the definition of $H(E_c, x)$ it can be concluded that its expected value equals the average $i$-state at which an individual dies. Integration by parts of this expectation yields an equation that relates this average $i$-state at death to the survival function $S(E_c, x)$:

$$\int_{x_b}^{x_m} x\, H(E_c, x)\, dx$$

$$= \int_{x_b}^{x_m} x\, \frac{d(E_c, x)}{g(E_c, x)} \exp\left( -\int_{x_b}^{x} \frac{d(E_c, \xi)}{g(E_c, \xi)}\, d\xi \right) dx$$

$$= \left. -x \exp\left( -\int_{x_b}^{x} \frac{d(E_c, \xi)}{g(E_c, \xi)}\, d\xi \right) \right|_{x=x_b}^{x=x_m} + \int_{x_b}^{x_m} \exp\left( -\int_{x_b}^{x} \frac{d(E_c, \xi)}{g(E_c, \xi)}\, d\xi \right) dx$$

$$= x_b + \int_{x_b}^{x_m} S(E_c, x)\, dx.$$

If the expected lifetime of an individual is of interest, the probability
density function of the age at which an individual dies can be derived
as follows (cf. the definition and derivation of $H$):

$$-\frac{dS\big(E_c,X(E_c,a)\big)}{da} = -\frac{dS\big(E_c,X(E_c,a)\big)}{dt}\frac{dt}{da}$$

$$= d\big(E_c,X(E_c,a)\big)\,S\big(E_c,X(E_c,a)\big)$$

$$= d\big(E_c,X(E_c,a)\big)\,\exp\left(-\int_{x_b}^{X(E_c,a)}\frac{d(E_c,\xi)}{g(E_c,\xi)}\,d\xi\right)$$

$$= d\big(E_c,X(E_c,a)\big)\,\exp\left(-\int_0^a d\big(E_c,X(E_c,\zeta)\big)\,d\zeta\right).$$

The resulting expression is analogous to the distribution of the state at
death given in function (10). It is a probability density function with an
integral equal to 1. Its expected value can again be derived by integration
by parts:

$$\int_0^\infty a\,d\big(E_c,X(E_c,a)\big)\,\exp\left(-\int_0^a d\big(E_c,X(E_c,\zeta)\big)\,d\zeta\right)\,da$$

$$= -a\,\exp\left(-\int_0^a d\big(E_c,X(E_c,\zeta)\big)\,d\zeta\right)\bigg|_{a=0}^{a=\infty}$$

$$+\int_0^\infty \exp\left(-\int_0^a d\big(E_c,X(E_c,\zeta)\big)\,d\zeta\right)\,da$$

$$= \int_0^\infty \exp\left(-\int_0^a d\big(E_c,X(E_c,\zeta)\big)\,d\zeta\right)\,da.$$

The last integral equals the average age individuals reach, given that the
death rate, $d(E_c,x)$, and the ($i$-state)-age relation, $X(E_c,a)$, are known.

As mentioned in the text, the first reproduction event is, from a proba-
bilistic point of view, comparable with the death of an individual. Therefore, it is relatively straightforward to see that the expected $i$-state at

which an individual gives birth for the first time is given by

$$\int_{x_b}^{x_m} x \, \frac{b(E_c, x)}{g(E_c, x)} \exp\left(-\int_{x_b}^{x} \frac{b(E_c, \xi)}{g(E_c, \xi)} \, d\xi\right) \, dx$$

$$= -x \exp\left(-\int_{x_b}^{x} \frac{b(E_c, \xi)}{g(E_c, \xi)} \, d\xi\right)\Bigg|_{x=x_b}^{x=x_m} + \int_{x_b}^{x_m} \exp\left(-\int_{x_b}^{x} \frac{b(E_c, \xi)}{g(E_c, \xi)} \, d\xi\right) \, dx$$

$$= x_b + \int_{x_b}^{x_m} \exp\left(-\int_{x_b}^{x} \frac{b(E_c, \xi)}{g(E_c, \xi)} \, d\xi\right) \, dx \, .$$

Like the arguments that led to the condition $S(E_c, x_m) = 0$, this derivation must also assume that individuals reproduce with probability 1 if they live long enough, which is biologically realistic. The exact assumption states that

$$\exp\left(-\int_{x_b}^{x_m} \frac{b(E_c, \xi)}{g(E_c, \xi)} \, d\xi\right) = 0 \, .$$

The expected age at which an individual reproduces for the first time is given by

$$\int_{0}^{\infty} a \, b(E_c, X(E_c, a)) \exp\left(-\int_{0}^{a} b(E_c, X(E_c, \zeta)) \, d\zeta\right) \, da$$

$$= -a \exp\left(-\int_{0}^{a} b(E_c, X(E_c, \zeta)) \, d\zeta\right)\Bigg|_{a=0}^{a=\infty}$$

$$+ \int_{0}^{\infty} \exp\left(-\int_{0}^{a} b(E_c, X(E_c, \zeta)) \, d\zeta\right) \, da$$

$$= \int_{0}^{\infty} \exp\left(-\int_{0}^{a} b(E_c, X(E_c, \zeta)) \, d\zeta\right) \, da \, .$$

The integrand in the first expression is the probability density function for an individual to reproduce for the first time at age $a$ (cf. the derivation for the expected lifetime of an individual, above). The expected age is, of course, equivalent to the "time till first reproduction."

## Appendix B

*Derivation of the Characteristic Equation of the* Daphnia *Model*

To derive the matrix $\mathbf{U}(\lambda)$ in the characteristic equation (63), the approximations (61) for the population state and food density in the neighborhood of the equilibrium are to be substituted into the model equations of Table 3. The equations are then linearized and solved as far as possible to obtain the elements of $\mathbf{U}(\lambda)$. In the following, function symbols subscripted with $F$ or $\ell$—for example, $d_F(F, \ell)$ and $d_\ell(F, \ell)$—refer to the partial derivatives of the function with respect to $F$ and $\ell$, respectively.

Substituting $m(t, a) = \tilde{m}(a) + \Delta_{\mathrm{m}}(a)\, e^{\lambda t}$ into the PDE (28) leads to the following derivation:

$$\frac{\partial \Delta_{\mathrm{m}}(a) e^{\lambda t}}{\partial t} + \frac{\partial \tilde{m}(a)}{\partial a} + \frac{\partial \Delta_{\mathrm{m}}(a) e^{\lambda t}}{\partial a}$$

$$= -d\left( \tilde{F} + \Delta_{\mathrm{F}} e^{\lambda t}, \tilde{L}(a) + \Delta_{\mathrm{L}}(a) e^{\lambda t} \right) \left( \tilde{m}(a) + \Delta_{\mathrm{m}}(a) e^{\lambda t} \right)$$

$$\approx -d\big( \tilde{F}, \tilde{L}(a) \big) \tilde{m}(a) \; - \; d_F\big( \tilde{F}, \tilde{L}(a) \big) \Delta_{\mathrm{F}} e^{\lambda t} \tilde{m}(a)$$

$$- \; d_\ell\big( \tilde{F}, \tilde{L}(a) \big) \Delta_{\mathrm{L}}(a) e^{\lambda t} \tilde{m}(a) \; - \; d\big( \tilde{F}, \tilde{L}(a) \big) \Delta_{\mathrm{m}}(a) e^{\lambda t}$$

$$\Rightarrow \quad \frac{\partial \Delta_{\mathrm{m}}(a)}{\partial a} = -(\lambda + \mu)\, \Delta_{\mathrm{m}}(a)$$

$$\Rightarrow \quad \Delta_{\mathrm{m}}(a) = \Delta_{\mathrm{B}}\, e^{-(\lambda+\mu)a} .$$

In the first step, the function $d(F, \ell)$ is replaced with its first-order Taylor expansion around the equilibrium values $\tilde{F}$ and $\tilde{L}(a)$. All higher-order terms involving the deviates $\Delta_{\mathrm{F}}$, $\Delta_{\mathrm{L}}(a)$, and $\Delta_{\mathrm{m}}(a)$ are neglected. These manipulations constitute the linearization step, since a linear equation for $\Delta_{\mathrm{m}}(a)$ results. In the second step, the partial derivatives of the function $d$ are replaced with the appropriate expressions from Table 1. In addition, the equilibrium condition (41) for $\tilde{m}(a)$ is used to simplify the equation. Both sides of the equation can be divided by $e^{\lambda t}$, which removes the time dependence. These manipulations lead to a simple differential equation for $\Delta_{\mathrm{m}}(a)$ that can be solved explicitly, as shown in the last equation. The symbol $\Delta_{\mathrm{B}}$ is used as shorthand for $\Delta_{\mathrm{m}}(0)$. Obviously, $\Delta_{\mathrm{m}}(a)$ is an explicit function of $\Delta_{\mathrm{B}}$ only, involving no other small deviates.

Analogous to the derivation that is given above, the approximation

$L(t, a) = \tilde{L}(a) + \Delta_L(a)\, e^{\lambda t}$ is substituted into the PDE (36):

$$\frac{\partial \Delta_L(a)e^{\lambda t}}{\partial t} + \frac{\partial \tilde{L}(a)}{\partial a} + \frac{\partial \Delta_L(a)e^{\lambda t}}{\partial a}$$

$$= g\left(\tilde{F} + \Delta_F e^{\lambda t}, \tilde{L}(a) + \Delta_L(a)e^{\lambda t}\right)$$

$$\approx g\big(\tilde{F}, \tilde{L}(a)\big) + g_F\big(\tilde{F}, \tilde{L}(a)\big)\Delta_F e^{\lambda t} + g_\ell\big(\tilde{F}, \tilde{L}(a)\big)\Delta_L(a)e^{\lambda t} \quad \text{(B1)}$$

$$\Rightarrow \quad \frac{\partial \Delta_L(a)}{\partial a} = -(\lambda + \gamma)\Delta_L(a) + \gamma \ell_m h'(\tilde{F})\Delta_F$$

$$\Rightarrow \quad \Delta_L(a) = \frac{\gamma \ell_m h'(\tilde{F})}{\lambda + \gamma}\left(1 - e^{-(\lambda + \gamma)a}\right)\Delta_F .$$

In the first step of this derivation, the equation is linearized by using the Taylor expansion of the function $g(F, \ell)$, neglecting higher-order terms. In the second step, the partial derivatives of $g$ are replaced with their appropriate expression (see Table 1), and the equilibrium condition (38) is used for simplification. A linear, time-independent differential equation for $\Delta_L(a)$ results, which can be solved to obtain an explicit expression for $\Delta_L(a)$ as a function of $\Delta_F$ only. The function $h'(\tilde{F})$ in this equation (and all following ones) is the derivative of the *Daphnia* functional response $h(F)$, evaluated at the equilibrium food density, $\tilde{F}$.

Substituting the relations $F(t) = \tilde{F} + \Delta_F e^{\lambda t}$, $L(t, a) = \tilde{L}(a) + \Delta_L(a)e^{\lambda t}$, and $m(t, a) = \tilde{m}(a) + \Delta_m(a)e^{\lambda t}$ into the ODE (34) for the food density gives

$$\frac{d\Delta_F e^{\lambda t}}{dt}$$

$$= R\left(\tilde{F} + \Delta_F e^{\lambda t}\right)$$

$$- \int_0^\infty I\left(\tilde{F} + \Delta_F e^{\lambda t}, \tilde{L}(a) + \Delta_L(a)e^{\lambda t}\right)\left(\tilde{m}(a) + \Delta_m(a)e^{\lambda t}\right) da$$

$$\approx R(\tilde{F}) + R'(\tilde{F})\Delta_F e^{\lambda t} \quad \text{(B2)}$$

$$- \int_0^\infty I\big(\tilde{F}, \tilde{L}(a)\big)\tilde{m}(a)\, da - \int_0^\infty I_F\big(\tilde{F}, \tilde{L}(a)\big)\Delta_F e^{\lambda t}\tilde{m}(a)\, da$$

$$- \int_0^\infty I_\ell\big(\tilde{F}, \tilde{L}(a)\big)\Delta_L(a)e^{\lambda t}\tilde{m}(a)\, da - \int_0^\infty I\big(\tilde{F}, \tilde{L}(a)\big)\Delta_m(a)e^{\lambda t}\, da$$

$$\Rightarrow \quad \left(R'(\tilde{F}) - \lambda\right)\Delta_\mathrm{F} - \int_0^\infty \nu h'(\tilde{F})\left(\tilde{L}(a)\right)^2 \Delta_\mathrm{F}\tilde{m}(a)\,da$$

$$- \int_0^\infty 2\nu h(\tilde{F})\tilde{L}(a)\Delta_\mathrm{L}(a)\tilde{m}(a)\,da - \int_0^\infty \nu h(\tilde{F})\left(\tilde{L}(a)\right)^2 \Delta_\mathrm{m}(a)\,da$$

$$= 0\,.$$

The function $R'(\tilde{F})$ is the derivative of the algal growth function $R(F)$, evaluated at the equilibrium food density, $\tilde{F}$. This derivation is analogous to the previous two. The linearization, which involves the partial derivatives of the *Daphnia* ingestion rate $I(F,\ell)$, and the simplification step, which uses the equilibrium condition (45), now leads to an algebraic equality. The first three terms of this equality are all linear expressions of $\Delta_\mathrm{F}$, since $\Delta_\mathrm{L}(a)$ has been derived above as an explicit function of $\Delta_\mathrm{F}$. The coefficients of $\Delta_\mathrm{F}$ in these three terms together constitute the element $U_{22}(\lambda)$ in the matrix $\mathbf{U}(\lambda)$. The last term in the equality is a linear function of $\Delta_\mathrm{B}$ by virtue of the relation between $\Delta_\mathrm{m}(a)$ and $\Delta_\mathrm{B}$ derived above. This term constitutes the element $U_{21}(\lambda)$ in the matrix $\mathbf{U}(\lambda)$.

The remaining two elements of the matrix $\mathbf{U}(\lambda)$ are obtained from the linearization of the boundary condition (32). This boundary condition contains the current duration of the juvenile period $A_\mathrm{j}(t)$ as a lower bound in the integral. If the state of the system is perturbed away from the equilibrium state, the juvenile delay changes as well. At every time, however, $A_\mathrm{j}(t)$ fulfills the condition $L\bigl(t, A_\mathrm{j}(t)\bigr) = \ell_\mathrm{j}$. By substituting the approximations $L(t,a) = \tilde{L}(a) + \Delta_\mathrm{L}(a)\,e^{\lambda t}$ and $A_\mathrm{j}(t) = \tilde{A}_\mathrm{j} + \Delta_\mathrm{A}\,e^{\lambda t}$ into this equality, a relation between $\Delta_\mathrm{A}$ and $\Delta_\mathrm{L}(\tilde{A}_\mathrm{j})$ can be derived:

$$
\begin{aligned}
\ell_\mathrm{j} &= L(t, \tilde{A}_\mathrm{j} + \Delta_\mathrm{A}\,e^{\lambda t}) \\[4pt]
&= \tilde{L}(\tilde{A}_\mathrm{j} + \Delta_\mathrm{A}\,e^{\lambda t}) + \Delta_\mathrm{L}(\tilde{A}_\mathrm{j} + \Delta_\mathrm{A}\,e^{\lambda t})\,e^{\lambda t} \\[4pt]
&\approx \tilde{L}(\tilde{A}_\mathrm{j}) + \frac{\partial \tilde{L}(\tilde{A}_\mathrm{j})}{\partial a}\Delta_\mathrm{A}\,e^{\lambda t} + \Delta_\mathrm{L}(\tilde{A}_\mathrm{j})\,e^{\lambda t}
\end{aligned}
\tag{B3}
$$

$$\Rightarrow \quad \Delta_\mathrm{A} = \frac{-\Delta_\mathrm{L}(\tilde{A}_\mathrm{j})}{g(\tilde{F},\ell_\mathrm{j})}\,.$$

The linearization step involves the Taylor expansion of the function $\tilde{L}(a)$ around the equilibrium value $a = \tilde{A}_\mathrm{j}$. Recognizing that $\ell_\mathrm{j} = \tilde{L}(\tilde{A}_\mathrm{j})$ and $\partial\tilde{L}(\tilde{A}_\mathrm{j})/\partial a = g(\tilde{F},\ell_\mathrm{j})$ following equation (38), $\Delta_\mathrm{A}$ can be expressed in terms of $\Delta_\mathrm{L}(\tilde{A}_\mathrm{j})$ and is hence also linearly related to $\Delta_\mathrm{F}$.

Linearization of the boundary condition (32) involves substitution of all approximate values given by (61):

$$\tilde{m}(0) + \Delta_{\mathrm{B}}\, e^{\lambda t}$$

$$= \int_{\tilde{A}_j + \Delta_A e^{\lambda t}}^{\infty} b\left(\tilde{F} + \Delta_F e^{\lambda t}, \tilde{L}(a) + \Delta_L(a)e^{\lambda t}\right)\left(\tilde{m}(a) + \Delta_m(a)e^{\lambda t}\right) da$$

$$\approx \int_{\tilde{A}_j}^{\infty} b\big(\tilde{F}, \tilde{L}(a)\big)\tilde{m}(a)\, da \;+\; \int_{\tilde{A}_j}^{\infty} b_F\big(\tilde{F}, \tilde{L}(a)\big)\tilde{m}(a)\Delta_F e^{\lambda t}\, da$$

$$+ \int_{\tilde{A}_j}^{\infty} b_\ell\big(\tilde{F}, \tilde{L}(a)\big)\tilde{m}(a)\Delta_L(a)e^{\lambda t}\, da \qquad\qquad\text{(B4)}$$

$$+ \int_{\tilde{A}_j}^{\infty} b\big(\tilde{F}, \tilde{L}(a)\big)\Delta_m(a)e^{\lambda t}\, da \;-\; b(\tilde{F}, \ell_j)\tilde{m}(\tilde{A}_j)\Delta_A e^{\lambda t}$$

$$\Rightarrow \quad \Delta_{\mathrm{B}} - b(\tilde{F}, \ell_j)\tilde{m}(\tilde{A}_j)\frac{\Delta_L(\tilde{A}_j)}{g(\tilde{F}, \ell_j)}$$

$$- \int_{\tilde{A}_j}^{\infty} r_m h'(\tilde{F})\big(\tilde{L}(a)\big)^2 \tilde{m}(a)\Delta_F\, da$$

$$- \int_{\tilde{A}_j}^{\infty} 2 r_m h(\tilde{F})\tilde{L}(a)\tilde{m}(a)\Delta_L(a)\, da$$

$$- \int_{\tilde{A}_j}^{\infty} r_m h(\tilde{F})\big(\tilde{L}(a)\big)^2 \Delta_m(a)\, da \;=\; 0\,.$$

The linearization step in the derivation above is complicated by the fact that $\tilde{A}_j + \Delta_A\, e^{\lambda t}$ occurs in the lower bound of the integral. The perturbation of this integral bound can be approximated to first order by the term $-b(\tilde{F}, \ell_j)\tilde{m}(\tilde{A}_j)\Delta_A e^{\lambda t}$. The remaining steps to linearize and simplify (using the equilibrium condition for $\tilde{m}(0)$) are identical to the derivation (B2). The resulting equality contains the expressions for the elements $U_{11}(\lambda)$ and $U_{12}(\lambda)$ in $\mathbf{U}(\lambda)$: the first and last term of the equality are both linear functions of $\Delta_{\mathrm{B}}$ on the basis that $\Delta_m(a)$ is related to $\Delta_{\mathrm{B}}$. The coefficients of these terms constitute the element $U_{11}(\lambda)$.

Because $\Delta_L(a)$ is a function of $\Delta_F$, all remaining terms in the equality are proportional to $\Delta_F$. The three coefficients together form the basis of the element $U_{12}(\lambda)$. This completes the derivation of the characteristic equation (63).

## Literature Cited

Bell, G. I., and E. C. Anderson. 1967. Cell growth and division. I. A mathematical model with applications to cell volume distributions in mammalian suspension cultures. *Biophysical Journal* 7: 329–351.

Caswell, H. 1989. *Matrix Population Models: Construction, Analysis, and Interpretation.* Sinauer, Sunderland, Mass.

DeAngelis, D. L., and L. J. Gross. 1992. *Individual-Based Models and Approaches in Ecology: Populations, Communities and Ecosystems.* Chapman & Hall, New York.

de Roos, A. M. 1988. Numerical methods for structured population models: The Escalator Boxcar Train. *Numerical Methods for Partial Differential Equations* 4: 173–195.

———. 1989. Daphnids on a train: Development and application of a new numerical method for physiologically structured population models. Ph.D. diss. Leiden University, Leiden.

de Roos, A. M., and J. A. J. Metz. 1991. Towards a numerical analysis of the Escalator Boxcar Train. Pp. 91–113 *in* J. A. Goldstein, F. Kappel, and W. Schappacher, eds., *Differential Equations with Applications in Biology, Physics, and Engineering.* Lecture Notes in Pure and Applied Mathematics 133. Marcel Dekker, New York.

de Roos, A. M., O. Diekmann, and J. A. J. Metz. 1988. The Escalator Boxcar Train: Basic theory and an application to *Daphnia* population dynamics. Technical Report AM–R8814, CWI, Amsterdam.

de Roos, A. M., E. McCauley, and W. G. Wilson. 1991. Mobility versus density-limited predator-prey dynamics on different spatial scales. *Proceedings of the Royal Society of London B* 246: 117–122.

de Roos, A. M., J. A. J. Metz, E. Evers, and A. Leipoldt. 1990. A size dependent predator-prey interaction: Who pursues whom? *Journal of Mathematical Biology* 28: 609–643.

de Roos, A. M., O. Diekmann, and J. A. J. Metz. 1992. Studying the dynamics of structured population models: A versatile technique and its application to *Daphnia*. *American Naturalist* 139: 123–147.

Diekmann, O., and J. A. J. Metz. 1993. On the reciprocal relationship between life histories and population dynamics. Technical Report AM–R9302, CWI, Amsterdam.

———. 1995. On the reciprocal relationship between life histories and population dynamics. Pp. 263–279 *in* S. A. Levin, ed., *Frontiers in Mathematical Biology.* Lecture Notes in Biomathematics 100. Springer-Verlag, Berlin.

Diekmann, O., M. Gyllenberg, J. A. J. Metz, and H. Thieme. 1992. The "cumulative" formulation of (physiologically) structured population models. Technical Report AM–R9205, CWI, Amsterdam.

———. 1994. The "cumulative" formulation of (physiologically) structured population models. Pp. 145–154 *in* Ph. Clément and G. Lumer, eds., *Evolution Equations, Control Theory and Biomathematics.* Lecture Notes in Pure and Applied Mathematics 155. Marcel Dekker, New York.

Gurney, W. S. C., and R. M. Nisbet. 1985. Fluctuation periodicity, generation separation, and the expression of larval competition. *Theoretical Population Biology* 28: 150–180.

Gurney, W. S. C., E. McCauley, R. M. Nisbet, and W. W. Murdoch. 1990. The physiological ecology of *Daphnia*: A dynamic model of growth and reproduction. *Ecology* 71: 716–732.

Kooijman, S. A. L. M. 1986. Population dynamics on basis of budgets. Pp. 266–297 *in* J. A. J. Metz and O. Diekmann, eds., *The Dynamics of Physiologically Structured Populations.* Lecture Notes in Biomathematics 68. Springer-Verlag, Berlin.

Kooijman, S. A. L. M., and J. A. J. Metz. 1984. On the dynamics of chemically stressed populations: The deduction of population consequences from effects on individuals. *Ecotoxicology and Environmental Safety* 8: 254–274.

McCauley, E., and W .W. Murdoch. 1987. Cyclic and stable populations: Plankton as paradigm. *American Naturalist* 129: 97–121.

———. 1990. Predator-prey dynamics in rich and poor environments. *Nature* 343: 455–457.

McCauley, E., W. W. Murdoch, R. M. Nisbet, and W. S. C. Gurney. 1990. The physiological ecology of *Daphnia*: Development of a model of growth and reproduction. *Ecology* 71: 703–715.

McKendrick, A. G. 1926. Application of mathematics to medical problems. *Proceedings of the Edinburgh Mathematical Society* 44: 98–130.

Metz, J. A. J., and A. M. de Roos. 1992. The role of physiologically structured population models within a general individual-based modeling perspective. Pp. 88–111 *in* D. L. DeAngelis and L. J. Gross, eds., *Individual-Based Models and Approaches in Ecology: Populations, Communities and Ecosystems.* Chapman & Hall, New York.

Metz, J. A. J., and O. Diekmann. 1986. *The Dynamics of Physiologically Structured Populations.* Lecture Notes in Biomathematics 68. Springer-Verlag, Berlin.

Metz, J. A. J., A. M. de Roos, and F. van den Bosch. 1988. Population models incorporating physiological structure: A quick survey of the basic concepts and an application to size-structured population dynamics in waterfleas. Pp. 106–126 *in* B. Ebenman and L. Persson, eds., *Size-Structured Populations.* Springer-Verlag, Berlin.

Murdoch, W. W., and E. McCauley. 1985. Three distinct types of dynamic behaviour shown by a single planktonic system. *Nature* 316: 628–630.

Murphy, L. F. 1983. A nonlinear growth mechanism in size structured population dynamics. *Journal of Theoretical Biology* 104: 493–506.

Nisbet, R. M., W. S. C. Gurney, E. McCauley, and W. W. Murdoch. 1989. Structured population models: A tool for linking effects at individual and population levels. *Biological Journal of the Linnean Society* 37: 79–99.

Paloheimo, J. E., S. J. Crabtree, and W. D. Taylor. 1982. Growth model of *Daphnia. Canadian Journal of Fisheries & Aquatic Sciences* 39: 598–606.

Press, W. H., B. P. Flannery, S. A. Teukolsky, and W. T. Vetterling. 1988. *Numerical Recipes in C: The Art of Scientific Computing.* Cambridge University Press.

Rosenzweig, M. L. 1971. Paradox of enrichment: Destabilization of exploitation ecosystems in ecological time. *Science* 171: 385–387.

Roughgarden, J. 1979. *Theory of Population Genetics and Evolutionary Ecology: An Introduction.* Macmillan, New York.

Sharpe, F. R., and A. J. Lotka. 1911. A problem in age-distributions. *Philosophical Magazine* 21: 435–438.

Sinko, J. W., and W. Streifer. 1967. A new model for age-size structure of a population. *Ecology* 48: 910–918.

van den Bosch, F., and A. M. de Roos. In press. The dynamics of infectious diseases in orchards with roguing and replanting as control strategy. *Journal of Mathematical Biology.*

van den Bosch, F., A. M. de Roos, and W. Gabriel. 1988. Cannibalism as a life boat mechanism. *Journal of Mathematical Biology* 26: 619–633.

von Bertalanffy, L. 1957. Quantitative laws in metabolism and growth. *Quarterly Review of Biology* 32: 217–231.

VonFoerster, H. 1959. Some remarks on changing populations. *in* F. Stohlman, ed., *The Kinetics of Cellular Proliferation.* Grune & Stratton, New York.

# CHAPTER 6

# Nonlinear Matrix Equations and Population Dynamics

## J. M. Cushing

Classical models for the dynamics of biological populations, such as the famous logistic and Lotka-Volterra equations, involve population-level statistics and parameters. In the "structured" population models of this book, individual members of the population are classified in some manner, usually by means of certain physiological characteristics, and the distribution of individuals based upon this classification is dynamically modeled.

A popular approach to modeling structured populations is to make both time and the structuring variable(s) discrete. This approach leads to a system of difference equations or, in the most commonly used form, a matrix equation for the population distribution. The chapter by Caswell (Chapter 2) discusses linear matrix models in detail. A linear theory cannot be used for long-term predictions for growing populations because it implies unlimited growth. If regulatory effects of some kind are taken into account (such as the adverse effects of population density on fertility, survival, and growth rates), then the matrix equations become nonlinear.

The goal of this chapter is to present a general theory for the asymptotic dynamics of nonlinear matrix equations as they apply to the dynamics of structured populations. The point of view taken is that of bifurcation theory. By this is meant that changes in dynamics are studied as a function of a single model parameter.

Models for population dynamics are generally replete with parameters describing various kinds of vital rates and other modeling coefficients. In principle, any parameter appearing in a particular application can be used as a bifurcation parameter. That is, the dynamics implied by the model equations can in principle be studied as a function of any one of the parameters appearing in the equations, and an investigator may have good reasons for choosing one particular model parameter over another in a specific application. In this chapter, however, I show how a biologically significant parameter (the inherent net reproductive number of the population) can always be defined, introduced into the model equations, and successfully used as a bifurcation parameter in a completely general setting.

I hope to show how several rewards can be gained from the point of view taken here. First, the approach establishes general results concerning the existence and stability of equilibrium distributions, which are then available for any particular application. Second, it makes available powerful analytical techniques for obtaining results about the asymptotic dynamics of what might otherwise be intractable model equations. Finally, the approach serves to organize one's study of any particular model in terms of a general, biologically meaningful parameter. Moreover, even when one or more of the general results fail to apply to a specific model (because of the failure of some required hypothesis or other), the approach often gives insights into what exceptional phenomena occur and what analytical steps should next be taken.

First, the approach is applied to linear matrix equations. This is done to present the point of view of bifurcation theory in such a way that a straightforward extension can be made to nonlinear equations. Next, this extension is made, and results concerning the existence and stability of equilibria are given. The following section illustrates these equilibrium results with some applications.

## 1 Linear Matrix Models

The earliest matrix models structure a population into age classes of a common length and then follow the number of individuals in each age class over discrete time steps of the same length.

Let $x_i(t)$ denote the number of individuals in the $i$th age class at time $t$, $1 \le i \le m$, and suppose that the time unit is taken, without loss in generality, to be 1. If $\tau_{i+1,i} \in (0,1]$ is the fraction of individuals in age class $i$ that survives to age class $i+1$, for $i = 0, 1, \ldots, m-1$, and if $f_{1j} \ge 0$ is the number of offspring produced by an individual in age class $j$, then (as in Chapter 2),

$$\mathbf{x}(t+1) = \mathbf{A}\mathbf{x}(t), \quad t = 0, 1, 2, \ldots$$

where

$$\mathbf{x}(t) = \begin{bmatrix} x_1(t) \\ x_2(t) \\ \vdots \\ x_m(t) \end{bmatrix}$$

$$\mathbf{A} = \begin{bmatrix} f_{11} & f_{12} & \cdots & f_{1,m-1} & f_{1m} \\ \tau_{21} & 0 & \cdots & 0 & 0 \\ 0 & \tau_{32} & \cdots & 0 & 0 \\ \vdots & \vdots & \ddots & \vdots & \vdots \\ 0 & 0 & \cdots & \tau_{m,m-1} & 0 \end{bmatrix} \tag{1}$$

$$f_{1j} \geq 0 \text{ (not all 0)}, \qquad 0 < \tau_{j+1,j} \leq 1.$$

This model assumes no immigration or emigration, and it assumes that no individual ever reaches age $m+1$. Sometimes the final class, $i = m$, is defined to be the class of individuals of age greater than $m$, with survival to all ages permitted, in which case a fraction $\tau_{mm}$ appears in the lower right-hand corner of $\mathbf{A}$.

Note that this projection matrix can be additively decomposed as $\mathbf{A} = \mathbf{F} + \mathbf{T}$, where $\mathbf{F} = [f_{ij}]$ is a fertility matrix and $\mathbf{T} = [\tau_{ij}]$ is a class-transition matrix.

For a population structured by any means into $n$ classes (or stages), a nonnegative projection matrix $\mathbf{A}$ can be constructed whose entries are nonnegative and represent births, deaths, and transitions between classes and which, by multiplication on the left, takes the population's class-distribution vector at time $t$ to its distribution vector at time $t + 1$ (see Chapter 2, by Caswell):

$$\mathbf{x}(t+1) = \mathbf{A}\mathbf{x}(t), \quad t = 0, 1, 2, \ldots \tag{2}$$

In general, such a matrix can be additively decomposed into a fertility and a class-transition matrix:

$$\mathbf{A} = \mathbf{F} + \mathbf{T} \geq 0,$$

where

$$\mathbf{F} = [f_{ij}], \quad f_{ij} \geq 0, \tag{3}$$

$$\mathbf{T} = [\tau_{ij}], \quad \tau_{ij} \in [0,1], \quad \sum_{i=1}^{m} \tau_{ij} \leq 1, \tag{4}$$

where $f_{ij}$ is the number of $i$-class offspring produced by a $j$-class individual (per unit of time) and $\tau_{ij}$ is the fraction of $j$-class individuals that lie in class $i$ after one unit of time (or the probability that a $j$-class individual survives and moves to class $i$ in one unit of time). Writing $\mathbf{A} \geq 0$ for a matrix (or a vector) means that every entry in $\mathbf{A}$ is greater than or equal to 0.

If the projection matrix $\mathbf{A}$ is constant (i.e., if the fertilities $f_{ij}$ and the transition probabilities $\tau_{ij}$ are all constants, unchanging in time), the dynamics are described using the methods in Chapter 2, by Caswell. Let the dominant eigenvalue of matrix $\mathbf{A}$ be called $\lambda_1$. The population becomes extinct if $\lambda_1 < 1$ and grows geometrically if $\lambda_1 > 1$. Mathematically, we say that the "trivial equilibrium" $\mathbf{x} = \mathbf{0}$ is (*globally*) *asymptotically stable* in the first case and is *unstable* in the second case. In other words, the trivial equilibrium state loses its stability as the dominant eigenvalue of $\mathbf{A}$ increases through the critical value of 1.

When the dynamics of population models are studied as a function of the entries in the projection matrix $\mathbf{A}$, the critical value $\lambda_1 = 1$ delineates the boundary between the radically different dynamic cases of population extinction and population survival. As shown below, this delineation carries through to nonlinear models as well, but in a more complicated way.

The asymptotic dynamics implied by the linear model (2) may also be characterized by another fundamental biological quantity. This, the "net reproductive number," has a more explicit biological meaning than $\lambda_1$ and is often a more convenient quantity with which to analyze matrix models, as we see below.

Consider the age-structured case and its Leslie matrix (1). The product

$$\tau_{21}\tau_{32} \cdots \tau_{j,j-1} = \prod_{k=0}^{j-1} \tau_{k+1,k}$$

is the probability of living to age $j$. (For notational convenience, $\tau_{10} = 1$, by definition.) The expected number of offspring that an individual of age $j$ will produce in one unit of time is therefore

$$f_{1j} \prod_{k=0}^{j-1} \tau_{k+1,k}.$$

Adding over all age classes yields the expected number of offspring per individual over its lifetime. This quantity, defined as

$$n = \sum_{j=1}^{m} f_{1j} \prod_{k=0}^{j-1} \tau_{k+1,k} \, ,$$

is called the net reproductive number for the age-structured population.

To define the net reproductive number for the general linear model (2), two assumptions are made about the transition and fertility matrices $\mathbf{T}$ and $\mathbf{F}$.

First, we suppose that the inverse exists:

$$\mathbf{E} = (\mathbf{I} - \mathbf{T})^{-1}. \tag{5}$$

If we denote $\mathbf{E} = [e_{ij}]$, then $e_{ij}$ is the expected amount of time that an individual starting initially in class $j$ will spend in class $i$ during its lifetime.

A sufficient condition for the matrix $\mathbf{I} - \mathbf{T}$ to have an inverse is that the following inequalities hold:

$$\sum_{i=1}^{m} \tau_{ij} < 1 \quad \text{for all} \quad j = 1, 2, \dots, m \tag{6}$$

(i.e., the sum of the column entries in $\mathbf{T}$ must be strictly less than 1). Since $\sum_{i=1}^{m} \tau_{ij}$ is the total fraction of individuals leaving class $j$ in one unit of time, it is clear that this sum must satisfy the condition that $\sum_{i=1}^{m} \tau_{ij} \leq 1$. Therefore, the biological meaning of assumption (6) is that there is always some loss of individuals from every class over each unit of time (due, for example, to deaths).

Assumption (6) is sufficient for the existence of the inverse $\mathbf{E} = (\mathbf{I} - \mathbf{T})^{-1}$, but it is not necessary. This can be seen from the case of a Leslie matrix (1) for which

$$\mathbf{F} = \begin{bmatrix} f_{11} & f_{12} & \cdots & f_{1,m-1} & f_{1m} \\ 0 & 0 & \cdots & 0 & 0 \\ 0 & 0 & \cdots & 0 & 0 \\ \vdots & \vdots & \ddots & \vdots & \vdots \\ 0 & 0 & \cdots & 0 & 0 \end{bmatrix},$$

$$\mathbf{T} = \begin{bmatrix} 0 & 0 & \cdots & 0 & 0 \\ \tau_{21} & 0 & \cdots & 0 & 0 \\ 0 & \tau_{32} & \cdots & 0 & 0 \\ \vdots & \vdots & \ddots & \vdots & \vdots \\ 0 & 0 & \cdots & \tau_{m,m-1} & 0 \end{bmatrix}.$$

In this case, we find that

$$
\mathbf{I} - \mathbf{T} = \begin{bmatrix}
1 & 0 & \cdots & 0 & 0 \\
-\tau_{21} & 1 & \cdots & 0 & 0 \\
0 & -\tau_{32} & \cdots & 0 & 0 \\
\vdots & \vdots & \ddots & \vdots & \vdots \\
0 & 0 & \cdots & -\tau_{m,m-1} & 1
\end{bmatrix}
$$

is always invertible (its determinant equals 1), even if inequality (6) does not hold. In fact, the inverse of $\mathbf{I} - \mathbf{T}$ here is

$$
\mathbf{E} = [e_{ij}], \quad e_{ij} = \prod_{k=0}^{i-1} \tau_{k+1,k} \quad \text{for all} \quad j
$$

(a matrix with identical columns).

Returning to the general case, the second assumption is that there is a strictly dominant positive eigenvalue for the matrix defined by

$$
\mathbf{R} = \mathbf{EF}. \tag{7}
$$

Because $\mathbf{T} \geq 0$ and $(\mathbf{I} - \mathbf{T})^{-1} = \mathbf{I} + \mathbf{T} + \mathbf{T}^2 + \cdots$, it follows that $\mathbf{E} = (\mathbf{I} - \mathbf{T})^{-1} \geq 0$. Therefore, since $\mathbf{F} \geq 0$ by assumption (3), it follows that the product $\mathbf{R} = \mathbf{EF}$ is also nonnegative. Thus, assumption (7) holds if, for example, $\mathbf{R}$ is irreducible and primitive.

As an example, for a Leslie matrix, we find that

$$
\mathbf{R} = [r_{ij}], \quad r_{ij} = f_{1j} \prod_{k=0}^{i-1} \tau_{k+1,k}.
$$

Since every column of this matrix is a multiple of the same column, it follows that 0 is an eigenvalue of multiplicity $m - 1$. The remaining eigenvalue is easily calculated to be the net reproductive number, namely,

$$
\sum_{j=1}^{m} f_{1j} \prod_{k=0}^{j-1} \tau_{k+1,k}.
$$

This eigenvalue of $\mathbf{R}$ (not of $\mathbf{A}$!) is nonzero, and hence strictly dominant, if not all $f_{1j}$ are zero, that is, if at least one age class is fertile.

It can be shown in general for any matrix $\mathbf{A} = \mathbf{F} + \mathbf{T}$ satisfying assumptions (5) and (7) that the dominant eigenvalue of the matrix $\mathbf{R} = (\mathbf{I} - \mathbf{T})^{-1}\mathbf{F}$ is equal to the expected number of offspring per individual over its lifetime (Cushing & Zhou 1995). Given this biological meaning for $\mathbf{R}$, we make the following definition.

**Definition 1** *Consider the nonnegative matrix* $\mathbf{A} = \mathbf{F} + \mathbf{T}$, *where* $\mathbf{T}$ *and* $\mathbf{F}$ *satisfy conditions (3) and (4). If assumptions (5) and (7) hold, then the net reproductive number* $n$ *associated with the matrix equation (2) is defined to be the strictly dominant eigenvalue of the matrix* $\mathbf{R} = (\mathbf{I} - \mathbf{T})^{-1}\mathbf{F}$.

Assuming that projection matrix $\mathbf{A}$ in (2) has a strictly dominant eigenvalue $\lambda_1$, in addition to satisfying assumptions (4) and (7), what is the connection between the net reproductive number $n$ and the dominant eigenvalue $\lambda_1$? In partial answer to this question, it has been shown that

$$\lambda_1 > 1 \quad (\text{or } < 1 \text{ or } = 1)$$

if and only if

$$n > 1 \quad (\text{ or } < 1 \text{ or } = 1, \text{ respectively})$$

(Cushing & Zhou 1995). It follows that the trivial equilibrium is asymptotically stable if $n < 1$ and is unstable if $n > 1$ and that positive equilibria exist if and only if $n = 1$. Given the biological meaning of the net reproductive number, $n$, these statements are biologically reasonable.

Both the geometric growth (or decay) rate of the population, $\lambda_1$, and the net reproductive number, $n$, determine the stability properties of the trivial equilibrium (i.e., determine whether a population whose dynamics are described by the linear model (2) survives or becomes extinct). From a mathematical standpoint, it can sometimes be more convenient for certain kinds of questions concerning the dynamics of the population to use and study $n$ rather than $\lambda_1$. This is because an analytical formula for $n$ is often available when no such formula for $\lambda_1$ exists.

Here is an example. Consider an Usher (1972) matrix, for which

$$\mathbf{T} = \begin{bmatrix} \tau_{11} & 0 & \cdots & 0 & 0 \\ \tau_{21} & \tau_{22} & \cdots & 0 & 0 \\ \vdots & \vdots & \ddots & \vdots & \vdots \\ 0 & 0 & \cdots & \tau_{m,m-1} & \tau_{mm} \end{bmatrix},$$

$$\mathbf{F} = \begin{bmatrix} f_{11} & f_{12} & \cdots & f_{1,m-1} & f_{1m} \\ 0 & 0 & \cdots & 0 & 0 \\ \vdots & \vdots & \ddots & \vdots & \vdots \\ 0 & 0 & \cdots & 0 & 0 \end{bmatrix},$$

and hence, if $\tau_{ii} < 1$, the inverse $\mathbf{E} = (\mathbf{I} - \mathbf{T})^{-1}$ is a matrix with identical columns,

$$\mathbf{E} = [e_{ij}], \quad e_{ij} = \prod_{k=0}^{i-1} \frac{\tau_{k+1,k}}{1 - \tau_{k+1,k+1}} \quad \text{for all} \quad j .$$

It follows that

$$\mathbf{R} = \mathbf{EF} = [r_{ij}], \quad r_{ij} = f_{1j} \prod_{k=0}^{i-1} \frac{\tau_{k+1,k}}{1 - \tau_{k+1,k+1}} .$$

(Again, for notational convenience, $\tau_{10} = 1$.) Since all the columns of $\mathbf{R}$ are multiples of the same column, this matrix has 0 as an eigenvalue of multiplicity $m - 1$. The remaining eigenvalue is calculated to be

$$n = \sum_{j=1}^{m} f_{1j} \prod_{k=0}^{j-1} \frac{\tau_{k+1,k}}{1 - \tau_{k+1,k+1}} ,$$

which is positive, and hence dominant, provided that not all $f_{1j}$ are zero (i.e., there must be at least one fertile class). This is a generalization of the formula for the Leslie matrix. It may be used to study, for example, the sensitivities of the net reproductive number to changes in these vital parameters.

As a numerical example, consider the matrix

$$\mathbf{A} = \begin{bmatrix} 0.72 & 0 & 0 & 0.74 & 1.04 & 9.03 \\ 0.28 & 0.69 & 0 & 0 & 0 & 0 \\ 0 & 0.31 & 0.75 & 0 & 0 & 0 \\ 0 & 0 & 0.25 & 0.77 & 0 & 0 \\ 0 & 0 & 0 & 0.23 & 0.63 & 0 \\ 0 & 0 & 0 & 0 & 0.37 & 0 \end{bmatrix}$$

used by Usher in a study of the dynamics of a Scots pine forest (Usher 1966). The formula for $n$ above easily yields a net reproductive rate of 15.1 (to three significant digits). Thus, the trivial equilibrium is unstable, and this population is growing geometrically. It is not as easy to find the dominant eigenvalue, which is a root of the sixth-degree characteristic polynomial. (Numerically, we find that $\lambda_1 \approx 1.20$.)

One way to study or answer questions about the stability properties of linear model (2) is by means of the net reproductive number, $n$. All of the entries in the projection matrix $\mathbf{A}$, of course, contribute to the value of $n$ in a complicated way. It would be more convenient if $n$ appeared explicitly in the dynamic equation (2).

One way to introduce $n$ explicitly into the equation is to normalize the fertilities in the matrix $\mathbf{F}$ as follows. Suppose that the assumptions (5) and (7) for the transition and fertility matrices $\mathbf{T}$ and $\mathbf{F}$ hold, so that $n$ is well defined. Then define $\beta_{ij}$ to be the ratio of the class-specific fertility, $f_{ij}$, to the net reproductive rate, $n$. That is, define $\beta_{ij}$ such that $f_{ij} = n\beta_{ij}$. Then, $\mathbf{F} = n\mathbf{N}$, where these normalized fertilities are described by the nonnegative matrix $\mathbf{N} = [\beta_{ij}] \geq 0$. The linear matrix equation (2) then becomes

$$\mathbf{x}(t+1) = (n\mathbf{N} + \mathbf{T})\,\mathbf{x}(t), \quad t = 0, 1, 2, \ldots, \tag{8}$$

where now it is true that the matrix $\mathbf{Q} = (\mathbf{I} - \mathbf{T})^{-1}\mathbf{N}$ has a strictly dominant eigenvalue equal to 1.

As a simple example, the Leslie matrix (1) for the classical age-structured matrix equation can be rewritten

$$\mathbf{A} = n
\begin{bmatrix}
\beta_{11} & \beta_{12} & \cdots & \beta_{1,m-1} & \beta_{1m} \\
0 & 0 & \cdots & 0 & 0 \\
0 & 0 & \cdots & 0 & 0 \\
\vdots & \vdots & \ddots & \vdots & \vdots \\
0 & 0 & \cdots & 0 & 0
\end{bmatrix}$$

$$+
\begin{bmatrix}
0 & 0 & \cdots & 0 & 0 \\
\tau_{21} & 0 & \cdots & 0 & 0 \\
0 & \tau_{32} & \cdots & 0 & 0 \\
\vdots & \vdots & \ddots & \vdots & \vdots \\
0 & 0 & \cdots & \tau_{m,m-1} & 0
\end{bmatrix},$$

where the normalized fertilities $\beta_{ij}$ and the survival probabilities $\tau_{j+1,j}$ are assumed to satisfy the condition that

$$\sum_{j=1}^{m} \beta_{1j} \prod_{k=0}^{j-1} \tau_{k+1,k} = 1.$$

Given the changes in the stability properties of the trivial equilibrium $\mathbf{x} = \mathbf{0}$ of the general linear matrix equation (8) and hence in the fate of the population as $n$ increases through the critical value of 1, the point $(n, \mathbf{x}) = (1, \mathbf{0})$ is called a "bifurcation point." One way to represent these facts schematically or graphically is by means of a so-called bifurcation diagram in which some measure of the magnitude of equilibria is plotted against the "bifurcation parameter" $n$. For example, the sum of the components, that is, the total population size, could be plotted against $n$. Such a dia-

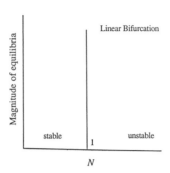

FIGURE 1. *If the magnitudes of the positive equilibria of general linear matrix equation (9) are plotted against the net reproductive number, n, then a vertical line is obtained at the critical value n = 1; at this line, the trivial equilibrium, $\mathbf{x} = \mathbf{0}$ (which exists for all n), loses its stability. This is because positive equilibria exist if and only if n = 1; there are, in fact, infinitely many positive equilibria given by $\mathbf{x} = c\mathbf{v}_1$ (c > 0). The graph of the trivial equilibrium coincides with the n-axis, and hence the intersection of these two continuous branches of equilibria where $(n, \mathbf{x}) = (1, \mathbf{0})$ constitutes a bifurcation.*

gram, for the linear matrix equation (2), consists of the vertical straight line obtained from the nontrivial equilibria at $n = 1$ and the horizontal straight line of trivial equilibria for all $n$ (which lies on the $n$-axis in the diagram and therefore cannot be seen). See Figure 1.

Furthermore, if the matrix $\mathbf{A} = \mathbf{F} + \mathbf{T}$ has a strictly dominant eigenvalue for all $n$ greater than 0 (for example, if it is irreducible and primitive for all $n$ greater than 0), then the trivial equilibrium is stable, the population becomes extinct for all $n$ less than 1, and the population grows geometrically without bound for all $n$ greater than 1. There is a loss of stability as $n$ increases through the bifurcation value, where $n = 1$. This is also indicated in the bifurcation diagram.

This bifurcation diagram is not the classical way to view and think about the dynamics of a linear matrix model (2). However, it is one way to view the dynamics implied by a linear model, since it depends on the vital parameters in its projection matrix. Moreover, this point of view lends itself very nicely to a generalization to nonlinear models, as we see in the next section.

## 2 Nonlinear Matrix Models

If the vital parameters appearing in the fertility matrix $\mathbf{F}$ and/or the transition matrix $\mathbf{T}$ are not all constants, but instead depend on population density, then projection matrix $\mathbf{A}$ in equation (2) depends on the entries in the distribution vector $\mathbf{x}$:

$$\mathbf{A}(\mathbf{x}) = \mathbf{F}(\mathbf{x}) + \mathbf{T}(\mathbf{x}). \tag{9}$$

The matrix equation for the dynamics of the distribution vector then becomes nonlinear:

$$\mathbf{x}(t+1) = \mathbf{A}[\mathbf{x}(t)]\,\mathbf{x}(t), \quad t = 0, 1, 2, \dots. \tag{10}$$

Here we assume that the entries in the projection matrix do not explicitly depend on time $t$; thus, this nonlinear matrix equation is autonomous. Because equation (10) is nonlinear, the asymptotic dynamics of its solutions as $t \to +\infty$ can be considerably more varied and complicated than that of the linear equation (2).

A first step toward understanding the dynamics of a nonlinear equation such as equation (10) is to determine its (nonnegative) equilibrium solutions and their stability properties. An equilibrium is a constant solution, $\mathbf{x}(t) = \mathbf{e} \in R^m$, of the algebraic equations

$$\mathbf{e} = \mathbf{A}(\mathbf{e})\mathbf{e}.$$

Clearly, $\mathbf{e} = \mathbf{0}$ is always an equilibrium of equation (10). The existence of other equilibria depends on the properties of the matrix $\mathbf{A}$ and presumably depends on the parameters in this matrix.

One way to study the equilibria of the nonlinear matrix equation (10) is to consider their existence and stability as a function of some selected parameter(s) in the projection matrix $\mathbf{A}$. The approach here, motivated by the approach to linear systems in the preceding section, is to define and introduce into the equation a single parameter $n$ for this purpose, namely, an "inherent" net reproductive number. Biologically, this number is the expected number of offspring per individual over its lifetime when the population level is low, that is, when density effects on vital rates are negligible. Once $n$ is introduced into the equation, bifurcation theory can be useful not only in organizing one's thinking about equilibria (and other asymptotic states), but in analyzing the dynamics of a general matrix equation. Certainly other approaches are possible, particularly in specific applications. However, one main advantage of the approach taken here is the generality of its applicability. It provides an overview of any particular application, in terms of a biologically meaningful parameter. Moreover, it provides analytical techniques that can yield information about equilibrium states for specific applications that might otherwise be quite intractable.

*Existence of Equilibria*

Consider the nonlinear matrix equation (10) with the projection matrix given by the additive decomposition in (9). The entries $\tau_{ij}(\mathbf{x})$ and $f_{ij}(\mathbf{x})$ in the transition and fertility matrices

$$\mathbf{T} = [\tau_{ij}(\mathbf{x})] \quad \text{and} \quad \mathbf{F} = [f_{ij}(\mathbf{x})]$$

are assumed to be twice-continuous differentiable functions of all $x_i$ on some open domain $D$ in $R_m$ that contains the origin, $\mathbf{x} = \mathbf{0}$. Thus, we assume that

$$\tau_{ij}, f_{ij} \in C^2(D; R^m),$$

$$0 \le \tau_{ij}(\mathbf{x}) \le 1, f_{ij}(\mathbf{x}) \ge 0 \quad \text{for all} \quad \mathbf{x} \in D. \tag{11}$$

Next, assume that the following inverse exists for all $\mathbf{x} \in D$:

$$\mathbf{E}(\mathbf{x}) = [\mathbf{I} - \mathbf{T}(\mathbf{x})]^{-1}. \tag{12}$$

Sufficient for this assumption to hold is that there is some loss at each time interval, in each class, at all population density levels; that is,

$$\sum_{i=1}^{m} \tau_{ij}(\mathbf{x}) < 1.$$

This inequality is not necessary for equation (12), however.

Suppose we define the matrix

$$\mathbf{R}(\mathbf{x}) = \mathbf{E}(\mathbf{x})\mathbf{F}(\mathbf{x}).$$

The final assumption is that

$R(\mathbf{0})$ has a strictly dominant, simple real eigenvalue $n > 0$, which is associated with a positive right eigenvector $\mathbf{v} > \mathbf{0}$ and a nonnegative left eigenvector $\mathbf{w} \ge \mathbf{0}$ ($\ne \mathbf{0}$). (13)

Without loss in generality we can assume that the scalar product $\langle \mathbf{v}, \mathbf{w} \rangle = 1$ (scalar products are defined in Chapter 2, by Caswell).

**Definition 2** *Consider the nonnegative matrix* $\mathbf{A}(\mathbf{x}) = \mathbf{F}(\mathbf{x}) + \mathbf{T}(\mathbf{x})$, *where* $\mathbf{T}(\mathbf{x})$ *and* $\mathbf{F}(\mathbf{x})$ *satisfy conditions (11). If assumptions (12) and (13) hold, then the inherent net reproductive number* $n$ *associated with matrix equation (2) is defined to be the strictly dominant eigenvalue of the matrix* $\mathbf{R}(\mathbf{0}) = [\mathbf{I} - \mathbf{T}(\mathbf{0})]^{-1} \mathbf{F}(\mathbf{0})$.

As for the linear equation, $n$ is introduced into the model equation (10) by normalizing the fertility rates $f_{ij}(\mathbf{x})$ relative to $n$. That is, the normalized fertility rates $\beta_{ij}(\mathbf{x})$ are defined by

$$f_{ij}(\mathbf{x}) = n\beta_{ij}(\mathbf{x})$$

and

$$\mathbf{F}(\mathbf{x}) = n\mathbf{N}(\mathbf{x}),$$

where

$$\mathbf{N}(\mathbf{x}) = [\beta_{ij}(\mathbf{x})] \geq 0.$$

This, together with assumption (13), implies that

$$\mathbf{Q}(\mathbf{0}) \quad \text{has dominant eigenvalue } 1, \tag{14}$$

where

$$\mathbf{Q}(\mathbf{x}) = [\mathbf{I} - \mathbf{T}(\mathbf{x})]^{-1} \mathbf{N}(\mathbf{x}) \geq 0.$$

With the explicit introduction of the net reproductive number, $n$, into the nonlinear matrix equation (10) in this manner, the projection matrix $\mathbf{A} = \mathbf{A}(\mathbf{x}, n)$ can be considered a function of $n$ as well as of $\mathbf{x}$, where

$$\mathbf{A}(\mathbf{x}, n) = n\mathbf{N}(\mathbf{x}) + \mathbf{T}(\mathbf{x}) = [a_{ij}(\mathbf{x}, n)].$$

Matrix equation (10) becomes

$$\mathbf{x}(t+1) = [n\mathbf{N}(\mathbf{x}) + \mathbf{T}(\mathbf{x})]\mathbf{x}(t), \quad t = 0, 1, 2, \ldots, \tag{15}$$

and the equilibrium equation

$$\mathbf{e} = [n\mathbf{N}(\mathbf{e}) + \mathbf{T}(\mathbf{e})]\mathbf{e}$$

can be rewritten as

$$\mathbf{e} = n[\mathbf{I} - \mathbf{T}(\mathbf{e})]^{-1}\mathbf{N}(\mathbf{e})\mathbf{e}$$

or

$$\mathbf{e} = n\mathbf{Q}(\mathbf{e})\mathbf{e}. \tag{16}$$

From this form of the equilibrium equations, the following general existence result for equilibria can be proved (Cushing 1995; see also Cushing 1988a).

**Theorem 1** *Under assumptions (11), (12), and (13), the matrix equation (10) has a "global" continuum of nontrivial "equilibrium pairs," $[n, \mathbf{x}(t)] = (n, \mathbf{e})$. At least near the bifurcation point, the equilibrium pairs $(n, \mathbf{e}) \neq (1, \mathbf{0})$ from the continuum are positive; that is, $\mathbf{e} > \mathbf{0}$.*

By "global" in this theorem is meant that either the continuum contains a pair $(n_0, \mathbf{0})$ where $n_0$ ($\neq 1$) is a characteristic value (i.e., the reciprocal of an eigenvalue) of $\mathbf{Q}(\mathbf{0})$ or the continuum connects to the boundary of $(0, +\infty) \times D$. This is referred to as the "Rabinowitz alternative" (Rabinowitz 1971). Incidentally, it is a fundamental fact of bifurcation theory that the only candidates for bifurcation points $(n, \mathbf{0})$ for nontrivial equilibria that bifurcate from the trivial equilibrium, $\mathbf{e} = \mathbf{0}$, are the characteristic values for $n$ of $\mathbf{Q}(\mathbf{0})$.

In the special case of linear equations, let $\mathbf{v}_1$ be the right eigenvalue of $\mathbf{A}$ corresponding to the dominant eigenvalue $\lambda_1$. The bifurcating continuum is just the set of equilibrium pairs $(n, \mathbf{e}) = (1, c\mathbf{v}_1)$ for positive constants $c$ greater than 0. Thus, the bifurcation of positive equilibria in this case is "vertical"; that is, the "spectrum" of $n$ values along the continuum consists of a single point where $n = 1$, leading to the vertical straight line in the bifurcation graph in Figure 1. For nonlinear equations, however, the graph of the bifurcating continuum is not in general vertical, and the spectrum of $n$ values is some interval containing $n = 1$. See Figure 2.

Although Theorem 1 does guarantee the existence of positive equilibria for nonlinear matrix equations under general circumstances, in specific applications one would like to know much more about the bifurcating branch. Are all of the equilibria on the continuum positive? What is the "spectrum" of $n$ values along the bifurcating continuum (i.e., for what values of $n$ does there exist a positive equilibrium)? For a given value of $n$ is there a unique positive equilibrium? Or can there exist several positive equilibria for some $n$? Are the equilibria from the continuum bounded or unbounded as a function of $n$? And, most important, what are the stability properties of the equilibria?

Detailed answers to these questions depend on the specific details of the matrix equation and its nonlinearities. Some useful things, however, can be said under rather general conditions.

In applications, the first part of the Rabinowitz alternative can often and easily be ruled out by the nature of the model equations. In any case, it is clear from the equilibrium equation (16) that the

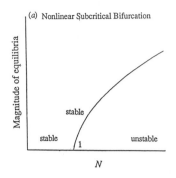
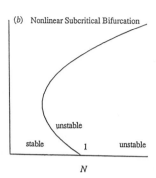

FIGURE 2. *If the magnitudes of the positive equilibria of general nonlinear matrix equation (16), whose existence is guaranteed by Theorem 1, are plotted against the inherent net reproductive number, $n$, then a continuous branch is obtained that bifurcates from the trivial equilibrium, $\mathbf{x} = \mathbf{0}$, at the critical value $n = 1$, where the trivial equilibrium loses stability. In this nonlinear case, the bifurcation is in general not vertical and is either* (a) *"supercritical" or* (b) *"subcritical." In the supercritical case, the bifurcating positive equilibria are (locally asymptotically) stable, at least for $n$ near 1. In the subcritical case, the bifurcating positive equilibria are unstable for $n$ close to 1.*

only equilibrium when $n = 0$ is $\mathbf{e} = \mathbf{0}$; consequently, the spectrum of $n$ values along the continuum guaranteed by Theorem 1 must be positive (negative values of $n$ make no biological sense, of course). Should the first alternative occur, however, it is known that $n_0$ must be of odd geometric multiplicity (Rabinowitz 1971).

With regard to the second part of the Rabinowitz alternative, in applications the domain $D$ under assumption (11) on which the problem is posed is usually unbounded, and in fact $D = R^m$. In this case, "connects to the boundary" means that the continuum of equilibrium pairs $(n, \mathbf{e})$ is unbounded in $R \times R^m$. This means that either the spectrum of $n$ values from the continuum is unbounded or the set of equilibrium $\mathbf{e}$ from the continuum is unbounded (or both).

If $q(\mathbf{e})$ is the eigenvalue of the matrix $\mathbf{Q}(\mathbf{e})$ for which $q(\mathbf{0}) = 1$, then $nq(\mathbf{e})$ is an eigenvalue of the matrix $\mathbf{R}(\mathbf{e}) = n\mathbf{Q}(\mathbf{e})$. The equilibrium form (16) of equation (15) implies that if $\mathbf{e} \neq \mathbf{0}$ is an equilibrium, then

$$nq(\mathbf{e}) = 1. \tag{17}$$

In principle, this scalar equation defines $\mathbf{e}$ as a function of $n$ (actually, it more conveniently defines $n = 1/q(\mathbf{e})$ as a function of $\mathbf{e}$). This equation can often be used in applications to derive properties of the bifurcating continuum of equilibria; see the examples below. Biologically, equation (17) states that the net reproductive rate at equilibrium necessarily equals 1.

In population models, of course, we would like to know if all nontrivial equilibria from the bifurcating continuum are positive (or, more mathematically, if the equilibria from the continuum remain in the positive cone of $R^m$). If they are not, the continuum would have to contain a pair $(n, \mathbf{e})$ with $n$ greater than 0 and a nonnegative, nontrivial equilibrium $\mathbf{e}$ that contains at least one zero component. This can often be ruled out by an examination of the projection matrix $\mathbf{A}$. In most applications, the only possible nonnegative, nontrivial equilibria are positive, and the second Rabinowitz alternative is ruled out. Such is the case, for example, if $\mathbf{Q}(0)$ has no positive real characteristic value other than when $n = 1$; or if it does, then eigenvectors associated with any other characteristic value are negative (which is true if $\mathbf{Q}(0)$ is irreducible; see Gantmacher 1960, p. 63).

Consider models in which all newborn individuals belong to class $i = 1$; each newborn has a nonzero probability of eventually reaching every class ($\tau_{i,i+1}(\mathbf{x}) > 0$ for $i \leq m-1$); and at least one class is inherently fertile ($f_{1i}(0) > 0$ for some $i$). For these kinds of model populations, it can be shown that the bifurcating continuum of equilibrium pairs is unbounded in $R_+ \times R_+$ and that all nontrivial equilibria along the continuum are positive (Cushing 1995).

An important example is the nonlinear matrix for which

$$
\mathbf{N}(\mathbf{x}) =
\begin{bmatrix}
\beta_1(\mathbf{x}) & \beta_2(\mathbf{x}) & \cdots & \beta_{m-1}(\mathbf{x}) & \beta_m(\mathbf{x}) \\
0 & 0 & \cdots & 0 & 0 \\
0 & 0 & \cdots & 0 & 0 \\
\vdots & \vdots & \ddots & \vdots & \vdots \\
0 & 0 & \cdots & 0 & 0
\end{bmatrix},
$$

$$
\mathbf{T}(\mathbf{x}) =
\begin{bmatrix}
0 & 0 & \cdots & 0 & 0 \\
\tau_1(\mathbf{x}) & 0 & \cdots & 0 & 0 \\
0 & \tau_2(\mathbf{x}) & \cdots & 0 & 0 \\
\vdots & \vdots & \ddots & \vdots & \vdots \\
0 & 0 & \cdots & \tau_{m-1}(\mathbf{x}) & \tau_m(\mathbf{x})
\end{bmatrix}, \quad (18)
$$

where there is a nonzero probability of surviving each of the first $m - 1$ time units (i.e., $\tau_i(\mathbf{x}) > 0$ for $1 \leq i \leq m - 1$) and where the

$m$th age class is fertile (i.e., $f_m(0) > 0$). Equation (18) describes a Leslie matrix if $\tau_m = 0$. In this case,

$$[\mathbf{I} - \mathbf{T}(\mathbf{x})]^{-1} =$$

$$
\begin{bmatrix}
1 & 0 & \cdots & 0 & 0 \\
\tau_1(\mathbf{x}) & 1 & \cdots & 0 & 0 \\
\tau_1(\mathbf{x})\tau_2(\mathbf{x}) & \tau_1(\mathbf{x}) & \cdots & 0 & 0 \\
\vdots & \vdots & \ddots & \vdots & \vdots \\
\dfrac{\prod_{j=1}^{m-1}\tau_j(\mathbf{x})}{1-\tau_m(\mathbf{x})} & \dfrac{\prod_{j=2}^{m-1}\tau_j(\mathbf{x})}{1-\tau_m(\mathbf{x})} & \cdots & \dfrac{\tau_{m-1}(\mathbf{x})}{1-\tau_m(\mathbf{x})} & \dfrac{1}{1-\tau_m(\mathbf{x})}
\end{bmatrix}
$$

exists and consequently assumption (12) holds. A calculation shows that

$$\mathbf{Q}(\mathbf{x}) = [\beta_1(\mathbf{x})\mathbf{v}(\mathbf{x}) \quad \beta_2(\mathbf{x})\mathbf{v}(\mathbf{x}) \quad \cdots \quad \beta_m(\mathbf{x})\mathbf{v}(\mathbf{x})] ,$$

where $\mathbf{v}(\mathbf{x})$ is the column vector

$$
\mathbf{v}(\mathbf{x}) = \begin{bmatrix}
1 \\
\tau_1(\mathbf{x}) \\
\tau_1(\mathbf{x})\tau_2(\mathbf{x}) \\
\vdots \\
\dfrac{\prod_{j=1}^{m-1}\tau_j(\mathbf{x})}{1-\tau_m(\mathbf{x})}
\end{bmatrix} > \mathbf{0} .
$$

Since all of the columns of $\mathbf{Q}(\mathbf{x})$ are multiples of the same positive vector $\mathbf{v}(\mathbf{x})$, the number 0 is an eigenvalue of multiplicity $m - 1$. The remaining eigenvalue, which is positive and simple, is given by the formula

$$q(\mathbf{x}) = \beta_1(\mathbf{x}) + \sum_{i=2}^{m-1} \beta_i(\mathbf{x}) \prod_{j=1}^{i-1} \tau_j(\mathbf{x}) + \beta_m(\mathbf{x})\frac{\prod_{j=1}^{m-1}\tau_j(\mathbf{x})}{1 - \tau_m(\mathbf{x})} , \quad (19)$$

where the normalization of the $\beta_i$ implies that $q(0) = 1$. As above, positive equilibria must satisfy equation (17); that is,

$$n\left[\beta_1(\mathbf{x}) + \sum_{i=2}^{m-1} \beta_i(\mathbf{x}) \prod_{j=1}^{i-1} \tau_j(\mathbf{x}) + \beta_m(\mathbf{x})\frac{\prod_{j=1}^{m-1}\tau_j(\mathbf{x})}{1 - \tau_m(\mathbf{x})}\right] = 1 .$$

The right and left eigenvectors of $\mathbf{Q}(0)$ (or equivalently $\mathbf{R}(0)$) in assumption (13) are

$$\mathbf{v} = \mathbf{v}(0) > 0$$

and

$$0 \neq \mathbf{w} = \begin{bmatrix} \beta_1(\mathbf{0}) \\ \vdots \\ \beta_i(\mathbf{0}) \\ \vdots \\ \beta_m(\mathbf{0}) \end{bmatrix} \geq 0.$$

Moreover, the bifurcating continuum of equilibrium pairs is unbounded in $R_+ \times R_+$, and all nontrivial equilibria lying on the continuum are positive.

It is atypical but possible for more than one positive equilibrium to be associated with a value for $n$. In this case, the bifurcation graph has "turns" in it, as drawn in Figure 2. (Moreover, the theorem does not claim that all positive equilibria must lie on the bifurcating continuum.) More is said about this possibility below.

*Stability of Equilibria*

The "linearization principle" is valid for equation (10). This principle states that the local stability properties of a hyperbolic equilibrium $\mathbf{e}$ are determined by those of the linearization of equation (10) at the equilibrium. (An equilibrium is called "hyperbolic" if none of the eigenvalues of $\mathbf{J}(\mathbf{e})$, $\lambda_i = \lambda_i(\mathbf{e})$, lies on the unit circle in the complex plane; i.e., if all $|\lambda_i(\mathbf{e})| \neq 1$.) The linearization at $\mathbf{e}$ is the linear matrix equation whose projection matrix is the Jacobian of the right-hand side of equation (10) evaluated at $\mathbf{e}$. If we denote the right-hand side of (10) by $\mathbf{f}(\mathbf{x}) = [f_i(\mathbf{x})]$, the Jacobian is the matrix $\mathbf{J}(\mathbf{x}) = [\partial f_i(\mathbf{x})/\partial x_j]$. The linearization principle implies that $\mathbf{e}$ is (locally asymptotically) stable if all of the eigenvalues $\lambda_i(\mathbf{e})$ lie inside the unit circle (i.e., $|\lambda_i(\mathbf{e})| < 1$ for all $i$) and that $\mathbf{e}$ is unstable if at least one of the eigenvalues lies outside the unit circle (i.e., $|\lambda_i(\mathbf{e})| > 1$ for at least one $i$).

"Locally asymptotically stable" means that $\mathbf{e}$ is stable and also that orbits that start close enough to $\mathbf{e}$ actually approach $\mathbf{e}$ as $t \to +\infty$. "Stable" means that when $\varepsilon > 0$, there is a $\delta$ greater than 0 such that $|\mathbf{x}(0)| < \delta$, implying that $|\mathbf{x}(t)| < \varepsilon$ $(t > 0)$. "Unstable" means "not stable."

Suppose that assumptions (11), (12), and (13) apply for equation (10) written in form (15), and consider the equilibrium $\mathbf{e} = \mathbf{0}$, for which the Jacobian matrix is $\mathbf{J}(\mathbf{0}) = \mathbf{A}(\mathbf{0})$. Following the preceding section, the equilibrium $\mathbf{e} = \mathbf{0}$ is (locally asymptotically) stable if $n < 1$ and is unstable if $n > 1$.

We say that the equilibrium $\mathbf{e} = \mathbf{0}$ "loses stability" as $n$ increases through the critical value $n = 1$. This fact has the same biological interpretation here, at least at low population densities, as it does for linear equations. Namely, if the expected number of offspring per individual over its lifetime is less than one (less than replacement), when population densities are low, then populations that are initially at a low level will become extinct. If this expected number is larger than one, however, then low-level populations will survive. Unlike the linear case, however, it may *not* be true when $n < 1$ that *all* populations decline to extinction.

Notice that in view of Theorem 1, the point $n = 1$ at which the equilibrium $\mathbf{e} = \mathbf{0}$ loses stability is a bifurcation point for nonzero equilibria. At this point the equilibrium $\mathbf{e} = \mathbf{0}$ is nonhyperbolic. The parameter values that yield nonhyperbolic equilibria are important because qualitative changes in asymptotic dynamics can occur at (and only at) such parameter values, often resulting in the appearance of new asymptotic states.

It was easy to see how the (local) stability of equilibrium $\mathbf{e} = \mathbf{0}$ depends on the inherent net reproductive number, $n$. In contrast, the stability properties of positive equilibria, especially those on the bifurcating continuum (as guaranteed by Theorem 1) are much more difficult to ascertain. These properties depend closely on the nonlinearities in the equation. However, one general result can be proved under the assumption that

$$1 \text{ is a strictly dominant eigenvalue of } \mathbf{N}(\mathbf{0}) + \mathbf{T}(\mathbf{0}). \qquad (20)$$

Note that $\mathbf{N}(\mathbf{0}) + \mathbf{T}(\mathbf{0})$ is just the projection matrix of (15) evaluated at the bifurcation point $(n, \mathbf{e}) = (1, \mathbf{0})$. That $n = 1$ is an eigenvalue of $\mathbf{N}(\mathbf{0}) + \mathbf{T}(\mathbf{0})$ follows from assumption (14).

With the addition of assumption (20), it can be shown that the stability properties of the positive equilibria lying on the bifurcating continuum depend, at least in the neighborhood of the bifurcation point $(n, \mathbf{e}) = (1, \mathbf{0})$, on the "direction of bifurcation" (Cushing 1995). Specifically, if the bifurcating pairs $(n, \mathbf{e})$ near $(1, \mathbf{0})$ exist when $n > 1$, then the equilibria $\mathbf{e} > \mathbf{0}$ are (locally asymptotically) stable; on the other hand, if these pairs exist when $n < 1$, then the equilibria $\mathbf{e} > \mathbf{0}$ are unstable. In the first case the bifurcation is said to be "to the right" (or "supercritical" or "stable"), and in the second case it is said to be "to the left" (or "subcritical" or "unstable"). See Figure 2.

Furthermore, the direction of bifurcation can be determined by a formula derived from so-called Liapunov-Schmidt expansions of the equilibrium branch. The result is that

$$\begin{cases} \text{the bifurcation is to the left (i.e., stable or supercritical)} \\ \text{if } n_1 > 0, \\ \\ \text{the bifurcation is to the right (subcritical or unstable)} \\ \text{if } n_1 < 0, \end{cases}$$

where

$$n_1 = -\left\langle \mathbf{w}, \left[\nabla_{\mathbf{x}}\langle a_{ij}(0,1), \mathbf{v}\rangle\right] \mathbf{v} \right\rangle,$$

$$\nabla_{\mathbf{x}} a_{ij}(\mathbf{x}, n) = \begin{bmatrix} \partial a_{ij}(\mathbf{x}, n)/\partial x_1 \\ \vdots \\ \partial a_{ij}(\mathbf{x}, n)/\partial x_m \end{bmatrix}$$

and $\mathbf{v}$ and $\mathbf{w}$ are the right and left eigenvectors of $\mathbf{R}(\mathbf{0})$ (see eq. 13). This formula allows the direction of bifurcation, and hence the local stability properties of the positive equilibria near the bifurcation point, to be calculated from the nonlinear entries in the projection matrix $\mathbf{A}(\mathbf{x}, n)$, at least "generically" (i.e., in most cases) when $n_1 \neq 0$. Note that for the linear case, when $\mathbf{A}$ is independent of $\mathbf{x}$ and the bifurcation is vertical (i.e., is neither to the left nor to the right), $n_1$ is zero.

A simplified formula holds when all newborns belong to the same class, say, to class $i = 1$ (without loss in generality). Then, there is a positive constant $c$ such that

$$n_1 = -c\langle \nabla_{\mathbf{x}} q(\mathbf{0}), \mathbf{v} \rangle, \qquad (21)$$

where $q(\mathbf{x})$ is given by equation (19) (Cushing 1988*a*, 1995).

This simplified formula is valid, for example, for the nonlinear model (18). The nonlinear Leslie model satisfies requirement (20), sufficient for stability to be determined by the direction of bifurcation, if 1 is strictly dominant as an eigenvalue of the constant Leslie matrix

$$\mathbf{N}(\mathbf{0}) + \mathbf{T}(\mathbf{0}) = \begin{bmatrix} \beta_1(\mathbf{0}) & \beta_2(\mathbf{0}) & \cdots & \beta_{m-1}(\mathbf{0}) & \beta_m(\mathbf{0}) \\ \tau_1(\mathbf{0}) & 0 & \cdots & 0 & 0 \\ 0 & \tau_2(\mathbf{0}) & \cdots & 0 & 0 \\ \vdots & \vdots & \ddots & \vdots & \vdots \\ 0 & 0 & \cdots & \tau_{m-1}(\mathbf{0}) & \tau_m(\mathbf{0}) \end{bmatrix}.$$

Necessary and sufficient conditions for this to hold are known (Impagliazzo 1980). A simple sufficient condition is that there are two adjacent age classes that are fertile. For example, $\beta_{m-1}(\mathbf{0}) > 0$ is sufficient, since it has been assumed all along that $\beta_m(\mathbf{0}) > 0$.

In the general case, since

$$\frac{\partial a_{ij}(\mathbf{0}, 1)}{\partial x_k} = \frac{\partial \beta_{ij}(\mathbf{0})}{\partial x_k} + \frac{\partial \tau_{ij}(\mathbf{0})}{\partial x_k},$$

it is the nonlinear dependence of the fertilities $\beta_{ij}$ (or, equivalently, $f_{ij}$) and the transition probabilities $\tau_{ij}$ that determine the direction of bifurcation and hence the local equilibrium stability near the point of bifurcation. The most common modeling assumptions in population-dynamics models, that fertility and survival are adversely affected by increased population density (and consequently that both $\beta_{ij}$ and $\tau_{ij}$, and hence $a_{ij}$, are decreasing in each component $x_k$), imply that $n_1 > 0$. Thus, the usual situation is that a stable (supercritical) bifurcation of positive equilibria occurs as $n$ increases through 1.

There are interesting cases, however, when unstable bifurcations to the left occur. Necessary for such an occurrence is that not all of the partial derivatives $\partial \beta_{ij}(\mathbf{0}, 1)/\partial x_k$, $\partial \tau_{ij}(\mathbf{0}, 1)/\partial x_k$ be nonnegative. One case is that of a so-called "Allee effect" (or "depensation"). Such an effect is present when the opposite of the usual regulatory density effects occurs on at least one of the vital rates, usually at low population densities only. Thus, at low population densities, birthrates might *increase* and/or death rates might *decrease* with increased population density for at least some age classes (see the application below; Allee 1931; Cushing 1988*b*; Dennis 1993). If these reversed effects are significant enough at *low* population densities, then $n_1$ can be negative and the bifurcating positive equilibria will be unstable, at least near the bifurcation point.

Normally it is assumed, however, that such Allee effects on the vital rates do not persist at *high* population densities, at which the usual adverse effects of density on the vital rates are assumed to hold. It turns out that this causes the bifurcating branch to "turn around" as in Figure 2*b*. Usually in this case, although not always, the "top" of the branch (i.e., the positive equilibria of larger magnitude) consists of stable equilibria. There are several things to note in this case. First of all, stable positive equilibria exist for values of the inherent net reproductive number, $n$, less than 1. This is in marked contrast to linear models, in which the population always becomes extinct when $n < 1$. Second, there are multiple equilibria and multiple stable equilibria when $n < 1$. Third, there is the potential for a *sudden* collapse of the population as $n$ decreases (unlike the case of a stable bifurcation when there is a continuous decrease to the trivial solution as $n$ decreases). Finally, a hysteresis effect is caused by the existence of multiple stable equilibria.

Another interesting case that can result in an unstable bifurcation is that of cannibalism (see Cushing 1991$a$; for an example of unstable bifurcation due to complicated nonlinear interactions and stage structure, see Cushing 1988$a$).

In a general setting, the loss of stability of the trivial equilibrium at a critical value, $n = 1$, results in the bifurcation of a global continuum of positive equilibria from the trivial equilibrium. Under suitable assumptions, the stability properties of these bifurcating equilibria near the bifurcation point are determined by the direction of bifurcation. However, it is not true that the stability properties of the positive equilibria near the bifurcation point necessarily persist globally along the continuum. Stable equilibria that result from a supercritical bifurcation can lose their stability farther along the continuum. Unstable equilibria that result from a subcritical bifurcation can gain stability farther along the continuum. Such changes in stability often result in the bifurcation of other kinds of stable attractors. The "generic" possibilities are described in the next subsection.

*Loss of Stability of a Positive Equilibrium*

The stability of a positive equilibrium, $\mathbf{x} = \mathbf{e} > \mathbf{0}$, of the nonlinear matrix equation (10) is determined by the eigenvalues $\lambda(\mathbf{e})$ of the Jacobian matrix. Loss of stability occurs if a change in a model parameter causes one or more of these eigenvalues to move outside of the unit circle. A critical parameter value, then, is one that places an eigenvalue exactly on the unit circle, that is, that causes $|\lambda(\mathbf{e})|$ to be equal to 1 for at least one eigenvalue. Such an eigenvalue has the complex form $\lambda_i(\mathbf{e}) = \exp[i\theta(\mathbf{e})]$ for some polar angle $\theta(\mathbf{e}) \in [0, 2\pi)$. Conversely, if the equilibrium $\mathbf{e}$ is unstable because an eigenvalue is outside the unit circle and a change in a model parameter causes this eigenvalue (and all other eigenvalues) to be inside the unit circle, then the equilibrium has been stabilized. Once again the critical parameter value is the one that places the eigenvalue exactly on the unit circle.

It turns out that the nature of the bifurcation that takes place at a critical parameter value, where a change in equilibrium stability occurs, depends on this polar angle $\theta(\mathbf{e})$ of the eigenvalue lying on the unit circle, that is, on exactly where the eigenvalue crosses the unit circle as it moves from the inside to the outside (or vice versa). The simplest occurrences, and the ones best understood,

are those in which there is either a single real eigenvalue at $\pm 1$
with the remaining $n - 1$ eigenvalues lying inside the unit circle
or a single complex conjugate pair lying on the unit circle with
the remaining $n - 2$ eigenvalues lying inside. Described below are
only the most commonly occurring types of bifurcations that can
occur in population models in these simplest cases (for a more
complete discussion, see Lauwerier 1986; Wiggins 1990). Relevant
discussions appear in the chapters by Caswell (2), Nisbet (4), and
de Roos (5).

*Eigenvalue +1.* Consider the case when an eigenvalue associated
with the linearization at an equilibrium **e** assumes the value $+1$
at a critical value of $n$ in the nonlinear matrix equation (15). The
bifurcation that occurs at such a critical point is associated, in one
way or another, with the existence of multiple equilibria.

For example, an eigenvalue of 1 occurs in the linearization of
equation (15) at $\mathbf{e} = \mathbf{0}$ when $n = 1$. This is an example of a
"transcritical bifurcation." Transcritical bifurcations involve the
intersection of two different branches of equilibria in such a way
that two equilibria exist for values of $n$ on both sides of the critical
value 1. For equation (15), at the point $(\mathbf{0}, 1)$, the branch consisting
of the trivial equilibria (which exists for all $n$) intersects a branch
of nontrivial equilibria $(\mathbf{e}, n)$ that exists for $n$ greater than and $n$
less than 1. Only one side, however, consists of positive equilibria,
$\mathbf{e} > \mathbf{0}$ (the side where $n > 1$ in the case of a stable supercritical
bifurcation and the side where $n < 1$ in the case of an unstable
subcritical bifurcation), and therefore the equilibria from the other
side of the intersecting branch in the discussion above are ignored.
It typically happens at a transcritical bifurcation that an "exchange
of stability" occurs, by which is meant that the branch containing
the stable equilibria for $n$ less than 1 contains unstable equilibria
for $n$ greater than 1, and vice versa (the stability having been
passed from one branch to the other as $n$ passes through 1).

As seen above, general nonlinear matrix models of the form (10)
possess a transcritical bifurcation and an exchange of stability at
the trivial equilibrium when the inherent net reproductive rate
increases through the critical value $n = 1$. Positive equilibria of
equation (15) can also experience bifurcations as $n$ changes. Besides transcritical bifurcations (which are not common for positive
equilibria of models of the form in eq. 15), "saddle-node" bifurcations can occur at positive equilibria when an eigenvalue of the
linearization equals $+1$ at a critical value of $n$. In a saddle-node
bifurcation, two different positive equilibria exist for $n$ on one side

of the critical value, move together as $n$ equals the critical value, and then disappear for $n$ on the other side of the critical value.

An example is illustrated in Figure 2$b$, where a subcritically bifurcating branch of positive equilibria "turns around" at the critical value when $n < 1$; a saddle-node bifurcation occurs at the "turn-around" point (the "nose" of the graph in Fig. 2$b$). Generally, in a saddle-node bifurcation one of the equilibria is stable (a node) and the other unstable (a saddle).

Another type of equilibrium bifurcation that can occur is called a "pitchfork" bifurcation. In this case a positive equilibrium loses stability at a critical value of $n$ on one side of which there exist two more positive equilibria. An exchange of stability takes place between these two equilibrium branches.

It is not always easy to determine, analytically, the kind of equilibrium bifurcation that occurs at an eigenvalue of $+1$. Formulas sufficient to determine the types above are known for two-dimensional models (see Wiggins 1990). For higher-dimensional cases, however, an additional complicated calculation (called a "center manifold calculation") is required in order to use these formulas. Numerical computer simulations are often useful in determining the kinds of bifurcations that occur.

*Eigenvalue $-1$.*   Suppose that a positive equilibrium of equation (15) loses stability at a critical value of $n$ where the linearization has an eigenvalue equal to $-1$ (all remaining $n-1$ eigenvalues remaining inside of the complex unit circle). In this case there exist, for $n$ on one side of the critical value or the other, non-equilibrium periodic solutions of equation (15) of period 2, so-called "2-cycles." The amplitude of these cycles grows from zero at the critical value of $n$; the 2-cycles either "pop out of" or "shrink into" the equilibrium as $n$ increases through the critical bifurcation value. This kind of bifurcation is called a "period-doubling" bifurcation.

When a period-doubling bifurcation occurs, there exist both a positive equilibrium and a 2-cycle (actually two 2-cycles, since a shift of a 2-cycle is still a 2-cycle) for $n$ on one side of the critical value or the other. However, the equilibria and the 2-cycles do not possess the same stability properties; if one is stable, the other is unstable (so another exchange of stability can be said to occur). In this sense the stability of the bifurcating 2-cycles is determined by the direction of bifurcation.

For example, if the positive equilibrium loses stability as $n$ increases through the critical value, then the 2-cycles are unstable if they exist for $n$ less than the critical value (a subcritical bifurcation) and stable if they exist for $n$ greater than the critical value (a supercritical bifurcation). Formulas exist for the calculation of a constant whose sign determines the direction of bifurcation and the stability of the 2-cycles; however, this calculation is notoriously complicated and not useful for examining parameter changes. Again, evidence of stable 2-cycles is usually gained through numerical simulations.

*Complex pair of eigenvalues of magnitude 1.* For one-dimensional difference equations, loss of stability can occur only in one of the two cases above. For higher-dimensional models, which are central to structured-population dynamics, other possibilities arise because the linearization at an equilibrium can possess complex eigenvalues, which leave the unit circle at points other than $\pm 1$.

In the simplest case, at the critical value of $n$ there is a pair of complex conjugate eigenvalues of magnitude 1,

$$\lambda(\mathbf{e}) = e^{\pm i\theta(\mathbf{e})}, \quad \theta(\mathbf{e}) \neq 0 \quad \text{or} \quad \pi,$$

while all the remaining eigenvalues lie inside the unit circle. For technical reasons, we assume that

$$\theta(\mathbf{e}) \neq 2\pi/3, \ 4\pi/3, \ \pi/4, \ 3\pi/4.$$

That is to say, we assume that

$$\lambda^k \neq 1 \quad \text{for} \quad k = 1, 2, 3, 4.$$

When these equalities hold, the eigenvalues are said to satisfy a "resonance" condition, and the resulting bifurcations are not well understood (Lauwerier 1986; Wiggins 1990). We exclude resonances because they are not "generic" (typical).

The result of a complex pair moving across the unit circle (but not at one of the first four roots of unity) is a "Naimark-Sacker" bifurcation to an invariant loop (sometimes called a "Hopf bifurcation"). An "invariant loop" is a closed curve $L$ in $m$-dimensional space with the property that if an orbit starts on $L$ then it forever stays on $L$. That is, if $x(0)$ lies on $L$, then the solution $x(t)$ of equation (15) lies on $L$ for all nonnegative $t$. This theorem states that such invariant loops exist for values of $n$ on one side of the critical value or the other. They "pop out of" the positive equilibrium at the critical value of $n$, in the sense that their amplitude shrinks to zero as $n$ passes through the critical value. Moreover, as in the case of a period doubling, the positive equilibrium and the loop are not "stable" at the same time, and hence, the "stability" of the bifurcating loops depends on the direction of bifurcation.

(The loop is called "stable" if all nearby orbits tend to the loop in $m$-dimensional space as $t \to +\infty$.) The direction of bifurcation can in principle be determined by the calculation of a certain quantity, but in practice this is tedious and often intractable. The plotting of two components of $x(t)$ by means of computer simulations can usually indicate the presence of stable invariant loops.

The use of the word "stable" with regard to the bifurcating invariant loops must be properly understood. Unlike the bifurcation results of the preceding subsections, the Naimark-Sacker bifurcation to an invariant loop says something about the geometry of the orbits (plots of solutions $x(t)$ in $m$-dimensional space) but nothing about the dynamics of solutions around the loop, which can be exactly periodic or, more typically, totally aperiodic. In the latter case, time series (i.e., plots of components of solutions $x(t)$) oscillate in a nearly periodic fashion, although their oscillation often looks irregular or "chaos-like" (but they are not technically chaotic in the sense of being sensitively dependent upon initial conditions; Wiggins 1990).

So far, the focus has been on the bifurcations that can occur when positive equilibrium states of equation (15) lose stability as the inherent net reproductive number, $n$, increases. We have seen that, when this happens, new kinds of "stable" entities come into existence, namely, new positive equilibria or oscillatory solutions (2-cycles or those on an invariant loop). This is certainly not the end of the story, however. As $n$ increases farther, these nonequilibrium states can lose their stability, and new kinds of bifurcations can occur. Probably the most familiar is the well-known period-doubling sequence of bifurcations that occurs when 2-cycles lose stability and stable 4-cycles come into existence, which then lose stability to stable 8-cycles, and so on. Often this cascade occurs until $n$ increases past a critical value, after which "chaos" occurs. The famous one-dimensional discrete logistic equation is the prototypical example. Stable invariant circles can lose their stability to chaos as $n$ increases as well. In this case, chaos is often preceded, however, by regions of "period locking," in which cycles of certain periods occur (and often period double).

The implications of complicated bifurcations and dynamics in discrete (and continuous) models and their relevance to biological populations are controversial. For example, claims that certain population data can be explained by the dynamics of simple difference-equation models are often poorly supported. Problems here include inaccurate or insufficient data sets, inadequate statistical tests for parameter estimation and model verification, use of simplistic models with little biological relevance, and neglect of stochastic effects. However, these issues are increasingly being addressed by more-careful studies, for which a thorough understanding of the deterministic dynamics of the model is certainly prerequisite.

## 3 Some Examples

This section presents some applications that illustrate how the analysis of nonlinear models can be carried out from the point of view of bifurcation theory.

The first application illustrates a typical, supercritical bifurcation of stable positive equilibria as $n$ increases through $n = 1$. In this example, as $n$ increases, the bifurcating continuum of positive equilibria ultimately loses stability, leading either to a bifurcation to stable 2-cycles or to a stable invariant loop, depending upon model parameter values.

The second application illustrates the possibility of a subcritical (unstable) bifurcation of positive equilibria where $n = 1$. This example also illustrates a saddle-node bifurcation at a subcritical value of the inherent net reproductive rate. This leads to an interval of $n$ values in which multiple stable equilibrium states exist and hysteresis effects occur.

The last application provides an example in which the fundamental result relating the direction of bifurcation to the stability of the bifurcating positive equilibria fails to hold. This example illustrates the necessity of assumption (20) for relating the direction of bifurcation to stability. It also shows that even when our general results fail, the bifurcation-theory viewpoint remains helpful. The manner in which the assumptions fail to hold often provides a clue as to what is happening at the bifurcation point and thus what analytical steps to take next. In this application, this clue ultimately leads to a full understanding of the dynamics at the primary bifurcation point.

### A Model for Flour Beetle Dynamics

Discrete models have been extensively used to help understand the complicated population dynamics of flour beetles of the genus *Tribolium* (Sokoloff 1974; Costantino & Desharnais 1991). The simplified model below tries to capture one important behavior of most species of *Tribolium*, the cannibalism of eggs and pupae by adults and larvae (see also the discussion by Botsford, Chapter 12). It is a simplification and precursor of a more complicated model used by Costantino et al. (1995) and Dennis et al. (1995).

In this model are two classes, adults and immatures. The fertility and transition matrices are given by

$$\mathbf{F} = \mathbf{F}(x_1, x_2) = \begin{bmatrix} 0 & be^{-c_E x_2} \\ 0 & 0 \end{bmatrix},$$

$$\mathbf{T} = \mathbf{T}(x_1, x_2) = \begin{bmatrix} 0 & 0 \\ e^{-c_P x_2} & 1 - \mu \end{bmatrix}.$$

Here the exponentials account for losses due to cannibalism, with the nonnegative coefficients $c_E$ and $c_P$ (not both 0) corresponding to cannibalism on eggs and pupae, respectively. The parameter $b$ is the immature recruitment rate in the absence of cannibalism, and $\mu$ is the adult mortality rate; $\mu \in (0, 1)$. The unit of time equals the maturation period.

Clearly, basic properties (11) hold on the whole $(x_1, x_2)$ plane where $D = R^2$, and the inverse exists for all $x_1$ and $x_2$:

$$\mathbf{E}(x_1, x_2) = \begin{bmatrix} 1 & 0 \\ e^{-c_P x_2}/\mu & 1/\mu \end{bmatrix}.$$

The matrix

$$\mathbf{R}(0,0) = \begin{bmatrix} 0 & b \\ 0 & b/\mu \end{bmatrix}$$

satisfies assumption (13) with

$$n = b/\mu, \quad \mathbf{v} = \begin{bmatrix} b \\ b/\mu \end{bmatrix}, \quad \mathbf{w} = \begin{bmatrix} 0 \\ \mu/b \end{bmatrix}.$$

This leads to the matrices

$$\mathbf{N}(x_1, x_2) = \begin{bmatrix} 0 & \mu e^{-c_E x_2} \\ 0 & 0 \end{bmatrix}, \quad \mathbf{Q}(x_1, x_2) = \begin{bmatrix} 0 & \mu e^{-c_E x_2} \\ 0 & e^{-(c_E + c_P)x_2} \end{bmatrix},$$

and to the matrix

$$\mathbf{N}(0,0) + \mathbf{T}(0,0) = \begin{bmatrix} 0 & \mu \\ 1 & 1 - \mu \end{bmatrix},$$

which has 1 as its strictly dominant eigenvalue. Hence, assumption (20) also holds.

We conclude that there exists an unbounded branch of positive equilibrium pairs that bifurcates from the trivial equilibrium at the critical value where $n = 1$ $(b/\mu = 1)$. Since the nonlinearities are decreasing functions of $x_2$, it follows that $n_1 > 1$ and that the bifurcation is supercritical and stable. (The quantity $n_1$ can also be explicitly calculated by the formula above, and whereby it is found that $n_1 = c_P b$.)

Equation (17) gives the relationship

$$n \exp[-(c_E + c_P)A_e] = 1$$

between the positive equilibria (specifically, the equilibrium adult number, $A_e > 0$) and the inherent net reproductive number, $n$. This scalar equation shows that the bifurcating continuum of positive equilibria is a logarithmic function of $n$. Thus, this fact supports the deductions that the bifurcation is supercritical, that the spectrum of $n$ values corresponding to positive equilibria is infinite (specifically, the half-line where $n > 1$), that the magnitudes of positive equilibria are unbounded as a function of $n$ greater than 1, and that for each $n$ greater than 1 there is a unique positive equilibrium (the equilibrium number of immatures is $I_e = n\mu \exp(-c_E A_e)$).

Stability does not persist along the continuum of positive equilibria as $n$ increases, however. It can be shown that a 2-cycle bifurcation can occur and an invariant loop bifurcation can occur, depending upon parameter values. The critical value of $n$ $(n_{cr} > 1)$ at which this loss of stability occurs can be calculated by computing the linearization of the equations at the positive equilibria and determining when the resulting $2 \times 2$ matrix has eigenvalues on the complex unit circle. This leads to the formula

$$n_{cr} = \begin{cases} e^{(1+1/\mu)(1+r)} & \text{when } r < r_{cr}, \\ e^{2(1-1/\mu)(1+r)/(1-r)} & \text{when } r > r_{cr}, \end{cases} \quad (22)$$

where

$$r = c_P/c_E \quad \text{and} \quad r_{cr} = (3 - \mu)/(1 + \mu).$$

A 2-cycle bifurcation occurs, where $n = n_{cr}$, when $r > r_{cr}$; and an invariant loop bifurcation occurs, where $n = n_{cr}$, when $r < r_{cr}$. See Figure 3 for a numerical example of the latter case.

## A Model with Allee Effect

We have previously pointed out that when increased population density reduces vital rates (as in the preceding example), we obtain a supercritical stable bifurcation at $n = 1$ in the model (10). A reversal of this relationship between population density and at

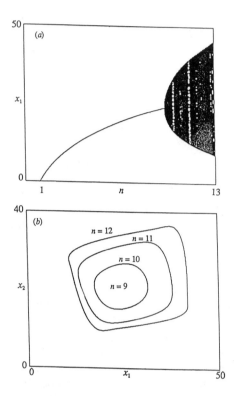

FIGURE 3. (a) *The bifurcation diagram for the* Tribolium *model when* $c_e = 0.10$, $c_p = 0.01$, *and* $\mu = 0.95$. *The (positive) asymptotic attractor is shown plotted against the inherent net reproductive number, n. The supercritical (and hence stable) bifurcation of equilibria is clearly seen occurring when $n = 1$. The loss of stability of these positive equilibria followed by a bifurcation to stable invariant loops is also seen to occur at larger values of n.* (b) *The attractors are shown in the phase plane generated when $n = 9$, at which a stable equilibrium occurs, and at $n = 10, 11$, and 12, at which stable invariant loops are seen. According to equation (22), a Naimark-Sacker bifurcation occurs at $n_{cr}$, just above 9.56.*

least one vital rate (fertility, survival, growth rates, etc.), at low population densities, may lead to a subcritical and unstable bifurcation. Allee effects or "depensation" effects are of this type (Allee 1931; Clark 1976; Dennis 1989).

To illustrate, consider an age-structured population, equation (18), in which fertility rates are density-dependent but survival rates are not. The fertility rates are assumed to be proportional to the rate $u$ at which a limited resource is consumed, and this consumption or uptake rate $u$ depends on population density. Specifically,

$$u = u[W(t)] > 0\,,$$

where

$$W(t) = \sum_{i=1}^{m} w_i x_i(t)$$

is a weighted total population size, $\sum_{i=1}^{m} w_i = 1$, and $w_i \geq 0$. The Allee effect is modeled by the assumption that

$$u'(0) > 0\,. \tag{23}$$

This means that as low population levels *increase* (i.e., as the weighted total population size $W$ increases from near zero), per capita resource consumption, and hence fertility, *also increases.*

However, we also assume that for high population densities, per capita resource consumption decreases with increased population density; specifically,

$$\lim_{W \to +\infty} u(W) = 0\,. \tag{24}$$

In the nonlinear Leslie matrix model (18), write

$$\beta_i(W) = b_i u(W)$$

(i.e., $f_i = n\beta_i(W) = nb_i u(W)$), yielding

$$q = q(W) = u(W) \left( b_1 + \sum_{i=2}^{m-1} b_i \prod_{j=1}^{i-1} \tau_j + b_m \frac{\prod_{j=1}^{m-1} \tau_j}{1 - \tau_m} \right)\,.$$

Assuming that $u(W)$ and the coefficients $b_i$ are normalized by $u(0) = 1$,

$$b_1 + \sum_{i=2}^{m-1} b_i \prod_{j=1}^{i-1} \tau_j + b_m \frac{\prod_{j=1}^{m-1} \tau_j}{1 - \tau_m} = 1\,,$$

such that $q(0) = 1$ (i.e., in the model, $n$ really equals the inherent net reproductive number). Thus, equation (17) becomes

$$nu(W) = 1. \tag{25}$$

From the general results, if $b_m > 0$, a global, unbounded continuum of positive equilibrium pairs $(n, \mathbf{x})$ bifurcates from $(1, \mathbf{0})$. The direction of bifurcation is determined by the sign of $n_1$ (if it is nonzero). Using (21) and the expression for $\mathbf{v}$ given after (18), the sign of $n_1$ is the same as the sign of

$$-\nabla_{\mathbf{x}} q(\mathbf{0}) \mathbf{v} = -u'(0) \sum_{i=1}^{m} \mathbf{w}_i \mathbf{v}_i$$

$$= -u'(0) \left( w_1 + \sum_{i=2}^{m-1} w_i \prod_{j=1}^{i-1} \tau_j + w_m \frac{\prod_{j=1}^{m-1} \tau_j}{1 - \tau_m} \right).$$

In other words, $n_1$ has the opposite sign from $u'(0) > 0$, and consequently, the direction of bifurcation is to the left (subcritical).

Under the added assumption that 1 is a strictly dominant eigenvalue of the Leslie matrix

$$\mathbf{N}(0) + \mathbf{T}(0) = \begin{bmatrix} b_1 & b_2 & \cdots & b_{m-1} & b_m \\ \tau_1 & 0 & \cdots & 0 & 0 \\ 0 & \tau_2 & \cdots & 0 & 0 \\ \vdots & \vdots & \ddots & \vdots & \vdots \\ 0 & 0 & \cdots & \tau_{m-1} & \tau_m \end{bmatrix},$$

the stability of the positive bifurcating equilibria near the bifurcation point is determined by the direction of bifurcation.

This example illustrates how the Allee assumption, inequality (23), causes an unstable primary bifurcation of equilibria at the critical value where $n = 1$.

An easier way to see that the bifurcation in this example is subcritical is to note that the bifurcating continuum of positive equilibria can be described mathematically by the pairs $(n, W)$, where $W > 0$, satisfying equation (25). Clearly, as $W$ increases from 0, $u'(0) > 0$, implying that $n$ must *decrease*. Moreover, as $W \to +\infty$ and hence $u(W) \to 0$, it follows from equation (25) that $n$ must also tend to $+\infty$. This means, among other things, that the bifurcating branch "turns around," as shown in Figure 2$b$. At this turning point there occurs a saddle-node bifurcation of positive equilibria, and one can usually expect that the "larger" equilibria lying on the upper part of the branch are stable, while those on the lower are unstable. This does not always occur, however, but when it does, one sees an interval of inherent net reproductive numbers less than 1 for which there exist two stable equilibrium states, the stable positive equilibrium and the trivial or extinction equilibrium. This in turn leads to hysteresis effects. (For a simple model displaying these phenomena, see Cushing 1988$b$.)

## A Size-Structured Model for Density-Dependent Growth

The following example (from Cushing & Li 1992) shows the importance of assumption (20) in drawing the conclusion that stability is related to the direction of bifurcation. In this example, this assumption fails to hold, although assumptions (11), (12), and (13) do hold and hence the existence of the bifurcation continuum of positive equilibria is ensured. In fact, in this example a supercritical but *un*stable bifurcation can occur. One point of this example is that even when the assumptions and conclusions of our bifurcation theorem fail to hold, this approach can give insights into the dynamics and suggest other avenues of analysis.

One question concerning intraspecific competition that has received considerable attention deals with competition between juvenile and adult members of a single population and whether such an interaction is a "stabilizing" or "destabilizing" influence on the population's dynamics. Generally, such competition of juveniles versus adults is considered destabilizing, although there can be exceptions and a great deal depends on how the notion of "destabilization" is measured. (Studies of this problem based upon discrete dynamic model equations can be found in Ebenman 1987, 1988$a,b$; Cushing & Li 1989, 1992; Loreau 1990. Studies using continuous models appear in May et al. 1974; Tschumy 1982; Cushing 1991$b$; Cushing & Li 1991.)

There are many ways of modeling competition between juveniles and adults. Most published models assume that competition reduces age-class survival rates and/or adult fertility rate. Many contributors to the book by Ebenman and Persson (1988) argued that body size, rather than chronological age, is the key individual variable. In this case, competition can significantly affect an individual's growth rate, size at maturation, etc.

Ebenman (1988$b$) studied a size-structured model for juvenile-versus-adult competition. His model is not analytically tractable, however, and he relied heavily on computer simulations. A simpler model is considered here that focuses on competition effects on the size at maturation and hence on adult fertility rates (Cushing & Li 1992).

Consider a population of juveniles and two adult size classes, one consisting of smaller and less fertile adults and the other of larger and more-fertile adults. It is assumed that after one unit of time a juvenile matures but that its size at maturity depends on the amount of competition (for, say, a limited food resource) that it experienced as a juvenile. Adults do not change size after maturity. The amount or intensity of competition experienced during maturation is assumed to be a function of the weighted population size,

$$W = x_1 + \gamma_1 x_2 + \gamma_2 x_3\,,$$

where the coefficients $\gamma_i$ measure the competitive effects of the adults on juvenile growth (relative to the self effects of juveniles). Density effects on survival and on adult fertility (except through adult size at maturation as described above) are ignored. If $x_1$ denotes the density of juveniles and $x_2$ and $x_3$ denote the densities of the smaller and larger adult classes, respectively, then the model is described by the difference equations

$$x_1(t+1) = b_1 x_2(t) + b_2 x_3(t)\,,$$
$$x_2(t+1) = \varphi[W(t)]\, x_1(t)\,,$$
$$x_3(t+1) = \{1 - \varphi[W(t)]\}\, x_1(t)\,,$$

where $b_2 > b_1 > 0$, $\varphi \in C(R_+, [0,1))$, and $\varphi'(W) > 0$ (Cushing & Li 1992). The inequalities $b_2 > b_1 > 0$ express the fact that larger individuals are more fertile. The function $\varphi(W)$, the fraction of juveniles growing to smaller size, incorporates the density dependence of juvenile growth. The mathematical assumption that this fraction is monotonically increasing as a function of $W$ reflects the biological assumption that, as the population density $W$ increases, a larger fraction of juveniles grow to the smaller adult size. (Here, in order to reduce the number of model parameters, juveniles are measured in units required to produce adults, considering a constant survivability over one unit of time. Thus, no mortality coefficient is needed in the equations for $x_2$ and $x_3$.)

This model has the form of equation (15) with

$$\mathbf{F} = \mathbf{F}(W) = \begin{bmatrix} 0 & b_1 & b_2 \\ 0 & 0 & 0 \\ 0 & 0 & 0 \end{bmatrix}\,,$$

$$\mathbf{T} = \mathbf{T}(W) = \begin{bmatrix} 0 & 0 & 0 \\ \varphi(W) & 0 & 0 \\ 1 - \varphi(W) & 0 & 0 \end{bmatrix} .$$

Assumptions (11) are fulfilled. There is an inverse,

$$\mathbf{E}(W) = [\mathbf{I} - \mathbf{T}(W)]^{-1} = \begin{bmatrix} 1 & 0 & 0 \\ \varphi(W) & 1 & 0 \\ 1 - \varphi(W) & 0 & 1 \end{bmatrix} ;$$

so assumption (12) is met. Regarding the final remaining assumption required for Theorem 1, the matrix

$$\mathbf{R}(0) = \mathbf{E}(0)\mathbf{F}(0) = \begin{bmatrix} 0 & b_1 & b_2 \\ 0 & b_1\varphi(0) & b_2\varphi(0) \\ 0 & b_1[1 - \varphi(0)] & b_2[1 - \varphi(0)] \end{bmatrix}$$

is seen to have a double eigenvalue 0 and the dominant simple eigenvalue

$$n = b_1\varphi(0) + b_2[1 - \varphi(0)] .$$

Writing the model equations in form (15) with

$$\mathbf{N}(W) = \begin{bmatrix} 0 & \beta_1 & \beta_2 \\ 0 & 0 & 0 \\ 0 & 0 & 0 \end{bmatrix} , \quad \mathbf{T}(W) = \begin{bmatrix} 0 & 0 & 0 \\ \varphi(W) & 0 & 0 \\ 1 - \varphi(W) & 0 & 0 \end{bmatrix} ,$$

where $\beta_i = b_i / [b_1\varphi(0) + b_2(1 - \varphi(0))]$, and applying Theorem 1, leads to an unbounded continuum of positive equilibria that bifurcates from the trivial equilibria at $n = 1$.

Moreover, a calculation shows that the quantity $n_1$ is a positive constant multiple of $\varphi'(0)(b_2 - b_1)$, and therefore, a supercritical bifurcation of positive equilibria occurs in this model system. For this to imply that the bifurcating positive equilibria are stable, requirement (20) must hold. However, the matrix

$$n\mathbf{N}(0) + \mathbf{T}(0) = \begin{bmatrix} 0 & n\beta_1 & n\beta_2 \\ \varphi(0) & 0 & 0 \\ 1 - \varphi(0) & 0 & 0 \end{bmatrix}$$

has eigenvalues 0, $-n$, and $+n$; consequently, $n = 1$ is not a dominant eigenvalue. As a result, assumption (20) fails to hold. This, of course, does not mean that the bifurcating equilibria are necessarily unstable, only that we do not know if the direction of bifurcation determines stability.

Since the eigenvalues of $n\mathbf{N}(0) + \mathbf{T}(0)$ lie inside the complex unit circle when $n < 1$ and outside the circle when $n > 1$, it follows that the trivial equilibrium loses stability at the bifurcation

point $(n, W) = (1, 0)$. However, when $n = 1$, this matrix has eigenvalue $-1$ in addition to $+1$, which suggests that in addition to a
(transcritical) bifurcation of the equilibrium branches there might
also occur a bifurcation of 2-cycles at $n = 1$! This is, in fact, true
(Cushing & Li 1992). When $n = 1$, there also bifurcates a global
branch of nonnegative "synchronous" 2-cycles (nonnegative, nontrivial 2-cycles in which juvenile and adult populations are never
simultaneously present).

Thus, when $n > 1$, there exists both a unique positive equilibrium and a unique nonnegative synchronous 2-cycle. At least near
the bifurcation point, one of these branches is stable and the other
is unstable, depending upon a "competition coefficient" defined by

$$\sigma = \gamma_1 \varphi(0) + \gamma_2 [1 - \varphi(0)]$$

(Cushing & Li 1992). If $\sigma < 1$, then the positive equilibrium branch
is stable and the synchronous 2-cycle branch is unstable; if $\sigma > 1$,
then the opposite is true.

It is interesting to note that in this model strong juvenile-versus-
adult competition implies that the population will stabilize into a
dynamic situation (a 2-cycle) in which adults and juveniles never
appear together. Thus, the population structure evolves, purely as
a result of the dynamic assumptions, to a state in which competition is altogether avoided, a kind of temporal niche.

## 4 Concluding Remarks

The familiar properties of linear matrix equations can be viewed
from the point of view of bifurcation theory using the net reproductive number, $n$, of the population as a bifurcation parameter.
From this perspective, the facts that the population dies out exponentially when $n < 1$, grows exponentially without bound when
$n > 1$, and possesses equilibrium states if and only if $n = 1$ can
be viewed as a continuum of (neutrally stable) equilibria vertically
bifurcating from the trivial equilibrium $\mathbf{x} = \mathbf{0}$ at the critical value,
$n = 1$, where the trivial equilibrium loses stability. Under rather
general circumstances, nonlinear matrix equations can in the same
way possess a continuum of equilibria that bifurcates from the equilibrium $\mathbf{x} = \mathbf{0}$ when $n = 1$, where $n$ is the inherent net reproductive
number. For nonlinear equations, the bifurcation is usually not vertical, and consequently, in contrast to linear equations, equilibria
exist for values of $n$ other than 1. Near the bifurcation point the

stability of the bifurcating equation depends on the direction of bifurcation.

The approach taken above to nonlinear matrix equations can also be applied to coupled systems of matrix equations as models for several interacting structured species. For such systems, the projection matrix for one species can depend on the density of the other species. General existence and stability results for equilibrium states can be obtained by techniques from bifurcation theory, by defining and using the inherent net reproductive number of one species, when its density is low and the remaining community is at equilibrium, as a bifurcation parameter. Stability results again depend on the direction of bifurcation. This approach can be viewed as the study of conditions under which a species can be added to or invade a community of other species as a function of its inherent net reproductive number. (For details, see Cushing 1995.) Moreover, the same approach can also be applied when the underlying community is not at a stable equilibrium but in a stable cycle. In this case, a bifurcating branch of stable cycles can be proved to exist, with stability again depending on the direction of bifurcation. These results have been used, for example, to show the cyclic coexistence of competing species under circumstances in which the principle of competitive exclusion (which is based upon equilibrium dynamics) would imply otherwise (Crowe 1991).

## Literature Cited

Allee, W. C. 1931. *Animal Aggregations.* University of Chicago Press.

Clark, C. W. 1976. *Mathematical Bioeconomics: The Optimal Management of Renewable Resources.* Wiley, New York.

Costantino, R. F., and R. A. Desharnais. 1991. *Population Dynamics and the 'Tribolium' Model: Genetics and Demography.* Monographs on Theoretical and Applied Genetics 13. Springer, Berlin.

Costantino, R. F., J. M. Cushing, B. Dennis, and R. A. Desharnais. 1995. Experimentally induced transitions in the dynamic behaviour of insect populations. *Nature* 375: 227–230.

Crowe, K. M. 1991. A discrete size-structured competition model. Ph.D. diss. University of Arizona, Tucson.

Cushing, J. M. 1988a. Nonlinear matrix models and population dynamics. *Natural Resource Modeling* 2(4): 539–580.

———. 1988b. The Allee effect in age-structured population dynamics. Pp. 479–505 *in* T. G. Hallam, L. J. Gross, and S. A. Levin, eds., *Mathematical Ecology.* World Scientific Press, Singapore.

———. 1991*a*. A simple model of cannibalism. *Mathematical Biosciences* 107(1): 47–72.

———. 1991*b*. Some delay models for juvenile vs. adult competition. Pp. 177–188 *in* S. Busenberg and M. Martelli, eds., *Differential Models in Biology, Epidemiology, and Ecology.* Springer-Verlag, Berlin.

———. 1995. Nonlinear matrix models for structured populations. Chapter 26 *in* O. Arino, D. Axelrod, M. Kimmel, and M. Langlais, eds., *Mathematical Population Dynamics III: Mathematical Methods and Modelling of Data.* Wuerz, Winnipeg.

Cushing, J. M., and J. Li. 1989. On Ebenman's model for the dynamics of a population with competing juveniles and adults. *Bulletin of Mathematical Biology* 51: 687–713.

———. 1991. Juvenile versus adult competition. *Journal of Mathematical Biology* 29: 457–473.

———. 1992. Intra-specific competition and density dependent juvenile growth. *Bulletin of Mathematical Biology* 54(4): 503–519.

Cushing, J. M., and Zhou Y. 1995. The net reproductive number and stability in linear structured population models. *Natural Resource Modeling* 8: 297–333.

Dennis, B. 1989. Allee effects: Population growth, critical density, and the chance of extinction. *Natural Resource Modeling* 3: 481–538.

Dennis, B., R. Desharnais, J. M. Cushing, and R. Costantino. 1995. Nonlinear demographic dynamics: Mathematical models, statistical methods, and biological experiments. *Ecological Monographs* 65: 261–281.

Ebenman, B. 1987. Niche difference between age classes and intraspecific competition in age-structured populations. *Journal of Theoretical Biology* 124: 25–33.

———. 1988*a*. Competition between age classes and population dynamics. *Journal of Theoretical Biology* 131: 389–400.

———. 1988*b*. Dynamics of age- and size-structured populations: Intraspecific competition. Pp. 127–139 *in* B. Ebenman and L. Persson, eds., *Size-Structured Populations: Ecology and Evolution.* Springer-Verlag, Berlin.

Ebenman, B., and L. Persson, eds. 1988. *Size-Structured Populations: Ecology and Evolution.* Springer-Verlag, Berlin.

Gantmacher, F. R. 1960. *The Theory of Matrices.* Vol. II. Chelsea, New York.

Impagliazzo, J. 1980. *Deterministic Aspects of Demography.* Springer, Berlin.

Lauwerier, H. A. 1986. Two-dimensional iterative maps. Pp. 58–95 *in* A. V. Holden, ed., *Chaos.* Princeton University Press, Princeton, N.J.

Loreau, M. 1990. Competition between age classes, and the stability of stage-structured populations: A re-examination of Ebenman's model. *Journal of Theoretical Biology* 144: 567–571.

May, R. M., G. R. Conway, M. P. Hassell, and T. R. E. Southwood. 1974. Time delays, density-dependence and single-species oscillations. *Journal of Animal Ecology* 43: 747–770.

Rabinowitz, P. H. 1971. Some global results for nonlinear eigenvalue problems. *Journal of Functional Analysis* 1(3): 487–513.

Sokoloff, A. 1974. *The Biology of 'Tribolium'*. Vol. 2. Oxford University Press.

Tschumy, W. O. 1982. Competition between juveniles and adults in age-structured populations. *Theoretical Population Biology* 21: 255–268.

Usher, M. B. 1966. A matrix approach to the management of renewable resources with special reference to forests. *Journal of Applied Ecology* 3: 355–367.

———. 1972. Developments in the Leslie matrix model. Pp. 29–60 *in* J. M. R. Jeffers, ed., *Mathematical Models in Ecology*. Blackwell, London.

Wiggins, S. 1990. *Introduction to Applied Nonlinear Dynamical Systems and Chaos*. Tests in Applied Mathematics, Vol. 2. Springer, Berlin.

PART II

# APPLICATIONS

CHAPTER 7

# The Relative "Importance" of Life-History Stages to Population Growth: Prospective and Retrospective Analyses

*Carol Horvitz, Douglas W. Schemske, and Hal Caswell*

Determining the importance of life-history events for population growth is a significant, if ill-defined, goal of population-dynamics research. Perturbation analyses, which explore the effects on population growth of changes in the vital rates, provide an approach to this problem. They have become a standard part of demographic practice. It is now rare to find a published report of population growth rate that does not investigate how that rate changes as the vital rates are perturbed, either actually (comparing different treatments, sites, species, etc.) or hypothetically (exploring the consequences of potential management strategies or of evolutionary changes). Applications include life-history theory (where it is important to know how the different vital rates influence fitness; see, e.g., Caswell & Werner 1978; Caswell 1985; Calvo & Horvitz 1990; Kalisz & McPeek 1992; Calvo 1993), conservation biology (where it is important to know how protecting different stages in the life cycle would affect population growth; see, e.g., Crouse et al. 1987; Menges 1990; Doak et al. 1994; Heppell et al. 1994; Schemske et al. 1994), ecotoxicology (where it is important to know how pollutants affect population growth; see, e.g., Caswell 1996a; Sibly 1996; Levin et al., in press), and assessing the accuracy of estimates of population growth rate (Lande 1988).

Unfortunately, as the use of demographic perturbation analyses has spread, so have some common misconceptions about them. Our goal in this chapter is to clarify some of the applications and interpretations of perturbation analyses. Much of the confusion stems from apparently reasonable but ambiguous questions, perhaps the most ill posed of which is, "which of the stages, or which of the vital rates, is most important to population growth?" The problem is that "importance" has several different meanings, each of which leads to a different perspective on the population.

It is useful to distinguish between *prospective* and *retrospective* answers to this question. Prospective analyses address the effects of potential future changes. Of all the changes in the vital rates, which would produce the biggest effect on population growth rate? No changes in any vital rates need have occurred to ask this question. It is even possible, and sometimes instructive, to ask about changes that are purely hypothetical ("if pigs had wings ..." or "if this species were to become a perennial instead of an annual ..."). Sensitivity and elasticity analyses (see Chapter 2 and Caswell 1978; Caswell et al. 1984; de Kroon et al. 1986) are prospective analyses.

A common misconception about prospective analyses is that a high sensitivity or elasticity for some parameter implies that an observed change in population growth rate was due to that parameter. This is a mistake; analyzing changes in population growth that have actually occurred is a retrospective, not a prospective, problem. Some change in the vital rates has actually occurred, leading to the change in population growth rate. We want to know how much of the change in population growth rate can be attributed to the changes in each of the vital rates. The vital rate that contributes the most to the variability in population growth is not necessarily the one to which population growth rate is most sensitive, nor the one that will make the biggest contribution to variability in population growth rate in another environment. Some vital rates to which population growth is very sensitive may never vary, and thus make no contribution to the variability.

The appropriate analytical tool for this question is the "decomposition analysis" of life-table-response experiments (Caswell 1989a, 1996a). If population growth rate is measured as a deviation from a reference value (e.g., a treatment relative to a control), then the treatment effect can be decomposed into contributions from each of the vital rates. If variability in population growth rate is expressed as a variance over a set of treatments, this variance can be decomposed into contributions from the variances and covariances of the vital rates.

Confusion about retrospective and prospective perturbation is analogous to some common misinterpretations of genetic-heritability analysis (Lewontin 1974). Heritability decomposes an observed phenotypic variance into genetic and environmental contributions. It is a retrospective calculation. High heritability does not mean that the trait is insensitive to the environment; that is a prospective conclusion. The trait might be, for example, sensitive to temperature, but the data were collected in a constant-temperature chamber. Similarly, low heritability does not mean that the trait is insensitive to changes in genetics. Perhaps the trait is affected by genotype, but the population studied was homozygous at the relevant loci.

We describe five analyses, two prospective and three retrospective, that address the question of how the vital rates affect population growth rate. Then we apply them to data on the Neotropical understory herb *Calathea ovandensis*. Comparison of the results sheds light on various aspects of the "importance" of the vital rates in this particularly well studied case.

## 1 Demographic Analyses

We use the linear time-invariant matrix model

$$\mathbf{n}(t+1) = \mathbf{A}\mathbf{n}(t)\,, \tag{1}$$

where $\mathbf{n}(t)$ is a vector giving the abundances of the stages in the population at time $t$, and $\mathbf{A}$ is the population-projection matrix, whose $ij$th entry $a_{ij}$ gives the contribution of an individual in stage $j$ to stage $i$ over one time step. See Caswell (Chapter 2, or 1989a) for details. The dominant eigenvalue $\lambda$ of the matrix gives the population growth rate. The associated right and left eigenvectors $\mathbf{w}$ and $\mathbf{v}$ give the stable stage distribution and the stage-specific reproductive values, respectively.

Sensitivity and elasticity formulas are derived in Chapter 2. The sensitivity of $\lambda$ to a change in the matrix entry $a_{ij}$ is given by

$$s_{ij} = \frac{\partial \lambda}{\partial a_{ij}} = \frac{v_i w_j}{\langle \mathbf{w}, \mathbf{v} \rangle}\,, \tag{2}$$

where $\langle \mathbf{w}, \mathbf{v} \rangle$ is the scalar product of $\mathbf{w}$ and $\mathbf{v}$. The elasticity, or proportional sensitivity, of $\lambda$ to a change in $a_{ij}$ is given by

$$e_{ij} = \frac{a_{ij}}{\lambda} \frac{\partial \lambda}{\partial a_{ij}}\,. \tag{3}$$

It is sometimes suggested that sensitivity or elasticity analysis is better, or more accurate, or less biased, or more biologically reasonable than the other. This is not so. The two give different pictures of the result of a perturbation. The pictures are each accurate, unbiased, and biologically meaningful. Population growth rate ($\lambda$) is a function of the entries of the population-projection matrix. Imagine plotting $\lambda$ as a function of one matrix, say $a_{ij}$, while holding the other entries constant. Plot $a_{ij}$ on the $x$-axis and $\lambda$ on the $y$-axis. The slope of the resulting curve is the sensitivity of $\lambda$ to changes in $a_{ij}$. If you were to repeat this operation for all the matrix entries, the result would be a set of curves with different slopes; those slopes would show how $\lambda$ responds to changes in all the matrix entries. Alternatively, imagine plotting $\lambda$ as a surface in a multidimensional space, as a function of all the $a_{ij}$. The sensitivities $s_{ij}$ give the gradient of this surface, showing in which directions it slopes steeply and in which directions it is relatively flat. This is the key to the evolutionary applications of sensitivities, which can be interpreted as selection gradients (Lande & Arnold 1983; Phillips & Arnold 1989).

Now, imagine the same exercise, but plot the logarithm of $\lambda$ as a function of the logarithm of $a_{ij}$. The slope of this line is the elasticity of $\lambda$ to changes in $a_{ij}$. The multidimensional analogue is the gradient of a surface that shows $\log \lambda$ as a function of the logarithm of the matrix entries. The elasticity $e_{ij} = 0$ if $a_{ij} = 0$. That is, if the wingspan of a pig increases by, say, 10 percent, there is no change in its flying ability. Transitions directly from newborn to large reproductive individuals (entries in the lower left corner of the matrix) usually have high sensitivities, even though such transitions may not occur. The sensitivity indicates correctly what *would* happen if that rate were increased from zero to a small positive number. The elasticity, however, is zero and can reveal nothing about such changes. This makes elasticity inappropriate for evolutionary questions such as asking how fitness would change if an annual were to become perennial, or seed dormancy were to be introduced where it does not exist, etc. The elasticity, in contrast, indicates correctly that increasing the matrix entry by some proportion has no effect on $\lambda$.

The elasticities sum to 1 across the whole matrix (Caswell 1986; de Kroon et al. 1986; Mesterton-Gibbons 1993) and can be interpreted as proportional contributions of the corresponding vital rates to $\lambda$ (see van Groenendael et al. 1994). There is some confusion between the correct statement "stage $x$ contributes $y$ percent to population growth rate" and the incorrect statement "stage $x$

explains $y$ percent of the *variation* in population growth rate," which cannot be determined from an elasticity analysis.

Is elasticity more meaningful than sensitivity? The choice of arithmetic or logarithmic axes is a matter not of right or wrong but of which graph reveals the aspects of the curve that are of interest. Logarithmic axes are useful because equal intervals correspond to equal proportions. If you are interested in patterns involving proportional changes in vital rates, use the slopes of the logarithmic plots (i.e., the elasticities). Equal intervals on arithmetic axes imply equal changes. If you are interested in rates of change on a linear scale (as evolutionary questions are, for example), then use sensitivities. Better yet, make both calculations and compare the results.

Remember that both sensitivity and elasticity are derivatives; they give the local slope of $\lambda$ (or $\log \lambda$) as a function of $a_{ij}$ (or $\log a_{ij}$). They may not accurately predict the result of large perturbations (although, in practice, they do a surprisingly good job). Using the second derivatives (Caswell 1996$b$) to take into account the curvature of $\lambda$ might improve the prediction. A more straightforward approach, however, is to vary parameters and calculate $\lambda$ numerically (see, e.g., Caswell & Werner 1978; Bierzychudek 1982; Horvitz & Schemske 1986; Martinez & Alvarez-Bullya 1986; Calvo 1993). For example, an effective but rare pollinator may increase plant reproductive success several-fold. A local-perturbation analysis near the usual low level of fruit production may not accurately predict the effect of the pollinator on $\lambda$, while a simulation that actually varies fruit set and calculates $\lambda$ will do so (Horvitz & Schemske, unpubl. data). Numerical perturbation analysis is also useful for investigating specific, biologically interpretable changes in several of the vital rates simultaneously. It should, however, be supplemented with analytical results.

## 2 Retrospective Analysis

A retrospective analysis begins with data on the vital rates and on $\lambda$, under two or more sets of environmental conditions. The goal of the analysis is to quantify the contribution of each of the vital rates to the variability in $\lambda$. A formal method for quantifying data of this kind has been called Life-Table-Response Experiments (LTRE's) (Caswell 1989$a$, $b$, 1996$a$, in press; see also Levin et al. 1987, in press; Walls et al. 1991; Brault & Caswell 1993). This usage defines "experiment" in a sense wide enough to include not only designed manipulations but also comparative observations.

Intuitively, variation in one matrix entry, say $a_{ij}$, makes a small contribution to the variation in $\lambda$ if $a_{ij}$ doesn't change much, or if $\lambda$ is not very sensitive to $a_{ij}$, or if both are true. Thus, the contribution of $a_{ij}$ to the variation in $\lambda$ involves the product of $s_{ij}$ and the observed variation in $a_{ij}$.

As in the familiar analysis of variance, LTRE analysis depends on whether the "treatments" are considered a random sample from some universe of possible conditions or a fixed set of treatments that are of interest in themselves and could be repeated if necessary. As in statistics, the distinction between random and fixed treatments is subtle, and one data set may well be interpreted differently by different people. In a random design, the results are characterized by the variance of $\lambda$, and the goal of the analysis is to decompose this variance into contributions from the variances of (and covariances among) the matrix entries (Brault & Caswell 1993; Caswell & P. Dixon, unpubl.). In a fixed design, the results are described in terms of the effect of each treatment on $\lambda$, measured relative to some baseline. The goal of the analysis is to decompose the treatment effects into contributions from the treatment effects on each of the vital rates. (There are also LTRE methods for quantitative treatments, analogous to regression models; see Caswell 1996*a*; Caswell & L.V. Martin, unpubl.)

*Random Treatments: Variance Decomposition*

Let $V(\lambda)$ denote the variance of $\lambda$ among treatments. Recall the formula for the variance of a linear combination of two random variables $x$ and $y$:

$$V(ax + by) = a^2 V(x) + b^2 V(y) + 2abC(x, y), \qquad (4)$$

where $C(x, y)$ is the covariance of $x$ and $y$. Approximating $\lambda$ as a linear function of the $a_{ij}$, and using the sensitivities $s_{ij}$ as the slopes, yields

$$V(\lambda) \approx \sum_{ij} \sum_{k\ell} C(ij, k\ell) s_{ij} s_{k\ell}, \qquad (5)$$

where $C(ij, k\ell)$ is the covariance of $a_{ij}$ and $a_{k\ell}$, and the sensitivities are calculated at the mean matrix. The covariances are calculated directly from the data on **A** for each environment. Each term in this summation is the contribution of one pair of vital rates to $V(\lambda)$. This calculation was first used to examine the variation in $\lambda$ among pods of killer whales (Brault & Caswell 1993).

Because $V(\lambda)$ depends on the covariances among pairs of vital rates, it is difficult to say which rate, as opposed to which pair of rates, makes the biggest contribution. One way to define the contribution of a single rate is to sum the contributions for all the covariances involving that rate:

$$\chi_{ij} = \sum_{k\ell} C(ij, k\ell) s_{ij} s_{k\ell}. \tag{6}$$

This sum includes the contribution to $V(\lambda)$ from the variance of $a_{ij}$ plus half the contributions from the covariances of $a_{ij}$ with the other rates. The other half of the covariance contributions is distributed among the contributions of the other matrix entries. We call this calculation of the contribution $\chi_{ij}$ the "covariance method."

*A special case: independent variation.* If it seems safe to assume that the vital rates vary independently, then the approximation for the variance reduces to

$$V(\lambda) \approx \sum_{ij} V(a_{ij}) s_{ij}^2, \tag{7}$$

where $V(a_{ij})$ is the observed variance in the $ij$th element of the matrix $\mathbf{A}$. Each term in this summation represents a contribution of one vital rate to $V(\lambda)$; by definition,

$$\chi_{ij}^{\mathrm{ind}} = V(a_{ij}) s_{ij}^2 . \tag{8}$$

As with the covariances in the preceding case, the variances here are calculated from the observed set of matrices and the sensitivities are calculated at the mean matrix. We call this the "variance method" of calculating the contribution $\chi_{ij}$ to $V(\lambda)$.

*Choice of a reference matrix.* The variance $V(\lambda)$ is the average of the squared deviations from the mean of the growth rates, and thus the sensitivities in the approximations (6) and (8) are calculated from the mean matrix. However, other matrices might serve as more-reasonable reference matrices. In the data set we analyze next, we use a summary matrix computed from the pooled data from all sites and years. This is equivalent to using a weighted mean with weights proportional to the sample sizes. In this case, $V(\lambda)$ is no longer strictly speaking a variance; it is the mean-squared deviation from the reference matrix, and $C(ij, k\ell)$ is no longer a covariance; it is the mean of the cross-product of the deviations of $a_{ij}$ and $a_{k\ell}$ from their values in the reference matrix.

## Fixed Treatments: Decomposing Treatment Effects

A fixed-effect analysis treats the matrices as representative of particular conditions, either experimental or natural (high vs. low nutrients in a one-way model, for example, or year and spatial location in a two-way model). The goal is to determine how much of
the main effect of each treatment level on $\lambda$ is contributed by each
of the vital rates. In a factorial design, this can be extended to
include the interaction effects of combinations of factors as well as
the main effects.

The analysis uses a linear approximation in which the sensitivities appear as slopes. The effect of a treatment on $\lambda$ depends on
its effect on each matrix entry and on the sensitivity of $\lambda$ to that
entry.

In order to analyze a data set in which plots and years appear
as fixed effects, we write a two-way model (as in an analysis of
variance) for population growth in plot $m$ and year $n$:

$$\lambda^{(mn)} = \lambda^{(..)} + \alpha_m + \beta_n + (\alpha\beta)_{mn}, \qquad (9)$$

where $\lambda^{(..)}$ is the growth rate calculated from $\mathbf{A}^{(..)}$, the grand mean
of all the matrices, and $\alpha_m$, $\beta_n$, and $(\alpha\beta)_{mn}$ are the plot, year, and
interaction effects. These effects are estimated by

$$\alpha_m = \lambda^{(m\cdot)} - \lambda^{(..)}, \qquad (10)$$

$$\beta_n = \lambda^{(\cdot n)} - \lambda^{(..)}, \qquad (11)$$

$$(\alpha\beta)_{mn} = \lambda^{(mn)} - \alpha_m - \beta_n - \lambda^{(..)}. \qquad (12)$$

Here, the growth rates $\lambda^{(m\cdot)}$ and $\lambda^{(\cdot n)}$ are the growth rates calculated from the mean over years of the matrices for plot $m$ and from
the mean over plots of the matrices for year $n$, respectively.

These treatment effects on $\lambda$ can be decomposed into contributions from the effects on each matrix element, as follows:

$$\alpha_m \approx \sum_{ij}(a_{ij}^{(m\cdot)} - a_{ij}^{(..)})s_{ij}, \qquad (13)$$

$$\beta_n \approx \sum_{ij}(a_{ij}^{(\cdot n)} - a_{ij}^{(..)})s_{ij}, \qquad (14)$$

$$(\alpha\beta)_{mn} \approx \sum_{ij}(a_{ij}^{(mn)} - a_{ij}^{(..)})s_{ij} \qquad (15)$$

(Caswell 1989a,b). Because the sensitivity structure may change
from one treatment to another, the sensitivity matrix is evaluated

at a matrix that is intermediate between the treatment being considered and its reference matrix. For $\alpha_m$, this intermediate matrix is $(\mathbf{A}^{(m\cdot)} + \mathbf{A}^{(\cdot\cdot)})/2$; for $\beta_n$, it is $(\mathbf{A}^{(\cdot n)} + \mathbf{A}^{(\cdot\cdot)})/2$; for $(\alpha\beta)_{mn}$, it is $(\mathbf{A}^{(mn)} + \mathbf{A}^{(\cdot\cdot)})/2$.

As in the random-effects case, it is possible to use matrices other than the mean as a reference matrix. In our example, instead of the overall mean, we use the summary matrix from the pooled data for $\mathbf{A}^{(\cdot\cdot)}$. To be consistent, we should also have used pooled data for the plot means $\mathbf{A}^{(m\cdot)}$ and the year means $\mathbf{A}^{(\cdot n)}$. We did not have the data accessible in this format at the time of these analyses, so we did not do this. Thus, we present the analysis with the cautionary note that our reference matrices are slightly inconsistent, but we are reasonably confident that this does not have a large effect on the qualitative results.

The decomposition analysis (eqs. 13–15) gives the contribution of each matrix entry to each treatment effect (i.e., to each plot effect, to each year effect, and to each plot-by-year interaction effect). It would be nice to have some way to compute a single number that measures the overall contribution of each matrix entry to each treatment factor. Some matrix entries are unaffected by the treatment and make a zero contribution to each level of the treatment. Others make small positive contributions at some levels and small negative contributions at other levels. The important matrix entries are those with large positive contributions at some treatment levels and large negative contributions at others. Taking the means of these contributions over the different levels of the factor does not work, because it is always approximately zero, just as in an analysis of variance the mean of the treatment effects is (exactly) zero.

This suggests using the variance, or perhaps the mean of the absolute values, of the contributions as a summary measure. In this chapter we use the means of the absolute values. This calculation is useful in simplifying the results of complex data sets with many levels of cross-classified factors. In our system, with 4 plots, 4 years, and 16 plot-year combinations (see below), taking these means yields a graph with a single line for the plot effects, a single line for the year effects, and a single line for the interaction effects. However, this simplification comes at a cost: presenting the results in this form obscures the contributions to the treatment and interaction effects, which are the original goal of the design (for complete factorial analyses, see Caswell 1989a; Walls et al. 1991).

TABLE 1. *Variance of λ among 16 Matrices for* Calathea ovandensis

| Approximation | Equation | Value |
|---|---|---|
| Variance from mean | standard | 0.0182 |
| Variance from pooled | 16 | 0.0187 |
| Covariance approximation | 5 | 0.0317 |
| Variance approximation | 7 | 0.0196 |

## 3 An Example: *Calathea ovandensis*

We now apply the analyses described above to a set of 16 matrices (four plots over four years) generated from a five-year study of a Neotropical understory herb, *Calathea ovandensis*. The life cycle was divided into eight stages: seeds, three vegetative stages (seedling, juvenile, and pre-reproductive) and small, medium, large, and extra-large reproductive plants. *C. ovandensis* has long-term seed dormancy. Seedlings have the highest mortality of all stages, and reproductives have low mortality. Plants that do not die sometimes shrink, sometimes remain the same size (stasis), and sometimes increase in size between seasons. Fertility is positively correlated with size (Horvitz & Schemske 1995).

Significant spatiotemporal variability is found in the vital rates and in the sensitivity structure (Horvitz & Schemske 1995). The population growth rates of the 16 plot-year matrices varied from 0.7356 to 1.2477, with an arithmetic mean of 0.9695 and a variance $V(\lambda) = 0.0182$ (Table 1). Nevertheless, the overall dynamics are well represented by a summary matrix $\mathbf{A}^{(\cdot\cdot)}$ that pooled all observations over the five-year period (Fig. 1). The dominant eigenvalue of this matrix is $\lambda^{(\cdot\cdot)} = 0.9923$ (Horvitz & Schemske 1995).

We characterize the variability in the population growth rate by the mean-squared deviation around the growth rate of the summary matrix, rather than as the variance around the mean. Let $\lambda^{(mn)}$ denote the growth rate in plot $m$ in year $n$. Then, our measure of variability is

$$V_s(\lambda) = \sum (\lambda^{(mn)} - \lambda^{(\cdot\cdot)})^2 / 16 . \tag{16}$$

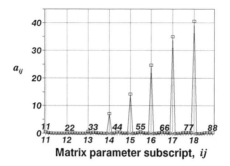

Figure 1. *Entries $a_{ij}$ of the pooled summary matrix (Horvitz &*
*Schemske 1995) for* Calathea ovandensis. *The 64 entries of the* $8 \times 8$
*matrix are arranged in column order: fates of seeds* $(a_{11}, \ldots, a_{81})$, *fates*
*of seedlings* $(a_{12}, \ldots, a_{82})$, *and so on. Subscripts for the entries of the*
*first row appear on the abscissa; subscripts for the diagonal elements*
*appear directly above the matrix entry.*

The variance from the summary matrix calculated in this way is
$V_s(\lambda) = 0.0187$ (Table 1).

*Prospective Analyses*

The sensitivities and elasticities of the summary matrix are shown
in Figure 2. The largest sensitivities are $s_{81}$, $s_{71}$, $s_{61}$, and $s_{51}$, in
that order. Additive perturbations thus exert the biggest effects if
they occur in the vital rates that represent extremely rapid growth
of newly germinated seedlings. Sensitivities to other seed transitions $(s_{41}$ and $s_{31})$ and to the rapid growth of established seedlings
$(s_{82}, s_{72}, s_{62}, s_{52}$ and $s_{42})$ are also high. Within each stage, stasis
and growth have higher sensitivities than fecundity.

The largest elasticity is $e_{55}$ (stasis of small reproductives), followed by $e_{54}$ (growth of pre-reproductives), $e_{44}$ (stasis of pre-reproductives), and $e_{11}$ (seed dormancy). Thus, proportional perturbations have the biggest impact if they affect the stasis of, and growth
to, small reproductives.

There is little correlation between the sensitivity and the elasticity patterns (Fig. 2). This is not unusual; the effects of additive

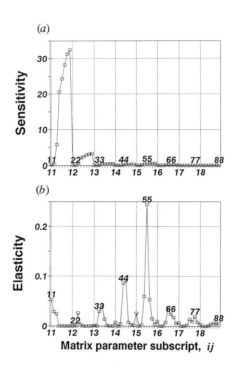

FIGURE 2. (a) *Sensitivities* ($s_{ij}$) *of* $\lambda$ *to the corresponding entries of the pooled summary matrix.* (b) *Elasticities* ($e_{ij}$) *of* $\lambda$ *to the corresponding entries of the pooled summary matrix. Axis labels as in Figure 1.*

and proportional perturbations are likely to differ.

*Retrospective Analyses: Random Effects*

The covariances of the matrix entries are shown in Figure 3a. The large positive values dominating the figure are covariances among the fecundities; a year or plot that is good for the reproduction of one size class tends to be good for the reproduction of all. The entries on the diagonal of this surface are the variances; these are

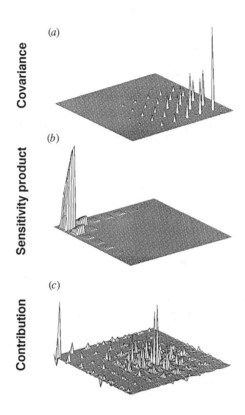

FIGURE 3. (a) *Covariances of the matrix elements* $a_{ij}$ *and* $a_{k\ell}$ *over 16 plot-year combinations. Variances appear on the diagonal.* (b) *The products* $s_{ij}s_{k\ell}$ *of pairs of sensitivities, calculated from the pooled summary matrix.* (c) *Contributions of each pair of matrix entries to* $V_s(\lambda)$, *from the product of the corresponding elements in* (a) *and* (b). *The order of the matrix elements in all three surfaces is the same as the abscissa in Figure 1.*

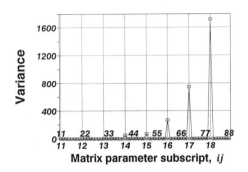

FIGURE 4. *The variances of the matrix elements $a_{ij}$ over 16 plot-year combinations. Axis labels as in Figure 1.*

plotted in Figure 4. The variances of fecundities dwarf all the other variances.

The products $s_{ij}s_{k\ell}$ of pairs of sensitivities are highest for the germination and rapid growth of seeds (Fig. 3*b*). Thus, in this data set there is a negative correlation between the sensitivity of $\lambda$ to a matrix entry and the variability in that matrix entry. The contributions to $V_s(\lambda)$ from each of the covariances is given by the element-wise product of Figures 3*a,b*; the result is shown in Figure 3*c*. Several large contributions from variance terms appear on the diagonal, and there are many small but not negligible contributions, both positive and negative, from the off-diagonal covariance terms. The sum of these values, as in equation (5), is 0.0317.

Taking sums across the rows (or down the columns) of the surface in Figure 3*c* gives the contributions $\chi_{ij}$ by the covariance method (eq. 6). The results are shown in Figure 5. This shows how $V_s(\lambda)$ is increased by positive covariances (including variances) of matrix entries and reduced by negative covariances among matrix entries. The largest of these contributions are from $a_{65}$ (growth of small reproductives), $a_{54}$ (growth of pre-reproductives), $a_{31}$ (rapid growth of newly germinated seeds), and $a_{45}$ (a negative contribution from covariances involving the shrinking of small reproductives). Negative contributions are also made by covariances involving $a_{44}$ and $a_{55}$ (stasis of pre-reproductives and small reproductives).

Figure 3 shows that covariances among the matrix entries cannot be ignored in this data set. Nevertheless, Figure 5*b* shows the

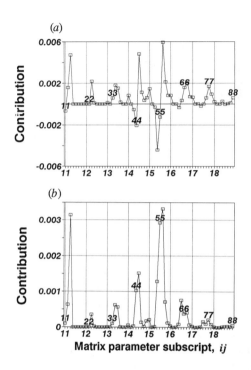

FIGURE 5. (a) *The total contribution* $(\chi_{ij})$ *of matrix entry* $a_{ij}$ *to the variance* $V_s(\lambda)$, *calculated by summing over the covariance contributions in Figure 3 according to equation (6).* (b) *The contributions* $(\chi_{ij}^{ind})$ *of the variance of matrix entries to* $V_s(\lambda)$ *assuming that the matrix entries vary independently, as in equation (8). Axis labels as in Figure 1.*

contributions $\chi_{ij}^{ind}$ calculated using the variance method (eq. 8); these values are just the diagonal elements of Figure 3c. They sum to 0.0196 (Table 1).

The largest contributions to $V_s(\lambda)$ in this analysis are from the variances of $a_{65}$ (growth of small reproductives), $a_{31}$ (growth of newly germinated seeds), $a_{55}$ (stasis of small reproductives), and $a_{54}$ (growth of small reproductives). Note that ignoring covariances leads to the conclusion that the variance of $a_{55}$ makes a large pos-

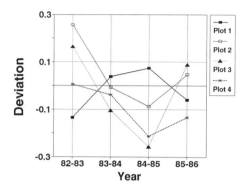

FIGURE 6. *Population growth* $\lambda$ *for each plot in each year, expressed as a deviation from the growth rate of the pooled summary matrix.*

itive contribution to $V_s(\lambda)$; in fact, the covariances involving $a_{55}$ make its net contribution to $V_s(\lambda)$ negative (Fig. 5*b*).

### Retrospective Analyses: Fixed Treatments

Using the fixed-effect factorial design, we can break down contributions to variability in $\lambda$ further, into independent effects of plots, years, and their interaction. Population growth rate is shown in Figure 6 as a function of plot and year. The plot and year effects $\alpha_m$ and $\beta_n$ are shown in Figures 7*a,b*. Plot 2 had the largest positive effect on $\lambda$, and plot 4 the largest negative effect. Year 1982–83 had the largest positive effect, and 1984–85 the largest negative effect. The plot and year effects are of similar magnitude, although years are slightly more influential than plots. The temporal pattern differed among plots (Fig. 6), suggesting a plot-by-year interaction, which appears in the interaction effects $(\alpha\beta)_{mn}$ in Figure 7*c*.

We do not show the decomposition of each of the plot, year, and interaction effects into contributions from each of the matrix entries. Instead, we show the means of the absolute values of the contributions of each matrix entry to the plot effects, to the year effects, and to the interaction effects (Fig. 8).

The matrix entries most influential in determining plot effects are $a_{65}$ (growth of small reproductives), $a_{44}$ (stasis of pre-reproductives), $a_{31}$ (germination and growth of seeds), and $a_{45}$ (shrinkage of small reproductives) (Fig. 8*a*).

The entries most influential in determining year effects are $a_{55}$

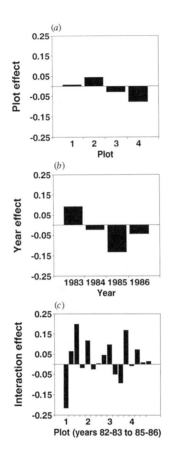

FIGURE 7. (a) *The main effects* $\alpha_m$, $m = 1, \ldots, 4$, *of plots, calculated from equation (13).* (b) *The main effects* $\beta_n$, $n = 1, \ldots, 4$, *of years, calculated from equation (14).* (c) *The interaction effects* $(\alpha\beta)_{mn}$, *calculated from equation (15). Years, from 1982 to 1985, are shown within each plot.*

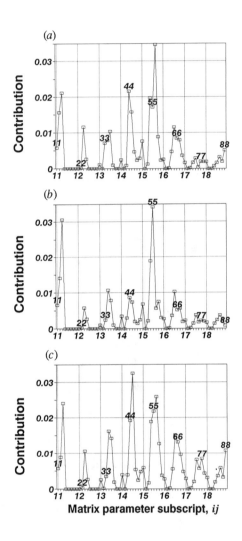

FIGURE 8. (a) *The mean, over 4 plots, of the absolute value of the contributions of each matrix element to the plot effect.* (b) *The mean, over 4 years, of the absolute value of the contributions of each matrix element to the year effect.* (c) *The mean, over 16 plot-year combinations, of the absolute values of the contributions of each matrix element to the plot-by-year interaction effect. Axis labels as in Figure 1.*

(stasis of small reproductives), $a_{31}$ (germination and growth of seeds), $a_{45}$ (regression of small reproductives), and $a_{21}$ (seed germination) (Fig. 8$b$).

The matrix entries with the largest influence on interaction effects are $a_{45}$ (growth of pre-reproductives), $a_{65}$ (growth of small reproductives), $a_{31}$ (germination and growth of seeds), and $a_{55}$ (stasis of small reproductives) (Fig. 8$c$).

## 4 Discussion

We show computations of seven different measures of the "importance" of individual matrix entries to population growth: two prospective and five retrospective. Here we discuss the matrix elements that rank among the top four in each type of analysis (Table 2). We also examine the correlations among the seven measures of importance (Table 3).

The sensitivity and elasticity analyses pick out completely different sets of matrix entries as the most important. Additive perturbations of the matrix elements in the last four locations of the first column, representing extremely rapid growth of newly germinated seedlings, have the biggest effect on $\lambda$. Proportional perturbations, in contrast, have the biggest impact when they involve the stasis and growth of pre-reproductive and small reproductive plants, or seed dormancy. Sensitivity is weakly, but negatively, correlated with indices generated by the other approaches (Table 3), in spite of the fact that sensitivity appears in the calculation of all the other indices. This correlation would change if it was calculated using only the sensitivities of the nonzero matrix entries, that is, those that are free to vary. The largest such sensitivity is $s_{31}$, and we note that $a_{31}$ also appears among the four most important entries in all five retrospective analyses.

Of the 64 matrix entries, only eight ($a_{55}$, $a_{54}$, $a_{44}$, $a_{11}$, $a_{65}$, $a_{31}$, $a_{45}$, and $a_{21}$) appear among the top four in importance according to elasticity or any of the retrospective analyses. None is important by all criteria. The growth of newly germinated seedlings ($a_{31}$) is important by all five retrospective indices; it contributes significantly to variance (according to both the covariance and variance methods), and its contribution to plot, year, and plot-by-year interaction effects is also important.

Only two matrix entries ($a_{55}$ and $a_{65}$, the stasis and growth of small reproductives) are important according to four indices. Entry $a_{55}$ has high elasticity and makes a large contribution to $V_s(\lambda)$ by the variance method. It has an important contribution to the year

TABLE 2. *The Indices (i, j) of the Four Most Important Matrix Entries*

| | | | Method | | | | | |
| --- | --- | --- | --- | --- | --- | --- | --- | --- |
| | Prospective | | Retrospective | | | | | |
| | | | Random Effect | | Fixed Effect | | | |
| Rank | Sensitivity | Elasticity | Covariance | Variance | Plot | Year | Plot-by-Year | |
| 1 | 8,1 | 5,5 | 6,5 (+) | 6,5 | 6,5 | 5,5 | 5,4 | |
| 2 | 7,1 | 5,4 | 5,4 (+) | 3,1 | 4,4 | 3,1 | 6,5 | |
| 3 | 6,1 | 4,4 | 3,1 (+) | 5,5 | 3,1 | 4,5 | 3,1 | |
| 4 | 5,1 | 1,1 | 4,5 (−) | 5,4 | 4,5 | 2,1 | 5,5 | |

NOTE.— Indices measured by two prospective and five retrospective methods. For the contributions to $V_s(\lambda)$ calculated according to the covariance method, the sign of the effect is also shown.

TABLE 3. *Correlations among the Two Prospective and Five Retrospective Measures of the Importance of the Matrix Entries to Population Growth*

| | Prospective | | Retrospective | | | | |
| --- | --- | --- | --- | --- | --- | --- | --- |
| | | | Random Effects | | Fixed Effects | | |
| | Sensitivity | Elasticity | Variance | Covariance | Plot | Year | Plot-by-Year |
| Sensitivity | 1.00 | −0.13 | −0.07 | −0.08 | −0.17 | −0.13 | −0.21 |
| Elasticity | −0.13 | 1.00 | 0.01 | 0.71 | 0.63 | 0.76 | 0.69 |
| Covariance | −0.07 | 0.01 | 1.00 | 0.46 | 0.41 | 0.14 | 0.47 |
| Variance | −0.08 | 0.71 | 0.46 | 1.00 | 0.88 | 0.81 | 0.84 |
| Plot effect | −0.17 | 0.64 | 0.41 | 0.88 | 1.00 | 0.71 | 0.89 |
| Year effect | −0.13 | 0.76 | 0.14 | 0.81 | 0.71 | 1.00 | 0.74 |
| Plot-by-year | −0.21 | 0.69 | 0.47 | 0.84 | 0.89 | 0.73 | 1.00 |

effect and to the plot-by-year interaction effect (Table 2). Entry $a_{65}$ makes a large contribution to $V_s(\lambda)$ and to the plot and plot-by-year interaction effects.

Shrinkage of small reproductives ($a_{45}$) and the growth of pre-reproductives ($a_{54}$) are important according to three of the indices. Shrinkage of small reproductives has high elasticity and makes important contributions to $V_s(\lambda)$ by the variance method and to the plot-by-year interaction. The growth of pre-reproductives makes an important contribution to $V_s(\lambda)$ and to the plot and year effects.

Stasis of pre-reproductives ($a_{44}$) has high elasticity and makes an important contribution to the plot effect. Seed dormancy and seed germination ($a_{11}$ and $a_{21}$) have high elasticity and make an important contribution to the year effect, respectively.

Table 3 shows the correlations between the various measures of "importance," underscoring the different insights produced by each analysis. Sensitivity and elasticity are nearly independent of contributions to $V_s(\lambda)$ calculated using equations (5) and (6), the appropriate variance decomposition for this data set (correlation coefficients $-0.07$ and $0.01$, respectively). Thus, treating the 16 matrices as a random sample characterized by the variance of $\lambda$, the prospective and retrospective analyses clearly address different questions. Elasticity, however, is positively correlated with contributions to $V_s(\lambda)$ by the variance method.

Sensitivity is negatively correlated with the results of all the other analyses. Elasticity has positive correlations with most other analyses, and large ones with the plot (0.64), year (0.77), and plot-by-year interaction (0.69) effects.

The random- and fixed-effects analyses are positively correlated, but not always strongly. Contributions to $V_s(\lambda)$ from the covariance approach has correlations of 0.41, 0.14, and 0.47 with contributions to the plot, year, and plot-by-year interaction effects. Within the fixed-effect design, there are strong correlations between contributions to plot and year effects (0.72), plot and interaction effects (0.89), and year and interaction effects (0.74). This suggests that the plot and year effects tended to be mediated by variation in the same set of vital rates (Table 2).

Finally, we note that $\chi_{ij}^{\text{ind}}$ is highly correlated with elasticity and with contributions to all three fixed effects. Thus, a random-effects, retrospective analysis that is inappropriate for this data set does a good job of predicting the results of both a fixed-effect, factorial, analysis and an elasticity analysis.

There is no single best method for determining the "importance" of the vital rates. It is important to define more precisely what we mean by importance, and to address specific, well-posed questions, such as the following.

What are the relative sensitivities of $\lambda$ to incremental changes in the vital rates?

What are the relative proportional sensitivities of $\lambda$ to proportional changes in the vital rates?

Over an observed sample of environmental conditions, how do the variances of, and covariances among, the vital rates contribute to the variance of $\lambda$?

How much does each of the vital rates contribute to the observed effect on $\lambda$ of each of a set of fixed treatments?

In the case analyzed here, the prospective elasticity analysis identifies one vital rate, seed dormancy, that is not important according to the retrospective methods. Similarly, the factorial analysis identifies seed germination as making an important contribution to year effects; this vital rate is not identified as important by any other method. Despite these differences, both the transition to and the fates of small reproductives are identified as key events for *Calathea ovandensis*, both in predicting the outcome of hypothetical perturbations and in explaining the observed variation in population growth rate.

## Acknowledgments

We thank all the students and other faculty in the structured-models course (1993) for stimulating discussions, especially Daniel Promislow, Shripad Tuljapurkar, Glenda Wardle, and Phil Dixon, and to Michael Neubert for comments on the manuscript. This research was supported by National Science Foundation grants DEB-8206993 and DEB-8415666 to C.H. and D.W.S. and DEB-9211945 to H.C. This is Contribution 510 of the Program in Tropical Biology at the University of Miami and Contribution 9226 of the Woods Hole Oceanographic Institution.

## Literature Cited

Bierzychudek, P. 1982. The demography of jack-in-the-pulpit, a forest perennial that changes sex. *Ecological Monographs* 52: 335–351.

Brault, S., and H. Caswell. 1993. Pod-specific demography of killer whales (*Orcinus orca*). *Ecology* 74: 1444–1454.

Calvo, R. N. 1993. Evolutionary demography of orchids: Intensity and frequency of pollination and the cost of fruiting. *Ecology* 74: 1033–1042.

Calvo, R. N., and C. C. Horvitz. 1990. Pollinator limitation, cost of reproduction, and fitness in plants: A transition matrix demographic approach. *American Naturalist* 136: 499–516.

Caswell, H. 1978. A general formula for the sensitivity of population growth rate to changes in life history parameters. *Theoretical Population Biology* 14: 215–230.

———. 1985. The evolutionary demography of clonal reproduction. Pp. 187–224 *in* J. B. C. Jackson, L. W. Buss, and R. E. Cook, eds., *Population Biology and Evolution of Clonal Organisms*. Yale University Press, New Haven, Conn.

———. 1986. Life cycle models for plants. *Lectures on Mathematics in the Life Sciences* 18: 171–233.

———. 1989*a*. The analysis of life table response experiments. I. Decomposition of treatment effects on population growth rate. *Ecological Modelling* 46: 221–237.

———. 1989*b*. *Matrix Population Models: Construction, Analysis and Interpretation*. Sinauer, Sunderland, Mass.

———. 1996*a*. Demography meets ecotoxicology: Untangling the population level effects of toxic substances. Pp. 255–292 *in* M. C. Newman and C. H. Jagoe, eds., *Ecotoxicology: A Hierarchical Treatment*. Lewis, Boca Raton, Fla.

———. 1996*b*. Second derivatives of population growth rate: Calculation and applications. *Ecology* 77: 870–879.

———. In press. Analysis of life table response experiments. II. Alternative parameterizations for size- and stage-structured models. *Ecological Modelling*.

Caswell, H., and P. A. Werner. 1978. Transient behavior and life history analysis of teasel (*Dipsacus sylvestris* Huds.). *Ecology* 59: 53–66.

Caswell, H., R. J. Naiman, and R. Morin. 1984. Evaluating the consequences of reproduction in complex salmonid life cycles. *Aquaculture* 43: 123–134.

Crouse, D. T., L. B. Crowder, and H. Caswell. 1987. A stage based population model for loggerhead sea turtles and implications for conservation. *Ecology* 68: 1412–1423.

de Kroon, H. J., A. Plaiser, J. van Groenendael, and H. Caswell. 1986. Elasticity: The relative contribution of demographic parameters to population growth rate. *Ecology* 67: 1427–1431.

Doak, D., P. Kareiva, and B. Klepetka. 1994. Modeling population viability for the desert tortoise in the western Mojave desert. *Ecological Applications* 4: 446–460.

Heppell, S. S., J. R. Walters, and L. B. Crowder. 1994. Evaluating management alternatives for red-cockaded woodpeckers: A modeling approach. *Journal of Wildlife Management* 58: 479–487.

Horvitz, C. C., and D. W. Schemske. 1986. Seed dispersal and environmental heterogeneity in a Neotropical herb: A model of population and patch dynamics. Pp. 169–186 *in* A. Estrada and T. Fleming, eds., *Frugivores and Seed Dispersal.* Junk, Dordrecht, The Netherlands.

———. 1995. Spatiotemporal variation in demographic transitions of a tropical understory herb: Projection matrix analysis. *Ecological Monographs* 65: 155–192.

Kalisz, S., and M. A. McPeek. 1992. Demography of an age-structured annual: Resampled projection matrices, elasticity analyses and seed bank effects. *Ecology* 73: 1082–1093.

Lande, R. 1988. Demographic models of the northern spotted owl (*Strix occidentalis caurina*). *Oecologia* 75: 601–607.

Lande, R., and S. J. Arnold. 1983. The measurement of selection on correlated characters. *Evolution* 37: 1210–1226.

Levin, L. A., H. Caswell, K. D. DePatra, and E. L. Creed. 1987. Demographic consequences of larval development mode: Planktotrophy vs. lecithotrophy in *Streblospio benedicti. Ecology* 68: 1877–1886.

Levin, L., H. Caswell, T. Bridges, D. Cabrera, G. Plaia, and C. DiBacco. In press. Demographic responses of estuarine polychaetes to sewage, algal, and hydrocarbon additions: Life table response experiments. *Ecological Applications.*

Lewontin, R. C. 1974. The analysis of variance and the analysis of causes. *American Journal of Human Genetics* 26: 400–411.

Martinez, M., and E. Alvarez-Bullya. 1986. Seed dispersal, gap dynamics and tree recruitment: The case of *Cecropia obtusifolia* at Los Tuxtlas, Mexico. Pp. 333–346 *in* A. Estrada and T. Fleming, eds., *Frugivores and Seed Dispersal.* Junk, Dordercht, The Netherlands.

Menges, E. S. 1990. Population viability analysis for an endangered plant. *Conservation Biology* 4: 52–62.

Mesterton-Gibbons, M. 1993. Why demographic elasticities sum to one: A postscript to de Kroon et al. *Ecology* 74: 2467–2468.

Phillips, P. C., and S. J. Arnold. 1989. Visualizing multivariate selection. *Evolution* 43: 1209–1222.

Schemske, D. W., B. C. Husband, M. H. Ruckelshaus, C. Goodwillie, I. M. Parker, and J. G. Bishop. 1994. Evaluating approaches to the conservation of rare and endangered plants. *Ecology* 75: 584–606.

Sibly, R. M. 1996. Effects of pollutants on individual life histories and population growth rates. Pp. 197–224 *in* M. C. Newman and C. H. Jagoe, eds. *Ecotoxicology: A Hierarchical Treatment.* Lewis, Boca Raton, Fla.

van Groenendael, J., H. de Kroon, S. Kalisz, and S. Tuljapurkar. 1994. Loop analysis: Evaluating life history pathways in population projection matrices. *Ecology* 75: 2410–2415.

Walls, M., H. Caswell, and M. Ketola. 1991. Demographic costs of *Chaoborus*-induced defenses in *Daphnia pulex. Oecologia* 87: 43–50.

CHAPTER 8

# Life-History Evolution and Extinction

## Steven Hecht Orzack

The use of structured-population models for the analysis of life-history evolution has increased dramatically in the last fifteen years or so (see, e.g., Caswell 1989; Tuljapurkar 1990a; Charlesworth 1994). An important subset of these analyses assumes that the demographic components of a life history vary temporally. This branch of research in stochastic demography stems from work on population dynamics in variable environments (Lewontin & Cohen 1969; Cohen 1977, 1979a,b; Tuljapurkar & Orzack 1980; Tuljapurkar 1982a,b), and of course, it stems in a more general way from the pioneering work of Fisher (1930) and Cole (1954) on life-history evolution.

A central feature of analyses of structured populations subject to density-independent environmental fluctuations is the use of a random-matrix product to describe population change. Changes in numbers of different kinds of individuals are described by a recursion equation,

$$\mathbf{N}_{t+1} = \mathbf{X}_{t+1}\mathbf{N}_t = \mathbf{X}_{t+1}\mathbf{X}_t \ldots \mathbf{X}_1\mathbf{N}_0 \ , \tag{1}$$

where $\mathbf{N}_j$ denotes a vector (whose $i$th element represents the number of individuals of the $i$th life-history stage at time $j$) and $\mathbf{X}_j$ denotes a projection matrix appearing at time $j$. The presence of matrices defines this as a structured-population model, in that it implies that there is demographic heterogeneity. At any one time, some individuals undergo one kind of transition, say, from a younger to older age class, while others undergo another kind of transition, say, from a nonreproducing class to a reproducing class. This framework has obvious connections with previous work on random-matrix products (Furstenberg & Kesten 1960) and demography (e.g., Leslie 1945).

In this chapter, the matrices representing different environments are chosen according to a stationary stochastic process. This assumption renders mathematical analysis more tractable. Its relevance to the real world is another matter, although in many instances secular environmental trends occur over times much longer than are usually considered relevant to population dynamic analyses. Moments for the nonzero elements of the projection matrix are evaluated with respect to all environmental states. Of these, the mean and variance play an especially important role in analyses of population dynamics and life-history evolution.

## 1  The Distribution of Populations

An important feature of equation (1) is that a population has an asymptotic growth rate:

$$a = \frac{1}{t} \, E \left( \log \frac{M_t}{M_0} \right), \quad t \to \infty \,,$$

where $M_i$ is the total population size at time $i$ (Furstenberg & Kesten 1960; Cohen 1977; Tuljapurkar & Orzack 1980). This result has more than practical formal significance: Monte Carlo simulations indicate that the realized growth rate of a population approaches $a$ within "real" time, depending on the length and structure of the life history (Tuljapurkar & Orzack 1980).

The stochastic growth rate, $a$, has additional general importance. The long-run distribution of population size is lognormal, and $a\,t$ is the average value of the limiting distribution of the logarithm of population size (if one disregards extinction; see below) (Tuljapurkar & Orzack 1980). This result implies that the average population size for an assemblage of "replicate" populations at any one time is, in general, unrepresentative of the typical size. This discrepancy also occurs for nonstructured populations (Lewontin & Cohen 1969). Given the additional assumption that environmental states have at most a Markovian dependence or are independent, this result applies as well to the temporal average of the size of a single population.

The lognormal distribution of population size is of fundamental ecological significance for a number of reasons (Tuljapurkar & Orzack 1980); for one, it can be used to show that common arguments for the irrelevance of density-independent population models to nature are not meaningful. One argument is that such populations must explode in size. Indeed, lognormality does imply that there is a necessary "clock-like" increase in the average population size, $E(M_t)$, even when the stochastic growth rate, $a$, is equal to zero. Nonetheless, because of the distinction between average and typical behavior associated with lognormality, many if not most populations do not "explode" in size but remain bounded in size or become extinct at finite times (as demonstrated in Orzack 1993; for a related point, see also Lande 1993). This is just the pattern exhibited by species in some of the better-studied natural communities such as those described by den Boer (1990). The point is that the boundedness of population size at finite times is not by itself evidence for density dependence or against density independence. What is required to settle the question of modes of population regulation in nature is information about the ecological mechanisms underlying observed population patterns. While hardly startling, such a statement contrasts markedly with common assertions that the issue can be settled by "logic" and that density dependence "must" be present in natural populations (see, e.g., Godfray & Hassell 1992). By themselves, such statements are not biologically meaningful.

## 2  How Does the Stochastic Growth Rate Relate to Life-History Evolution?

The relevance of the stochastic growth rate, $a$, to life-history evolution stems from both specific and general considerations. Tuljapurkar (1982$b$) showed that a mutant life history can invade a monomorphic population if the associated heterozygote has a higher stochastic growth rate than the resident. Associated numerical results (Orzack 1985) indicate that inequalities among the growth rates derived from such a diallelic-invasion analysis predict the outcome of selection more generally. For example, a mutant life history predicted to invade a monomorphic population on the basis of Tuljapurkar's result also goes on to sweep through the population if the associated homozygote has a higher stochastic growth rate than either of the other two genotypes. Alternatively, a balanced polymorphism can result if the heterozygote has a higher stochastic growth rate than either of the other two genotypes.

The more general results pertaining to the relevance of the stochastic growth rate, $a$, to life-history evolution are the sort commonly used to motivate a "phenotypic" analysis like that found in many optimality models and models in evolutionary quantitative genetics. In particular, one presumes that evolution leads to the fixation or near-fixation of the phenotype with the maximum fitness or growth rate (given some initial set of variants). To this extent, this approach assumes that genetic details such as epistasis, heterosis, and pleiotropy "will not get in the way" of the evolution of the trait. Although the truth of this assumption cannot be taken as self-evident, it appears to be true in particular instances (Orzack & Sober 1994), and in addition, it at least has the virtue of revealing how the trait "should" evolve under the influence of natural selection (see Eshel 1984).

## 3 The Calculation of the Stochastic Growth Rate

The dependence of the stochastic growth rate on the form of the life history and the nature of environmental variability can be determined exactly in some instances (see, e.g., Roerdink 1988 and below), but for any suitable life history and environment, it is possible to construct an approximation to $a$ using the general approach outlined by Tuljapurkar (1982$b$). At present, two kinds of approximation exist. Both include the contribution of the average life history in the form of the dominant eigenvalue of the average projection matrix. When environmental variability is uncorrelated, a second-order or "small-noise" approximation is composed additionally of the nonpositive contribution of one-period variances of vital rates. This approximation can be remarkably accurate even in highly variable environments (Lande & Orzack 1988; Orzack & Tuljapurkar 1989; see below). One can retain additional terms in order to account for higher-order moments of the environmental process, although the analytical simplicity of the approximation is somewhat obscured. A fourth-order approximation has been used in some instances (see, e.g., Tuljapurkar & Istock 1993; Tuljapurkar & Wiener, unpubl. ms.; and below). An important common feature of these approximations is that they stem from *dynamic* analysis, that is, one in which sample paths of the stochastic process, products of random matrices, are analyzed. A common alternative form of analysis, one in which a Taylor series is used to approximate the geometric mean growth rate of a population, is not dynamically based and fails to capture important qualitative features of the process (Tuljapurkar 1990$a$; Orzack 1993).

Negative or positive autocorrelation of environmental fluctuations can be accounted for by the use of additional terms in either of the dynamic approximations. Such fluctuations can have important qualitative effects on the direction of life-history evolution. These have only begun to be explored (Orzack 1985).

These dynamically based approximations can be used with any life history for which an ergodicity condition is satisfied, such that the initial distribution of individuals among the various groups in the population (e.g., ages) has no asymptotic effect on the population dynamics. Beyond this constraint, the approximation scheme can accommodate a great variety of assumptions about life-history structure and variability. To this extent, life-history investigators have an analytical power that is unparalleled in the history of the subject.

## 4  Life-History Evolution

The power of dynamic analysis can be seen by examining its application to one of the central questions in the analysis of life histories: the temporal dispersion of reproduction. More specifically, one can ask two closely connected questions. Under what circumstances will repeated reproduction evolve in a variable environment? If repeated reproduction does evolve, what form will it take? (See Orzack & Tuljapurkar 1989; Tuljapurkar 1990a; Orzack 1993.)

The general focus of these questions stems from the classic paper by Cole (1954). From a simple model, he concluded that the growth rates of a semelparous population and an iteroparous population are equal when the semelparous brood is one individual larger than the iteroparous brood. This raises an obvious question: why are iteroparous life histories so common? Important contributions to the answer to this question were made by Murphy (1968), Schaffer (1974), and Goodman (1984), with a consensus emerging that iteroparous life histories are superior because they allow an organism to avoid necessarily reproducing at times when offspring could not survive. Yet, none of these analyses can be regarded as reasonably general because all contain restrictive assumptions about life-history structure, population dynamics, or the nature of environmental variability.

Most of these restrictive assumptions can be avoided using the dynamic framework mentioned above. This can be best illustrated by carrying out a comparative analysis of a semelparous life history and of various iteroparous life histories. By definition, such life histories differ with respect to the length of life and the number of reproductive events during life. Of course, for such a comparison to be meaningful, it is necessary that some constraint be shared by the life histories, and in this instance, the total amount of net reproduction is held equal. As shown in Table 1, the total average net reproduction is identical (= 1.01) for all life histories, but they differ

TABLE 1. *Life Histories Sharing a Constraint on the Total Amount of Average Net Reproduction over the Lifetime*

| Life History | $\phi_i$ | $\alpha$ | $\omega$ | $\lambda_0$ | $T_0$ | $D$ |
|---|---|---|---|---|---|---|
| 1 | 1.010 | 1 | 1 | 1.0100 | 1.000 | 1.000 |
| 2 | 0.505 | 1 | 2 | 1.0067 | 1.498 | 0.500 |
| 3 | 0.2525 | 1 | 4 | 1.0040 | 2.495 | 0.250 |
| 4 | 0.16833 | 1 | 6 | 1.0029 | 3.492 | 0.167 |
| 5 | 0.12625 | 1 | 8 | 1.0022 | 4.488 | 0.125 |
| 6 | 0.101 | 1 | 10 | 1.0018 | 5.485 | 0.100 |
| 7 | 0.12625 | 3 | 10 | 1.0015 | 6.492 | 0.125 |
| 8 | 0.16833 | 5 | 10 | 1.0013 | 7.496 | 0.167 |
| 9 | 0.2525 | 7 | 10 | 1.0012 | 8.499 | 0.250 |
| 10 | 0.505 | 9 | 10 | 1.0010 | 9.500 | 0.500 |

NOTE.—$\phi_i$ is the average net reproduction at age $i$ and is constant from the first ($\alpha$) to the last ($\omega$) age at reproduction; $\lambda_0$ is the dominant eigenvalue of the average projection matrix; $T_0$ is the mean generation length associated with the average life history. $D$ $(= \Sigma(\phi_i/\lambda_0^i)^2)$ is a measure of the degree of iteroparity of the life history; it decreases as reproductive spread increases. (All values are rounded; for further details, see Orzack 1993.)

with respect to the number of reproductive events and the ages at first and last reproduction. For all of these life histories, if one assumes that temporal variation is uncorrelated, it is straightforward to show (Tuljapurkar 1982*b*; Orzack & Tuljapurkar 1989) that the "small-noise" approximation to the stochastic growth rate is

$$
a \approx \log \lambda_0 - \frac{1}{2} \left[ \sum_{\alpha}^{\omega} \left( \frac{\partial \log \lambda_0}{\partial \phi_i} \right)^2 \sigma_{\phi_i}^2 \right.
$$
$$
\left. + \sum_{\alpha}^{\omega} \sum_{\alpha}^{\omega} \frac{\partial \log \lambda_0}{\partial \phi_i} \frac{\partial \log \lambda_0}{\partial \phi_j} \mathrm{cov}(\phi_i, \phi_j) \right], \quad (2)
$$
$$
i \neq j,
$$

where $\alpha$ and $\omega$ are the ages at first and last reproduction, $\phi_i$ and $\sigma_{\phi_i}^2$ are the mean and variance, respectively, of the net reproduction at age $i$ ($\phi_i$ is the mean value of $\ell_i m_i$, where $\ell_i$ is the survivorship to age $i$ and $m_i$ is the fertility at age $i$), $\text{cov}(\phi_i, \phi_j)$ is the covariance between the fluctuations of net reproduction at ages $i$ and $j$ at a given time, and $\lambda_0$ is the dominant eigenvalue of the average projection matrix.

Beyond the assumptions that environmental fluctuations are stationary and "small," this equation encompasses all types of environmental fluctuations experienced by components of the life history. There is good agreement between the predictions of equation (2) and stochastic growth rates observed in Monte Carlo simulations (Orzack & Tuljapurkar 1989). Additional assumptions about the type of fluctuation can in some instances make the predictions more accurate. For example, when environmental fluctuations are assumed to be lognormally distributed, equation (2) is exact for a scalar life history; that is,

$$a = \log \lambda_0 - \frac{\sigma^2}{2}$$

if one uses an additional formula, $\sigma^2 = \log(1 + C^2)$, where $C$ is the coefficient of variation of $\lambda_0$ (Johnson & Kotz 1970; Lande & Orzack 1988).

A central feature of these applications of equation (2) has been the determination of the life histories that have identical stochastic growth rates, that is, that are selectively neutral with respect to one another. One such determination for the life histories in Table 1 is shown in Figure 1, where it is assumed that the coefficients of variation of the net reproduction values do not vary with age. Such a set of life histories comprise an *indifference* curve. Such a set displays the potential for nonselective evolution of the trait. Presumably, mutations may cause a variety of life histories to accumulate in a population as long as they have the same stochastic growth rate as the resident phenotype. Subsequent random drift or population subdivision may result in the fixation of different life histories in different daughter populations. The likelihood of this occurring is not a matter resolved by appeal to claims that life-history evolution must be determined only by natural selection. Any trait has the potential for nonselective evolution in particular circumstances. This is well known in the context of other traits whose evolution, like that of life-history traits, would at first glance appear to necessarily require natural selection. One of the best-known examples is the sex ratio produced by an individual in an

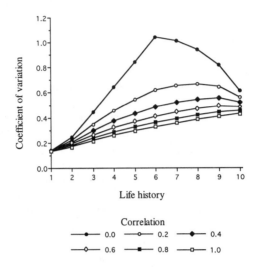

FIGURE 1. *Indifference curves for the life histories in Table 1 for a range of values of the correlation between the fluctuations in net reproduction at ages $i$ and $j$. $a = 0.0$. Values of the coefficient of variation are derived from equation (2).*

infinite panmictic population. Assuming that sons and daughters are of equal energetic cost to parents, there is no selection on the sex ratio when the population has equal numbers of males and females (Kolman 1960).

Indifference curves describe the potential for neutral life-history evolution. Beyond this, they underscore the need for an understanding of the relationship between environmental dynamics and life-history dynamics. For example, major changes in life-history structure need not result in changes in the relationship between environmental variability and growth rate. Consider life histories 4 and 10, which have very different reproductive schedules (see Table 1). Yet, when the vital-rate fluctuations at any one time are uncorrelated, almost identical levels of environmental variability result in the same stochastic growth rate for each (see curve for zero correlation in Fig. 1).

The important insight obtained here is that environmental variability does not necessarily lead to a selective advantage for iteroparity or for iteroparous life histories with increasingly dispersed reproduction. This can be seen in Figure 2, which displays the re-

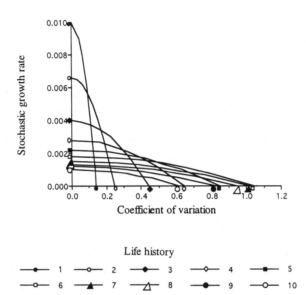

FIGURE 2. *The relationship between the coefficient of variation and stochastic growth rate for the life histories in Table 1 when the fluctuations in net reproduction at ages i and j are uncorrelated. Values of the stochastic growth rate are derived from equation (2).*

lationship between stochastic growth rate and environmental variability for the life histories in Table 1. Observe that the semelparous life history has a higher stochastic growth rate at lower levels of environmental variability; its high value of $\lambda_0$ more than compensates for its "susceptibility" to environmental variability. As variability increases, the longer life cycles associated with iteroparity become more important as a means of discounting the deleterious effect of variability. Yet, the near equality in Figure 2 of stochastic growth rates for life histories 4, 5, and 6 at intermediate levels of environmental variability reflects the particularity of the advantage of iteroparity. In this instance, life history 6, the iteroparous life history with the most evenly distributed reproduction, does best at maintaining a nonnegative growth rate in the face of high environmental variability. In contrast, if vital-rate fluctuations at any one time are completely correlated (not shown), life history 10 does best because it has very concentrated reproduction near the end of life, allowing the individual to avoid the deleterious consequences of environmental variability. The main point is not that the notion of iteroparity as an adaptation to environmental variability is wrong. Clearly, it can be such an adaptation in some circumstances. But talk about the general selective consequences of environmental variability is *ambiguous*. These results demonstrate that there is no unique adaptive response to environmental variability. We can now gain a more precise, quantitative understanding of the relationship between life-history structure and the magnitude and nature of environmental variability. The complexity of the resulting picture is at least partially compensated for by having the ability to accommodate a broad variety of assumptions about the structures of the environment and of the life history.

## 5  The Flavors of Reproductive Delay

The power of the dynamic framework described above is also revealed by an exercise in "comparative" demography (Orzack 1993). I exploited the fact that structurally distinct life histories, such as one with a diapause stage and one with iteroparity, still share a common feature: the "tension" between present and future reproduction. This is well known, of course, but we lack a full understanding of the dynamic similarities and differences among such life histories. The life histories I compare are called "iteroparity," "biennial," "prereproductive delay," and "diapause" (see Table 2). All of these life histories have a delay in the transition of individuals in to or out of the reproductive class or classes. The repeated "reproduction" caused by this delay ensures that the ergodicity condition is satisfied, even for the biennial and prereproductive-delay life histories, which have only one class of individuals producing true newborns. As a result, one can apply the analytical tools described above to all of these delay life histories.

Important aspects of the population dynamics of the biennial, prereproductive-delay, and diapause life histories were presented by Roerdink (1988, 1989), Tuljapurkar (1990$b$), and Tuljapurkar and Istock (1993). These authors determined the stochastic growth rate exactly for a specific type of environmental variability experienced by the reproductive rate (e.g., $m_2$ in Table 2) and also showed that some constant degree of reproductive delay is advantageous in a variable environment. In their analysis, Tuljapurkar and Istock also demonstrated that a nonstructured approach to understanding the dynamics of diapause does not reveal an important result of their structured analysis: there is a broad peak for the stochastic

TABLE 2. *Life Histories with Different Kinds of Reproductive Delay*

| Phenotype | Projection Matrix | Type(s) of Individuals in Life-History Class | |
| --- | --- | --- | --- |
| | | $N_1$ | $N_2$ |
| iteroparity, $f \equiv$ age 1 reproductive fraction | $\begin{bmatrix} fm_1 & (1-f)m_2 \\ s_1 & 0 \end{bmatrix}$ | age 1 reproductive | age 2 reproductive |
| biennial, $f \equiv$ flowering fraction | $\begin{bmatrix} 0 & fm_2 \\ s_1 & (1-f)s_2 \end{bmatrix}$ | age 1 non-reproductive | age 2 ... reproductive and non-reproductive |
| pre-reproductive delay, $f \equiv$ reproductive fraction | $\begin{bmatrix} (1-f) & m_2 \\ fs_1 & 0 \end{bmatrix}$ | ages 1, 2, ... non-reproductive | ages 2 ... reproductive |
| diapause, $f \equiv$ direct-developing fraction | $\begin{bmatrix} 0 & (1-f) \\ m_1 s_1 & fm_2 \end{bmatrix}$ | age 2 reproductive | age 1 reproductive and non-reproductive |

NOTE.—For each life history, $f$ is the average value of the "reproductive" fraction, $m_1$ and $m_2$ are the fertilities of class-1 and class-2 individuals, $s_1$ and $s_2$ are the survivorships associated with the transitions between class 1 and class 2 and between class 2 and class 2, respectively. (From Orzack 1993, with corrections.)

growth rate as a function of reproductive delay, implying that very different life histories can have very similar growth rates. A similar broad peak occurs for the other delay life histories. This relationship reflects a potential for neutral differentiation of phenotypes that is similar to the potential for neutral differentiation discussed above in the context of iteroparity.

To examine the dynamic consequences of variation in the delay itself when environmental fluctuations are not temporally correlated, I derived a second-order approximation to the stochastic

growth rate of each of the four life histories (Orzack 1993). In order
to assess the differences among the life histories in their responses
to environmental variability, vital rates were chosen such that the
life histories had identical values of $\lambda_0$ (= 1.0001). Owing to a programming error, some of the indifference curves shown in Figure
8 of the 1993 paper are incorrect. A correct analysis reveals some
remarkable features of these life histories.

The first is their potential for dynamic similarity. In particular, for a given survival rate, $s_1$, the stochastic growth rates for
the iteroparous and diapause life histories are *identical*, provided
that the average values of the fertilities, $m_1$ and $m_2$, are equal (see
Table 2). For the iteroparous life history, the second-order approximation to $a$ is

$$a \approx \log \lambda_0 - \frac{1}{2}\left[ \left(\frac{\partial \log \lambda_0}{\partial fm_1}\right)^2 \sigma^2_{fm_1} + \left(\frac{\partial \log \lambda_0}{\partial (1-f)m_2}\right)^2 \sigma^2_{(1-f)m_2} \right.$$
$$\left. -2\frac{\partial \log \lambda_0}{\partial fm_1}\frac{\partial \log \lambda_0}{\partial (1-f)m_2}\mathrm{cov}[fm_1, (1-f)m_2] \right],$$

where $f$ is the average reproductive fraction (which, by definition,
is greater than or equal to zero and less than or equal to one). But
this expression equals

$$\log \lambda_0 - \frac{\sigma^2_f}{2\lambda_0^2}\left(\frac{\lambda_0}{2\lambda_0 - fm_1}m_1 - \frac{s_1}{2\lambda_0 - fm_1}m_2\right)^2. \qquad (3)$$

For the diapause life history, the second-order approximation to
$a$ is

$$a \approx \log \lambda_0 - \frac{1}{2}\left[ \left(\frac{\partial \log \lambda_0}{\partial (1-f)}\right)^2 \sigma^2_{1-f} + \left(\frac{\partial \log \lambda_0}{\partial fm_2}\right)^2 \sigma^2_{fm_2} \right.$$
$$\left. -2\frac{\partial \log \lambda_0}{\partial (1-f)}\frac{\partial \log \lambda_0}{\partial fm_2}\mathrm{cov}\big[(1-f), fm_2\big] \right],$$

which, in turn, equals

$$\log \lambda_0 - \frac{\sigma^2_f}{2\lambda_0^2}\left(\frac{s_1 m_1}{2\lambda_0 - fm_2} - \frac{\lambda_0}{2\lambda_0 - fm_2}m_2\right)^2, \qquad (4)$$

which is equivalent to the expression (3) for the iteroparous life
history when $m_1 = m_2$. It is straightforward to show that this
identity holds regardless of the order of the approximation used.

In a similar way, one can find dynamic similarity between the
other delay life histories. In particular, one can show that the

second-order approximation to the stochastic growth rate for the biennial life history is

$$a \approx \log \lambda_0 - \frac{\sigma_f^2}{2\lambda_0^2} \left( \frac{s_1}{2\lambda_0 - (1-f)s_2} m_2 - \frac{\lambda_0}{2\lambda_0 - (1-f)s_2} s_2 \right)^2.$$

(5)

This expression is identical to the approximation for the prereproductive-delay life history,

$$a \approx \log \lambda_0 - \frac{\sigma_f^2}{2\lambda_0^2} \left( \frac{\lambda_0}{2\lambda_0 - (1-f)} - \frac{m_2}{2\lambda_0 - (1-f)} s_1 \right)^2$$ (6)

when $s_2 = 1.0$. As above, this identity does not depend on the order of the approximation.

Thus, dynamic analysis simultaneously reveals how similar and different these life histories are. Iteroparity and diapause can share a dynamic response to environmental variability, but one that is different from the response shared by the biennial and prereproductive-delay life histories. The above identities have not been noted previously. In one sense, they are not surprising given the similarities in the nature of their life-history classes (see Table 2). On the other hand, the life histories in each pair do differ from one another in some important ways, such as whether individuals can skip reproduction and whether individuals must make the transition between the first and second life-history classes.

By definition, each identity implies that the associated indifference curves are identical, as shown in Figure 3, where the curve for the biennial life history is included as well. These indifference curves reveal that the life histories differ in regard to which reproductive fraction is associated with the least sensitivity to environmental variability. The iteroparous and diapause life histories are least sensitive when the average reproductive fraction is smallest $(= 0.1)$, whereas the biennial life history has the opposite pattern, with the least sensitivity being associated with the largest average reproductive fraction $(= 0.9)$. In addition, the life histories differ in sensitivity to change in the average value, with the iteroparous and diapause life histories benefiting much more from a decrease from 0.5 to 0.1 than from a decrease from 0.9 to 0.5. In contrast, the biennial life history is relatively insensitive to all changes in the average delay fraction.

These curves imply that the direction of selection acting on reproductive delay differs in populations characterized by these

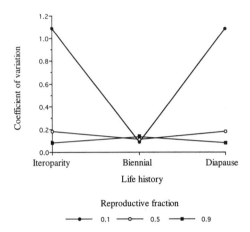

FIGURE 3. *Indifference curves for three of the life histories in Table 2 for various values of the average reproductive fraction* $f$. $a = 0.0, \lambda_0 = 1.0001, s_1$ *(and $s_2$, if present) = 0.8. Fertility, $m_1$ or $m_2$, is determined by $\lambda_0$ and the vital rates; $m_1$ and $m_2$ are equal if both are present. Values of the coefficient of variation are derived from equations (3)–(5).*

different life histories. Natural selection favors small reproductive fractions in populations with iteroparous and diapause life histories. Natural selection in a biennial population has the opposite tendency. The same is true of a population with a prereproductive-delay life history. This distinction in regard to the implied direction of life-history evolution can be understood by examining the sensitivity of a life history to a change in its components. This sensitivity is embodied in partial derivatives such as $\partial \log \lambda_0 / \partial f m_1$. In an environment with uncorrelated fluctuations, small sensitivities are better because they more effectively discount the negative effect of a given amount of variability. But the life histories differ in regard to how changing the average reproductive fraction, $f$, affects the sensitivities. In the case of iteroparity, for example, the relevant sensitivities, $\partial \log \lambda_0 / \partial f m_1$ and $\partial \log \lambda_0 / \partial (1 - f) m_1$, increase as the reproductive fraction increases. In contrast, the relevant sensitivities for the biennial life history, $\partial \log \lambda_0 / \partial f m_2$ and $\partial \log \lambda_0 / \partial (1 - f) s_2$, decrease as the reproductive fraction increases.

The second remarkable feature involving these delay life histories relates to the biennial and prereproductive-delay life histories. In particular, these life histories can be much more resistant to environmental fluctuations than are the other delay life histories. It is revealing in this context to use a higher-order approximation to the stochastic growth rate. For example, for the prereproductive-delay life history, it is straightforward to show (using the approach outlined in Tuljapurkar 1990$a$) that a fourth-order approximation is

$a \approx$

$$
\begin{aligned}
&\log \lambda_0 - \frac{\sigma_f^2}{2\lambda_0^2}\left(\frac{\lambda_0}{2\lambda_0 - (1-f)} - \frac{m_2}{2\lambda_0 - (1-f)}s_1\right)^2 \\
&+ \frac{\sigma_f^3}{3\lambda_0^3}\left(\frac{\lambda_0}{2\lambda_0 - (1-f)} - \frac{m_2}{2\lambda_0 - (1-f)}s_1\right)^3 \\
&- \frac{\sigma_f^4}{\lambda_0^4}\left(\frac{\lambda_0}{2\lambda_0 - (1-f)} - \frac{m_2}{2\lambda_0 - (1-f)}s_1\right)^2 \\
&\times \left[\frac{1}{4}\left(\frac{\lambda_0}{2\lambda_0 - (1-f)} - \frac{m_2}{2\lambda_0 - (1-f)}s_1\right)^2\right. \\
&\quad - \frac{1}{1-\lambda_1/\lambda_0}\left(\frac{\lambda_0}{2\lambda_0 - (1-f)} - \frac{m_2}{2\lambda_0 - (1-f)}s_1\right) \\
&\quad \times \left(\frac{\lambda_1}{2\lambda_1 - (1-f)} - \frac{m_2}{2\lambda_1 - (1-f)}s_1\right) \\
&\quad \left. + \frac{1}{2[1-(\lambda_1/\lambda_0)^2]}\left(\frac{\lambda_1}{2\lambda_1 - (1-f)} - \frac{m_2}{2\lambda_1 - (1-f)}s_1\right)^2\right],
\end{aligned}
\tag{7}
$$

where $\lambda_1$ ($= 1 - f - \lambda_0$) is the second eigenvalue of the average projection matrix. This expression also applies to the biennial life history if $s_2 = 1.0$.

When $\lambda_0 = 1.0001$ and $s_1 = s_2 = 0.8$, the second-order approximations are reasonably accurate for the other life histories when the average reproductive fraction is 0.5, for example (see Fig. 4). Fourth-order approximations to the stochastic growth rates for these life histories are not more accurate, at least as judged by their agreement with the simulation results (not shown). But for the same reproductive fraction, the prereproductive-delay life history is so resistant to environmental variability that huge coefficients of variation are required to produce a stochastic growth rate of 0.0. As a result, the accuracies of the second- and fourth-order approximations for the prereproductive-delay life history are impossible to judge for this reproductive fraction since the predicted coefficients of variation surpass those possible for a random variable bounded between 0 and 1. However, this life history is more

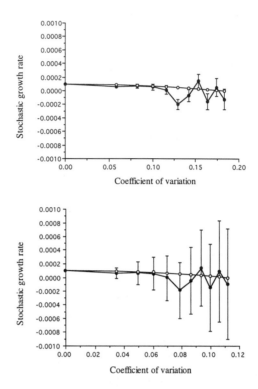

FIGURE 4. *Predicted* (open circles) *and observed* (closed circles) *stochastic growth rates for the iteroparous and diapause life histories* (top) *and for the biennial life history* (bottom) *for a range of coefficients of variation when the average reproductive fraction, $f$, equals 0.5; $a = 0.0$, $\lambda_0 = 1.0001$, $s_1 = s_2 = 0.8$. Fertility, $m_1$, is determined by $\lambda_0$ and the vital rates; $m_1$ and $m_2$ are equal if both are present. Second-order predictions are derived from equations (3)–(5). Each observed value is the mean value of $\log(M_t/M_{t-1})$ for a sample path of 10,000 time units, in which realized values of the delay fraction are sampled independently at each time step from a two-state distribution having a mean value of 0.5 and a variance determined by the given coefficient of variation; the 95 percent confidence interval for each mean is shown. The initial population vector is $(5.0, 5.0)$.*

sensitive to environmental variation when $\lambda_0$ is larger, the consequence being that smaller coefficients of variation are needed to reduce the stochastic growth rate to 0.0, for example. As a result, for higher values of $\lambda_0$, very low reproductive fractions ($f = 0.01$)

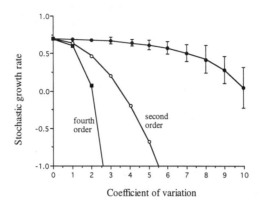

FIGURE 5. *Predicted and observed stochastic growth rates for the prereproductive-delay life history for a range of values of the coefficient of variation when the delay fraction, $f$, equals 0.01; $a = 0.0$, $\lambda_0 = 2.0$, $s_1 = 0.8$. Fertility, $m_1$, is determined by $\lambda_0$ and the vital rates. Second-order predictions are derived from equation (6); fourth-order predictions from equation (7). Each observed value is the mean value of $\log(M_t/M_{t-1})$ for a sample path of 10,000 time units, in which realized values of the delay fraction are sampled independently at each time step from a two-state distribution having a mean value of 0.01 and a variance determined by the given coefficient of variation; the 95 percent confidence interval for each mean is shown. The initial population vector is $(5.0, 5.0)$.*

result in feasible coefficients of variation. Even so, the resulting numerical analysis indicates that both approximations are close to being accurate only for a coefficient of variation of 1.0 (see Fig. 5). Despite their overall quantitative inaccuracy, both approximations correctly reflect the fact that this life history and the biennial life history can be much less sensitive to environmental variation than the other delay life histories These life histories are distinct from the others in containing at least one stage in which individuals may remain forever. This reservoir of individuals of varied ages can dramatically dampen the effects of environmental variation.

Further work needs to be done in regard to understanding the dynamics of the biennial and prereproductive-delay life histories. It is especially intriguing that as $\lambda_0$ increases, the ordering of the analytical approximations reverses, with the predicted coefficients of variation from the second-order approximation being smaller

than those from the fourth-order approximation for small values
of $\lambda_0$ and being larger for large values of $\lambda_0$ (not shown). Gaining
a better understanding of these life histories and the development
of accurate approximations to their stochastic growth rates are
obvious goals for future work.

It is worth mentioning that the differences among the dynamics
of these life histories serve as a reminder of the importance of
combining analytical and numerical work, since they are revealed
only by Monte Carlo simulations.

## 6  The Geometry of Reproductive Delay

The delay life histories just described encompass much of the observed variety of such life histories. Nonetheless, they have a common feature: delay occurs at only one stage in the life history.
Much of interest can be learned from a more general analysis, one
in which delay occurs in both life-history stages. Such a life history
is essential to consider in the context of comparative life-history
analysis. After all, differential proliferation of a life history with
one kind of delay could occur because the demographic "anatomy"
of the life history makes it easier to evolve than others, as opposed
to its being inherently superior in a selective sense. Understanding
the degree to which this is true is made possible by considering
a population with an average developmental fraction $f_L$ and an
average reproductive fraction $f_A$ (see Table 3). Such a life history encompasses all of the delay life histories considered above
and can be loosely thought of as a "hybrid" of the biennial and
prereproductive-delay life histories. When the reproductive rate is
temporally variable, the two kinds of delay are interchangeable if
they have equal costs and if any dependence among the environmental states (such as autocorrelation) is the same for forward and
backward sample paths.

Of most importance is Tuljapurkar and Wiener's demonstration
(unpubl. ms.) that entrainment may occur in the direction of life-
history evolution. In particular, consider the surface defined by
values of stochastic growth rate as a function of the two fractions.
This surface is asymmetrical, such that, for example, when one
fraction is zero, for a given total change in fraction, the selective
gradient is steeper when changing the nonzero fraction as opposed
to increasing the other fraction away from zero. To this extent,
the initial direction of life-history evolution could determine the
subsequent direction.

TABLE 3. *Life Histories with Larval and Adult Delays*

| Phenotype | Projection Matrix | Type(s) of Individuals in Life-History Class | |
|---|---|---|---|
| | | $N_1$ | $N_2$ |
| Mixed larval and adult delays | $\begin{bmatrix} s_\mathrm{L}(1 - f_\mathrm{L}) & m f_\mathrm{A} \\ f_\mathrm{L} & s_\mathrm{A}(1 - f_\mathrm{A}) \end{bmatrix}$ | ages $1, 2, \ldots$ non-reproductive | ages $2 \ldots$ reproductive and non-reproductive |
| Pure larval delay $f_\mathrm{A} = 1$ | $\begin{bmatrix} s_\mathrm{L}(1 - f_\mathrm{L}) & m \\ f_\mathrm{L} & 0 \end{bmatrix}$ | ages $1, 2, \ldots$ non-reproductive | ages $2 \ldots$ reproductive |
| Pure adult delay $f_\mathrm{L} = 1$ | $\begin{bmatrix} 0 & m f_\mathrm{A} \\ 1 & s_\mathrm{A}(1 - f_\mathrm{A}) \end{bmatrix}$ | age $1$ non-reproductive | ages $2 \ldots$ reproductive and non-reproductive |

NOTE.—$s_\mathrm{L}$ and $s_\mathrm{A}$ are the larval and adult survivorships, $f_\mathrm{L}$ is the average developmental fraction, $f_\mathrm{A}$ is the average reproductive fraction, $m$ is the fertility of class-2 individuals.

As in the case of the delay life histories, an important analysis concerns the dynamic consequences of variability in the developmental and reproductive fractions. A second-order approximation to the stochastic growth rate of a life history with two delay fractions is

$$
\begin{aligned}
a \approx \log \lambda_0 - \frac{1}{2} \Bigg[ & \left( \frac{\partial \log \lambda_0}{\partial s_\mathrm{L}(1 - f_\mathrm{L})} \right)^2 \sigma^2_{s_\mathrm{L}(1-f_\mathrm{L})} + \left( \frac{\partial \log \lambda_0}{\partial f_\mathrm{L}} \right)^2 \sigma^2_{f_\mathrm{L}} \\
& - 2 \frac{\partial \log \lambda_0}{\partial s_\mathrm{L}(1 - f_\mathrm{L})} \frac{\partial \log \lambda_0}{\partial f_\mathrm{L}} \mathrm{cov}\big[ s_\mathrm{L}(1 - f_\mathrm{L}), s_\mathrm{L} \big] \\
& + \left( \frac{\partial \log \lambda_0}{\partial m f_\mathrm{A}} \right)^2 \sigma^2_{m f_\mathrm{A}} + \left( \frac{\partial \log \lambda_0}{\partial s_\mathrm{A}(1 - f_\mathrm{A})} \right)^2 \sigma^2_{s_\mathrm{A}(1-f_\mathrm{A})} \\
& - 2 \frac{\partial \log \lambda_0}{\partial m f_\mathrm{A}} \frac{\partial \log \lambda_0}{\partial s_\mathrm{A}(1 - f_\mathrm{A})} \mathrm{cov}\big[ m f_\mathrm{A}, s_\mathrm{A}(1 - f_\mathrm{A}) \big] \Bigg],
\end{aligned}
$$

where $s_L$ and $s_A$ are the larval and adult survival rates and the same general notation is used as in the case of the delay life histories. Fluctuations in the developmental and reproductive fractions are assumed to be independent of one another. But this expression equals

$$
\log \lambda_0 - \frac{1}{2\lambda_0^2} \left[ \sigma_{f_L}^2 \left( s_L \frac{\partial \lambda_0}{\partial s_{rmL}(1 - f_L)} - \frac{\partial \lambda_0}{\partial f_L} \right)^2 \right.
$$
$$
\left. + \sigma_{f_A}^2 \left( m \frac{\partial \lambda_0}{\partial m f_A} - s_A \frac{\partial \lambda_0}{\partial s_A(1 - f_A)} \right)^2 \right]
$$
$$
= \log \lambda_0 - \frac{1}{2\lambda_0^2} \left[ \sigma_{f_L}^2 \left( s_L \frac{\lambda_0 - s_A(1 - f_A)}{2\lambda_0 - s_L(1 - f_L) - s_A(1 - f_A)} \right. \right. \tag{8}
$$
$$
\left. - \frac{m f_A}{2\lambda_0 - s_L(1 - f_L) - s_A(1 - f_A)} \right)^2
$$
$$
+ \sigma_{f_A}^2 \left( m \frac{f_L}{2\lambda_0 - s_L(1 - f_L) - s_A(1 - f_A)} \right.
$$
$$
\left. \left. - s_A \frac{\lambda_0 - s_L(1 - f_L)}{2\lambda_0 - s_L(1 - f_L) - s_A(1 - f_A)} \right)^2 \right].
$$

This expression is reasonably accurate as judged from a comparison of predicted and observed stochastic growth rates, especially when the average reproductive fraction is 0.5 (see Fig. 6). In fact, Monte Carlo simulations indicate that the second-order approximation is reasonably accurate even for coefficients of variation of 6 or 7 (see Fig. 7) and for fractions close to 1.0, at which there is a well-known singularity in the relationship between stochastic growth rate and delay (see, e.g., Roerdink 1988).

When larval ($s_L$) and adult ($s_A$) survival rates are equal and delay fractions are equal as well, the structure of equation (8) indicates that the stochastic growth rate is symmetrical with respect to the effects of delay; that is, it is affected equally by variability in the developmental fraction and by variability in the reproductive fraction. This exchangeability is a special case of Result 4 from Tuljapurkar and Wiener (unpubl. ms.).

An indifference surface defining the dependence of the coefficient of variation on the two fractions is shown in Figure 7. The symmetry of the surface means that the dynamic consequences of delay are interchangeable. The life histories most resistant to

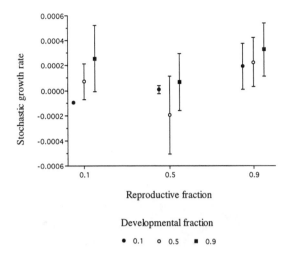

FIGURE 6. *Observed stochastic growth rates for life histories with two delays; $a = 0.0$, $\lambda_0 = 1.0001$, $s_L = s_A = 0.8$. Fertility, $m$, has a value determined by $\lambda_0$ and the vital rates. Each observed value is the mean value of $\log(M_t/M_{t-1})$ for a sample path of 5,000 time units, in which realized values of the developmental and reproductive fractions are sampled independently at each time step from a two-state distribution having a given mean value ($f_L$ or $f_A$) and a variance determined by the given coefficient of variation; the 95 percent confidence interval for each mean is shown. The initial population vector is $(5.0, 5.0)$.*

environmental variability are those with large developmental and large reproductive fractions. Populations with such life histories can withstand levels of environmental variability that cause other populations to have highly negative stochastic growth rates. But a large developmental or reproductive fraction by itself does little to accentuate the ability of a population to withstand environmental fluctuations. In addition, the shallowness of the indifference surface demonstrates that populations with very distinct life histories can exhibit similar responses to environmental variability. As in the examples discussed above, this similarity implies that there is a substantial potential for neutral life-history evolution in this instance. In general, this potential depends on the magnitude of $\lambda_0$, since this affects the slope of the indifference surface.

Given the advantage of life histories with two delays for sustaining growth in highly variable environments, it is telling that such

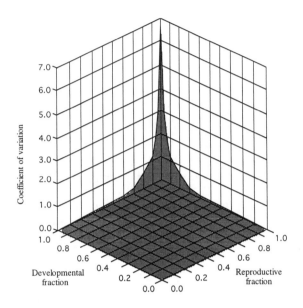

FIGURE 7. *Indifference surface for life histories with a developmental fraction ($f_L$) and reproductive fraction ($f_A$); $a = 0.0$, $\lambda_0 = 1.0001$, $s_L = s_A = 0.8$. Fertility, $m$, has a value determined by $\lambda_0$ and the vital rates. Values of the coefficient of variation are derived from equation (8). The points $(0,0)$, $(0,1)$, $(1,0)$, and $(1,1)$ are inaccessible.*

life histories are rare at best. It may be that such life histories cannot evolve because of developmental constraints, but compelling data in this regard are not available.

These analyses, whether they concern the "flavors" of reproductive delay or the consequences of developmental and reproductive fractions, show how much remains to be explored with regard to the stochastic dynamics of life histories. Their dynamics can be complex despite the relatively simple structure of the life history. This is best exemplified by the surprising behavior of the prereproductive-delay life history. It is likely that more surprises await us.

## 7  Population Extinction

Equality of stochastic growth rates is the neutral basis of the indifference curves discussed above. Yet the life histories associated with an indifference curve may differ with respect to another important metric of evolutionary success: a population's probability of extinction.

The extinction dynamics of nonstructured and structured populations subject to density-independent environmental stochasticity are analytically well understood, at least to the extent that diffusion analysis can be taken to describe the dynamics adequately. In this context, the standard assumptions underlying diffusion analysis amount to assuming that the population undergo only local changes in size (as opposed to undergoing catastrophes) and that the stochastic growth rate and its variance be temporally invariant. The latter assumption in particular implies that demographic stochasticity is not accounted for in the analysis, and to this extent, it implies that populations under consideration are at least moderate in size except perhaps during a relatively fast decline to extinction.

The asymptotic lognormality of population size implies that a Wiener process is a good descriptor of changes in population size (Tuljapurkar & Orzack 1980). Accordingly, the probability that a population starting from any initial size is a given size at a particular time is given by the first-passage-time probability for this process, given an absorbing barrier. From this quantity one can calculate finite-time and ultimate probabilities of extinction as well as the moments of the extinction-time distribution (Lande & Orzack 1988). In a strict sense, use of the Wiener process implies that changes in the logarithm of population size be temporally independent and normally distributed. Yet, both assumptions are violated for a structured population studied in finite time. Nonetheless, the usefulness of the Wiener process is borne out by Monte Carlo simulations (Lande & Orzack 1988; Orzack 1993, unpubl. ms.), with finite-time extinction probabilities being predicted accurately as long as the distance between initial population size and the extinction boundary is large enough that approximate lognormality is attained before extinctions start to occur. This accuracy is a testimonial to the fact that extinction dynamics in these simulations are in fact determined primarily by the stochastic growth rate and its variance (as is implied by the use of the diffusion equation) and by the initial population size. Since the stochastic growth rate and its variance can be obtained from time-series data (see, e.g., Lande & Orzack 1988; Dennis et al. 1991), as opposed to necessarily requiring the enumeration of the structure of the life history, the analysis can be almost scalar in character. The only "concession" to the structured nature of the problem is that differences in the future reproductive potential of individuals in different classes need to be accounted for in the estimate of the initial population size. After all, structure in the demography inherently implies that individuals may differ in their effects on future population growth. These differences can be accurately accounted for by calculating an adjusted initial population size as determined by the total reproductive value in the initial population (the dot product of the left eigenvector of the average projection matrix and the initial population vector; Lande & Orzack 1988).

## 8  Evolution Within and Between Populations

One key aspect of extinction dynamics is their connection to evolutionary dynamics within populations. The role of the stochastic growth rate in life-history evolution within populations has been described above. It also plays an important role in determining extinction dynamics. For example, the cumulative probability of extinction up to time $t$ is

$$G(t|x_0) = \Phi\left(\frac{-x_0 - at}{\sigma\, t^{1/2}}\right)$$
$$+ \exp\left(\frac{-2ax_0}{\sigma^2}\right)\left[1 - \Phi\left(\frac{x_0 - at}{\sigma\, t^{1/2}}\right)\right],$$

where $\Phi(y)$ is the normal probability integral

$$\Phi(y) = \frac{1}{2\pi}\int_{-\infty}^{y}\exp\left(\frac{-z^2}{2}\right)dz\,,$$

$\sigma^2$ is the variance of the stochastic growth rate, $a$, and $x_0$ is the distance on the logarithmic scale between adjusted initial population size and the extinction boundary (Lande & Orzack 1988). This is a remarkable instance of a good analytical understanding of the relationship between a determinant of within-population success and a determinant of between-population success.

A number of features of the extinction dynamics of structured populations underscore the importance of accounting for environmental and demographic dynamics. The probability of ultimate extinction, $G(\infty|x_0)$, is 1.0 when $a \leq 0$ and is $\exp(-2ax_0/\sigma^2)$ when $a > 0$. Suppose that $a$ is greater than zero. Since $\sigma^2$ is approximately equal to $2(\log \lambda_0 - a)$, life histories with higher average growth rates ($\lambda_0$) have a *higher* probability of ultimate extinction, assuming that the life histories have identical stochastic growth rates and identical initial population sizes (Orzack 1993).

This point is of special applied importance for assessing the extinction threat to an endangered species, because deterministic theory has often been taken to imply that increased average growth rate necessarily reduces the threat of extinction.

The second feature concerns finite-time extinction probabilities. Consider for simplicity the case $a = 0$. Here, the cumulative probability of extinction is

$$G(t|x_0) = \Phi\left(\frac{-x_0}{\sigma\, t^{1/2}}\right) + 1 - \Phi\left(\frac{x_0}{\sigma\, t^{1/2}}\right),$$

$$= 2\left[1 - \Phi\left(\frac{x_0}{\sigma\, t^{1/2}}\right)\right]$$

$$\approx 2\left[1 - \Phi\left(\frac{x_0}{(\log \lambda_0 t)^{1/2}}\right)\right].$$

It can be seen that life histories with a higher value of $\lambda_0$ have higher finite-time extinction probabilities for the same initial population size. This result is true even when $a \neq 0$. As noted above in regard to ultimate extinction probabilities, such a counterintuitive relationship is potentially of great applied significance.

## 9  Life-History Evolution and Extinction

The life histories associated with an indifference curve may or may not be neutral with respect to ultimate and finite-time extinction probabilities. When the stochastic growth rate is greater than zero, ultimate extinction probabilities may differ because of their dependence on $\lambda_0$ and on $x_0$. For example, consider the life histories shown in Table 1. When $a = 0.001$ and ten newborn individuals initiate a population, the semelparous life history (life history 1) has an ultimate extinction probability of 0.771. In contrast, life history 6 has an ultimate extinction probability of 0.028, and life history 10 has an ultimate extinction probability of $<1 \times 10^{-8}$. Such differences dramatically demonstrate that populations characterized by short-lived life histories can become extinct at much higher rates than, for example, long-lived populations with iteroparous life histories. The differential extinction of such life histories is one possible answer to Cole's question of why such life histories are uncommon. The important general point is that selection within populations is not the only possible answer to this question.

Of course, one can define indifference curves with respect to stochastic growth rate and ultimate extinction probability. Perhaps the most interesting case in this regard is when life histories share

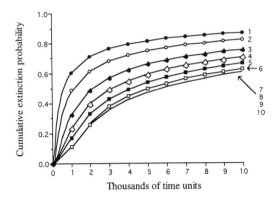

FIGURE 8. *Transient extinction probabilities for the life histories shown in Table 1; $a = 0.0$ and the extinction boundary is $\Sigma N_i = 1.0$. The adjusted initial population size for each life history is based on the assumption that the initial population vector is $(5.0, 5.0)$. These curves are independent of the covariance between vital rates at any one time. (From Orzack 1993.)*

a stochastic growth rate of 0.0. Populations with such a stochastic growth rate eventually must become extinct, despite being in ecological "balance" at least as judged from their expected average growth in the absence of an extinction threshold. Yet such life histories are not neutral in the *transient* sense; that is, their extinction probabilities at finite times differ.

Consider again the life histories shown in Table 1. Finite-time extinction probabilities are shown in Figure 8. Two features of these curves are important. The first is that there is no simple relationship between differences in life-history structure and differences in transient extinction probabilities. Compare, for example, life histories 2, 3, 4, and 5 with life histories 7, 8, 9, and 10. These sets have the same range of differences in the number of reproductive events during life (see Table 1). Yet, at finite times, the life histories in the first set differ appreciably in extinction probabilities, whereas those in the second set are very similar.

A similar comparison of finite-time extinction probabilities can be made with regard to the various delay life histories. Despite the marked difference in the structure of these life histories and their differences in response to environmental variability, when $f = 0.5$, their extinction probabilities are always within a few percent of one

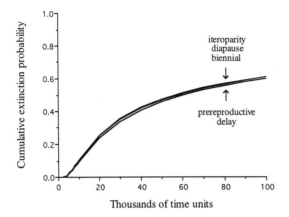

FIGURE 9. *Transient extinction probabilities for the life histories in Table 2; $a = 0.0$, $f = 0.5$, $\lambda_0 = 1.0001$, $s_1 = s_2 = 0.8$. Fertility, $m_1$, has a value determined by $\lambda_0$ and the vital rates; $m_1$ and $m_2$ are equal if both are present. The extinction boundary is $\Sigma N_i = 1.0$. The adjusted initial population size for each life history is based on the assumption that the initial population vector is $(5.0, 5.0)$. "Iteroparity," "biennial," and "diapause" have similar extinction probabilities that cannot be distinguished in this graph; their order is reflected in the order of the labels.*

another, as shown in Figure 9. In this instance, the iteroparous population tends to disappear at a higher rate than populations with the other life histories. Extinction probabilities are not strongly affected by a change in the delay fraction in this case, although its quantitative effect can be greater if $\lambda_0$ is larger.

It is important to stress that life histories with equal stochastic growth rates do *not* generally have identical extinction dynamics. Differences in reproductive-value schedules and the resulting differences in adjusted initial population sizes (gven identical initial "raw" population sizes) cause the diapause life history to have higher finite-time extinction probabilities when the reproductive fraction is small. When the reproductive fraction is large, this inequality is reversed and the iteroparous life history has a higher extinction rate (as shown in Fig. 9). The extinction calculations shown here are valid for the prereproductive-delay life history in the sense that they depend only on there being some level of environmental variability that causes the stochastic growth rate to be zero. Neither the vital rate(s) affected by variability nor the magnitude of the variability need be specified.

In a similar way, the symmetry of the indifference surface involving developmental and reproductive fractions does not carry over to extinction dynamics. Given the life histories underlying the data in Figure 7, one can show that a given amount of delay is associated with a higher extinction probability when it occurs in the reproductive fraction as opposed to the developmental fraction. For example, a population with a reproductive fraction of 0.5 has a probability of 0.607 of being extinct after 100,000 time units; a population with a developmental fraction of 0.5 has a probability of 0.602. In this instance, the difference in extinction probabilities is small, but in general it need not be. The important evolutionary point is that the exchangeability of stochastic growth rates does not result in exchangeability of extinction dynamics. This again underscores the potential difference between within-population and between-population dynamics. Life-history analyses have traditionally focused almost exclusively within populations (but see Holgate 1967). As should be clear, much can be learned from a broader view.

## 10  Future Directions

All the analyses described here, whether they relate to life-history evolution within populations, to population regulation, or to extinction, represent just partial glimpses of dynamic phenomena in nature. This is discouraging only if one forgets how quickly our understanding of ecological and evolutionary dynamics of structured populations has improved and how much unprecedented power lies in the new analytical tools developed in stochastic demography.

### Literature Cited

Caswell, H. 1989. *Matrix Population Models.* Sinauer, Sunderland, Mass.

Charlesworth, B. 1994. *Evolution in Age-Structured Populations.* Cambridge University Press.

Cohen, J. 1977. Ergodicity of age structure in populations with Markovian vital rates, III. Finite-state moments and growth rate; an illustration. *Advances in Applied Probability* 9: 462–475.

———. 1979*a*. Long-run growth rates of discrete multiplicative processes in Markovian environments. *Journal of Mathematical Analysis and Applications* 69: 243–257.

———. 1979*b*. Comparative statics and stochastic dynamics of age-structured populations. *Theoretical Population Biology* 16: 159–171.

Cole, L. C. 1954. The population consequences of life history phenomena. *Quarterly Review of Biology* 19: 103–137.

den Boer, P. J. 1990. Density limits and survival of local populations in 64 carabid species with different powers of dispersal. *Journal of Evolutionary Biology* 3: 19–48.

Dennis, B., P. L. Munholland, and J. M. Scott. 1991. Estimation of growth and extinction parameters for endangered species. *Ecological Monographs* 61: 115–143.

Eshel, I. 1984. Evolutionary stable strategies and viability selection in Mendelian populations. *Theoretical Population Biology* 22: 204–217.

Fisher, R. A. 1930. *The Genetical Theory of Natural Selection.* Clarendon Press, Oxford.

Furstenberg, H., and H. Kesten. 1960. Products of random matrices. *Annals of Mathematical Statistics* 31: 457–469.

Godfray, H. C. J., and M. P. Hassell. 1992. Long time series reveal density dependence. *Nature* 359: 673–674.

Goodman, D. 1984. Risk spreading as an adaptive strategy in itero- parous life histories. *Theoretical Population Biology* 25: 1–20.

Holgate, P. 1967. Population survival and life history phenomena. *Journal of Theoretical Biology* 14: 1–10.

Johnson, N. L., and S. Kotz. 1970. *Distributions in Statistics: Continuous Univariate Distributions–1.* Wiley, New York.

Kolman, W. 1960. The mechanism of natural selection on the sex ratio. *American Naturalist* 94: 373–377.

Lande, R. 1993. Risks of population extinction from demographic and environmental stochasticity and random catastrophes. *American Naturalist* 142: 911–927.

Lande, R., and S. H. Orzack. 1988. Extinction dynamics of age-struc- tured populations in a fluctuating environment. *Proceedings of the National Academy of Sciences (USA)* 85: 7418–7421.

Leslie, P. H. 1945. On the use of matrices in certain population mathematics. *Biometrika* 33: 213–245.

Lewontin, R. C., and D. Cohen. 1969. On population growth in a randomly varying environment. *Proceedings of the National Academy of Sciences (USA)* 62: 1056–1060.

Murphy, G. 1968. Pattern in life history and the environment. *American Naturalist* 102: 391–403.

Orzack, S. H. 1985. Population dynamics in variable environments V. The genetics of homeostasis revisited. *American Naturalist* 125: 550–572.

———. 1993. Life history evolution and population dynamics in variable environments: Some insights from stochastic demography. Pages 63–104 in J. Yoshimura and C. W. Clark, eds., *Adaptation in Stochastic Environments.* Springer-Verlag, New York.

Orzack, S. H., and E. Sober. 1994. Optimality models and the test of adaptationism. *American Naturalist* 143: 361–380.

Orzack, S. H., and S. Tuljapurkar. 1989. Population dynamics in variable environments VII. The demography and evolution of iteroparity. *American Naturalist* 133: 901–923.

Roerdink, J. B. T. M. 1988. The biennial life strategy in a random environment. *Journal of Mathematical Biology* 26: 199–215.

———. 1989. The biennial life strategy in a random environment. [Supplement.] *Journal of Mathematical Biology* 27: 309–319.

Schaffer, W. M. 1974. Optimal reproductive effort in fluctuating environments. *American Naturalist* 108: 783–790.

Tuljapurkar, S. D. 1982*a*. Population dynamics in variable environments II. Correlated environments, sensitivity analysis and dynamics. *Theoretical Population Biology* 21: 114–140.

———. 1982*b*. Population dynamics in variable environments III. Evolutionary dynamics of *r*-selection. *Theoretical Population Biology* 21: 141–165.

———. 1990*a*. *Population Dynamics in Variable Environments*. Lecture Notes in Biomathematics 85. Springer-Verlag, New York.

———. 1990*b*. Delayed reproduction and fitness in variable environments. *Proceedings of the National Academy of Sciences (USA)* 87: 1139–1143.

Tuljapurkar, S. D., and C. Istock. 1993. Environmental uncertainty and variable diapause. *Theoretical Population Biology* 43: 251–280.

Tuljapurkar, S. D., and S. H. Orzack. 1980. Population dynamics in variable environments I. Long-run growth rates and extinction. *Theoretical Population Biology* 18: 314–342.

Tuljapurkar, S. D., and P. Wiener. Escape in time: Stay young or age gracefully? Unpubl. ms.

CHAPTER 9

# Population Dynamics of *Tribolium*

## Robert A. Desharnais

The study of population dynamics is complicated by the difficulty of obtaining experimental evidence that supports the predictions of mathematical models. Biologists are faced with numerous challenges such as sampling error, environmental fluctuations, and lack of replication in the study of populations in the field. One approach to this problem is to examine experimental populations under controlled laboratory conditions. Flour beetles of the genus *Tribolium* are an excellent experimental system for evaluating predictions of demographic models.

The flour beetle has a long history in population biology. Royal Norton Chapman (1928) introduced *Tribolium* as an experimental insect for the study of population growth. Age structure in flour beetles was first investigated by John Stanley (1932), a student of Chapman. Thomas Park (1948) began an extensive series of experimental investigations on species competition involving *T. castaneum* and *T. confusum*. Reviews of the vast literature that has developed on *Tribolium* can be found in the papers by King and Dawson (1972), Mertz (1972), and Bell (1982) and in the books by Sokoloff (1972, 1974, 1977) and Costantino and Desharnais (1991).

There are several reasons for the popularity of *Tribolium* as an animal model for the study of populations. Cultures can be maintained indefinitely on a simple medium of flour and yeast. The insects undergo a holometabolous development from egg to adult in four to six weeks. Adults are about 3 mm in length; hundreds of insects can be maintained on 8–20 grams of flour. An accurate census of the population is obtained by sifting the medium and counting the life stages. Large numbers of replicate cultures can be kept

in an unlighted incubator under constant conditions of temperature and humidity. Several species, mutants, and genetic strains
are available. Much is already known about the basic anatomy,
taxonomy, physiology, development, genetics, and ecology of this
beetle (Sokoloff 1972, 1974, 1977).

This chapter focuses on one specific area of population research
involving *Tribolium*: nonlinear demographic dynamics. The emphasis is on approaches and results appropriate to the theme of
"structured-population models" found throughout this book. A
comprehensive review of the literature is not attempted; this chapter deals mostly with the work of my colleagues and me (for more
in-depth treatment of these subjects, see Costantino and Desharnais 1991).

The chapter is divided into four parts. It begins with a brief description of the life-stage interactions that occur in flour beetle cultures. The remaining three sections illustrate different approaches
that have been used to model these interactions. The first (used
in Desharnais & Liu 1987) is based on the classic Leslie matrix
model for age-structured populations. The next section describes
the integral-equation model of Hastings (1987) and Hastings and
Costantino (1987) for the cannibalism of eggs by larvae; this is
referred to as the "egg-larval submodel." The last section deals
with the more recent results of Costantino et al. (1995), Cushing et al. (in press), and Dennis et al. (1995), which are based on
the "LPA model," a system of three difference equations for the
dynamics of the larval, pupal, and adult life stages. Each of these
approaches represents a different trade-off between biological complexity and mathematical tractability.

## 1 Life-Stage Interactions

One of the most compelling reasons for using *Tribolium* in the
study of populations is that it provides a fascinating example of
nonlinear demographic dynamics. Laboratory populations maintained under constant environmental conditions usually exhibit
dramatic fluctuations in density and age structure. These fluctuations cannot be characterized as stochastic; they are the result
of strong behavioral and physiological interactions among the life
stages—the most important being cannibalism.

The life-stage interactions that drive the dynamics of *Tribolium*
populations are summarized in Figure 1. The open arrows represent the life cycle, which, for *T. castaneum* at 34°C, has a duration

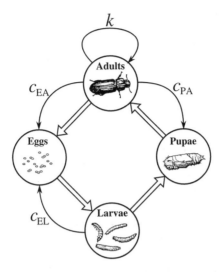

FIGURE 1. *Life-stage interactions in* Tribolium.

of approximately 28 days. The single arrows represent the interactions. The arrows labeled $c_{EA}$ and $c_{PA}$ represent the cannibalism of eggs and pupae, respectively, by adults. The arrow labeled $c_{EL}$ represents the cannibalism of eggs by larvae. Pupal cannibalism by larvae occurs at a much lower rate; for simplicity, it can be ignored. The fecundity of females decreases with crowding; this is represented by the arrow labeled $k$.

An important consequence of these life-stage interactions is that *Tribolium* populations rarely reach the "carrying capacity" of their laboratory "habitat." As the number of adult beetles increases, the effects of cannibalism and reduced fecundity cause decreases in adult recruitment. As an example, consider adults eating pupae, ignoring for the moment the other life-stage interactions. If $b$ denotes the number of pupae produced per adult, then the number of potential adult recruits is $bN$, where $N$ is the adult number. However, these pupae must survive cannibalism by adults. Ignoring subscripts, let $c$ denote the probability that a given adult will find and eat a given pupa in some fixed time interval. The probability that the pupa is not eaten by the adult is $1 - c$. If $N$ adults are present during the time interval, then the probability that the pupa avoids being eaten by any of the adults is $(1 - c)^N$, which, if $c$ is small, is approximately equal to $e^{-cN}$. The total number of recruits into the adult population becomes $bNe^{-cN}$, which decreases toward zero as $N \to \infty$. The same argument holds for the other life-stage interactions. A common observation is that the most cannibalistic species and genetic strains of *Tribolium* are the ones with the lowest population densities (Park et al. 1965; Costantino & Desharnais 1991). This negative exponential function appears throughout the *Tribolium* literature.

## 2 Leslie Matrix Model

An earlier model (Desharnais & Liu 1987) of the demographic dynamics of *Tribolium* populations was based on the projection-matrix approach introduced into biology by Bernardelli (1941), Lewis (1942), and Leslie (1945). In this model, the beetle life span is divided into $\omega$ age classes, where each age class spans a single day. The population is represented by a $\omega \times 1$ vector $\mathbf{n}(t)$, whose elements are the densities of each age class at time $t$. A Leslie matrix $\mathbf{M}(t)$ is used to project the age structure forward one unit of time (one day). The dynamics are given by the matrix equation

$$\mathbf{n}(t+1) = \mathbf{M}(t)\,\mathbf{n}(t), \tag{1}$$

which, in expanded form, is

$$\begin{pmatrix} n_1(t+1) \\ n_2(t+1) \\ n_3(t+1) \\ \vdots \\ n_\omega(t+1) \end{pmatrix} = \begin{pmatrix} b_1(t) & b_2(t) & \cdots & b_{\omega-1}(t) & b_\omega(t) \\ s_1(t) & 0 & \cdots & 0 & 0 \\ 0 & s_2(t) & \cdots & 0 & 0 \\ \vdots & \vdots & \ddots & \vdots & \vdots \\ 0 & 0 & \cdots & s_{\omega-1}(t) & 0 \end{pmatrix} \begin{pmatrix} n_1(t) \\ n_2(t) \\ n_3(t) \\ \vdots \\ n_\omega(t) \end{pmatrix}. \tag{2}$$

The $b_i(t)$'s and $s_i(t)$'s are age-specific values for fecundities and survival probabilities, respectively, which, in general, vary in time.

The strategy of the 1987 model was to represent the vital rates of the Leslie matrix by functions that capture the basic biology described in Figure 1 using as few parameters as possible. Parameters were estimated from data, and the model simulations were compared with experimental results. Inferences about the effects of life-stage interactions on the demographic dynamics of populations were obtained using numerical simulations to explore the model's behavior in various regions of parameter space. (A more general description of this model can be found in Costantino & Desharnais 1991.)

The first step is to group the age classes into life stages. Based on estimates of the life-stage durations for the corn oil sensitive (*cos*) mutant of *T. castaneum* (Moffa 1976), the life stages can be approximated using the following sets of integers: $E = \{1, 2, 3\}$ for eggs, $L = \{4, 5, \ldots, 23\}$ for larvae, $P = \{24, 25, 26\}$ for pupae, and $A = \{27, 28, \ldots, \omega = 300\}$ for adults. The densities of each stage are obtained by summing the age vector over these sets:

$$N_J(t) = \sum_{i \in J} n_i(t), \qquad J \in \{E, L, P, A\}.$$

For example, $N_L(t)$ and $N_A(t)$ are the number of larvae and adults, respectively.

Fecundity rates depend on the age of females and are reduced by the effects of crowding. The data of Moffa (1976) suggest that, for the *cos* strain, the fecundity of female adults decreases linearly with age. Rich (1956) showed that the fecundity rates of *T. confusum* females decrease as the density of adults increases. These results lead to the parameterization of the fecundity terms (Desharnais & Liu 1987) using

$$b_i(t) = \begin{cases} \left[\alpha - \beta(i - \varepsilon)\right] \exp\left[-kN_A(t)\right] & \text{for } i \in A, \\ & i \leq \varepsilon + \text{int}\,(\alpha/\beta), \quad (3) \\ 0 & \text{otherwise}, \end{cases}$$

where $\alpha$ is the maximum fecundity rate per adult (one-half the number of eggs laid per female per day), $\beta$ is the slope of the linear decrease in fecundity with age, $k$ is a parameter describing the sensitivity of fecundity to the effects of crowding, $\varepsilon = \min(A)$ is the age at which a beetle enters the adult stage, and "int" is the integer function. These fecundity values comprise the first row of the Leslie matrix in equation (2).

Survival probabilities involve both "natural" (nonpredatory) mortality and death due to cannibalism. In the absence of cannibalism, the survival probabilities of eggs and pupae are usually high (Moffa 1976); for simplicity, no natural mortality is assumed for these two life stages. Larvae and adults are assigned natural mortality rates of $\mu_L$ and $\mu_A$, respectively. The rate at which a larva consumes eggs depends on the age of the larva; older, larger larvae are more voracious cannibals of eggs. Let $c_{EL}(j)$ denote the cannibalistic rate of a larva of age $j$. Using the data of Park et al. (1965) for four genetic strains of *T. castaneum*, $c_{EL}(j)$ can be approximated as a linearly increasing function of age: $c_{EL}(j) = c'_{EL}(1 + j - \zeta)$, where $c'_{EL}$ is the slope of the linear increase and $\zeta = \min(L)$, the age at which a beetle enters the larval stage (Desharnais & Liu 1987). It is assumed that the rates $c_{EA}$ and $c_{PA}$, at which an adult canni-balizes eggs and pupae, respectively, are independent of the age of the adult. With these assumptions, the survival probabilities are

$$
s_i(t) = \begin{cases}
\exp\left[-c'_{EL} \sum_{j \in L} (1 + j - \zeta)n_j(t) \right. \\
\qquad \left. -c_{EA} N_A(t)\right] & \text{for } i \in E\,, \\
\exp(-\mu_L) & \text{for } i \in L\,, \\
\exp\left[-c_{PA} N_A(t)\right] & \text{for } i \in P\,, \\
\exp(-\mu_A) & \text{for } i \in A\,, \\
& \quad i \neq \omega\,.
\end{cases}
\tag{4}
$$

These probabilities form the subdiagonal of the Leslie matrix in equation (2).

Data from several sources are used to estimate the parameters in (3) and (4) (Desharnais & Liu 1987). Table 1 lists these parameters, their estimates and standard errors, and a reference to the source of the data. Whenever possible, data on the *cos* strain of *T. casta-neum* were used. However, the estimate of the rate of egg cannibal-ism by larvae was obtained from the data of Park et al. (1965) for four different genetic strains of *T. castaneum*, and the estimates of the parameters for crowding and egg cannibalism by adults were obtained from the data of Rich (1956) for the species *T. confusum*. (For details on how the numerical estimates and standard errors were computed, see Desharnais & Liu 1987; Costantino & Deshar-nais 1991.)

It is possible to make inferences about the rate of population growth under "density-independent" conditions. If one sets the pa-rameters $k$, $c_{EL}$, $c_{EA}$, and $c_{PA}$ equal to zero, the vital rates in the Leslie matrix $\mathbf{M}$ become constants, and one obtains the classic lin-ear model of geometric growth. The net reproductive rate is given by $R_0 = \sum_{i=1}^{\omega} s_{i-1} b_i$, where $s_0 = 1$. Assuming a 1:1 sex ratio, this equals half the number of eggs a female beetle is expected to produce in her lifetime. The daily rate of population increase, $\lambda_0$, is the dominant eigenvalue of the matrix $\mathbf{M}$, and the intrinsic rate of increase per day, $r_0$, is the natural logarithm of this eigenvalue. For the estimates in Table 1, we obtain $R_0 = 185.5$, $\lambda_0 = 1.140$, and $r_0 = 0.131$ per day. A female is expected to produce approx-

TABLE 1. *Parameter Estimates for the Leslie Matrix Model*

| Parameter | Description | Estimate (±SE) | Source of Data |
|---|---|---|---|
| $\mu_L$ | larval mortality | $0.0251 \pm 0.0020$ | Desharnais & Liu (1987) |
| $\mu_A$ | adult mortality | $0.0130 \pm 0.0009$ | Desharnais & Costantino (1980) |
| $\alpha$ | maximum fecundity | $7.96 \pm 0.20$ | Moffa (1976, p. 51) |
| $\beta$ | decrease in fecundity with age | $0.0664 \pm 0.0028$ | Moffa (1976, p. 51) |
| $k$ | decrease in fecundity from crowding | $0.00164 \pm 0.00006$ | Rich (1956, Table IV) |
| $c'_{EL}$ | larval cannibalism of eggs (slope) | $0.000760 \pm 0.000047$ | Park et al. (1965, Table 10) |
| $c_{EA}$ | adult cannibalism of eggs | $0.00252 \pm 0.00016$ | Rich (1956, Table IV) |
| $c_{PA}$ | adult cannibalism of pupae | $0.00558 \pm 0.00026$ | Jillson & Costantino (1980) |

imately 371 eggs in her lifetime. Once a constant age distribution is obtained, the population density increases by 14 percent each day—a prolific rate of growth. It is clear why flour beetles are such an important pest of stored grain and food products. These conclusions are consistent with other estimates for *T. castaneum* cultured at the same temperature and humidity (Sokoloff 1974, Table 11.22).

Predictions were obtained for the equilibrium densities and population stability. A unique equilibrium age vector $\mathbf{n}^* > \mathbf{0}$ exists if and only if $R_0 > 1$ (Desharnais & Liu 1987). Determining the stability of this equilibrium requires calculating a "stability matrix" $\mathbf{S}$ whose elements are the coefficients of a "linearization" of the model

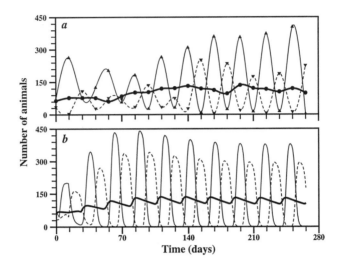

FIGURE 2. *Densities of adults* (heavy lines), *large larvae plus pupae*
(dashed lines), *and small larvae* (thin lines) *for* (a) *control replicate "A"*
*(Desharnais & Liu 1987) and* (b) *the Leslie matrix model.* (a) circles,
triangles, *Census points;* curves, *interpolating cubic splines.*

in the neighborhood of $n^*$. The eigenvalues of $S$ determine the
stability of the equilibrium; if the modulus of the dominant eigenvalue is less than one, the equilibrium is stable. Estimates of the
parameters have been used to provide expressions for computing
the elements of $n^*$ and $S$ (Desharnais & Liu 1987; Costantino & Desharnais 1991). When grouped into life stages, the predicted equilibrium densities are 430 eggs, 134 larvae, 10 pupae, and 88 adults,
which are within the ranges normally observed for this species.
However, this equilibrium is unstable; the dominant eigenvalues
are the complex conjugates $\lambda_1, \lambda_2 = 1.00 \pm 0.23\,i$ with a modulus of
1.03. The largest subdominant eigenvalues are $\lambda_3, \lambda_4 = 0.99 \pm 0.02\,i$
with a modulus of 0.99. All the remaining eigenvalues are complex
and unique. The linearized stability analysis predicts an unstable
equilibrium.

The model's demographic dynamics show good qualitative agreement with those observed in the laboratory. Figure 2 is a plot of
the life-stage densities for control replicate "A" of an earlier experiment (Desharnais & Costantino 1980) and the Leslie matrix
model. For this comparison, the larval age class is divided into two

TABLE 2. *Bifurcation Values for the Leslie Matrix Model*

| Model Parameter | Bifurcation Value | Ratio to Estimated Value | Effect of Increasing the Parameter |
|---|---|---|---|
| $\mu_L$ | 0.0788 | 3.14 | stabilize |
| $\alpha$ | 0.718 | 0.0902 | destabilize |
| $k$ | 0.0211 | 12.9 | stabilize |
| $c'_{EL}$ | 0.000146 | 0.192 | destabilize |
| $c_{EA}$ | 0.00896 | 3.56 | stabilize |
| $c_{PA}$ | 0.187 | 33.5 | stabilize |

groups of equal duration. The number of pupae is combined with the number of large larvae. Cubic splines interpolate the census data, which were collected every two weeks. Large oscillations in the number of immatures are evident. Smaller oscillations in the number of adults are predicted by the model; these low-amplitude oscillations are not obvious in the census data. With respect to the magnitudes of the life-stage oscillations, there is good agreement between the model and the census data. The same observation can be made for additional data on four control populations and nine populations subjected to demographic perturbations (Desharnais & Liu 1987). Overall, the model does a good job of capturing the qualitative behavior of the life-stage densities.

Numerical analyses of the Leslie matrix model provide information on the effects of each parameter on demographic stability. Varying each parameter singly, while keeping the remaining parameters at their estimated values, and employing a simple bisection searching algorithm, "bifurcation points" can be located at which the equilibrium loses stability (Desharnais & Liu 1987). The results are presented in Table 2. Oscillations persist for all values of $\beta$ and $\mu_A$, so these parameters are absent from the table. Increasing the rate of reproduction, $\alpha$, or the rate at which larvae eat eggs, $c_{EL}$, has a destabilizing effect; increasing the remaining parameters stabilizes the model. The ratios of the bifurcation values to the estimated values suggest that large modifications of the parameters would be required to stabilize the model; the laboratory populations lie well within the unstable region of parameter space.

A significant drawback of the nonlinear Leslie matrix model is that it is analytically intractable. A local stability analysis requires the calculation of the eigenvalues of a $300 \times 300$ matrix. Laborious numerical simulations are required to gain some modest information on the effects of the parameters on population growth; a systematic mapping of parameter space is not practical. So although the model includes several biologically realistic characteristics, it is limited in its ability to provide insight into the demographic dynamics of *Tribolium*. The next section describes a different modeling approach that sacrifices some biological realism in exchange for mathematical tractability.

## 3 Egg-Larval Submodel

In the Leslie matrix model of the preceding section, it was found that the cannibalism of eggs by larvae is destabilizing; this lifestage interaction drives the huge oscillations found in the immature age classes. By contrast to the dynamics of the immatures, the fluctuations in adult numbers are relatively small. A continuoustime model that focuses on the dynamics of the egg and larval life stages is called the "egg-larval submodel" (Hastings 1987; Hastings & Costantino 1987; Costantino & Desharnais 1991).

The egg-larval submodel is a special case of the McKendrick–von Foerster equation. This treatment of population growth, which was made popular by von Foerster (1959), is built on the work of Sharpe and Lotka (1911), Lotka (1925), and McKendrick (1926). If $n(x,t)$ denotes the number of individuals of age $x$ at time $t$, the aging and death of the population is given by

$$\frac{\partial}{\partial t}n(x,t) + \frac{\partial}{\partial x}n(x,t) = -\mu(x,t)\,n(x,t)\,, \qquad (5)$$

where $\mu(x,t)$ is the mortality rate for an individual of age $x$ at time $t$. Reproduction is expressed as a boundary condition, $n(0,t) = B(t)$, where $B(t)$ is the total birthrate of the population. Defining $b(x,t)$ as the number of offspring born to an individual of age $x$ at time $t$, the total birthrate is computed using

$$B(t) = \int_0^\omega n(x,t)\,b(x,t)\,dx\,, \qquad (6)$$

where $\omega$ is the maximum attainable age. To complete the model, one must also specify an initial age distribution $n(x,0)$. In integral form, the model can be written as

$$
n(x,t) = \begin{cases} n(x-t,0)\exp\left(-\displaystyle\int_{x-t}^{x}\mu(s,s+t-x)\,ds\right) & \text{for } t < x \le \omega, \\[2ex] n(0,t-x)\exp\left(-\displaystyle\int_{0}^{x}\mu(s,s+t-x)\,ds\right) & \text{for } x \le t . \end{cases}
$$

$$(7)$$

The first expression is for the mortality of members of the initial population; the second is for individuals born after time $t = 0$. Webb (1985) provided detailed derivations and analyses of models of the McKendrick–von Foerster type.

The egg-larval submodel for flour beetles is based on two simplifying assumptions. First, it is assumed that the numbers of adults are constant and that these adults produce a steady supply of new eggs. The recruitment of new eggs into the population can be written as $n(0,t) = B$, where $B$ is a constant representing the net rate at which new eggs are produced by adults. The second simplifying assumption is that the rate at which a larva eats eggs is independent of the age of the larva. This assumption is unrealistic because the cannibalistic voracity of larvae increases with size. Unfortunately, the analysis of the model becomes intractable with age-dependent rates of cannibalism. As a rough approximation, it will be assumed that all larvae eat eggs at a constant rate $c$. With these assumptions, the mortality rates of eggs and larvae are given by

$$
\mu(x,t) = \begin{cases} \mu_{\mathrm{E}} + cN_{\mathrm{L}}(t) & \text{for } 0 < x \le D_{\mathrm{E}} , \\[1ex] \mu_{\mathrm{L}} & \text{for } D_{\mathrm{E}} < x \le D_{\mathrm{E}} + D_{\mathrm{L}} , \end{cases}
$$

$$(8)$$

where $D_{\mathrm{E}}$ and $D_{\mathrm{L}}$ are the durations of the egg and larval stages, respectively, $\mu_{\mathrm{E}}$ and $\mu_{\mathrm{L}}$ are the "natural" mortality rates for eggs and larvae, and the total number of larvae is given by

$$
N_{\mathrm{L}}(t) = \int_{D_{\mathrm{E}}}^{D_{\mathrm{E}}+D_{\mathrm{L}}} n(x,t)\,dx .
$$

$$(9)$$

Since pupae do not contribute to the mortality of eggs or larvae, they are excluded from the model. Assuming an initial population of adults only, substitution of (8) into (7) gives

$$
n(x,t) =
\begin{cases}
B \exp\left( -\mu_{\mathrm{E}} - c \int_0^x N_{\mathrm{L}}(t-s)\,ds \right) \\
\qquad\qquad \text{for } 0 \le x < D_{\mathrm{E}} \;, \\[2mm]
n(D_{\mathrm{E}}, t - x + D_{\mathrm{E}}) \exp\left( -\mu_{\mathrm{L}}(x - D_{\mathrm{E}}) \right) \\
\qquad\qquad \text{for } D_{\mathrm{E}} \le x < D_{\mathrm{E}} + D_{\mathrm{L}} \;.
\end{cases}
\tag{10}
$$

The first equation is for eggs, which must survive both natural mortality and cannibalism, and the second is for the natural mortality of larvae. Using (10) in (9) yields an integral equation for the total number of larvae:

$$
N_{\mathrm{L}}(t) = B \exp(-\mu_{\mathrm{E}} D_{\mathrm{E}}) \int_0^{D_{\mathrm{L}}} \exp\left( -\mu_{\mathrm{L}} y - c \int_0^{D_{\mathrm{E}}} N_{\mathrm{L}}(t - s - y)\,ds \right) dy \,,
\tag{11}
$$

where $y = x - D_{\mathrm{E}}$ is the age of larvae from the time of hatching. This single integral equation can be used to investigate the dynamics of the model.

Hastings (1987) derived results for the equilibrium and local stability of (11) for the case where $\mu_{\mathrm{L}} = 0$. He showed that a unique equilibrium exists whenever $B > 0$. This equilibrium is stable provided that

$$
cB < \gamma \exp(\mu_{\mathrm{E}} D_{\mathrm{E}} + \gamma D_{\mathrm{L}} D_{\mathrm{E}}) \,,
\tag{12}
$$

where $\gamma$ is given by

$$
\gamma = \theta^2 / \left[ 2 - \cos(D_{\mathrm{E}}\theta) - \sin(D_{\mathrm{L}}\theta) \right]
\tag{13}
$$

and $\theta = 2\pi/(D_{\mathrm{E}} + D_{\mathrm{L}})$. Figure 3 is a plot of the stability boundaries for several egg-stage durations.

Hastings (1987) also derived results that describe the behavior of the model near the stability boundary. As the rate of cannibalism, $c$, or the birthrate, $B$, increases, the model undergoes a subcritical Hopf bifurcation at the point where the inequality (12) no longer holds. This means that in the neighborhood of the boundary, a stable equilibrium is surrounded by an unstable orbit. Initial conditions inside the unstable orbit approach the stable equilibrium; initial conditions outside the unstable orbit approach another attractor. The unstable orbit defines a local domain of attraction for the stable equilibrium. This implies the existence of multiple attractors for some subset of parameter space.

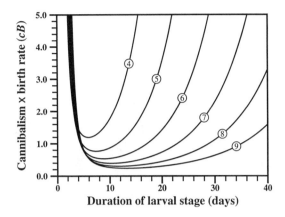

FIGURE 3. *Stability boundaries of the egg-larval submodel from equation (12) with $\mu_E = 0$. Each curve is for a different duration of the egg stage (circled numbers). Parameter values below the curve result in a stable equilibrium.*

Figure 4 shows two of the numerical results from a discretized analogue of the egg-larval submodel (Costantino & Desharnais 1991). The inner trajectory approaches a stable equilibrium point, and the outer trajectory approaches a stable limit cycle. Two stable attractors—a point equilibrium and a loop—coexist for the same parameter values. In fact, for the parameter values in this figure, multiple attractors are found for egg production rates in the range from $B = 134$ to $B = 424$. From a biological point of view, this is a significantly large region of parameter space.

The parameters of the egg-larval submodel can be estimated from experimental data. Park et al. (1961, 1964, 1965) published life-history and census data for four genetic strains of *T. confusum* (strains *bI–bIV*) and four genetic strains of *T. castaneum* (strains *cI–cIV*). From their results, estimates and standard errors were calculated for the parameters $D_E$, $D_L$, $c$, and $B$ of the egg-larval submodel (for details, see Costantino & Desharnais 1991). Table 3 shows the dynamics predicted by the egg-larval submodel. Numerical simulations were used to determine the smallest value of the product $cB$ at which a stable cycle persists; below this value, the equilibrium point is globally stable. Expressions (12) and (13) provide the bifurcation point for $cB$ where the equilibrium loses stability. The interval bounded by these two values is a region of parameter space containing multiple attractors; a stable point and

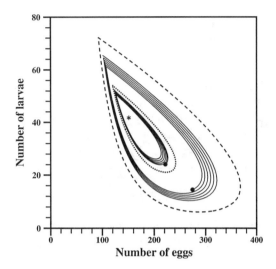

FIGURE 4. *Simulations of the egg-larval submodel for* $D_E = 4$ *days,* $D_L = 18$ *days,* $B = 150$ *eggs/day,* $c = 0.025$*, and* $\mu_E = \mu_L = 0$*. The solid curves are trajectories with two different initial conditions* (circles)*. The inner trajectory spirals toward the stable equilibrium* (asterisk)*. The outer trajectory spirals toward a stable cycle* (dashed loop)*. An unstable cycle* (dotted loop) *defines the domain of attraction of the stable equilibrium.*

a stable orbit coexist. For values of $cB$ above the bifurcation point, only a stable orbit is found. The estimated value of the product $cB$ allows a classification of the predicted dynamics for the eight genetic strains. As this table indicates, all three possibilities are predicted: three of the strains have a globally stable point attractor, four of the strains are in the interval of multiple attractors, and a single strain, $cI$, is predicted to have only a stable orbit. This suggests that a rich spectrum of dynamic behaviors can be found in biologically relevant regions of parameter space.

   The qualitative predictions of the egg-larval submodel are valid for less-restrictive assumptions. Although no larval mortality was assumed ($\mu_L = 0$), when this restriction is relaxed somewhat, a subcritical Hopf bifurcation is still predicted (Hastings 1987). A similar model with all four life stages has also been analyzed (Hastings & Costantino 1987; Costantino & Desharnais 1991). Expressions (12) and (13) give approximate stability bounds, and a region of multiple attractors still exists. Adding age-dependent egg canni-

TABLE 3. *Parameter Estimates and Dynamics Predicted by the Egg-Larval Submodel for Eight Genetic Strains of* Tribolium

| Genetic Strain | $D_E$ | $D_L$ | Onset of Cycles $(cB)$ | Bifurcation Point $(cB)$ | Est'd Value $(cB)$ | Predicted Dynamics |
|---|---|---|---|---|---|---|
| *bI* | 5.37 | 18.96 | 1.60 | 2.60 | 0.75 | stable point |
| *bII* | 5.43 | 20.11 | 1.62 | 2.93 | 1.67 | multiple attractors |
| *bIII* | 5.63 | 18.89 | 1.41 | 2.06 | 1.01 | stable point |
| *bIV* | 5.60 | 19.92 | 1.48 | 2.45 | 1.54 | multiple attractors |
| *cI* | 4.09 | 17.58 | 3.13 | 8.51 | 30.32 | stable orbit |
| *cII* | 4.17 | 25.11 | 3.56 | 38.52 | 2.94 | stable point |
| *cIII* | 4.32 | 19.02 | 2.83 | 8.35 | 5.07 | multiple attractors |
| *cIV* | 4.19 | 20.93 | 3.28 | 14.96 | 3.75 | multiple attractors |

balism by larvae complicates the results (Costantino & Desharnais 1991). In general, a subcritical Hopf bifurcation does not always exist. However, if larval cannibalism rates increase quickly as larvae grow, then multiple attractors are found. Unfortunately, detailed information on age-specific cannibalism rates is needed to make predictions for any particular population.

## 4 Life-Stage Models

Recent efforts have concentrated on an interdisciplinary research program that integrates model derivation and analysis, parameter estimation and model verification, and design and implementation of biological experiments. The goal is to provide solid experimental evidence for a variety of nonlinear dynamic behaviors in flour beetle populations. Mathematical and statistical analyses are used to make a priori predictions concerning the range of dynamic behaviors; *Tribolium* experiments are designed to reveal transitions from one type of dynamic behavior to another. This section describes recent results and current efforts of the author and his collaborators (Costantino et al. 1995), which are based on this approach.

*Discrete Generations*

The simplest "life-stage" model of population growth is a discrete-generation model involving a single reproductive life stage. If $N(t)$ denotes the number of individuals in the reproductive life stage, then the population dynamics can be described using a simple difference equation of the form $N(t+1) = bN(t)f\big[N(t)\big]$, where $b$ is the per capita reproductive rate and $f(N)$ is a "density-regulating factor" that depends on population size. A large number of functional forms for $f(N)$ have appeared in the literature (see, e.g., May & Oster 1976). For *Tribolium*, the function $f(N) = \zeta N^{\alpha-1}e^{-cN}$ attenuates population growth rates at high densities and allows for an "Allee effect" (Allee 1931) at low densities when $\alpha > 1$. The population growth model becomes

$$N(t+1) = \beta N(t)^\alpha \exp\big[-cN(t)\big], \tag{14}$$

where $\beta = b\zeta$. A graph of the right-hand side of (14) as a function of $N$ shows a "one-humped curve" that is common to many population growth models. Typically, as the reproductive rate increases, these models undergo transitions from a stable equilibrium, to a period-doubling cascade (stable cycles of period 2, 4, 8, ...), to chaos, with bifurcations occurring at calculable critical points.

Parameter estimation and model evaluation make use of a stochastic analogue of the deterministic population model. At population sizes typical of flour beetle cultures, variability due to environmental fluctuations outweighs the component due to demographic fluctuations (Dennis & Costantino 1988). A characteristic of models with environmental variability is that noise is additive on a logarithmic scale (Dennis et al. 1991). Applying these ideas, the stochastic version of (14) is

$$N(t+1) = \beta N(t)^\alpha \exp\big[-cN(t) + Z(t)\big], \tag{15}$$

where $Z(t)$ has a normal distribution with a zero mean and variance $\sigma^2$. It is assumed that there is no serial autocorrelation in the random components; that is, $Z(0), Z(1), Z(2), \ldots$ are uncorrelated. This formulation preserves the deterministic model as the conditional expectation of $\ln N(t+1)$ given $N(t)$:

$$\mathrm{E}\big[\ln N(t+1) \mid N(t)=n\big] = \ln\big[\beta n^\alpha \exp(-cn)\big]. \tag{16}$$

Each conditional "one-step" transition can be treated as an independent observation for the purposes of parameter estimation and model evaluation.

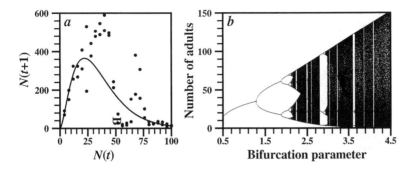

FIGURE 5. *Results of the pilot study of the discrete-generation model (14). (a) The numbers of adults recovered plotted against the numbers of adults from the preceding generation; the smooth curve is based on equation (14) with the maximum-likelihood parameter estimates. (b) Predicted bifurcation diagram for the populations with $\beta$ as the bifurcation parameter.*

Flour beetles can be cultured to mimic the discrete-generation life history implied by (14). Adults are placed in fresh medium and allowed to oviposit for a fixed interval of time called the "breeding interval." At the end of the breeding interval, all adults are removed and the medium containing eggs is returned to the vials. Five weeks later, the flour is sifted and the next generation of adults is counted. The immatures that do not reach adulthood are discarded. The new adults are placed in fresh medium to initiate another generation. By altering the duration of the breeding interval, one can control the rate of reproduction.

A pilot study was conducted to investigate the potential of this experimental protocol for studying discrete-generation population dynamics. A fixed number of adults of the *sooty* strain of *T. castaneum* was placed into vials containing 20 grams of medium. Adult numbers ranged from 4, 8, 12, ..., to 100 for a total of 25 treatments. Each adult density was repeated. After a breeding interval of seven days, the adults were removed and the offspring were allowed to develop for five weeks.

The results of the pilot study are plotted in Figure 5. The data are the equivalent of a one-dimensional map of adult numbers in two consecutive generations (Fig. 5a). The model (15) was fitted to the data using the method of maximum likelihood. Parameter estimates of $\beta = 10.8$, $\alpha = 1.68$, and $c = 0.076$ were obtained. Figure 5b shows the predicted dynamics as a bifurcation diagram with $\beta$ as the bifurcation parameter. The ordinate represents the asymptotic values of adult number for a given value of $\beta$, with the remaining parameters fixed at their estimated values. For small values of $\beta$, a stable equilibrium is predicted. As $\beta$ increases, this equilibrium bifurcates into a stable 2-cycle. The 2-cycle gives way to a 4-cycle, 8-cycle, etc., until acyclic chaotic dynamics appear. Within the chaotic region are intervals of "period locking." Note that the estimated value of $\beta = 10.8$ places the experimental populations well within the chaotic region.

This protocol opens the opportunity of documenting experimentally transitions in dynamic behavior. By shortening the breeding interval, one can decrease the value of $\beta$. With an array of experimental treatments, it should be possible to cover the sequence of dynamic behavior from stable equilibria, to period doublings, to chaos. Such experiments are currently under way.

## Overlapping Generations

A structured model is required to account for population dynamics in the case of overlapping generations. These are coupled difference equations or matrix equations that describe the dynamics of two or more life stages or groupings of life stages (Cushing 1988; Caswell 1989). Dennis et al. (1995) have proposed the use of three coupled difference equations to describe the dynamics of larvae, pupae, and adults in *Tribolium* cultures:

$$L(t+1) = bA(t)\exp\left[-c_{EA}A(t) - c_{EL}L(t)\right], \qquad (17a)$$

$$P(t+1) = L(t)(1 - \mu_L), \qquad (17b)$$

$$A(t+1) = P(t)\exp\left[-c_{PA}A(t)\right] + A(t)(1 - \mu_A). \qquad (17c)$$

Here $L(t)$ refers to the number of feeding larvae, $P(t)$ refers to the number of nonfeeding larvae, pupae, and callow adults, and $A(t)$ refers to the number of reproductive adults. The unit of time is taken to be the maturation interval for feeding larvae, so that after one unit of time either a larva dies or it survives and pupates. This unit of time is also the cumulative time spent as a nonfeeding larva (sometimes called a "prepupa"), pupa, and callow adult. The quantity $b$ ($> 0$) is the number of larval recruits per adult per unit of time in the absence of cannibalism. The fractions $\mu_L$ and $\mu_A$ are the larval and adult probabilities, respectively, of dying from causes other than cannibalism. The exponential nonlinearities account for the cannibalism of eggs by both adults and larvae and the cannibalism of pupae by adults (Fig. 1). It is assumed that the only significant source of pupal mortality is cannibalism by adults. Dennis et al. (1995) referred to this model as the "LPA model."

Adding noise on a logarithmic scale produces the following stochastic model:

$$L(t+1) = bA(t)\exp\left[-c_{\mathrm{EA}}A(t) - c_{\mathrm{EL}}L(t) + Z_1(t)\right], \quad (18a)$$

$$P(t+1) = L(t)(1 - \mu_{\mathrm{L}})\exp\left[Z_2(t)\right], \quad (18b)$$

$$A(t+1) = \left\{P(t)\exp\left[-c_{\mathrm{PA}}A(t)\right] + A(t)(1 - \mu_{\mathrm{A}})\right\} \\ \times \exp\left[Z_3(t)\right]. \quad (18c)$$

The random vector $\mathbf{Z}(t) = (Z_1(t), Z_2(t), Z_3(t))$ is assumed to have a trivariate normal distribution with mean vector of $\mathbf{0}$ and a covariance matrix of $\boldsymbol{\Sigma}$. Including covariance terms in the off-diagonal elements of $\boldsymbol{\Sigma}$ allows for the possibility of correlations in the growth fluctuations among the life stages during the same time intervals. However, we assume the correlations between time intervals to be small by comparison; that is, $\mathbf{Z}(0), \mathbf{Z}(1), \mathbf{Z}(2), \ldots$ are uncorrelated.

The dynamics of the deterministic model (17) are preserved by the stochastic model (18) as conditional expectations on a logarithmic scale:

$$\mathrm{E}\left[\ln L(t+1)\right] = \ln\left[ba\exp(-c_{\mathrm{EA}}a - c_{\mathrm{EL}}\ell)\right], \quad (19a)$$

$$\mathrm{E}\left[\ln P(t+1)\right] = \ln\left[\ell(1 - \mu_{\mathrm{L}})\right], \quad (19b)$$

$$\mathrm{E}\left[\ln A(t+1)\right] = \ln\left[p\exp(-c_{\mathrm{PA}}a) + a(1 - \mu_{\mathrm{A}})\right], \quad (19c)$$

given that $L(t) = \ell$, $P(t) = p$, and $A(t) = a$. This result allows an explicit connection between the mathematical model and the population time-series data. Given the numbers of each life stage at time $t$ and estimates for the model parameters, one can predict the expected numbers of each life stage at time $t+1$ (two weeks later). Using the stochastic model (18), one can choose a set of parameter values that maximizes the joint probability of the one-time-step transitions in the observed time series. The model is connected to the data by one-step forecasts, not by continued iteration.

Dennis et al. (1995) have applied this maximum-likelihood procedure to earlier data on the *cos* genetic strain of *T. castaneum* (Desharnais & Costantino 1980). They used the 4 control cultures for parameter estimation and the 9 cultures subjected to demographic perturbations for model evaluation. Sophisticated diagnostics were conducted on the residuals. The LPA model (17) did a remarkably good job of predicting the dynamics of all 13 populations. Partic-

TABLE 4. *Parameter Estimates for the LPA Model*

| Parameter | Estimate | 95% Confidence Interval |
| --- | --- | --- |
| $b$ | 11.7 | (6.2, 22.2) |
| $\mu_A$ | 0.11 | (0.07, 0.15) |
| $\mu_L$ | 0.51 | (0.43, 0.58) |
| $c_{EA}$ | 0.011 | (0.004, 0.180) |
| $c_{EL}$ | 0.009 | (0.008, 0.011) |
| $c_{PA}$ | 0.018 | (0.015, 0.021) |

ularly impressive was the fact that a single set of parameter values from the control cultures was able to describe the dynamics of the 9 demographically manipulated cultures, even though none of the data from these manipulated populations was used to obtain the parameter estimates.

Maximum-likelihood parameter estimates for the LPA model are given in Table 4. The 95 percent confidence intervals are based on profile likelihoods (McCullagh & Nelder 1989). When these parameter values are substituted into (17), the model simulations approach a stable 2-cycle.

Figure 6 shows the one-step predictions (Desharnais & Costantino 1980, control replicate "A"). Keep in mind that a single set of parameter values is used for all the predictions. The model does a particularly good job of prediction in the region of strong oscillations. Dennis et al. (1995) provided similar plots for all 13 populations.

Equilibrium densities and stability boundaries can be obtained numerically for the LPA model. Although no closed-form solution exists for the equilibrium densities of each life stage, it can be shown that a single unique nontrivial equilibrium exists provided that $b > \mu_A/(1 - \mu_L)$ (Dennis et al. 1995). This equilibrium can be found numerically by locating the unique real root of a simple nonlinear equation. The $3 \times 3$ linearized stability matrix $\mathbf{S}$ can be expressed in terms of the model's parameters and equilibrium densities. Given parameter values and the corresponding equilibrium life-stage densities, the eigenvalues of $\mathbf{S}$ can be computed. If the moduli of these three eigenvalues are all less than unity, the equilibrium is stable. This process can be repeated for different combinations of parameter values to map out the stability boundaries in parameter space.

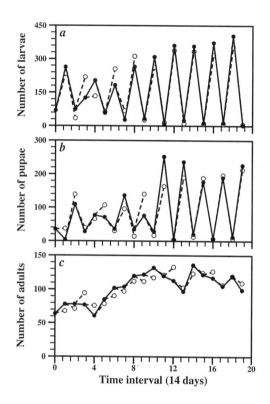

FIGURE 6. *Densities of* (a) *larvae,* (b) *pupae, and* (c) *adults (Desharnais & Costantino 1980, control replicate "A").* Solid circles, *Census points;* open circles, *one-step forecasts.*

Figure 7 shows the stability boundaries of the LPA model as functions of the adult mortality rate, $\mu_A$, and the rate of egg cannibalism by larvae, $c_{EL}$. For this figure, the remaining parameters of the LPA model were set at their estimated values (Table 4). Crossing the stability boundary to the left of the large peak represents a bifurcation into a stable 2-cycle. Here, a single real dominant eigenvalue of **S** becomes equal to $-1$. Asymptotically, life-stage densities oscillate between two discrete values. Crossing the stability boundary to the right of this peak represents a bifurcation into an invariant loop. Here **S** has a pair of complex conjugate eigenvalues of modulus one. Asymptotically, life-stage densities move aperiodically around a closed loop. The solid circle in Figure 7 shows the location of the *cos* genetic strain of *T. castaneum* in

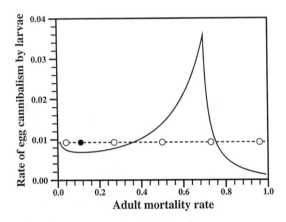

FIGURE 7. *Stability boundaries for the LPA model (17) as a function of the adult mortality rate and the rate of egg cannibalism by larvae. The remaining parameters were set at the estimated values for the* cos *strain of* T. castaneum. Solid circle, *Location of the* cos *strain in parameter space;* open circles, *adult mortality treatments in an experimental study now under way.*

the unstable region of parameter space near the 2-cycle boundary. Its location in parameter space is consistent with the demographic dynamics observed in Figure 6.

The stability boundary in Figure 7 has stimulated new experiments. In a recent study (Costantino et al. 1995), adult mortality rates were manipulated by removing or adding adults at the time of census to make the total number of adults that died during an interval consistent with a predetermined value of $\mu_A$. In addition to a control (no manipulation), values of $\mu_A$ were chosen to be 0.04, 0.27, 0.50, 0.73, and 0.96 (Fig. 7, *open circles*). This experimental design was implemented using two different genetic strains of *T. castaneum*. There were four replicates for each combination of adult mortality rate and genetic strain.

The predicted outcomes from this experimental design are visualized in Figure 8. This bifurcation diagram was obtained by iterating the deterministic LPA model for different values of $\mu_A$, with the remaining parameters fixed at their estimated values (Table 4). This graph shows the asymptotic behavior of the LPA model as one moves along the dashed line in Figure 7. At very low adult

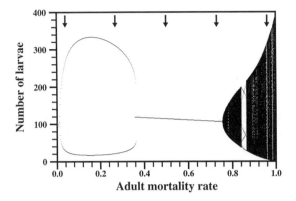

FIGURE 8. *A bifurcation diagram for the LPA model (17) with adult mortality rate as the bifurcation parameter. The remaining parameters were set at the estimated values for the* cos *strain of* T. castaneum. *Arrows show the adult mortality treatments of Costantino et al. (1995).*

mortalities, there is a stable point equilibrium, but this equilibrium soon bifurcates into a stable 2-cycle. This 2-cycle is followed by a narrow region of multiple attractors ($0.357 \leq \mu_A \leq 0.363$) where stable equilibria coexist with stable 2-cycles. As $\mu_A$ increases, positive equilibria persist, whereas the stable 2-cycles disappear. At high values of adult mortality, there is a bifurcation to an invariant loop. Adult-mortality treatments were chosen to sample these different dynamic behaviors (Fig. 7, *open circles*; Fig. 8, *arrows*). The goal is to provide convincing evidence for nonlinear population dynamics by documenting transitions in the demographic behavior of experimental populations. For the results of this study, see Costantino et al. (1995).

## 5 Concluding Remarks

The research results described in this chapter are joined by a common theme: the integration of theoretical and experimental approaches in order to study nonlinear dynamics in structured populations. In this regard, the *Tribolium* experimental system provides many opportunities. This animal model shows the potential for exhibiting many of the exotic behaviors so far identified only in theoretical models: discrete-point cycles, limit cycles, invariant loops, multiple attractors, strange attractors, chaos. Through a careful choice of species, genetic strains, environmental conditions, and husbandry protocols, one can "sample" parameter space experimentally, obtaining populations that cover the entire spectrum of dynamic behaviors. The possibility exists for "nudging" populations across boundaries within parameter space, documenting experimentally transitions in behavior. If multiple attractors are predicted, a variety of initial conditions could be used to seek out the attractors. If multiple attractors are found, perturbations could be used to shift a population from one domain of attraction to another. This work is only beginning. The prospects are exciting.

## Acknowledgments

The research described in this chapter was supported in part by U.S. National Science Foundation grants DMS-9206678, DMS-9306271, and DMS-9319073.

## Literature Cited

Allee, W. C. 1931. *Animal Aggregations.* University of Chicago Press.

Bell, A. E. 1982. The *Tribolium* model and animal breeding. *Second World Congress on Genetics and Applications to Livestock Production* 5: 26–42.

Bernardelli, H. 1941. Population waves. *Journal of the Burma Research Society* 31: 1–18.

Caswell, H. 1989. *Matrix Population Models.* Sinauer, Sunderland, Mass.

Chapman, R. N. 1928. Quantitative analysis of environmental factors. *Ecology* 9: 111–122.

Costantino, R. F., and R. A. Desharnais. 1991. *Population Dynamics and the 'Tribolium' Model: Genetics and Demography.* Springer-Verlag, New York.

Costantino, R. F., J. M. Cushing, B. Dennis, and R. A. Desharnais. 1995. Experimentally induced transitions in the dynamic behaviour of insect populations. *Nature* 375: 227–230.

Cushing, J. M. 1988. Nonlinear matrix models and population dynamics. *Natural Resource Modelling* 2: 539–580.

Cushing, J. M., B. Dennis, R. A. Desharnais, and R. F. Costantino. In press. An interdisciplinary approach to understanding nonlinear ecological dynamics. *Ecological Modelling.*

Dennis, B., and R. F. Costantino. 1988. Analysis of steady-state populations with the gamma abundance model and its application to *Tribolium. Ecology* 69: 1200–1213.

Dennis, B., P. L. Munholland, and J. M. Scott. 1991. Estimation of growth and extinction parameters for endangered species. *Ecological Monographs* 61: 115–143.

Dennis, B., R. A. Desharnais, J. M. Cushing, and R. F. Costantino. 1995. Nonlinear demographic dynamics: Mathematical models, statistical methods and biological experiments. *Ecological Monographs* 65: 261–281.

Desharnais, R. A., and R. F. Costantino. 1980. Genetic analysis of a population of *Tribolium*. VII. Stability: Response to genetic and demographic perturbations. *Canadian Journal of Genetics and Cytology* 22: 577–589.

Desharnais, R. A., and L. Liu. 1987. Stable demographic limit cycles in laboratory populations of *Tribolium castaneum*. *Journal of Animal Ecology* 56: 885–906.

Hastings, A. 1987. Cycles in cannibalistic egg-larval interactions. *Journal of Mathematical Biology* 24: 651–666.

Hastings, A., and R. F. Costantino. 1987. Cannibalistic egg-larva interactions in *Tribolium*: An explanation for the oscillations in population numbers. *American Naturalist* 130: 36–52.

Jillson, D., and R. F. Costantino. 1980. Growth, distribution, and competition of *Tribolium castaneum* and *Tribolium brevicornis* in fine-grained habitats. *American Naturalist* 116: 206–219.

King, C. E., and P. S. Dawson. 1972. Population biology and the *Tribolium* model. *Evolutionary Biology* 5: 133–227.

Leslie, P. H. 1945. On the use of matrices in certain population mathematics. *Biometrika* 33: 183–212.

Lewis, E. G. 1942. On the generation and growth of a population. *Sankhya* 6: 93–96.

Lotka, A. J. 1925. *Elements of Physical Biology*. Williams & Wilkins, Baltimore.

May, R. M., and G. F. Oster. 1976. Bifurcations and dynamic complexity in simple ecological models. *American Naturalist* 110: 573–599.

McCullagh, P., and J. A. Nelder. 1989. *Generalized Linear Models*. Chapman & Hall, London.

McKendrick, A. G. 1926. Applications of mathematics to medical problems. *Proceedings of the Edinboro Mathematical Society* 54: 98–130.

Mertz, D. B. 1972. The *Tribolium* model and the mathematics of population growth. *Annual Review of Ecology and Systematics* 3: 51–78.

Moffa, A. M. 1976. Genetic Polymorphism and Demographic Equilibrium in *Tribolium castaneum*. Ph.D. diss. University of Rhode Island, Kingston.

Park, T. 1948. Experimental studies of interspecies competition. I. Competition between populations of the flour beetles *Tribolium confusum* Duval and *Tribolium castaneum* Herbst. *Ecological Monographs* 18: 265–308.

Park, T., D. B. Mertz, and K. Petrusewicz. 1961. Genetic strains of *Tribolium*: Their primary characteristics. *Physiological Zoology* 34: 62–80.

Park, T., P. H. Leslie, and D. B. Mertz. 1964. Genetic strains and competition in populations of *Tribolium*. *Physiological Zoology* 37: 97–162.

Park, T., D. B. Mertz, W. Grodzinski, and T. Prus. 1965. Cannibalistic predation in populations of flour beetles. *Physiological Zoology* 38: 289–321.

Rich, E. L. 1956. Egg cannibalism and fecundity in *Tribolium*. *Ecology* 37: 109–120.

Sharpe, F. R., and A. J. Lotka. 1911. A problem in age-distribution. *Philosophy Magazine* 21: 435–438.

Sokoloff, A. 1972. *The Biology of 'Tribolium.'* Vol. 1. Oxford University Press.

———. 1974. *The Biology of 'Tribolium.'* Vol. 2. Oxford University Press.

———. 1977. *The Biology of 'Tribolium.'* Vol. 3. Oxford University Press.

Stanley, J. 1932. A mathematical theory of the growth of populations of the flour beetle, *Tribolium confusum* Duval. *Canadian Journal of Research* 6: 632–671.

von Foerster, H. 1959. Some remarks on changing populations. Pp. 382–407 *in* F. Stohlman, ed., *The Kinetics of Cellular Proliferation*. Grune & Stratton, New York.

Webb, G. F. 1985. *Theory of Nonlinear Age-Dependent Population Dynamics*. Marcel Dekker, New York.

CHAPTER 10

# Evolutionary Dynamics of Structured Populations

## Jochen Kumm, Sido D. Mylius, and Daniel Promislow

Traditional theory in population genetics largely ignores the internal structure of a population and treats populations as homogeneous with respect to age, size, developmental stage, or other individual states. Although this is often a legitimate assumption for organisms with discrete generations, many species have overlapping generations, for which within-population structure should be taken into account. The last few decades have seen considerable interest in extending evolutionary models to include age structure. Charlesworth (1994) gave an account of the major results that have been obtained by applying ecological and demographic models of age-structured populations to population genetics and evolutionary theory. Although much progress has been made in this arena, there are still numerous problems in which neither age nor stage structure has been addressed.

In this chapter, we explore some of these problems and examine the effect of population structure on simple population-genetic and game-dynamic models. We start by investigating the relationship between density-dependent population regulation and population genetics in the presence of age structure. Earlier work by Asmussen and Feldman (1977), Asmussen (1979a), and Rougharden (1979) explored the interaction of population dynamics under density-dependent population regulation, with a simple, single-locus genetic system and no age structure. We extend this work by incorporating stage structure and show that, although stage structure typically acts to stabilize temporal dynamics, it can also lead to previously unidentified, and complex, dynamics in both population numbers and allele frequencies. We investigate the effect of selection on different life-history stages and show that the introduction of stage structure into previous models yields results that differ from those of classical theory.

We also extend Dawkins' (1976) "battle of the sexes" by introducing a population with mating structure. Previous models of the battle of the sexes specify a dynamic system on the basis of the payoff matrices of the evolutionary population game. Here the war of the sexes is modeled by expressing the costs of raising offspring and performing courtship in terms of a time delay during which individuals cannot mate anew. Numerical analysis of the system shows that the dynamics of this model are qualitatively different from those of the "homogeneous" models of the battle of the sexes. Evolutionary predictions of a game-theory model, in which the costs and benefits of a strategy are summarized by a single entry of a payoff matrix, may be quite different from a model that explicitly considers the mechanisms controlling mating in a structured population.

## 1  Population Genetics and the Dynamics of Stage-Structured Populations

In the past two decades, a number of theoretical papers have asked whether changes in population size and growth rate affect genetic diversity within populations. Although population dynamics and population genetics are generally studied independently, Roughgarden (1971) introduced a theoretical framework in which population size and gene frequencies change simultaneously. He assumed that natural selection is density-dependent, that is, the viability of individuals of a given genotype depends on population size. Also, as in traditional genetic models, he assumed that populations are composed of randomly mating individuals with discrete, non-overlapping generations. Assuming that an individual's genotype at a locus with two alleles $(A_1, A_2)$ affects both its fertility and its demand for resources (i.e., the carrying capacity), the fitness of an individual of genotype $A_i A_j$ is given by

$$w_{ij} = 1 + r_{ij}(1 - N/K_{ij}), \qquad (1)$$

where $r_{ij}$ is the growth rate and $K_{ij}$ is the carrying capacity of individuals of the $A_i A_j$ genotype (Roughgarden 1971, 1979). Roughgarden showed that trajectories in the phase space of gene frequency and population size depend on $r_{ij}$ and $K_{ij}$ and converge to equilibria. When $A_i A_i$ homozygotes have the highest carrying capacity, only the $A_i$ allele is present in the population at equilibrium. When heterozygotes have a higher carrying capacity than both homozygotes, both alleles can be maintained in the population. When the heterozygotes have lower carrying capacity than both homozygotes, either one of the alleles may exclude the other from the population at equilibrium, depending on the size and genetic makeup of the population at the outset. When the environment is stable, Roughgarden (1979) and Charlesworth (1980) showed that, in this model, the invasion of a new allele depends only on its carrying capacity and not on the intrinsic growth rate, $r$.

Similar models have been used to explore a variety of phenomena, for example, the maintenance and stability of genetic polymorphisms (Asmussen 1979a), the effects of competition among genotypes (Adler 1990; Cushing & Li 1992; Gatto 1993), and the role of fluctuating environments on density-dependent selection (Roughgarden 1979). Other models (e.g., May 1974) have shown that a great variety of dynamic behaviors can be observed even in simple systems when selection is a nonlinear function of population size. Asmussen and Feldman (1977; Asmussen 1979a,b) showed how density-regulated population dynamics can lead to stable polymorphisms, periodicity, and even chaos. They used simple single-locus, two-allele models with discrete, non-overlapping generations. Recent work (Hastings 1992, 1993) is beginning to explore the effects of geographic structure on such systems.

In recent years, a sizeable literature has examined the effects of adding structure to models in both population dynamics and population genetics. For example, a recent paper (Botsford 1992) demonstrates that the inclusion of a small amount of age structure in models of density-regulated populations can dramatically increase the stability of the population size over time. For a genetic example, Ellner and Hairston (1994) have shown that age structure greatly increases the range of circumstances under which fluctuating environments can maintain genetic variation.

Here we examine the genetics of a stage-structured population with density-dependent population growth. Individuals are born, mature, and age, moving from life stage to life stage, and the genetic characteristics and the population size within each stage may change as a result of selection and density-dependent effects. This introduces nonlinearities that can lead to complex dynamics. It is for this reason that interactions between the ecological and genetic characteristics of a population may lead to new results, even in simple models.

In this chapter we focus on examples that capture biologically relevant patterns. We show that this model can be used to study the effects of age structure on the rate of genetic change; adding age structure to a density-dependent model can favor genotypes with low growth rates; and genotypic variation changes with time among age classes.

Although the results here are limited in scope, we hope to motivate further study of the interaction between the dynamics, genetics, and structure of populations.

*Methods*

We consider individuals with a two-stage life cycle schematically presented in Figure 1. Individuals are born as larvae ($L$) and develop into reproducing individuals ($D$). The latter produce gametes ($G$) depending on their fecundity $F$. Larvae are formed from the total pool of gametes. Maturation of a larva occurs with probability $s_L$, and following reproduction, adults may live on to reproduce again with probability $\alpha$. When adults reproduce more than once, their offspring join different cohorts of larvae, so that $\alpha$ is a measure of cohort mixing within this structured population. Throughout this chapter we use the parameter $\alpha$ or the term adult survival to measure the effects of repeated reproduction (as did Botsford 1992).

For a single locus with alleles $A_1$ and $A_2$ in a diploid organism, three genotypes, $A_1A_1$, $A_1A_2$, and $A_2A_2$, may be present in each life stage. Generally, we indicate parameters or variables specific to genotype $A_iA_j$ by the superscript $ij$. For example, the probability of survival from larval stage to mature stage for genotype $A_iA_j$ is $s_L^{ij}$, and the number of larvae of genotype $A_iA_j$ is $L^{ij}$. Thus, at any time $t$, the vector

$$N_t = \begin{bmatrix} L^{11} & L^{12} & L^{22} & D^{11} & D^{12} & D^{22} & G^1 & G^2 \end{bmatrix}_t \quad (2)$$

characterizes the population. $D^{ij}$ denotes the number of adults of each genotype, and $G^k$ denotes the number of gametes carrying allele $A_k$.

The dynamics of the model depend on the the entries of an expanded Leslie matrix $\mathbf{Q}_t$, which describe the survival, reproduction, and fertility of individuals in the population. This matrix has entries for every life stage, genotype, or gamete and describes the

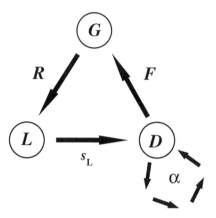

FIGURE 1. *A simple life cycle for a stage-structured population. Here* $s_L$ *is the probability that a larva survives to the mature stage, and* $\alpha$ *is the probability that an adult survives to reproduce a second time. F is the fecundity of the population, the average number of gametes produced per adult. R describes the formation of larvae from gametes and is explained in detail in the text.*

transitions between life stages. Let $R_t^{ij,k}$ denote the probability that, at time $t$, the genotype $A_i A_j$ is produced by a mating that involves gamete $A_k$. Assuming random mating among gametes, we have, for example, that $R_t^{11,1}$ is given by $\frac{1}{2} G_t^1 \left( G_t^1 + G_t^2 \right)^{-1}$. Then, $\alpha^{ij}$ is the probability that a mature individual of genotype $A_i A_j$ survives to reproduce again. $F_t^{ij}$ denotes the fecundity of a mature individual of genotype $A_i A_j$ at time $t$. The expanded Leslie matrix is

$$
\mathbf{Q}_t = \begin{bmatrix}
0 & 0 & 0 & 0 & 0 & 0 & R_t^{11,1} & R_t^{11,2} \\
0 & 0 & 0 & 0 & 0 & 0 & R_t^{12,1} & R_t^{12,2} \\
0 & 0 & 0 & 0 & 0 & 0 & R_t^{22,1} & R_t^{22,2} \\
s_L^{11} & 0 & 0 & \alpha^{11} & 0 & 0 & 0 & 0 \\
0 & s_L^{12} & 0 & 0 & \alpha^{12} & 0 & 0 & 0 \\
0 & 0 & s_L^{22} & 0 & 0 & \alpha^{22} & 0 & 0 \\
0 & 0 & 0 & F_t^{11} & \frac{1}{2}F_t^{12} & 0 & 0 & 0 \\
0 & 0 & 0 & 0 & \frac{1}{2}F_t^{12} & F_t^{22} & 0 & 0
\end{bmatrix} . \quad (3)
$$

The product of the vector $\mathbf{N}$ describing the population at time $t$ and the expanded Leslie matrix yields a new population vector

**N** for time $t + 1$. Adult fecundity and survival within life stages may vary between genotypes because of selection or density dependence. In our model, adult fecundity is described by the exponential logistic

$$F_t^{ij} = F \exp\left[r_{ij}\left(1 - D_t/K_{ij}\right)\right], \qquad (4)$$

where $F$ is constant, genotype-independent fecundity, $r_{ij}$ and $K_{ij}$ are the rates of growth and of carrying capacity, respectively, of the genotype $ij$, and $D_t$ is the total number of reproductive individuals in the population at time $t$. We choose the exponential logistic form because the simpler logistic equation can lead to biologically impossible results (Roughgarden 1971; Asmussen & Feldman 1977).

Genotypic differences in fecundity are reflected in the relative number of gametes produced by each genotype. Let $p$ and $q$ denote the frequencies of $A_1$ and $A_2$ gametes produced. Assume that each gamete fuses with one other gamete at random, and, in the youngest age class, the three genotypes $A_1A_1$, $A_1A_2$, and $A_2A_2$ are present in Hardy-Weinberg proportions $p^2$, $2pq$, and $q^2$. The actual number of offspring produced by each genotype in each generation is simply one-half the number of gametes produced by that genotype.

## Results

*Slower genetic change with stage structure.* Population-genetic models for age-structured populations have shown that pools of dormant seeds can slow the rate of evolution (Templeton & Levin 1979). This theory has been substantiated in the field by Tonsor et al. (1993). Dormant propagules can be thought of as representatives of an organism's genetic past, as can iteroparous adults.

Our model shows that iteroparous adults can slow the effect of selection on maturing larvae. We examine the time it takes for an allele $A_1$, whose carriers have moderately higher survival probability $s_L$, to become fixed (in this case, from $p = 0.50$ to $p = 0.99$), as a function of adult survival. With one reproductive class in a two-stage model, increased adult survival leads to a greater-than-exponential increase in the amount of time it takes for $A_1$ to reach a frequency of 0.99 (Fig. 2).

This pattern remains virtually unaltered if the larval-stage individuals reproduce. Thus, cohort mixing, whether at the level of dormant seeds or reproductive adults, slows change in allele frequency.

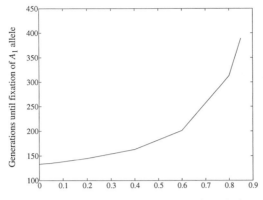

FIGURE 2. *The relationship between adult survival (i.e., the probabililty that adults may reproduce more than once) and the number of generations required for allele $A_1$ to reach a frequency of 0.99.*

*Differences among stages.*     We assume that some fraction of adults in generation $t$ survive to reproduce again in the subsequent population. The number of adults in generation $t + 1$ is the sum of adults who matured from larvae and the number of survivors from the preceding generation. Once an adult has lived through two generations, however, we assume that it dies.

Two equations describe the number of $A_i A_j$ individuals:

$$D_{t+1}^{ij} = s_{\mathrm{L}}^{ij} L_t^{ij} + \alpha D_t^{ij}, \tag{5}$$

$$L_{t+1}^{ij} = \sum_{\substack{v=i \\ w=i}}^{j} (G_{t+1}^v G_{t+1}^w) \bigg/ 2 \sum_i^2 G_{t+1}^i. \tag{6}$$

A subtle, but important, consequence of adult survival of this sort is that the equilibrium size of a population is no longer the carrying capacity $K$. The recursion for the total population size $N_t$ can be written as

$$N_{t+1} = s_{\mathrm{L}}^{ij}(1 + \alpha) \exp\left[r_{ij}(1 - N_t/K)\right] N_t. \tag{7}$$

From this, the equilibrium population size is

$$\tilde{N} = K\,(1 + \Delta)\,, \tag{8}$$

where $\Delta = \ln\left[s_{\mathrm{L}}^{ij}(1 + \alpha)\right]/r_{ij}$.

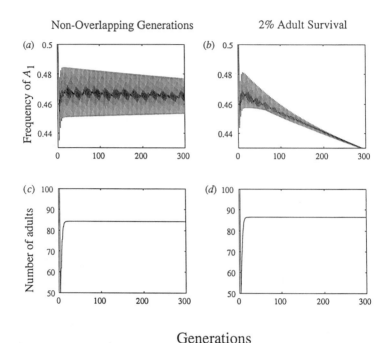

FIGURE 3. *The frequency of allele $A_1$ and the number of adults: when all adults reproduce only once, allele A is maintained at an intermediate frequency in the population; a small amount of adult survival (2%) leads to the elimination of allele $A_1$ from the population.*

$\Delta$ is greater than zero and the equilibrium population size exceeds $K$ when $s_L^{ij}(1+\alpha) > 1$ or when $\alpha > 1/s_L^{ij} - 1$. Thus, when a significant percentage of adults survives to reproduce a second time, population size may exceed the carrying capacity. As a consequence, genes that alter the characteristics of density dependence may be subject to selection.

In Asmussen's (1979a) models, either stable or unstable population dynamics occur, and gene frequencies may be stable or unstable. Thus, there are four possible combinations of genetic and population dynamics. We found all of these in our simulations. Below, we focus on the case for which gene frequencies exhibit periodic dynamics, and total population size is either stable or fluctuating.

Botsford (1992) showed that in a simple density-dependent population model, increasing adult survival from 0 to 5 percent can change the population dynamics from chaotic to periodic. Is the same true of fluctuations in gene frequencies? When there is zero adult survival (i.e., non-overlapping generations), we find the same pattern as did Asmussen (1979$a$). Adding a small amount of adult survival, however, led to regular cycles that first appeared at much higher values of $r$ than were seen for non-overlapping generations. Chaotic patterns appeared only with high values of either $r$ or $F$. This effect is illustrated by the example in Figure 3, where the $A_2A_2$ genotype has the highest value of the parameter $K$ but also has a higher mortality in the juvenile stage. For some parameter values and with no adult survival, adult population size can reach a stable equilibrium while the gene frequencies in adults continue to fluctuate. With adult survival as low as 2 percent, these fluctuations in gene frequencies are rapidly damped, and the population moves to a stable polymorphism. Thus, a small amount of adult survival has a dramatic effect on the trajectory of gene frequencies.

In addition, in certain regions in our model both adult population size and gene frequencies fluctuate. But in some cases, gene frequencies are relatively constant among adults and fluctuate only among larvae (Fig. 4). Although gene frequencies among larvae exhibit periodic fluctuations, the pool of adults comprises larvae entering that stage, together with those individuals already in it. Thus, the surviving adults serve to buffer the effects of individuals entering that stage for the first time.

*Slow growth favored by density-dependent selection.* In a model that combined density dependence and genetic variation in a fluctuating environment, Roughgarden (1979) showed that selection on density-dependent fecundity can lead to the evolution of lower intrinsic rates of increase. In our model, density regulates fecundity, as in equation (4). Individuals differ in respect to the growth rate, $r$. Even in a constant environment, we find that a homozygous genotype with a lower value of $r$ goes to fixation. If the heterozygote has the lowest $r$, a stable polymorphism can be obtained. When adults reproduce more than once ($\alpha > 0$), the genotype with the lowest $r$ becomes fixed more rapidly. This is interesting in itself, since in other circumstances increasing adult survival lowers the rate at which the favored allele becomes fixed.

Why should the population with the lowest growth rate become fixed? Imagine two separate populations, $m$ and $n$, with equal carrying capacities, $K$ (as in eq. 4), and values of $r$ such that $r_m > r_n$.

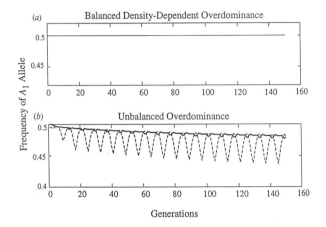

FIGURE 4. *The frequency of allele $A_1$ with unbalanced overdominance: solid line, in adults; dotted line, in larvae. Fluctuations in allele frequency in adults are subtle, and the allele gradually becomes less abundant, even as the adult population oscillates. With balanced overdominance, however, the frequency of $A_1$ remains steady.*

We wish to find the circumstances under which

$$F_{\mathrm{m}} = F \exp\left[r_{\mathrm{m}}\left(1 - \frac{N_t}{K}\right)\right] > F_{\mathrm{n}} = F \exp\left[r_{\mathrm{n}}\left(1 - \frac{N_t}{K}\right)\right]. \quad (9)$$

It is easy to show that as long as $N_t$ remains below the carrying capacity, fecundity in species $m$ is greater than in species $n$. However, this inequality is reversed when the population exceeds carrying capacity ($1 - N/K$ becomes negative). Thus, during the exponential growth phase, the genotype with highest $r$ grows fastest. But once the population exceeds carrying capacity, the genotype with the lowest $r$ has the highest relative fecundity, and thus the advantage.

When the equilibrium population size exceeds the carrying capacity (eq. 8), an allele associated with a lower growth rate, $r$, can invade even in a stable environment. The greater $\alpha$ is, the greater is the equilibrium population size, and the greater is the advantage of a low-$r$ allele.

One can understand intuitively why increasing adult survival has this effect. The population is regulated by a decrease in fecundity in response to increased numbers of adults. When adult survival increases, the proportion of adults in the population also increases, thereby enhancing density regulation. To illustrate, return to a simple two-stage life cycle, with juvenile survival $s_L$, adult fecundity $F$, and adult survival $\alpha$. The fraction $\phi$ of adults in the stable age distribution is

$$\phi = [(\alpha^2 + 4Fs_L)^{1/2} + \alpha]/2F. \tag{10}$$

If $F$ and $s_L$ are constant, this fraction of adults increases with $\alpha$, as does the total number of adults.

## Discussion

We present a simple model of a stage-structured population with density-dependent fecundity. Such models are less likely to generate oscillatory dynamics or genetic polymorphisms. This is largely due to cohort mixing.

One effect of cohort mixing is to slow the rate of genetic change in the population. Both adult survival and dormant seed pools act in this way. Although cohort mixing leads to apparent homogeneity in the life stage in which it occurs, fluctuations in gene frequencies may continue to occur at other life stages. This should be relevant to field sampling, since measures of genetic heterogeneity depend on which life stage is sampled, and results may be sensitive to "mixing" samples across life stages.

Another important consequence of cohort mixing is that the equilibrium population size (see eq. 8) differs from the carrying capacity controlling density dependence. As a result, density-dependent selection may favor lower growth rates than predicted by traditional models, but it does not maintain some plausible kinds of genetic polymorphisms.

## 2  A "Battle of the Sexes" with Pair Formation

The term "battle of the sexes" was introduced by Richard Dawkins in *The Selfish Gene* (1976). Currently, the phrase refers to a specific conflict described in his book: an evolutionary game between males and females about the costs of raising offspring. The conflict is rooted in Trivers' (1972) theory of parental investment and sexual selection, and has the following rationale.

In many species, raising offspring requires a considerable investment by the parents. One parent may find it tempting to reduce its investment, at the expense of the other. Often one of the sexes, typically the female, makes a larger investment in the offspring.

The male is then not as committed to the children as the female and is tempted to desert shortly after the mating, leaving the female with the task of raising the offspring. A female could prevent this by choosing a faithful partner, perhaps testing his fidelity by insisting on a long engagement period.

Dawkins used the following caricature of the possible types of behavioral strategies to help analyze this conflict. Males can be either helping or nonhelping, and females can be choosy or nonchoosy. (Actually, Dawkins and many others used the terms "philanderous" or "faithful" for males, and "coy" or "fast" for females.) Choosy females insist that their partner performs a prolonged courtship before mating; nonchoosy females do not. Helping males help the female in raising the offspring, and they court if the (choosy) female insists; nonhelping males do not court and leave immediately after conception. As a result, choosy females do not mate with nonhelping males.

Dawkins constructed a game-theory model of this scenario (sensu Maynard Smith & Price 1973), assigning fixed numerical values to the various costs and benefits. These can be represented by different payoff matrices for males and females. An asymmetrical conflict (Maynard Smith 1982) corresponds to the bimatrix games of classical game theory; there are two separate populations, males and females, with different strategy sets and payoff functions.

Dawkins' analysis leads to a totally mixed equilibrium of the game, at which all types of players are present and it does not pay any player to deviate from the equilibrium strategy. He claimed that this is an evolutionarily stable strategy (ESS). Schuster and Sigmund (1982) refuted this on the grounds that it lacks certain stability properties. The male and female payoffs for a "rare-mutant" strategy in a resident population at the Dawkins equilibrium are independent of the frequencies of the mutants. Hence, any mutant strategy is an alternative best reply to the mixed equilibrium, which therefore satisfies the first (Nash equilibrium) condition of the definition of an ESS (see Maynard Smith 1982). But if the mutant males and females are either more faithful and more choosy or less faithful and less choosy, the mutants fare better against themselves than do the residents against the mutants. Hence, the mixed-equilibrium point does not satisfy the second condition for an ESS. Selten (1980) has shown that for asymmetrical games in general, only pure strategies can be ESS's. The battle of the sexes is sometimes referred to as one of the most simple biological games without an ESS (Maynard Smith & Hofbauer 1987).

*Dynamic Models*

Another way to illustrate the instability of the mixed equilibrium is from the point of view of a dynamic system. Because the payoffs for the male strategies depend only on the state of the female population, and the payoffs of the female strategies depend only on the male population, there is no penalty for a one-sided deviation from the equilibrium frequencies. If a fluctuation increases the proportion of helping males, the payoff for each male strategy stays the same. But in this situation, it pays the females to become less choosy; then, nonhelping males are at an advantage. But if the frequency of nonhelping males is high, choosy females fare better; with a lot of choosy females around, helping males increase, and we are back at the beginning. However, there is no guarantee of a return to equilibrium, and there is a tendency to oscillatory behavior, for which the static game-theory approach is not sufficient.

Taylor and Jonker (1978) introduced a class of ordinary differential equations, known as game-dynamic equations, to model the dynamics of games. They assumed that the rate of increase of each type of player is equal to the difference between its expected payoff and the average payoff. For the battle of the sexes, using Dawkins' (1976) payoff matrices, Schuster and Sigmund (1981) derived corresponding dynamic equations. The solutions of these are neutrally stable oscillations around the Nash equilibrium, which is the time average of all these orbits (see Fig. 5). This behavior is similar to that of models of predator-prey dynamics of the Lotka-Volterra type.

Especially for games without an ESS, small modifications of the underlying dynamics can change the qualitative behavior considerably. With somewhat different equations in continuous time, the system converges to an asymptotically stable equilibrium (Maynard Smith 1982; Hofbauer & Sigmund 1988). With discrete-time game dynamics, however, the equilibrium is always unstable (Eshel & Akin 1983), in agreement with Maynard Smith's (1982) remark that any time delay destabilizes the equilibrium solution.

Note also that in the game-dynamic equations, just as in the static game-theory approach of Dawkins, "like begets like": male offspring inherit the strategy of the father, and female offspring inherit the strategy of the mother. All these models assume asexual reproduction, an awkward assumption for a biological war between the sexes. Incorporating diploid sexual inheritance in the Schuster-Sigmund model (Bomze et al. 1983) still gives rise to periodic oscil-

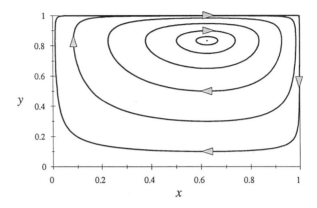

FIGURE 5. *Phase portrait of system (13). The fraction of helping males,*
*x, is plotted on the abscissa, and the fraction of choosy females, y, on*
*the ordinate.*

lations. Converting the game dynamics for the diploid model from
continuous time (Bomze et al. 1983) to discrete time (Maynard
Smith & Hofbauer 1987) changes the qualitative behavior from
neutrally stable oscillations to an unstable equilibrium surrounded
by a limit cycle (Maynard Smith 1982).

### A Model with Pair Formation

One of the (often implicit) assumptions in game-theory models is
that all consequences of the behavioral strategies can be expressed
in the payoffs. The fitness gain for an individual playing a certain strategy is then equal to the expected payoff for playing that
strategy in the population. In many cases this assumption is questionable. Our modification of the sex-war game expresses the costs
of raising offspring and performing courtship by means of a time
delay for the corresponding individuals in a pair-formation model.
We stay close to the model of Schuster and Sigmund (1981), the
first and simplest dynamic version of the game.

*Formulation of the model.*    Suppose that raising offspring takes $\tau_r$
time units, and performing courtship plus raising offspring takes $\tau_{cr}$
time units ($0 < \tau_r \leq \tau_{cr}$). The time needed for raising offspring does
not depend on whether the male helps the female. We assume that
when the female does the raising by herself, she produces $\kappa_1$ male

offspring and also $\kappa_1$ female offspring, whereas when the male helps her, $\kappa_2$ offspring of each sex are produced $(0 < \kappa_1 \leq \kappa_2)$. The difference between $\kappa_1$ and $\kappa_2$ is not in the original game (Dawkins 1976) and its descendants, but it is a natural extension.

Define $u_i(t)$ and $v_j(t)$ to be the densities at time $t$ of unmated males and females following strategies $i$ and $j$, respectively. The density of all unmated males is $u(t)$, where $u(t) = u_1(t) + u_2(t)$, and the density of all free females is $v(t)$, where $v(t) = v_1(t) + v_2(t)$. Furthermore, $w_{ij}(t, \tau) \, d\tau$ is either the density of type-$i$ males mated with type-$j$ females during a time interval of length $d\tau$ around $\tau$ time units before time $t$, or the density of (the same) type-$j$ females mated with (the same) type-$i$ males. Notice that, for $(i, j) = (1, 1)$, $0 < \tau < \tau_{\text{cr}}$, and for $(i, j) \in \{(1, 2), (2, 2)\}$, $0 < \tau < \tau_{\text{r}}$. Additionally, $w_{22}(t, \tau)$ denotes only the density of type-2 females mated with type-2 males because these males are always free.

We assume that mating between a type-$i$ male and a type-$j$ female occurs at a rate $m_{ij} = \alpha_{ij} u_i v_j / (u + v)$. Here, $\alpha_{ij}$ is a mating rate constant, equal to zero for a nonhelping-choosy mating and positive for all other combinations. For all but the nonhelping-choosy mating,

$$m_{ij}(t) = \alpha \frac{u_i(t) \, v_j(t)}{u(t) + v(t)}, \qquad ij \in \{11, 12, 22\}. \qquad (11)$$

It is only after raising the offspring that the individuals are free to mate again, along with the offspring.

For example, at time $t$, free males of type 1 disappear at a rate $m_{11}(t)$ by mating with females of type 1; free males reappear at a rate $w_{11}(t, \tau_{\text{cr}})$ following matings of this type that occurred at time $t - \tau_{\text{cr}}$. Assuming that "like begets like" and that individuals are mature immediately after raising, we can account for the male type-1 offspring by multiplying $w_{11}(t, \tau_{\text{cr}})$ by $(1 + \kappa_2)$.

Assuming a constant mortality rate $\mu$, the following partial differential equations describe the densities of mated individuals:

$$\frac{\partial w_{11}(t, \tau)}{\partial t} = -\frac{\partial w_{11}(t, \tau)}{\partial \tau} - \mu \, w_{11}(t, \tau), \qquad 0 < \tau < \tau_{\text{cr}},$$

$$\frac{\partial w_{12}(t, \tau)}{\partial t} = -\frac{\partial w_{12}(t, \tau)}{\partial \tau} - \mu \, w_{12}(t, \tau), \qquad 0 < \tau < \tau_{\text{r}}, \qquad (12)$$

$$\frac{\partial w_{22}(t, \tau)}{\partial t} = -\frac{\partial w_{22}(t, \tau)}{\partial \tau} - \mu \, w_{22}(t, \tau), \qquad 0 < \tau < \tau_{\text{r}},$$

with the side conditions that

$$w_{ij}(t,0) = m_{ij}(t), \qquad ij \in \{11, 12, 22\}. \qquad (13)$$

Considering all possible combinations of matings, together with reproduction and mortality, the following system of ordinary differential equations describes the densities of free individuals:

$$\frac{du_1(t)}{dt} = -m_{11}(t) + (1 + \kappa_2)\, w_{11}(t, \tau_{\mathrm{cr}}) - m_{12}(t)$$
$$+ (1 + \kappa_2)\, w_{12}(t, \tau_{\mathrm{r}}) - \mu\, u_1(t),$$

$$\frac{du_2(t)}{dt} = \kappa_1\, w_{22}(t, \tau_{\mathrm{r}}) - \mu\, u_2(t),$$

$$\frac{dv_1(t)}{dt} = -m_{11}(t) + (1 + \kappa_2)\, w_{11}(t, \tau_{\mathrm{cr}}) - \mu\, v_1(t), \qquad (14)$$

$$\frac{dv_2(t)}{dt} = -m_{12}(t) + (1 + \kappa_2)\, w_{12}(t, \tau_{\mathrm{r}}) - m_{22}(t)$$
$$+ (1 + \kappa_1)\, w_{22}(t, \tau_{\mathrm{r}}) - \mu\, v_2(t).$$

The fractions of helping males and choosy females in the population at time $t$, $x(t)$ and $y(t)$ respectively, are

$$x(t) = \frac{u_1(t) + W_{11}(t) + W_{12}(t)}{M(t)}, \quad y(t) = \frac{v_1(t) + W_{11}(t)}{F(t)}, \qquad (15)$$

where the densities of mated individuals, either males or females (except $W_{22}$, which represents only females) at time $t$, are

$$W_{ij}(t) = \int_0^{\tau_{\bullet}} w_{ij}(t, \tau)\, d\tau, \qquad ij \in \{11, 12, 22\} \qquad (16)$$

—here and below, $\tau_{\bullet} = \tau_{\mathrm{cr}}$ if $ij = 11$ and $\tau_{\bullet} = \tau_{\mathrm{r}}$ if $ij \in \{12, 22\}$— and the total densities of males and females, respectively, are

$$M(t) = u_1(t) + u_2(t) + W_{11}(t) + W_{12}(t),$$
$$F(t) = v_1(t) + v_2(t) + W_{11}(t) + W_{12}(t) + W_{22}(t). \qquad (17)$$

The appendix shows how the manifold on which male and female densities are equal is invariant and attracting. Thus, if the densities of males and females are initially equal, they remain equal. Finally, notice that there is no density dependence in the model; in the long run, the values of the parameters (notably $\alpha$ and $\mu$) produce an exponentially growing or an exponentially decaying population.

*An equivalent model with delay-differential equations.* We can easily convert the system of equations (12)–(14) into an equivalent system of delay-differential equations. Defining

$$u_{1w}(t) = W_{11}(t) + W_{12}(t),$$
$$v_{1w}(t) = W_{11}(t),$$
$$v_{2w}(t) = W_{12}(t) + W_{22}(t),$$
(18)

for mated individuals of type-1 males and for type-1 and type-2 females, respectively, and adding these three variables to equations (14), we obtain

$$\frac{du_1(t)}{dt} = -m_{11}(t) + (1 + \kappa_2)\, n_{11}(t, \tau_{cr}) - m_{12}(t)$$
$$+ (1 + \kappa_2)\, n_{12}(t, \tau_r) - \mu\, u_1(t),$$

$$\frac{du_2(t)}{dt} = \kappa_1\, n_{22}(t, \tau_r) - \mu\, u_2(t),$$

$$\frac{dv_1(t)}{dt} = -m_{11}(t) + (1 + \kappa_2)\, n_{11}(t, \tau_{cr}) - \mu\, v_1(t),$$

$$\frac{dv_2(t)}{dt} = -m_{12}(t) + (1 + \kappa_2)\, n_{12}(t, \tau_r) - m_{22}(t)$$
$$+ (1 + \kappa_1)\, n_{22}(t, \tau_r) - \mu\, v_2(t), \qquad (19)$$

$$\frac{du_{1w}(t)}{dt} = m_{11}(t) - n_{11}(t, \tau_{cr}) + m_{12}(t)$$
$$- n_{12}(t, \tau_r) - \mu\, u_{1w}(t),$$

$$\frac{dv_{1w}(t)}{dt} = m_{11}(t) - n_{11}(t, \tau_{cr}) - \mu\, v_{1w}(t),$$

$$\frac{dv_{2w}(t)}{dt} = m_{12}(t) - n_{12}(t, \tau_r) + m_{22}(t)$$
$$- n_{22}(t, \tau_r) - \mu\, v_{2w}(t),$$

where

$$n_{ij}(t, \tau_\bullet) = m_{ij}(t - \tau_\bullet)\, e^{-\mu\tau_\bullet} \qquad (20)$$

is the mating rate between type-$i$ males and type-$j$ females at time $t - \tau_\bullet$, multiplied by the survival probability over the time interval $(t - \tau_\bullet, t)$.

We can integrate this system of seven delay-differential equations numerically using the SOLVER program (Blythe et al. 1990$a$). The three mating rates $m_{ij}(t)$ are stored as history variables (Blythe et al. 1990$b$) in the SOLVER equations. Initial conditions of the system are set by entering $(x_0, y_0)$ and the total population size. The distribution over free and mated individuals at and before $t = 0$ is calculated by assuming that 99 percent of the individuals are mated. Total numbers of males and females are equal. Unless stated otherwise, $\alpha = 1$ and $\mu = 0.1$.

Notice that the time-delayed system (19) cannot keep track of the types of the partners of the mated individuals. From the definitions (18) there is sufficient information to calculate $x$ and $y$ using equations (15). With respect to $x$ and $y$, the systems (12)–(14) and (19) are equivalent, as long as the initial conditions are equivalent.

*Results*

In this subsection we discuss the qualitative behavior of the model in $(x, y)$-space, since we are interested mainly in the dynamics of the frequencies of the male and female behavioral strategies.

At the corners of $\Sigma^2 = \{x, y \mid 0 \leq x, y \leq 1, x + y = 1\}$ are always four trivial equilibria, where the variables $x$ or $y$ are equal to 0 or 1. It is laborious but straightforward to show analytically that on the boundary of $\Sigma^2$ the following holds. If $x = 0$, then $\dot{y} < 0$. For $x = 1$, $\dot{y} < 0$ if $\tau_{cr} > \tau_r$, and $\dot{y} = 0$ if $\tau_{cr} = \tau_r$. If $y = 1$, then $\dot{x} > 0$. On the border $y = 0$ and in the interior, the situation is more complicated. From here on, we discuss the qualitative behavior of $x$ and $y$ as it appears from numerical simulations of the model with delay-differential equations (19).

The simulations show that the system can have at least two nontrivial equilibria. We employ, as typical, the parameter values $(\tau_r, \tau_{cr}) = (10, 10.5)$ and $(\kappa_1, \kappa_2) = (1, 2)$. Figure 6 shows that on the border $y = 0$ is an asymptotically stable equilibrium at $(x, y) \approx (0.28, 0)$, connected through the "unstable manifold" of an interior "saddle point" at $(0.34, 0.01)$ with a (totally mixed) stable interior equilibrium at $(0.92, 0.42)$. All trajectories in this figure started in the upper left corner. The irregular shape of the orbits is due to imperfect initialization of the system.

Note that the dynamic systems (12)–(14) and (19) have infinite dimensions. Here, the dynamics of $x$ and $y$ are a two-dimensional projection of these systems, which strictly speaking rules out discussions of saddle points, stable and unstable manifolds, separatrices, etc. Nevertheless, it is convenient to continue to do so. But the fact that the behavior in $(x, y)$-space is reminiscent of a two-dimensional system suggests that this system can be simplified considerably.

The stable manifold of the saddle separates the basins of attraction of the interior and the boundary equilibrium. The basin of attraction of the interior equilibrium lies inside the interior of

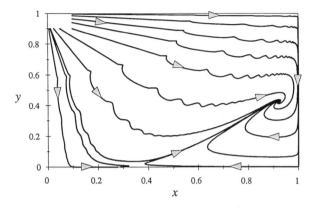

FIGURE 6. *Phase portrait of system (22), with* $(\tau_r, \tau_{cr}, \kappa_1, \kappa_2) =$ $(10, 10.5, 1, 2)$. *Again,* $x$ *is the fraction of helping males, and* $y$ *is the fraction of choosy females in the population. For explanation, see the text.*

$\Sigma^2$ and is surrounded by that of the boundary equilibrium, except that they both touch the corner $(x, y) = (0, 1)$. The interior basin of attraction almost touches the boundary $x = 1$, where the boundary basin is extremely narrow. The system can make large excursions through $(x, y)$-space before settling down. An example is the uppermost curve in Figure 6, which starts at $(0.1, 0.997)$ and converges to the interior equilibrium. Trajectories starting with $y$ even closer to 1 follow approximately the same route, but on the boundary side of the stable manifold of the saddle, finally converging to the boundary equilibrium.

The effect of varying the courtship period $(\tau_{cr} - \tau_r)$ is shown in Figure 7. Again, all trajectories start in the upper left corner, except for the curves in panel $a$, where the courtship period is zero. For a very short courtship period, the proportion of helping males in the interior equilibrium is almost 1 (see panel $b$).

Convergence to the interior equilibrium is, in this case, oscillatory: curves repeatedly pass extremely close to the boundary $x = 1$ while approaching the equilibrium. When $\tau_r = \tau_{cr}$ (no courtship; panel $a$), the interior equilibrium is absorbed by the boundary $x = 1$, which turns into a line of equilibria. Above $y \approx 0.52$, the boundary is attracting in the $x$ direction; below this value it is repelling. Nonhelping males are able to invade if the proportion of choosy females is smaller than 0.52; but after an initial increase,

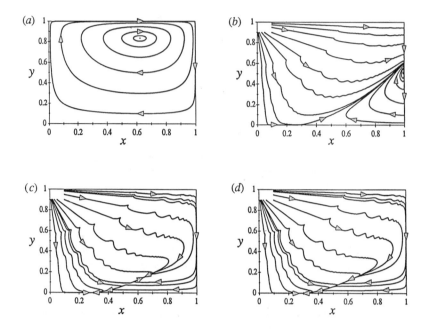

FIGURE 7. *Phase portraits of system (22), with varying courtship dura-*
*tions and $\tau_r = 10$, $\kappa_1 = 1$, $\kappa_2 = 2$: (a) zero, $\tau_{cr} = 10$; (b) very short,*
*$\tau_{cr} = 10.05$; (c) long, $\tau_{cr} = 11.2$; (d) slightly longer, $\tau_{cr} = 11.25$. Again,*
*x is the fraction of helping males, and y is the fraction of choosy females*
*in the population.*

they decrease in frequency again because the number of choosy females increases. The system settles on a boundary point with only
helping males, but with a higher proportion of choosy females than
before; from there nonhelping males are not able to invade again.

Increasing the courtship period (panels *b* and *c*) decreases the
frequencies of helping males and choosy females in the interior
equilibrium, and the basin of attraction for the interior equilibrium
shrinks until the equilibrium hits the saddle point and the two
disappear near $\tau_{cr} \approx 11.22$. For higher $\tau_{cr}$ values, the boundary
equilibrium is the only nontrivial one, and all trajectories from the
interior of $\Sigma^2$ converge to it (panel *d*).

The effect of a decrease in $\tau_r$ is similar to an increase in $\tau_{cr}$,
except that the boundary equilibrium moves to higher $x$ values,
whereas its location is independent of $\tau_{cr}$. At $\tau_r = \tau_{cr} = 10.5$,
the boundary equilibrium is located about where $x \approx 0.26$. At
$\tau_r \approx 9.42$ the interior equilibrium and the saddle annihilate each

other and only the boundary equilibrium remains. With a decrease in $\tau_r$, the boundary equilibrium moves to the right until it coincides with $(1,0)$ at $\tau_r \approx 2.8$.

Varying $\kappa_1$ yields approximately the same scenario as varying $\tau_r$, except that the location of the boundary equilibrium now spans the whole range of $0 \leq x \leq 1$. When $\kappa_1 \approx 1.08$ and higher, the boundary equilibrium is located at $(0,0)$; when $\kappa_1 \approx 0.47$ and lower, at $(1,0)$; and for $\kappa_1$ between these values, a nontrivial boundary equilibrium exists. Additionally, decreasing $\kappa_1$ lets the $y$ value of the interior equilibrium drop (!), whereas its $x$ value remains almost constant. Here, the disappearance of the interior equilibrium and the saddle point takes place at higher values of $x$. The interior equilibrium emerges again through a "saddle-node bifurcation" where $\kappa_1 \approx 1.19$. When $\kappa_1 \approx 0.55$, the interior equilibrium and the saddle seem to annihilate each other at the boundary equilibrium, which remains. Varying $\kappa_2$ amounts to the same thing as varying $\kappa_1$; it is the difference between the two that counts.

Choosy females can persist (and an interior equilibrium point exists) if the advantage of both parents' raising the offspring together is neither too small nor too large and if the length of the courtship period is not too large relative to the time needed for raising the offspring. The window of $\tau_\bullet$ values for a totally mixed equilibrium point to exist is wider for intermediate differences between $\kappa_1$ and $\kappa_2$, and vice versa. Outside this cone of values for $\tau_r$, $\tau_{cr}$, $\kappa_1$, and $\kappa_2$, the population consists of exclusively nonchoosy females. Even then, there can be a stable polymorphism in male strategies, depending on the exact parameter values.

### Discussion

An important difference between the game-dynamic model and the present models is seen in the behavior that arises when there are only nonhelping males around (at $x = 0$). In this case, choosy females do not mate, and they cannot reproduce. Without reproduction, their density must decrease at the mortality rate $\mu$. Nonchoosy females have positive growth terms because of matings with nonhelping males. Consequently, the proportion of choosy females decreases. In the game-dynamic model, however, the joint representation of reproduction and parental investment in one payoff value gives the choosy females a zero growth rate and the nonchoosy females a negative growth rate. This causes the proportion of choosy females to increase (and, if modeled, the total population size to decrease).

Another interesting difference is the existence of the boundary equilibrium (at $y = 0$). In a population of only nonchoosy females, helping and nonhelping males can coexist because of the difference in their reproductive outputs. This is not true in the game-dynamic models. Helping males can invade a nonhelping population if they raise sufficiently more offspring by helping the females, and nonhelping males can invade a helping population provided that the reproductive advantage for helping males is not too large and that the period needed for raising the offspring is sufficiently long. This observation shows that the raising period assists the helping males to keep females out of the mating market and unable to mate with nonhelping males.

The local behavior near the interior equilibrium point resembles the global behavior of the game-dynamic model more closely: clockwise convergence and oscillations. Both can be explained by the argument given in the subsection "Dynamic Models," clarifying the lack of stability of the mixed-equilibrium point: with more helping males around, nonchoosy females do better; with more nonchoosy females, nonhelping males are at an advantage, and so on.

If there is an interior equilibrium for the mating-delay models, it is a stable one, weakening Maynard Smith's (1982) claim that any time delay is destabilizing. Without a gain in reproductive output for a helping male ($\kappa_1 < \kappa_2$), there is no interior equilibrium; apparently, this condition more than compensates for destabilization due to time delays.

There are difficulties in formulating a more mechanistic model of the game, while keeping the model close to the Schuster-Sigmund model. This form maintains a necessary equality between the rates of mating of males and females. However, the Schuster-Sigmund model involves only quadratic terms of the form $u_i v_j$. The form used here is convenient, but it does not have a clear biological derivation; such a derivation would be a useful addition to dynamic models.

The densities acting as state variables in systems (12)–(14) cannot simply be scaled to frequencies. In the work of Taylor and Jonker (1978), Schuster and Sigmund (1981), and Hofbauer and Sigmund (1988), such a scaling of the variables is equivalent to a change in time scale, and only the information about the total population size is lost. For our system with delays, the densities are necessary in the delay terms, while a change in time scale also affects the time delay.

In the model presented here, paired individuals die individually, together with their would-be offspring (see eqs. 12, 14). The widowed partner, provided it stays alive, continues to raise its offspring (possibly courting a dead body). More-detailed ways of accounting for the mortality in pairs lead to more-complicated equations, but the global behavior of the model remains unaltered.

A striking drawback of this model (and those of Schuster & Sigmund 1981; Maynard Smith 1982; Eshel & Akin 1983) is that it discusses the dynamics of the battle of the sexes without taking sexual inheritance into account, thus implying that the organisms are haploid. An improvement would be to treat a diploid organism, for example, with two autosomal loci, one of which is expressed only in males and the other only in females (as in Bomze et al. 1983; Maynard Smith & Hofbauer 1987).

## Appendix

*Male and Female Densities in the Battle of the Sexes*

The manifold on which male and female densities are equal is attracting, that is,

$$\lim_{t \to \infty} [M(t) - F(t)] = 0.$$

Using the definitions in equation (14),

$$\begin{aligned}
\frac{d}{dt} &[M(t) - F(t)] \\
&= \frac{d}{dt} [u_1(t) + u_2(t) + W_{11}(t) + W_{12}(t)] \\
&\quad - \frac{d}{dt} [v_1(t) + v_2(t) + W_{11}(t) + W_{12}(t) + W_{22}(t)] \\
&= -\mu [u_1(t) + u_2(t)] + \mu [v_1(t) + v_2(t)] \\
&\quad + m_{22}(t) - w_{22}(t, \tau_\mathrm{r}) - \frac{d}{dt} W_{22}(t).
\end{aligned}$$

Because

$$\frac{dW_{ij}(t)}{dt} = -w_{ij}(t, \tau_\bullet) + m_{ij}(t) - \mu W_{ij}(t),$$

this is equal to

$$-\mu [u_1(t) + u_2(t)] + \mu [v_1(t) + v_2(t)] + \mu W_{22}(t).$$

Adding and subtracting $-\mu [W_{11}(t) + W_{12}(t)]$ leads to

$$\frac{d}{dt} [M(t) - F(t)] = -\mu [M(t) - F(t)],$$

and, therefore,

$$M(t) - F(t) = [M(0) - F(0)] \ e^{-\mu t} \ .$$

This proves that the manifold of equal male and female densities is invariant and, furthermore, that after disturbance the system returns to this manifold.

## Acknowledgments

The work by S.M. was supported by the Life Sciences Foundation (SLW), subsidized by the Netherlands Organization for Scientific Research (NWO). D.P. thanks R. Montgomerie, P. Taylor, and P. Abrams for their helpful advice during preparation of this manuscript.

## Literature Cited

Adler, F. R. 1990. Coexistence of two types on a single resource in discrete time. *Journal of Mathematical Biology* 28: 695–713.

Asmussen, M. A. 1979*a*. Density-dependent selection II. The Allee effect. *American Naturalist* 114: 796–809.

———. 1979*b*. Regular and chaotic cycling in models of ecological genetics. *Theoretical Population Biology* 16: 172–190.

Asmussen, M. A., and M. W. Feldman. 1977. Density dependent selection 1: A stable feasible equilibrium may not be attainable. *Journal of Theoretical Biology* 64: 603–618.

Blythe, S., W. S. Gurney, P. Maas, and R. M. Nisbet. 1990*a*. *Introduction & Installation Guide for* SOLVER *(rev. 4)*. User Guide *40*. Applied Physics Industrial Consultants, University of Strathclyde, Glasgow, Scotland.

———. 1990*b*. *Programming & Model Building Guide for* SOLVER *(rev. 4)*. User Guide *41*. Applied Physics Industrial Consultants, University of Strathclyde, Glasgow, Scotland.

Bomze, I., P. Schuster, and K. Sigmund. 1983. The role of Mendelian genetics in strategic models on animal behavior. *Journal of Theoretical Biology* 101: 19–38.

Botsford, L. W. 1992. Further analysis of Clark's delayed recruitment model. *Bulletin of Mathematical Biology* 54: 275–293.

Charlesworth, B. 1980. *Evolution in Age-Structured Populations*. Cambridge University Press.

———. 1994. *Evolution in Age-Structured Populations*. 2nd ed. *Cambridge Studies in Mathematical Biology*, Vol. 1. Cambridge University Press.

Cushing, J. M., and J. Li. 1992. Intra-specific competition and density dependent juvenile growth. *Bulletin of Mathematical Biology* 4: 503–519.

Dawkins, R. 1976. *The Selfish Gene.* Oxford University Press.

Ellner, S., and N. G. Hairston. 1994. The role of overlapping generations in maintaining genetic variation in a fluctuating environment. *American Naturalist* 143: 403–417.

Eshel, I., and E. Akin. 1983. Coevolutionary instability of mixed Nash solutions. *Journal of Mathematical Biology* 18: 123–133.

Gatto, M. 1993. The evolutionary optimality of oscillatory and chaotic dynamics in simple population models. *Theoretical Population Biology* 43: 310–336.

Hastings, A. 1992. Age dependent dispersal is not a simple process: Density dependence, stability, and chaos. *Theoretical Population Biology* 41: 388–400.

———. 1993. Complex interactions between dispersal and dynamics: Lessons from coupled logistic equations. *Ecology* 74: 1362-1372.

Hofbauer, J., and K. Sigmund. 1988. *The Theory of Evolution and Dynamical Systems: Mathematical Aspects of Selection.* Cambridge University Press.

May, R. M. 1974. *Stability and Complexity in Model Ecosystems.* Princeton University Press, Princeton, N.J.

Maynard Smith, J. 1982. *Evolution and the Theory of Games.* Cambridge University Press.

Maynard Smith, J., and J. Hofbauer. 1987. The "battle of the sexes": A genetic model with limit cycle behavior. *Journal of Theoretical Biology* 32: 1–14.

Maynard Smith, J., and G. Price. 1973. The logic of animal conflicts. *Nature* 246: 15–18.

Roughgarden, J. 1971. Density-dependent natural selection. *Ecology* 52: 453–468.

———. 1979. *Theory of Population Genetics and Evolutionary Ecology.* Macmillan, New York.

Schuster, P., and K. Sigmund. 1981. Coyness, philandering and stable strategies. *Animal Behaviour* 29: 186–192.

Selten, R. 1980. A note on evolutionarily stable strategies in asymmetrical animal conflicts. *Journal of Theoretical Biology* 84: 93–101.

Taylor, P., and L. Jonker. 1978. Evolutionarily stable strategies and game dynamics. *Mathematical Biosciences* 70: 65–90.

Templeton, A. R., and D. A. Levin. 1979. Evolutionary consequences of seedbanks. *American Naturalist* 114: 232–249.

Tonsor, S. J., S. Kalisz, J. Fisher, and T. P. Holtsford. 1993. A life-history based study of population genetic structure: Seed bank to adults in *Plantago lanceolata. Evolution* 47: 833–843.

Trivers, R. 1972. Parental investment and sexual selection. Pp. 136–179 *in* B. Campbell, ed., *Sexual Selection and the Descent of Man.* Aldine, Chicago.

CHAPTER 11

# The Effect of Overlapping Generations and Population Structure on Gene-Frequency Clines

*Oscar E. Gaggiotti, Carol E. Lee,*
*and Glenda M. Wardle*

Many natural populations exhibit clines or spatial patterns in gene frequency or phenotype (Roughgarden 1979). The maintenance of these clines is understood as a balance between the diversifying process of spatial variation in selection intensities and the homogenizing process of gene flow or migration. A quantitative analysis of the amount of gene flow necessary to prevent genetic differentiation between two populations is also directly related to theories of speciation, in particular, that of parapatric speciation (Slatkin 1973). In this chapter, we extend previous models of clines to examine the effects of population structure and overlapping generations in clines maintained by the balance between selection and gene flow.

Although some spatially discrete models of selection and migration have been used in the study of clines (e.g., Nagylaki 1977; Moody 1979, 1981), most of the studies have used the more tractable continuous counterparts (e.g., Slatkin 1973; Nagylaki 1978, 1989; Peletier 1978; Nagylaki & Moody 1980; Fife & Peletier 1981; Keller 1984).

All previous models considered non-overlapping generations. However, the life cycle of many organisms is composed of discrete life-history stages, and selection, migration, and survival are of-

ten age- or stage-specific. For instance, some studies give evidence
for stage-specific selection in natural populations. In this regard,
Koehn et al. (1980) showed that the differential mortality of larval
recruits of the bivalve mussel *Mytilus edulis* results in differences
in allele frequencies between larvae and adults. Differences in migration pattern among life-history stages are common. Many stationary bottom-dwelling marine invertebrates have free-swimming
larvae; conversely, some invertebrates, such as hydrozoan jellyfish,
have a sessile phase (in this case, as attached polyps) followed by
a free-swimming adult stage (Kozloff 1990). Finally, stage-specific
survival is well documented. In many animal species, mortality declines with age from the larval to adult stages (Deevey 1947); and
in marine invertebrates, mortality rates are generally higher for
larvae than for adults (Rumrill 1990).

Here we present a model that incorporates overlapping generations and stage structure for a linear series of demes (subpopulations) that exchange migrants with adjacent demes. We investigate
the effect of different selection regimes (larval and adult viability
and adult fecundity) on both unstructured and structured models
for a range of selection intensities and migration rates. The life
cycle consists of three stages—gametes, larvae, and adults—with
selection operating through differences in fecundity, larval survival,
or adult survival. The main objective is to show that the consideration of stage structure in models of clines can give us new insights
into the important evolutionary problems of speciation and maintenance of genetic variation. In particular, adding structure to the
model leads to a smoothing out of the cline (i.e., a decrease in the
steepness of the cline). In other words, there is an increase in the
effect of gene flow for a given selection intensity. In unstructured
models (with non-overlapping generations), selection is more effective at countering the effects of gene flow because all individuals
are exposed to selection and they all die after reproduction. In
contrast, when generations overlap, stage-specific selection acts on
only a fraction of the population, either larvae or adults. Moreover,
since adult individuals can survive and reproduce for longer periods
of time, there is a storage of genotypes at this stage. This storage
effect provides a mechanism for the smoothing out of the cline. If
selection is assumed to act on only a particular stage, either larvae
or adults, the degree to which the cline is smoothed depends on the
stage on which selection acts. If selection affects the adult stage,
the less favored adult genotypes are subject to selection in every
time interval and the storage of genotypes is reduced, leading to
steeper clines.

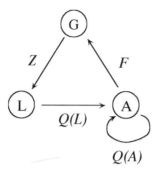

FIGURE 1. *Stage-classified life cycle for gametes, larvae, and adults.*
$Q(L)$, *larval survival;* $Q(A)$, *adult survival;* $F$, *fecundity, in terms of
number of gametes produced per adult;* $Z$, *fusion of gametes to produce
eggs.*

## 1 The Model

*Assumptions*

Consider a sexually reproducing diploid population subdivided into
a linear array of panmictic demes (linear stepping-stone model;
Malecot 1948; Kimura & Weiss 1964) that exchange larvae independently of genotype. The life cycle is represented by three stages:
gametes, larvae, and adults (see Fig. 1). The survival from larva
to adult is represented by $Q(L)$; an adult either survives until the
next iteration, represented by the self-loop $Q(A)$, or it dies. For
reasons that become clear below, reproduction is separated into
the production of gametes ($F$) and the fusion of gametes ($Z$).
Reproduction occurs within each deme and is followed by larval
dispersal. We consider three selection regimes: (a) fecundity selection, in which selection acts through differences in adult fecundity;
(b) larval-viability selection, in which selection acts through differences in larval survival; and (c) adult-viability selection, in which
selection acts through differences in adult survival. We restrict our
analysis to a single locus with two alleles, $B_1$ and $B_2$.

Migration is assumed to occur only between adjacent demes at
rate $m$; that is, a single deme in the habitat receives immigrants
from both neighboring demes at the rate $m/2$ per generation. In
addition, the effects of mutation and random genetic drift are
ignored, so that gene frequencies after mating are given by the
Hardy-Weinberg relation.

## The Equations

In order to obtain simple equations to describe the dynamics of the system, we separate reproduction into two steps, the production of gametes and the fusion of gametes. The model can thus be written as the product of three matrices and a vector:

$$\mathbf{N}_{t+1} = \mathbf{M} \times \mathbf{G} \times \mathbf{D}_t \times \mathbf{N}_t .$$

The product of vector $\mathbf{N}_t$ and matrix $\mathbf{D}_t$ represents the first step, during which selection is applied and the adults produce gametes; this matrix is time-dependent because it contains the adult fecundities, which are assumed to be density-dependent (see below). The product of the resulting vector, $\mathbf{N}_t^*$, and matrix $\mathbf{G}$ represents the second step, during which the fusion of gametes takes place. Finally, the product of the resulting vector, $\mathbf{N}_t^{**}$, and matrix $\mathbf{M}$ represents the step during which larvae migrate.

The full matrix model is given by

$$
\begin{bmatrix}
\mathbf{N}(1) \\
\mathbf{N}(2) \\
\mathbf{N}(3) \\
\vdots \\
\mathbf{N}(n-1) \\
\mathbf{N}(n)
\end{bmatrix}_{t+1}
=
$$

$$
\begin{bmatrix}
\mathbf{M}_{1\leftarrow 1} & \mathbf{M}_{1\leftarrow 2} & 0 & \cdots & 0 & 0 \\
\mathbf{M}_{2\leftarrow 1} & \mathbf{M}_{2\leftarrow 2} & \mathbf{M}_{2\leftarrow 3} & \cdots & 0 & 0 \\
0 & \mathbf{M}_{3\leftarrow 2} & \mathbf{M}_{3\leftarrow 3} & \cdots & 0 & 0 \\
\vdots & \vdots & \vdots & \ddots & \vdots & \vdots \\
0 & 0 & 0 & \cdots & \mathbf{M}_{n-1\leftarrow n-1} & \mathbf{M}_{n-1\leftarrow n} \\
0 & 0 & 0 & \cdots & \mathbf{M}_{n\leftarrow n-1} & \mathbf{M}_{n\leftarrow n}
\end{bmatrix}
$$

$$
\times
\begin{bmatrix}
\mathbf{G}(1) & 0 & 0 & \cdots & 0 & 0 \\
0 & \mathbf{G}(2) & 0 & \cdots & 0 & 0 \\
0 & 0 & \mathbf{G}(3) & \cdots & 0 & 0 \\
\vdots & \vdots & \vdots & \ddots & \vdots & \vdots \\
0 & 0 & 0 & \cdots & \mathbf{G}(n-1) & 0 \\
0 & 0 & 0 & \cdots & 0 & \mathbf{G}(n)
\end{bmatrix}
$$

$$\times \begin{bmatrix} \mathbf{D}(1) & 0 & 0 & \cdots & 0 & 0 \\ 0 & \mathbf{D}(2) & 0 & \cdots & 0 & 0 \\ 0 & 0 & \mathbf{D}(3) & \cdots & 0 & 0 \\ \vdots & \vdots & \vdots & \ddots & \vdots & \vdots \\ 0 & 0 & 0 & \cdots & \mathbf{D}(n-1) & 0 \\ 0 & 0 & 0 & \cdots & 0 & \mathbf{D}(n) \end{bmatrix}_t$$

$$\times \begin{bmatrix} \mathbf{N}(1) \\ \mathbf{N}(2) \\ \mathbf{N}(3) \\ \vdots \\ \mathbf{N}(n-1) \\ \mathbf{N}(n) \end{bmatrix}_t .$$

The components of the subvector $\mathbf{N}_t(k)$ give the number of larvae, $\mathbf{N}_t^{ij}(L,k)$, and adults, $\mathbf{N}_t^{ij}(A,k)$, of genotype $ij$ and the number of gametes carrying allele $i$, $\mathbf{N}_t^i(k)$, for deme $k$:

$$\mathbf{N}_t(k) = \begin{bmatrix} \mathbf{N}_t^{11}(L,k) & \mathbf{N}_t^{12}(L,k) & \mathbf{N}_t^{22}(L,k) & \mathbf{N}_t^{11}(A,k) \end{bmatrix}$$

$$\mathbf{N}_t^{12}(A,k) \quad \mathbf{N}_t^{22}(A,k) \quad \mathbf{N}_t^1(k) \quad \mathbf{N}_t^2(k) \big]^{\mathrm{T}} .$$

The matrix $\mathbf{D}_t$ is composed of submatrices

$\mathbf{D}_t(k) =$

$$\begin{bmatrix} 0 & 0 & 0 & 0 & 0 & 0 & 0 & 0 \\ 0 & 0 & 0 & 0 & 0 & 0 & 0 & 0 \\ 0 & 0 & 0 & 0 & 0 & 0 & 0 & 0 \\ Q^{11}(L,k) & 0 & 0 & Q^{11}(A,k) & 0 & 0 & 0 & 0 \\ 0 & Q^{12}(L,k) & 0 & 0 & Q^{12}(A,k) & 0 & 0 & 0 \\ 0 & 0 & Q^{22}(L,k) & 0 & 0 & Q^{22}(A,k) & 0 & 0 \\ 0 & 0 & 0 & F_t^{11}(k) & {}^{1\!}/_{2}\,F_t^{12}(k) & 0 & 0 & 0 \\ 0 & 0 & 0 & 0 & {}^{1\!}/_{2}\,F_t^{12}(k) & F_t^{22}(k) & 0 & 0 \end{bmatrix},$$

which, multiplied by the corresponding subvector $\mathbf{N}_t(k)$, gives the number of gametes with alleles $B_1$ and $B_2$ produced by deme $k$. The submatrices above contain all the demographic information for deme $k$ ($k = 1, 2, \ldots, n$) and genotype $ij$, namely, the transitions from larva to adult, $Q^{ij}(L,k)$, the transition from adult to adult, $Q^{ij}(A,k)$, and the fecundities (in terms of number of gametes), $F_t^{ij}(k)$. Note that only one of the demographic parameters (either larval survival, adult survival, or fecundity) is a function of genotype and deme, depending on the pattern of selection. In order to allow individual demes to reach an equilibrium population size, we incorporate density dependence into adult fecundity using an exponential logistic equation,

$$F_t^{ij}(k) = F^{ij}(k) \exp\left\{r\left[1 - P_t(k)/K\right]\right\},$$

where $F^{ij}(k)$ is the density-independent, genotype-dependent fecundity coefficient, $r$ is the population growth rate, $K$ is the carrying capacity of each deme, and $P_t(k)$ is the total number of individuals in deme $k$ at time $t$.

The matrix **G** is composed of submatrices

$$\mathbf{G}(k) = \begin{bmatrix} 0 & 0 & 0 & 0 & 0 & 0 & \dfrac{N^1(k)}{2\left[N^1(k) + N^2(k)\right]} & 0 \\[2ex] 0 & 0 & 0 & 0 & 0 & 0 & \dfrac{N^2(k)}{2\left[N^1(k) + N^2(k)\right]} & \dfrac{N^1(k)}{2\left[N^1(k) + N^2(k)\right]} \\[2ex] 0 & 0 & 0 & 0 & 0 & 0 & 0 & \dfrac{N^2(k)}{2\left[N^1(k) + N^2(k)\right]} \\[2ex] 0 & 0 & 0 & 1 & 0 & 0 & 0 & 0 \\ 0 & 0 & 0 & 0 & 1 & 0 & 0 & 0 \\ 0 & 0 & 0 & 0 & 0 & 1 & 0 & 0 \\ 0 & 0 & 0 & 0 & 0 & 0 & 0 & 0 \\ 0 & 0 & 0 & 0 & 0 & 0 & 0 & 0 \end{bmatrix},$$

which, when multiplied by the corresponding subvector $\mathbf{N}_k^*$, give the number of larvae of genotype $ij$ produced by deme $k$, and the number of adults remaining in that deme.

The matrix **M** is composed of two types of submatrices. The submatrix

$$\mathbf{M}_{k \leftarrow k} = \begin{bmatrix} 1-m & 0 & 0 & 0 & 0 & 0 & 0 & 0 \\ 0 & 1-m & 0 & 0 & 0 & 0 & 0 & 0 \\ 0 & 0 & 1-m & 0 & 0 & 0 & 0 & 0 \\ 0 & 0 & 0 & 1 & 0 & 0 & 0 & 0 \\ 0 & 0 & 0 & 0 & 1 & 0 & 0 & 0 \\ 0 & 0 & 0 & 0 & 0 & 1 & 0 & 0 \\ 0 & 0 & 0 & 0 & 0 & 0 & 1 & 0 \\ 0 & 0 & 0 & 0 & 0 & 0 & 0 & 1 \end{bmatrix}$$

multiplied by the corresponding vector $\mathbf{N}_k^{**}$ gives the number of larvae that remain in deme $k$. The submatrix

$$\mathbf{M}_{k\leftarrow k-1} = \mathbf{M}_{k\leftarrow k+1} = \begin{bmatrix} m/2 & 0 & 0 & 0 & 0 & 0 & 0 & 0 \\ 0 & m/2 & 0 & 0 & 0 & 0 & 0 & 0 \\ 0 & 0 & m/2 & 0 & 0 & 0 & 0 & 0 \\ 0 & 0 & 0 & 0 & 0 & 0 & 0 & 0 \\ 0 & 0 & 0 & 0 & 0 & 0 & 0 & 0 \\ 0 & 0 & 0 & 0 & 0 & 0 & 0 & 0 \\ 0 & 0 & 0 & 0 & 0 & 0 & 0 & 0 \\ 0 & 0 & 0 & 0 & 0 & 0 & 0 & 0 \end{bmatrix}$$

multiplied by the corresponding subvector $\mathbf{N}^{**}_{k-1}$ or $\mathbf{N}^{**}_{k+1}$ gives the number of larvae immigrating from demes $k-1$ and $k+1$ into deme $k$. The remaining submatrices, $\mathbf{M}_{k\leftarrow k-1}$ and $\mathbf{M}_{k\leftarrow k+1}$, for $i > 1$, are zero matrices because migration occurs only between adjacent demes.

*The Selection Gradient*

As previously stated, we consider three patterns of selection: (a) fecundity selection, (b) larval-viability selection, and (c) adult-viability selection. If we number the demes from 1 to $n$, starting at one extreme of the geographic range, we can define the larval survival for the different genotypes as follows:

$$\begin{aligned} B_1B_1, \quad & Q^{11}(L,k) = [1 - sc(k)]Q_L, \\ B_1B_2, \quad & Q^{12}(L,k) = Q_L, \\ B_2B_2, \quad & Q^{22}(L,k) = [1 + sc(k)]Q_L, \end{aligned} \tag{1}$$

where $c(k)$ describes the spatial variation in selection intensity ($k = 1, 2, \dots, n$ indicates the position of the deme along the species' range), $s$ measures the strength of selection, and $Q_L$ is a constant. We can obtain similar expressions for adult survival and (density-independent) fecundity for the different genotypes by replacing $Q^{ij}(L,k)$ and $Q_L$ in equation (1) with $Q^{ij}(A,k)$ and $Q_A$, and $F^{ij}(k)$ and $F$, respectively. The selection-gradient function, $c(k)$, has the form

$$c(k) = \begin{cases} 1 & \text{for } k \geq \frac{n+1+\Delta}{2}, \\ (2k - n + 1)/\Delta & \text{for } \frac{n+1-\Delta}{2} \leq k \leq \frac{n+1+\Delta}{2}, \\ -1 & \text{for } k \leq \frac{n+1-\Delta}{2}, \end{cases} \tag{2}$$

where $n$ is the total number of demes considered and $\Delta$ is the number of demes in the transition zone. The selection gradient

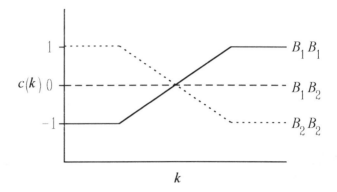

FIGURE 2. *Selection gradient as described by equation (2). For small k, the $B_1B_1$ genotype has a selective advantage; for large $k$, the $B_2B_2$ genotype has the selective advantage; in the transition zone, of width $\Delta$, the change in fitness occurs.*

is illustrated in Figure 2. For small $k$, the genotype $B_1B_1$ has a selective advantage measured by the quantity $s$; for large $k$, $B_2B_2$ has this selective advantage; and in between is a transition zone of width $\Delta$. This formulation is a discrete-space version of that used by May et al. (1975).

### Results

The main objective of the model is to determine whether the introduction of population structure can alter the outcome of earlier models of clines with discrete (non-overlapping) generations. To achieve this goal, we consider a model with non-overlapping generations, and compare the results from this model with those from the overlapping-generations models. The unstructured model assumes that the life cycle is divided into two stages, gametes and adults; all other assumptions remain unchanged. Simulations were run until the system reached equilibrium, which was assumed to occur when $|p_t^{ij} - p_{t+1}^{ij}| \leq 10^{-5}$ for all $i$ and $j$; the initial numbers of individuals were 15 adults of each genotype for each deme (see the figure legends for more details).

   Following previous analyses (see Slatkin 1973; May et al. 1975), we present the results in terms of three parameters that describe the effect of gene flow, the width of the cline, and the "environmental grain" (sensu Levins 1968, see below). The quantity that characterizes the main effect of gene flow is the root-mean-square migration length, which in our case (and that of Endler 1973) is given by $\ell = m^{1/2}$, where $m$ is the migration rate. We define the width, $W$, of the gene-frequency cline as the distance between the points where $p = 0.1$ and $p = 0.9$ ($p$ is the frequency of allele $B_1$). The quantity that sets the environmental grain is the characteristic length of the spatial variation in the frequencies of alleles (Slatkin 1973), which is given by $\ell_c = \ell s^{-1/2} = (m/s)^{1/2}$. As Slatkin pointed out, a population cannot respond to changes in selection that occur over distances smaller than the characteristic length, whereas it can respond to changes that occur over distances greater than $\ell_c$. Clearly, $\ell_c$ is related to the concept of environmental grain; in a fine-grained environment, spatial variation in the environment occurs on a short scale, so that the population is unable to respond. In a coarse-grained environment, the opposite is true.

In unstructured models, the relationship between the characteristic length, $\ell_c$, to $\Delta$ (the transition zone in the selection gradient) determines the width of the resulting cline, $W$ (Slatkin 1973; May et al. 1975). For $\ell_c > \Delta$, $W = \ell_c$ and the cline width is smoothed (i.e., greater than $\Delta$); for $\ell_c < \Delta$, $W = \Delta(\ell_c/\Delta)^{2/3}$ and the cline width is less than $\Delta$.

Figure 3 shows the equilibrium clines produced by the various modes of selection and population structure for different values of the ratio $m/s$. The upper panel is for $m = 0.1$, $s = 0.01$, $\Delta = 8$, and $\ell_c = 3.6$. Here, $\ell_c < \Delta$, and therefore, as expected from the equations above, the unstructured model gives a narrow cline with $W = 4.3$, which is less than $\Delta$. By comparison, the structured models produce smoother clines (less steep and wider) than those of the unstructured model. The difference between the unstructured and structured models is less pronounced when $\ell_c \ll \Delta$ (lower panel; $m = 0.01$, $s = 0.1$, $\Delta = 8$, and $\ell_c = 0.32$).

Figure 4 shows $W$ (expressed as the dimensionless ratio $W/\Delta$) plotted against the corresponding values of $\ell_c$ (measured as the dimensionless variable $\lambda/\Delta = (\ell_c/\Delta)^{2/3}$). A shallow slope indicates that the persistence of the cline is not sensitive to the balance between selection and migration and therefore can be easily maintained despite high gene flow. Conversely, a steep slope indicates that the cline is easily smoothed by the increasing effect of gene flow. The slope of the curve for the unstructured model is small, indicating that the effect of gene flow in this model is small; this result is consistent with Endler's (1973) results. The slope for the structured models, in contrast, is more pronounced and is larger for selection acting on adult fecundity or larval survival than for selec-

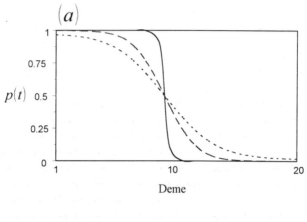

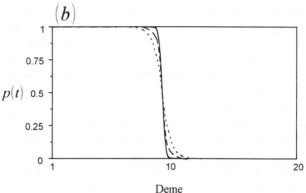

FIGURE 3. *Equilibrium clines (frequency, $p(t)$, of $B_1$ at equilibrium) produced by the various modes of selection and two different values of the ratio $m/s$.* Solid line, *Unstructured population with selection acting on fecundity or survival;* dashed line, *structured population with selection acting on adult survival;* dotted line, *structured, selection on fecundity or larval survival. Here $n = 20$, $\Delta = 8$, $F = 1$, $r = 1.4$, $K = 600$, $Q_L = 0.3$, $Q_A = 0.8$.* (a), $m = 0.1$, $s = 0.01$; (b) $m = 0.01$, $s = 0.1$.

tion acting on adult survival. In our structured models, selection acts on only a fraction of the individuals, and the adults can survive and reproduce for longer periods of time. Thus, there is a storage of genotypes in the adult stage, which provides a mechanism for the smoothing of the cline. The balance between selection and migration is altered so that the relative effect of migration increases. Note that the effect of migration is less pronounced when selection

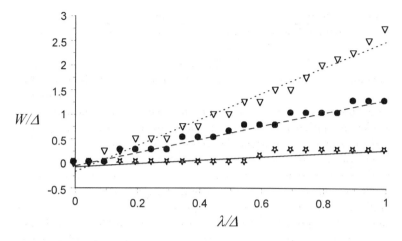

FIGURE 4. *Plot of W (expressed as the dimensionless ratio $W/\Delta$) against the corresponding values of $\ell_c$ (measured as the dimensionless variable $\lambda/\Delta = (\ell_c/\Delta)^{2/3}$). Stars, Unstructured population; circles, structured population with selection acting on adult survival; triangles, structured, selection on fecundity or larval survival. Here $n = 40$, $\Delta = 8$, $F = 1$, $r = 1.1$, $K = 600$, $Q_L = 0.3$, $Q_A = 0.8$.*

acts on adult survival than when it acts on fecundity or larval survival. This difference arises because, when selection affects adults, the less favored adult genotypes are subject to selection at every time interval and the storage of genotypes is reduced, leading to steeper clines. Note also that both fecundity selection and larval selection lead to the same result. Because the storage of genotypes occurs at the adult stage, fecundity selection and larval-viability selection are equivalent. That is, both fecundity and larval-viability selection result in a reduction in the number of larvae carrying the less favored allele that are able to survive to adulthood.

Different demographic parameters affect the outcome of the models. An extensive number of simulations show that as long as $K$ and $r$ are independent of genotype, they have no bearing on the comparison of the models. This is true for equilibrium conditions only; clearly, if recurrent catastrophes depress the population size of each deme to some low value, the outcome of the model would depend on $r$ and $K$.

Further simulations show that changes in density-independent fecundity and larval-survival coefficients, $F$ and $Q_L$, respectively, do not affect the outcome of the model, whereas adult survival, $Q_A$, does so only in the case of selection acting on adult survival

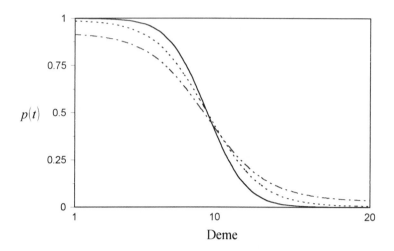

FIGURE 5. *Effect of the genotype-independent adult-survival coefficient,* $Q_A$, *on the shape of the cline. Here* $n = 20$, $\Delta = 8$, $F = 1$, $r = 1.4$, $K = 600$, $Q_L = 0.3$, $m = 0.1$, $s = 0.01$. *Solid line,* $Q_A = 0.8$; *dotted line,* $Q_A = 0.6$; *dashed line,* $Q_A = 0.4$.

(Fig. 5). The larval stage is short-lived, with larvae either dying or developing into adults; no storage of genotypes occurs at this stage. Thus, regardless of the selection pattern, changes in the coefficients of genotype-independent fecundity and larval survival have little effect on the genetic structure of the population. In contrast, changes in the coefficient of genotype-independent adult survival, $Q_A$, can affect the genetic structure because of the cumulative effect across time. However, when adults are not subject to selection (fecundity or larval-survival selection patterns), all adult genotypes have the same survival probability, and genetic structure is not affected. Only under adult-survival selection can changes in $Q_A$ affect the outcome of the model; Figure 5 shows this effect. The cline becomes steeper for increasing values of $Q_A$ because the proportion of adults within each deme increases, leading to an increase in the proportion of individuals experiencing selection in any given deme and to greater differences in gene frequencies among demes.

Regarding the dynamics of the system, the simulations show that, as expected, the behavior of the system is extremely rich, with limit cycles, $n$-point cycles ($n = 2, 4, \ldots$), and chaotic behavior. Moreover, some demes reach equilibrium, while others show unstable behavior. The parameters most important in determining the stability of the system appear to be the selection coefficient,

$s$, and the intrinsic population growth rate, $r$. The model is quite flexible, and clearly a careful investigation of the stability of the system and particularly the interaction between the population dynamics and the genetic structure would be an interesting direction for future work.

## 2 Discussion

The results of this simulation study show that stage structure has an important effect on clines maintained by the balance between selection and gene flow. Adding stage structure results in a cline that is wider and less steep than in the unstructured case. The smoothing effect of migration is more pronounced in organisms with overlapping generations because selection acts on only a fraction of the individuals and adults survive to reproduce repeatedly. The storage of genotypes in the adult stage provides the mechanism for the increased effect of gene flow on the width of the cline. The storage effect observed in the present analysis is similar to that observed by Ellner and Hairston (1994), who presented a model for stabilizing selection with a temporally fluctuating optimum and concluded that organisms with overlapping generations can maintain genetic variation in a temporally fluctuating environment. According to them, it is the storage of genotypes in long-lived stages on which the fluctuating selection does not act that provides the mechanism for the maintenance of polymorphism.

It should be noted that there is a time-scale difference between the unstructured and structured models. In the model for non-overlapping generations, $m$ and $s$ represent per generation rates, whereas in the model for overlapping generations, the same parameters represent per year rates. One could correct for this difference by multiplying $m$ and $s$ by the generation time of the overlapping-generations model. This correction would give the per generation mutation and migration rates for the overlapping-generations model. As mentioned in the preceding section, however, the results of the models depend on the ratio $m/s$ and not on the absolute values of $m$ and $s$. The above-mentioned correction amounts to multiplying both $m$ and $s$ by a constant, which cancels out of the ratio $m/s$. Our simulations using different sets of migration and selection coefficients with the same $m/s$ ratios lead to the same results, regardless of the magnitudes of $m$ and $s$. This confirms our assertion that the time-scale difference has no bearing on the comparison of the models.

Previous models considered only non-overlapping generations

and were not appropriate for studying the effect of selection acting on different life-history stages. Our formulation indicates that
selection acting through differences in adult survival among genotypes leads to steeper clines because selection affects the longest-
lived stage, reducing the storage effect.

Our model assumes that migration occurs only between adjacent
demes and therefore is more similar to that of Endler (1973) than to
other models (e.g., Slatkin 1973; May et al. 1975). Endler's results
indicate that clines are rather insensitive to the effects of gene flow,
and Slatkin asserted that those results are the product of assuming
that migration occurs only between adjacent demes. Our analysis,
however, shows that the cline can be sensitive to gene flow even if
migration is restricted to adjacent demes.

The study of clines maintained by a balance between selection
and migration is directly related to theories of speciation, in particular to that of parapatric speciation. A cline in gene frequency
is established if selection favors different alleles in two adjacent, or
parapatric, populations. With sufficiently strong selection on loci
that contribute to reproductive isolation, the populations can dif-
ferentiate into reproductively isolated species (Endler 1977; Lande
1982; Barton & Charlesworth 1984). Our results suggest that parapatric speciation would be more easily achieved in species with non-
overlapping generations because selection affects all individuals in
each generation. In the case of species with stage structure, parapatric speciation is more likely to occur if selection acts through
differences in adult survival than if it acts through differences in
fecundity or larval survival. This is because the adult stage is the
longest-lived one, and therefore, selection acting at this stage is
more efficient at purging the less favored genes.

## Literature Cited

Barton, N. H., and B. Charlesworth. 1984. Genetic revolutions, founder
    effects, and speciation. *Annual Review of Ecology and Systematics* 15:
    133–164.

Deevey, E. 1947. Life tables for natural populations of animals. *Quarterly
    Review of Biology* 22: 283–314.

Ellner, S., and N. G. Hairston. 1994. Role of overlapping generations in
    maintaining genetic variation in a fluctuating environment. *American
    Naturalist* 143: 403–417.

Endler, J. A. 1973. Gene flow and population differentiation. *Science
    (Washington, D.C.)* 179: 243–250.

———. 1977. *Geographic Variation, Speciation and Clines*. Princeton University Press, Princeton, N.J.

Fife, P. C., and L. A. Peletier. 1981. Clines induced by variable selection and migration. *Proceedings of the Royal Society of London, B* 214: 99–123.

Keller, J. B. 1984. Genetic variability due to geographical inhomogeneity. *Journal of Mathematical Biology* 20: 223–230.

Kimura, M., and W. H. Weiss. 1964. The stepping stone model of genetic structure and the decrease of genetic correlation with distance. *Genetics* 49: 561–576.

Koehn, R. K., R. I. E. Newell, and R. Immermann. 1980. Maintenance of an aminopeptidase allele frequency cline by natural selection. *Proceedings of the National Academy of Sciences (USA)* 77: 5385–5389.

Kozloff, E. N. 1990. *Invertebrates*. Saunders, New York.

Lande, R. 1982. Rapid origin of sexual isolation and character divergence in a cline. *Evolution* 36: 213–223.

Levins, R. 1968. *Evolution in Changing Environments*. Princeton University Press, Princeton, N.J.

Malecot, G. 1948. *Les mathématiques de l'hérédité*. Masson, Paris.

May, R. M., J. A. Endler, and R. E. McMurtrie. 1975. Gene frequency clines in the presence of selection opposed by gene flow. *American Naturalist* 109: 659–676.

Moody, M. E. 1979. Polymorphism with migration and selection. *Journal of Mathematical Biology* 8: 73–109.

———. 1981. Polymorphism with selection and genotype-dependent migration. *Journal of Mathematical Biology* 11: 245–267.

Nagylaki, T. 1977. *Selection in One- and Two-Locus Systems*. Springer-Verlag, Berlin.

———. 1978. Clines with asymmetric migration. *Genetics* 88: 813–827.

———. 1989. The diffusion model for migration and selection. Pp. 55–75 *in* A. Hastings, ed., *Models in Population Biology*. American Mathematical Society, Providence, R.I.

Nagylaki, T., and M. E. Moody. 1980. Diffusion model for genotype-dependent migration. *Proceedings of the National Academy of Sciences (USA)* 77: 4842–4846.

Peletier, L. A. 1978. A nonlinear eigenvalue problem occurring in population genetics. Pp. 170–187 *in* P. Benilan and J. Robert, eds., *Journées d'Analyse Non Linéare*. Proceedings, Beasancon, France, 1977. Springer-Verlag, Berlin.

Roughgarden, J. 1979. *Theory of Population Genetics and Evolutionary Ecology: An Introduction*. Macmillan, New York.

Rumrill, S. S. 1990. Natural mortality of marine invertebrate larvae. *Ophelia* 32: 163–198.

Slatkin, M. 1973. Gene flow and selection in a cline. *Genetics* 75: 733–756.

CHAPTER 12

# Dynamics of Populations with Density-Dependent Recruitment and Age Structure

## Louis W. Botsford

The combination of age or size structure and density-dependent recruitment has a defining influence on the dynamics of many plant and animal populations. Over the past 40 years, population biologists have combined mathematical models and field data in attempts to better understand the dynamics of such populations. Here I present a tutorial review of the interplay between mathematical and empirical biology that has shaped our understanding of populations with these characteristics.

Most investigations have sought the relationships between life-history parameters and population behavior. They have attempted to determine the characteristics of age structure and density dependence that lead to population stability, persistence, chaotic behavior, or shifts in equilibria. From a practical point of view, knowing these relationships allows us to formulate and test hypotheses regarding the causes of observed behavior such as cycles in abundance or collapse to low levels. They also allow us, as theoreticians, to begin to assemble a general understanding of the kinds of behavior to be expected from populations with different life-history characteristics. We are improving our ability to answer such questions as why populations of the same species in different locations behave differently (see, e.g., Murdoch & McCauley 1985) and why some populations cycle with a period roughly twice the mean age while others cycle with a period slightly greater than the mean age (see, e.g., Gurney et al. 1983).

Many models of populations have included age or size structure and density-dependent recruitment, and investigators have
employed several different approaches to understanding their behavior. Early efforts depended primarily on simulation (see, e.g.,
Ricker 1954; Pennycuick et al. 1968; Usher 1972), whereas later
researchers analyzed the stability of linearized versions of models (Allen & Basasibwaki 1974; Beddington 1974; Frauenthal 1975;
Rorres 1976; Botsford & Wickham 1978; Levin & Goodyear 1980).
Recently, needed attention has been paid to explicitly nonlinear
aspects of behavior (Guckenheimer et al. 1977; Botsford 1992*a*;
Higgins et al., in press). These models have been developed in a
variety of different fields, and their developers have not always been
aware of existing, parallel results in other fields. Early results in
the fisheries literature focused on variability and cycles in fish populations (e.g., Ricker 1954; Allen & Basasibwaki 1974; Botsford &
Wickham 1978); other studies were concerned with demographic
cycles in human populations through the Easterlin effect (e.g., Keyfitz 1972; Lee 1974; Frauenthal 1975; Rorres 1976); and still others
appeared in the general ecological literature (e.g., Pennycuick et
al. 1968; Usher 1972; Beddington 1974).

To describe these many developments in a common context, I
first present a fairly general model of populations with age structure and density-dependent recruitment, along with conditions for
equilibrium and local stability of a linearized version of the model.
I then use this model to describe several examples of population
studies involving age structure and density-dependent recruitment.
In presenting analyses of linearized models, I do not mean to pre-
sume that populations necessarily ever go to or hover about an
equilibrium or that such an analysis reveals all aspects of population behavior. Rather, I describe results in these terms because
for models with age structure and density-dependent recruitment,
(1) local stability of a linearized model is a good initial indicator
of nonlinear population behavior on generational time scales over
a wide range of parameter space, and (2) the dependence of equilibrium levels on life-history parameters appears to reflect slower
population changes in a meaningful way. The results of linearized
analyses are fairly robust indications of the way in which population behavior depends on life-history patterns. Following several
examples of this behavior, I describe progress thus far on specifically nonlinear aspects of the behavior of these populations. I point
out some of the differences in behavior due to the size depen-

dence of vital rates, rather than age dependence, and also indicate some progress in extending our understanding of single, local
populations to the understanding of the behavior of many intercon-
nected density-dependent, age-structured populations distributed
over space.

By way of definitions, here I take recruitment to be the number
of progeny in the next time instant after reproduction (as opposed
to the number entering the reproductive population or the number entering the harvested population). For density dependence in
recruitment, I do not here discuss Allee effects and their influence
on population dynamics (i.e., here density dependence is compensatory). I also do not focus explicitly on the population processes
leading to extinction; unless a population is low because of habitat
degradation, density dependence is typically not important in such
problems.

## 1 The Beginning

A rewarding pastime is to read (or re-read) an old, pivotal classic
in one's field. The result for me has often been amazement at the
similarity between the description of topical issues then and current
views in population biology. Differences are frequently merely a
matter of semantics, using different words to describe what are
essentially, biologically the same issues. This is certainly the case
for W. E. Ricker's "Stock and recruitment" (1954). Ricker defined
a relationship between recruitment to a population and the size
of the adult stock that had produced it (Fig. 1). To put both
axes in the same units, he defined both stock and recruitment at
the same point in the life history (i.e., number of eggs produced,
in his case), so that the population would exactly replace itself
for recruitment on a line of slope 1.0 passing through the origin,
and equilibria would lie on this line. He used this relationship to
simulate population behavior for both semelparous and iteroparous
populations with the following results.

For a semelparous population,

1. when the slope of the stock-recruitment curve was less than $-1$,
   cycles of period 2 would arise.

   For iteroparous populations,

2. the slope at which cycles occurred was less than $-1$;

3. the period of the cycles was twice the median time from egg

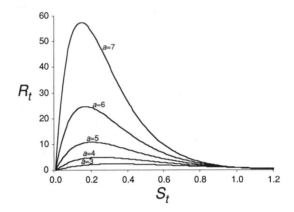

FIGURE 1. *Ricker's (1954) stock-recruitment relationship, normalized so that the equilibrium (for a semelparous model) is always 1.0, $R_t = S_t e^{a(1-S_t)}$. Note that as a (ln of discrete-time population growth rate at zero density) increases, both the rate of increase in recruitment at low stock densities and the rate of decrease in recruitment at high densities are greater in magnitude. The former reflects greater per capita reproduction, while the latter reflects greater per capita density dependence.*

production of adults to egg production of the next generation;

4. the amplitude of these cycles was greater as (a) maturation delay increased or (b) the number of age classes decreased; and

5. harvesting reduced the amplitude of these cycles.

If one allows for some ambiguity between stability and the amplitude of the cycles, and for the fact that Ricker did not examine behavior for slopes much steeper than the one in which cycles initially appeared, the seeds of much of our current knowledge were contained in his results, 40 years ago.

Other early simulation results contained elements of our current understanding of how these populations work. For example, Pennycuick et al.'s simulations (1968) of populations with density-dependent fecundity and survival showed that lags in reproduction and higher fecundities lead to cycles. Usher's $3 \times 3$ Leslie matrix (1972) with density-dependent first-year survival went from equilibrium to cycles to extremely erratic (possibly chaotic) behavior as the slope of the density dependence becomes more negative. Menshutkin (1964) simulated cannibalistic behavior directly to show the effect of harvest on cyclic fluctuations in a perch population.

## 2  A Model of Populations with Age Structure and Density-Dependent Recruitment

A reasonably general and biologically realistic way of describing density-dependent recruitment in an age-structured population is

$$R_t = B_t f(C_t) = \sum_{a=1}^{a_{\max}} b_a n_{a,t} f \left( \sum_{a=1}^{a_{\max}} c_a n_{a,t} \right) , \qquad (1)$$

where $R_t$ is recruitment at time $t$, $B_t$ is reproduction at time $t$, $C_t$ is effective population size at time $t$, $f(.)$ is recruitment survival, $n_{a,t}$ is the number of individuals at age $a$ at time $t$, and $b_a$ and $c_a$ are weighting functions. The weighting function $b_a$ reflects fecundity at age $a$, and the function $c_a$ reflects the relative influence on the density dependence in recruitment. This model describes recruitment as the product of the usual expression for egg production ($B_t$) and a density-dependent survival function ($0 \le f \le 1$), which describes the way in which older individuals affect recruitment. It allows reproduction and the density dependence in recruitment to depend on adult age structure in different ways (i.e., through the weighting functions $b_a$ and $c_a$). This expression applies to a closed population, one in which all recruits originate from reproduction within the population. For many populations, larvae or early juvenile stages enter the population from elsewhere rather than being produced by the population, and the population must be considered open. The appropriate model is then

$$R_t = B_c f(C_t) = B_c f \left( \sum_{a=1}^{a_{\max}} c_a n_{a,t} \right) , \qquad (2)$$

where $B_c$ is a constant.

The form of the function $f(.)$ varies depending on the mechanism being described, but it is typically monotonically decreasing from 1.0 to 0. Examples (Fig. 2) include (1) exponential decay, which is the form corresponding to the Ricker stock-recruitment model (Ricker 1954; Levin & Goodyear 1980) and has been used to describe cannibalism (Hastings & Costantino 1987, 1991); (2) linear decline to zero, which is the form typically used to describe adult-juvenile competition for a resource such as space (Roughgarden et al. 1985; Bence & Nisbet 1989); and (3) a constant level up to a certain effective population size, then linear decline, which is a form used to describe density-dependent recruitment in the Dungeness crab and which would reflect the situation described

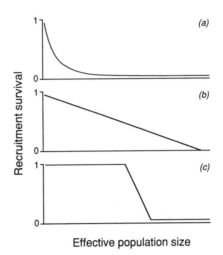

FIGURE 2. *Three frequently used recruitment-survival functions:* (a) $f(C_t) = e^{-aC_t}$, *which could reflect mortality due to random cannibalistic encounters between adults and pre-recruits;* (b) $f(C_t) = 1 - \beta C_t$ *(with $\beta$ a constant), which could represent the amount of space available for recruitment; and* (c) $f(C_t) = 1$ *(no density-dependent mortality) up to a point at which it declines linearly with specified slope. The third reflects a density-vague mechanism whereby density dependence is in effect only at high densities.*

as density-vague (i.e., no density dependence is seen at moderate to low population levels; rather, it occurs only at high abundance; Strong 1986).

This model does not explicitly include sex, but it is relatively easy to do so provided one can assume that reproduction is independent of the number of males. The remainder of this model is just the relationship between the number of individuals at each age at one time in terms of the number of individuals at the next lower age at the preceding time. Assuming fixed juvenile and adult survival, where $a \geq 2$,

$$n_{a+1,t+1} = p_a n_{a,t} , \tag{3a}$$

$$n_{1,t} = R_t . \tag{3b}$$

From these,

$$n_{a+1,\,t+1} = \sigma_a R_{t-a} , \tag{3c}$$

where $\sigma_a = \Pi_{i=1}^{a} p_i$.

Equilibrium conditions for these two models can be obtained by letting $R_t$ go to a constant value $R_e$. These are

$$1 = \Phi_b f(R_e \Phi_c), \tag{4a}$$

$$R_e = B_c f(R_e \Phi_c), \tag{4b}$$

for the closed and open population models, respectively. In these expressions we have defined total reproduction and effective population size at equilibrium:

$$\Phi_b = \sum_{a=1}^{a_{max}} \phi_b(a) \quad \text{and} \quad \Phi_c = \sum_{a=1}^{a_{max}} \phi_c(a) \tag{5}$$

as the sum, over age, of functions that reflect the influence of the adult population at each age on reproduction and density-dependent recruitment, respectively:

$$\phi_b(a) = b_a \sigma_a \quad \text{and} \quad \phi_c(a) = c_a \sigma_a. \tag{6}$$

These are termed *influence functions* because, as we see below, they describe the influence of past recruitment, $a$ time units ago, on current recruitment.

The expressions for equilibrium (eqs. 4 a,b) are related to standard analogues in linear populations; $\Phi_b$ is the same as $R_0$, total reproduction in the lifetime of an individual female; and $b_a \sigma_a$ in $\phi_b$ is the same as the notation $\ell_a m_a$ in linear models (e.g., the Leslie matrix). Similarly, $\Phi_c$ has a different weighting function and could be considered the total lifetime effect of each individual on density-dependent recruitment. Note that since $f \leq 1.0$, $\Phi_b$ (i.e., $R_0$) must always be greater than or equal to 1.0 for an equilibrium to exist. In contrast to the linear case, because of density dependence, an $R_0$ greater than 1.0 can lead to a constant population rather than an increasing population.

Local stability can be determined by exploring the dynamic behavior of a small variation in $R_t$ about equilibrium, $\triangle R_t = R_t - R_e$. Standard procedures involving a Taylor series approximation of $f(.)$ lead to the following expressions for a closed population,

$$\triangle R_t = \sum_{a=1}^{a_{max}} \left[ \tilde{\phi}_b(a) + K \tilde{\phi}_c(a) \right] \triangle R_{t-a},$$

and for an open population,

$$\triangle R_t = \sum_{a=1}^{a_{max}} K \tilde{\phi}_c(a) \triangle R_{t-a}. \tag{7}$$

where $\tilde{\phi}_b(a) = \phi_b(a)/\Phi_b$ and $\tilde{\phi}_c(a) = \phi_c(a)/\Phi_c$ are normalized influence functions and $K = f'(C_e)C_e/f(C_e)$, where $C_e$ is the effective population size at equilibrium. Note that $K$ has the form of a normalized slope of the recruitment survival function at equilibrium, which is also known as an elasticity (i.e., the relative change in one quantity with respect to another). In both cases, variation in recruitment is a sum over past values of recruitment weighted by the total influence function. The influence function is composed of reproductive and density-dependent influence functions for a closed population but only the density-dependent influence function for an open population. The effects on stability of the shape of the age dependence and the magnitude of the age dependence are separated into the normalized influence functions and the value of $K$, respectively. Shapes of influence functions vary widely depending on the mechanism underlying the density dependence in recruitment. Examples corresponding to hypothetical recruitment mechanisms for Dungeness crab include one linking recruitment to metabolic demand (weight$^{0.8}$), one reflecting fecundity and female mortality rates, and one based on cannibalism data and adult mortality rates (Fig. 3). Also shown is a theoretical influence function from the delayed-recruitment model that includes no influence on recruitment up to a certain age and then a constant effect at each age, with the number at each age decaying geometrically (Fig. 3$d$; Clark 1976; Botsford 1992$a$).

Substitution of $\triangle R_t = \lambda^t$ leads to the following characteristic equations:

$$
\begin{array}{cc}
\text{Closed Population} & \text{Open Population}
\end{array}
$$

$$
\sum_{a=1}^{a_{\max}} \left[ \tilde{\phi}_b(a) + K\tilde{\phi}_c(a) \right] \lambda^{-a} = 1, \quad \sum_{a=1}^{a_{\max}} K\tilde{\phi}_c(a)\lambda^{-a} = 1. \tag{8}
$$

Solutions to these characteristic equations for the dominant eigenvalue with various parametric values indicate population behavior. If all terms are greater than zero (e.g., if $K > 0$), the dominant eigenvalue is positive and real, from the Perron-Frobenius theorem, indicating either geometric approach to or divergence from equilibrium. However, $K$ is not greater than zero in a compensatory population. The dominant eigenvalue for models with compensatory density dependence (i.e., negative $K$) is typically complex, indicating decaying cycles when $|\lambda| < 1$. When $|\lambda|$ becomes greater than 1.0, the model undergoes a Hopf bifurcation, and typically, model solutions approach an invariant loop, a cycle that is either periodic or quasiperiodic. An exception to this is a semel-

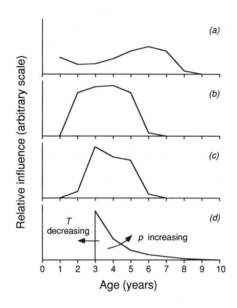

FIGURE 3. *Three examples of influence functions used for Dungeness crab and an idealized influence function. (a) Density dependence in recruitment proportional to the metabolic demand of older individuals. (b) Recruitment reflecting density-dependent fecundity. (c) Recruitment under conditions of cannibalism, as demonstrated by an analysis of gut contents. (d) The influence function for a delayed-recruitment model with the age at first reproduction of $T = 3$ years and adult mortality of $p = 0.6$ $year^{-1}$. The arrows indicate the stabilizing direction of changes in the age at first reproduction and adult survival (see Table 2).*

parous population, which undergoes period-doubling bifurcations (see, e.g., May & Oster 1976; Levin 1981).

The conditions under which this kind of model becomes locally unstable are typically described in terms of the slope of the recruitment survival function or the stock-recruitment function. These conditions can be determined by various analytical methods (Levin & Goodyear 1980; Bergh & Getz 1988) or by direct numerical solution for roots of the characteristic equation. Applications of various analytical methods to a variety of similar models, in addition to many simulations, have led to several reasonably general conclusions relating population behavior to life-history characteristics: (1) as $K$ decreases from zero, at some point the population becomes locally unstable and begins to cycle in recruitment; (2) the point at

which this occurs becomes more positive as the ratio of the width of
the age distribution (i.e., the total influence function) to the mean
age declines; (3) the period of the cycles is approximately twice
the mean age of the influence function. It is important to realize
that these are not strict generalizations; that is, some exceptions
may exist. The generalizations, which synthesize many different results, are based on the application of various methods of analysis
to different specific models of populations with age structure and
density-dependent recruitment. For example, the first and second
characteristics are consistent with Ricker's (1954) results and in
part with Usher's (1972) results.

   The dependence of stability on the width of the age structure
has also been shown through approximate analytical methods (e.g.,
Botsford & Wickham 1978; Levin 1981). This result is of particular
importance in fisheries because of the potential for size-selective
harvesting to cause populations to cycle (Botsford & Wickham
1978). That the period of the cycles is approximately twice the
mean age follows from many simulations and a simple analytical
approximation (see, e.g., Botsford & Wickham 1978). A more accurate approximation was derived for the delayed-recruitment model
(Bergh & Getz 1988).

   Analytical studies indicate that the necessary and sufficient condition for stability of the semelparous case of the characteristic
equation (e.g., eq. 8) is a sufficient condition for the iteroparous
case (see, e.g., Levin 1981; Bergh & Getz 1988). In this sense, the
semelparous case has the least stable age structure. This is consistent with the second general conclusion, since it is the minimum
value of the ratio of the width to the mean of the age distribution. From equation (8), this condition is $K < -1$ for the closed
population and $K < -2$ for the open population.

   These results indicate that population stability can be viewed as
an essential tension between (1) the destabilizing effect of compensatory density dependence represented by $K$ and (2) the stabilizing effect of age structure represented by $K'$, the value of $K$ below
which the population becomes locally unstable. Steeper recruitment survival curves make populations less stable, and vice versa.
The "age structure" of a population, or more accurately, the relative influence on recruitment at each age, determines how resilient
the population is to density dependence in recruitment. The influence of a large number of age classes close to the origin tends
to make a population stable, and vice versa. Adding age structure
to a semelparous population is stabilizing. Local instability and

TABLE 1. *Values of $f'(N_e)$ for Which the Population Becomes Locally Unstable about the Equilibrium, Producing Cycles*

| $T$ | $p$ 0.0 | 0.2 | 0.4 | 0.6 | 0.8 | 1.0 |
|---|---|---|---|---|---|---|
| 0 | $-1.00$ | $-1.20$ | $-1.40$ | $-1.60$ | $-1.80$ | $-2.00$ |
| 1 | $-1.00$ | $-1.00$ | $-1.00$ | $-1.00$ | $-1.00$ | $-1.00$ |
| 2 | $-1.00$ | $-0.91$ | $-0.82$ | $-0.74$ | $-0.68$ | $-0.62$ |
| 3 | $-1.00$ | $-0.86$ | $-0.74$ | $-0.63$ | $-0.53$ | $-0.45$ |
| 4 | $-1.00$ | $-0.84$ | $-0.70$ | $-0.56$ | $-0.45$ | $-0.35$ |
| 5 | $-1.00$ | $-0.83$ | $-0.67$ | $-0.52$ | $-0.39$ | $-0.29$ |

NOTE.—Computed from Clark 1976.

cycles result when the density dependence in recruitment survival is stronger than the inherent stabilizing effect of the age structure.

These results are demonstrated by what is probably the simplest form of an age-structured model, the delayed-recruitment model (Clark 1976; Botsford 1992a). In this model, the number of adults in one year is the sum of new recruits to the adult population and the fraction of adults surviving from the preceding year:

$$N_{t+1} = f(N_{t-T}) + pN_t. \tag{9}$$

The former is recruitment to the population $T$ years ago, where $T$ is the age at first reproduction, and the latter is assumed to be a constant fraction of last year's abundance, $p$. The function $f$ represents the nonlinear dependence of recruitment on adult stock (both reproduction and density dependence).

Through linearized analysis, Clark (1976) obtained the characteristic equation

$$\lambda^{T+1} - p\lambda^T - f'(N_e) = 0, \tag{10}$$

which leads to the following sufficient condition for stability:

$$f'(N_e) > -(1+p). \tag{11}$$

He then determined necessary and sufficient conditions for stability in terms of values of $f'(N_e)$ at which the population begins to produce cyclic behavior (Table 1).

This model can also be written explicitly in terms of density-

TABLE 2. *Values of K for Which the Population Becomes Locally Unstable about the Equilibrium, Producing Cycles*

| $T$ | $p$ 0.0 | 0.2 | 0.4 | 0.6 | 0.8 | 1.0 |
|---|---|---|---|---|---|---|
| 0 | −1.00 | −1.50 | −2.33 | −4.00 | −9.00 | ∞ |
| 1 | −1.00 | −1.20 | −1.67 | −2.50 | −5.00 | ∞ |
| 2 | −1.00 | −1.13 | −1.37 | −1.86 | −3.33 | ∞ |
| 3 | −1.00 | −1.08 | −1.23 | −1.57 | −2.65 | ∞ |
| 4 | −1.00 | −1.05 | −1.16 | −1.41 | −2.23 | ∞ |
| 5 | −1.00 | −1.04 | −1.12 | −1.31 | −1.97 | ∞ |

NOTE.—Computed from Clark 1976.

dependent recruitment, similar to our original general model:

$$R_t = f\left(\sum_{a=T}^{\infty} R_{t-a} p^{t-a}\right) . \tag{12}$$

The resulting equilibrium condition is

$$R_e = f\left[R_e/(1-p)\right] , \tag{13}$$

the normalized influence function is

$$\begin{aligned} \tilde{\phi}(a) &= (1-p)\,p^{a-T} \quad &\text{where } a \geq T, \\ \tilde{\phi}(a) &= 0 &\text{where } a < T, \end{aligned} \tag{14}$$

and the normalized slope (or the elasticity) of the recruitment survival function is

$$K = f_e'/(1-p) . \tag{15}$$

The characteristic equation (eq. 8) then follows, and the values of $K$ for which the population becomes locally unstable (i.e., $K'$) can be obtained (Table 2).

The implications for the effects of life-history parameters on stability in Table 2 are consistent with those outlined above under the more general age-structured model. For iteroparous populations, long maturation delays and low adult survivals, both of which increase the ratio of the mean age of the influence function to its width, are destabilizing (Fig. 3*d*). However, Table 1 differs from Table 2 in some respects (e.g., increasing $p$ is not always stabilizing) because the general formulation derived here focuses on the effects of age structure on stability. Age structure is normalized;

hence, the effects of changes in parameter values on equilibrium are treated separately. We see the value of this below, where some kinds of harvesting change equilibrium values whereas others do not. An important point here is that one must be careful in making general statements about whether a specific change in a certain life-history characteristic is stabilizing. The answer often depends on what is being held constant as the characteristic of interest changes. In both formulations, the semelparous case (i.e., $p = 0$) is the least stable, and this does not vary with maturation delay. For the semelparous case only, changing $T$ merely rescales time.

## 3 Examples of Applications

Many populations have age structure and density-dependent recruitment, and cycles in recruitment or general levels of variability are issues of interest. Below are a few of the examples in which the results described thus far have been used to understand better the variability in such populations.

*Dungeness Crab,* Cancer magister

For the past 30 years, investigators have been attempting to determine the cause(s) of cycles in abundance in the more northern fisheries for the Dungeness crab (*Cancer magister*) (Fig. 4). Potential causes include a predator-prey relationship (with man or chinook salmon as the predator) and an influence of a cyclic environmental factor, but we focus here on a third possibility, density-dependent recruitment (for reviews and recent results, see Botsford et al. 1989, 1994; Botsford & Hobbs 1995).

Researchers have used a model and the results obtained above regarding the relationship between life-history characteristics and stability to evaluate whether locally unstable cycles could be caused by each of three potential mechanisms of density-dependent recruitment: (1) density-dependent fecundity; (2) cannibalism; and (3) an egg-preying worm (Botsford & Wickham 1978; McKelvey et al. 1980). The initial models were versions of those presented above, in some cases made more realistic by including the influence of variability in growth patterns (Botsford 1984) and the effect of a delayed response of fishing mortality rate to changing abundance (Botsford et al. 1983). Including variability was stabilizing, whereas a delayed response in fishing mortality rate was destabilizing, and both lengthened the period of potential cycles.

In a model with density-dependent fecundity as the mechanism

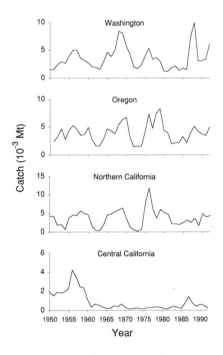

FIGURE 4. *Catch (in megatons) of Dungeness crab for four areas along the coast. Because of the intense fishing pressure in this size-selective fishery, catch is a good proxy for recruitment of juveniles four years earlier. Note that abundance at the three northern locations is cyclic, whereas the most southern population collapsed to low levels in the late 1950's.*

underlying density-dependent recruitment, the influence function is the fraction of females surviving to each age, weighted by fecundity (McKelvey et al. 1980). The period of the cycles matches the observed period, but there is no way to test other characteristics or even to show that density-dependent fecundity exists. A potential problem with modeling density-dependent fecundity as depending solely on females is that density-dependent fecundity commonly results from competition for food resources, which are depleted by both males and females.

Cannibalism was initially modeled using a standard expression for metabolic demand (i.e., $w^{0.8}$) as the weighting function $c_a$ (Botsford & Wickham 1978). This yielded a cycle period of 7 or 8 years, which became 8 or 9 years when the effects of lagged changes in effort were added, still short of the observed period of approximately 10 years. More-recent data on the age dependence of cannibalism show that most cannibalism is by adult females, leading to an influence function that is almost identical to that determined from fecundity (Botsford & Hobbs 1995). Replacing $w^{0.8}$ by the empirically determined weighting function, however, still produces a period lower than that produced by an influence function involving fecundity only. The period of the cycles proves to be sensitive to the exact shape of the influence function; it can vary by a year or two, even as the influence function varies by an empirically imperceptible amount (for details, see Botsford & Hobbs 1995). This behavior is somewhat disturbing as regards the precision with which mechanisms can be discerned from the cycle period.

The egg-preying worm is characterized as having density-dependent recruitment because worm density depends on adult crab density; however, a slightly different model was necessary (Hobbs & Botsford 1989). Recruitment survival, $g$, depends on worm abundance:

$$R_t = g\left(W_{t-1}\right)B_t\,, \tag{16}$$

where worm abundance is proportional to crab eggs not eaten in the preceding year:

$$W_t = \left[1 - g\left(W_{t-1}\right)\right]B_t = B_t - R_t\,. \tag{17}$$

This model results in the same characteristic equation as the closed population above with a slightly different density-dependent influence function (for details, see Hobbs & Botsford 1989). For this mechanism, recruitment survival data are available; hence, the actual value of $K$ can be determined (Fig. 5). The slope of the recruitment survival function is not steep enough for the egg predator to be the sole cause of the cycles. Analysis of the combined effect of both the egg predator and the original form of density-dependent recruitment (i.e., recruitment survival calculated as the product $f\left[C_t\right]g\left[w_{t-1}\right]$), however, shows that although the egg-preying worm alone cannot cause the cycles, it does contribute to instability if another density-dependent recruitment mechanism is present. The presence of the worm (i.e., when $K_w$, the slope of the recruitment survival function $g$, is $-1.1$ rather than 0) moves the value of $K_c$, the normalized slope of $f$, at which the population becomes unstable, from about $-3$ to about $-2$ (Fig. 6). This analysis also yields the possible combinations of values of $K_c$ and

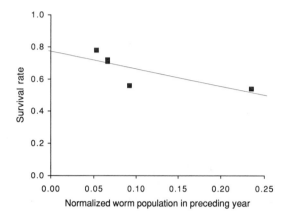

FIGURE 5. *An example of a density-dependent recruitment survival function: survival of Dungeness crab eggs through the period of egg predation at various densities of an egg-preying worm. Worm density depends on adult crab density. (Redrawn from Hobbs & Botsford 1989.)*

$K_{\rm w}$ that would produce both a sufficient magnitude of the dominant eigenvalue (i.e., greater than 0.8) and cycles of the observed period. The tongue of possible values is shown in Figure 6.

### *Barnacle,* Balanus glandula

The barnacle *Balanus glandula* is a sessile intertidal invertebrate with age structure and density-dependent recruitment. This species has been studied intensively for the past 10 years or so in central California to determine the factors that shape the species composition of intertidal communities (Gaines & Roughgarden 1985; Roughgarden et al. 1985). Because larvae are released into a large planktonic pool where they reside for several weeks, the population is modeled as an open population. The time increment in this discrete-time model is weeks. Recruitment to the rocky substrate is assumed to be proportional to the amount of free space available for settlement, $F_t$:

$$R_t = s\,F_t, \tag{18}$$

where $s$ is a constant reflecting the rate at which potential recruits (planktonic larvae) become available for settlement. The

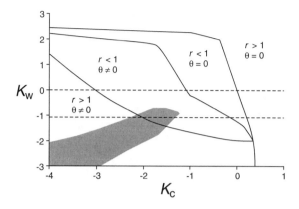

FIGURE 6. *Stability of a Dungeness crab population faced with two density-dependent recruitment mechanisms, cannibalism and egg predation by a worm (Hobbs and Botsford 1989). The axes are the elasticities of the dependence on worm density $(K_w)$ and the dependence on cannibalism $(K_c)$. The population is locally unstable and cyclic when the magnitude of the dominant eigenvalue of the linearized model, $r$, is greater than 1.0 with angle $\theta \neq 0$. The horizontal lines correspond to the worm's not being present $(K_w = 0)$ and the slope shown in Figure 5. The shaded area corresponds to combinations of values that would produce the cycles with the observed period (Fig. 4). (Redrawn from Gaines & Roughgarden 1985.)*

total amount of space is the sum of free space and occupied space,

$$A = F_t + \sum_{a=1}^{a_{max}} c_a n_{a,t}, \qquad (19)$$

where $c_a$ is the amount of space occupied by an individual of age $a$. Substituting this into equation (18) leads to an expression of the form of equation (2), in which $B_c = s$, $f(C_t) = A - C_t$, and

$$C_t = \sum_{a=1}^{a_{max}} c_a n_{a,t}.$$

From the definition following equations (7), the normalized slope (i.e., the elasticity) of $f(C_t)$ at equilibrium is

$$K = -C_e / (A - C_e). \qquad (20)$$

From this and the equilibrium condition, $F_e = A - C_e$ and $K = 1 - A/F_e$. From the sufficient condition for stability of an open

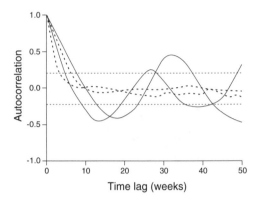

Figure 7. *Autocorrelation functions computed from recruitment time series to determine the relative cyclicity of populations with different rates of larval inflow (Gaines & Roughgarden 1985). The populations with higher inflow* (solid lines) *are more cyclic than those with lower inflow* (dashed) *as predicted by the analysis. (Redrawn from Gaines & Roughgarden 1985.)*

population (i.e., $K' \geq -1$), it follows that the population can be unstable when

$$F_e/A \leq \tfrac{1}{2}. \tag{21}$$

This is Roughgarden et al.'s (1985) rule regarding 50 percent free space: the population is stable when the equilibrium level of free space is greater than 50 percent of the available space. Available space declines with an increase in the rate at which propagules arrive (i.e., $B_c = s$) increases. Gaines and Roughgarden (1985) tested this result by comparing the degree of cyclicity at sites with different propagule-arrival rates. Computed autocorrelation functions indicated that populations are more cyclic when the arrival rate is high (Fig. 7). However, subsequent estimation of the influence function (i.e., the product of size at age and survival to each age) indicates that the life-history patterns of these populations does not cause cyclic behavior, and Possingham et al. (1994) have proposed an alternative interpretation of these data.

*Cannibalism*

The study of cannibalism, a common mechanism of density-dependent recruitment, has advanced our understanding of such recruitment in age-structured populations in general. It is now generally recognized that cannibalism is widespread among animal species (for reviews, see Fox 1975; Polis 1981). Several population-dynamic questions remain regarding the behavior of cannibalistic populations, two of which we discuss here: (1) whether cannibalism can regulate populations and, if so, under what conditions, and (2) whether cannibalism can cause populations to cycle, and under what conditions. A third major question not discussed here is how cannibalism contributes to population persistence through the lifeboat phenomenon (Polis 1981; Gabriel 1985).

Answering these questions involves describing how aspects of cannibalistic behavior at the individual level contribute to population dynamics. Important characteristics include the relative voracity of predators of different ages and the age distribution of their prey, how the level of available food affects voracity, the nutritional benefit of cannibalism, the numerical response, and the dependence of cannibalism rate per individual on prey density (i.e., functional response) (for more-extensive reviews, see Dong & Polis 1992; Q. Dong, G. Polis, A. Hastings & Botsford, pers. comm.).

The model developed here could be used to evaluate how some of these characteristics influence population regulation and the possibility of population cycles. It would represent a cannibalistic population in which cannibalism occurs only over a short period in early life (i.e., during recruitment), there is no functional response (i.e., $f$ depends on predators only), and there is no numerical response by the prey. The results discussed thus far imply that cannibalism on young stages can regulate populations if its response to density is strong enough near equilibrium (i.e., if the equilibrium condition is satisfied and the equilibrium is stable). If this rate of change is too strong, cycles can occur, and they are more likely to occur when cannibalism is by few age classes. The period of the cycles must be roughly twice the delay from the age of prey (i.e., zero) to the average age of predators.

Models of cannibalism often describe the process as a consequence of predators randomly encountering prey. On this basis, mortality is proportional to predator population size over a short time period, leading to an exponential recruitment survival function (Fig. 2a). For a closed population, equation (2) becomes

$$R_t = B_t e^{-aC_t} , \tag{22}$$

where $a$ is a constant. From equations (3), the equilibrium recruitment level is then

$$R_e = \ln \Phi_b / a \Phi_c ; \tag{23}$$

and from equations (4), the elasticity of the recruitment survival function is

$$K = -a R_e \Phi_c , \tag{24}$$

which from equation (23) is

$$K = -\ln \Phi_b . \tag{25}$$

Because of the result in equation (25), some investigators have described the instability of populations with density-dependent recruitment (i.e., cycles) as a consequence of increasing per capita reproduction. It is important to realize, however, that the unstable behavior follows from the increased slope of the density dependence, not from reproduction. The only way to increase the slope of the density dependence in this model while maintaining equilibrium is to increase reproduction.

Diekmann et al. (1986) extended these results significantly in their evaluation of the effects of a functional response in predation. Recruitment in their continuous-time model was written as

$$R(t) = B(t) e^{-C(t) F[p(t)]} , \tag{26}$$

where $F[p(t)]$ is the density-dependent factor in a functional response (i.e., the factor by which proportional dependence is multiplied) at prey density $p(t)$. They provided a careful derivation of mortality due to cannibalism that included the effects of feeding behavior, yet resulted in an expression in which cannibalism occurred over an infinitesimal time, thus making analysis easier.

To determine how adding a functional response changes behavior, we can compare their results to those obtained thus far. Diekmann et al. plotted the results of their linearized stability analysis in terms of $\Phi_b$ and a parameter $q = -\alpha \Phi_c F_e^2 / F_e'$, where $\alpha$ is the inverse of the product of handling time and vulnerability (Fig. 8). Since $F_e'$ is the rate of change of the density-dependent part of feeding, a low value of $q$ corresponds to a large amount of density dependence in feeding (i.e., a strong functional response). For age structure in their examples, they assumed constant fecundity from age 1 to infinity and a constant effect on density dependence from $T_3$ to $T_4$. They tested the effects on local stability of (1) increasing the age at which cannibalism begins ($T_3$), keeping the maximum age ($T_4$) infinite (panels $a,b$), (2) varying the maximum age ($T_4$),

$q$

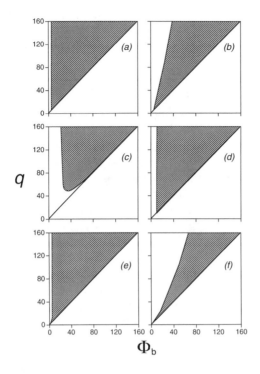

FIGURE 8. *The dependence of local stability on total lifetime reproduction,* $\Phi_b$, *and the strength of density dependence in feeding,* $q$, *for a model of cannibalism that includes a functional response (Diekmann et al. 1986). Solutions are locally unstable and cyclic in the hatched area. The area below the line* $\Phi_b = q + 1$ *does not have a stable equilibrium. The remaining area, generally to the left of the unstable area, is a stable equilibrium.* Top row, *Comparison of stability with two different minimum ages of cannibalism* (a, $T_3 = 2.0$; b, $T_3 = 0.25$), *with maximum age* ($T_4$) *equal to infinity.* Middle, *Stability with two different values of the maximum age of cannibalism* (c, $T_4 = 5.0$; d, $T_4 = 3.0$), *with the minimum age fixed at 1.0.* Bottom row, *The effect on stability of different mean ages of cannibalism* (e, $T_3 = 0.25$, $T_4 = 0.75$; f, $T_3 = 0.05$, $T_4 = 0.55$), *with the total cannibalistic interval constant. (Redrawn from Diekmann et al. 1986.)*

with the minimum age ($T_3$) held constant (panels c,d), and (3) varying both $T_3$ and $T_4$ while maintaining their difference constant (panels e,f). From the results regarding the dependence of stability on the shape of the influence function, for the model without a functional response (i.e., eq. 21), the stability boundary in Figure 8 would be a vertical line, which (1) moves to the left as $T_3$ increases, with $T_4 = \infty$, (2) moves to the left as $T_4$ decreases, with $T_3$ held constant, and (3) moves to the left as $T_3$ and $T_4$ increase, with their difference constant ($\Phi_c$ and $\Phi_b$ constant). Because a lack of functional response corresponds to the case where $F'_e = 0$ (i.e., no density dependence in feeding) in the definition of $q$, these results can be expected only at high values of $q$.

At high values of $q$, the model behaves as predicted by the model without a functional response. The two primary differences introduced by the addition of a functional response seem to be the curved nature of stability boundaries at low values of $q$ (i.e., it is not a vertical line) and the boundary on equilibrium at $q = \Phi_b - 1$ (i.e., the diagonal line in Fig. 8). In most cases, stability decreases (i.e., a lower value of $\Phi_b$ is required for instability) as $q$ decreases. When the reproductive influence function and the density-dependent influence function are similar, however, stability can increase with decreasing $q$, and in some cases, the population can be stable, then unstable, then stable as $\Phi_b$ increases (e.g., Fig. 8c,d; for further discussion, see Diekmann et al. 1986).

The second difference, an upper limit on $\Phi_b$ beyond which the population increases exponentially, is a direct consequence of the limitation in predation rate implied by the functional response. Biologically, as the density dependence in predation increases, this limitation prevents mortality due to cannibalism from being large enough to prevent an exponential increase in abundance. Mathematically, in the model without a functional response, when $\Phi_b$ increases, $R_e$ can increase to maintain an equilibrium. For this model, however, another factor (i.e., density dependence in the functional response) is present, which declines with $R_e$; hence, increasing $R_e$ cannot decrease $f$ to a value low enough to satisfy the equilibrium condition.

Another study of the population dynamics of cannibalism investigated the effects of cannibalism occurring over a nonzero time period in early life. Hastings (1987) and Hastings and Costantino (1987) modeled cannibalism in the flour beetle genus *Tribolium* as a random encounter of the egg stage by the larval stage. The egg stage was assumed to be in constant supply (i.e., an open population). The prey stage had nonzero duration, occurring over ages 0 to $A_E$, and the larval stage occurred over ages $A_E$ to $A_E + A_L$. Voracity and vulnerability were constant over these ranges.

We can evaluate the differences in behavior that a nonzero prey period introduces by comparing the results of Hastings and Costantino to those that would be expected from our simpler model with infinitesimal prey period. For density-dependent mortality due to random encounter, the recruitment-survival function declines exponentially with larval abundance at a rate proportional to the per capita consumption rate and the length of the egg stage. If there were no mortality other than cannibalism, equilibrium recruitment in the open-population model would be

$$R_e = B_c \exp(-cA_E A_L R_e) \, , \tag{27}$$

where $c$ is the per capita consumption rate. The normalized slope of $f(C_t)$ would be

$$K = -cA_E A_L R_e \, . \tag{28}$$

As $B_c$ increases, equilibrium recruitment increases, $K$ becomes more negative, and the population becomes less stable. This is similar to increasing the rate of propagule input in the barnacle example. Increasing $A_E$ exerts two effects, a decrease in $K$ and an increase in the mean lag between the ages of predator and prey. From our earlier results, both of these effects would be destabilizing. Increasing $A_L$ also affects stability, through the value of $K$ and the age structure, but in opposite ways. It decreases $K$, which is destabilizing, but it also increases the width of the cannibalistic influence function, which is stabilizing.

The effect of $B_c$ on stability is consistent with the local stability results obtained by Hastings and Costantino (1987) (Fig. 9); local stability is lost as $B_c$ increases. However, the effect of $A_E$ on stability is consistent over only part of the parameter space, the part in which stability boundaries have positive slope in Figure 9. It appears that the differences in behavior due to a nonzero prey period occur when $A_L$ is small, possibly as $A_L$ becomes similar to $A_E$. More analysis is needed to further elucidate this aspect of the effect of a nonzero prey stage. The period of cycles obtained by Hastings and Costantino was $A_E + A_L$, which is twice the distance between the mean age of prey $(A_E/2)$ and the mean age of predators $(A_E + A_L/2)$, a logical extension of the period being twice the mean age of predators.

Hastings and Costantino (1991) also examined the effect of adding age-dependent cannibalism rather than having all larvae be equally voracious (Fig. 9). Having cannibalism increase with larval age to a constant level, rather than being constant for all ages, is destabilizing. This effect is consistent with the standard model because the age dependence increases the ratio of mean to width of

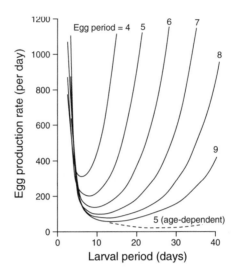

FIGURE 9. *Stability of a model of* Tribolium, *an open population with random-encounter cannibalism and a nonzero duration of the period of susceptibility to cannibalism (Hastings & Costantino 1987, 1991). With an increase in the rate at which pre-recruits are introduced into the population (i.e., egg production rate, $B_c$), the population becomes locally unstable and cyclic after crossing the line corresponding to the appropriate duration of the susceptible stage (i.e., egg period, $A_E$) and duration of the cannibalistic stage (i.e., larval period, $A_L$). Solid lines, Uniform cannibalism throughout the larval stage; dotted line, cannibalism increases gradually with age to a constant value. Note that the latter is much less stable. (Redrawn from Hastings & Costantino 1987, 1991.)*

the influence function. They also showed that the effect of increasing the rate of cannibalism (i.e., $c$ in eqs. 27 and 28) is destabilizing. This effect is biologically important because it implies that virtually all *Tribolium* populations should be locally unstable (for further details, see Hastings & Costantino 1991).

*Equilibria*

Examination of the effects of life-history parameters on equilibrium conditions can provide explanations of long-term shifts in population abundance. Here we discuss two examples: (1) why single-sex harvesting or hunting does not change density and hence leads to greater variability in recruitment than removals of both sexes; and (2) how density-dependent changes in growth rate can lead to permanent shifts in abundance levels (i.e., multiple equilibria).

The disadvantages of single-sex harvesting were first noted on an empirical basis in management of deer populations (Leopold et al. 1947). Single-sex harvesting of deer is a common management strategy based on the conservative notion that as long as the females of a polygamous species are not removed, persistence of a hunted species is guaranteed. In the first half of the twentieth century, wildlife managers began to notice that hunted deer populations frequently exhibit long-term (5- to 15-year) fluctuations in abundance termed irruptions (Caughley 1970). Deer populations would increase to high levels, food levels would decline because of heavy browsing, and then the deer population would decline; this cycle might repeat. In a classic study of this phenomenon, Leopold et al. (1947) pointed out that this was because females were not harvested, and hence there was no upper limit on the density of these populations.

Invertebrate fisheries, especially crustacean fisheries, also commonly employ single-sex harvesting by direct regulation or differential harvesting of males, either because of differential growth and a lower size limit or because of a restriction preventing the landing of ovigerous females (Botsford 1991). The basis for this strategy is the same as for deer: protection of the reproductive stock. The Dungeness crab fishery is an example of a single-sex harvest with highly variable recruitment (for others, see Botsford 1991).

The differences between male-only and both-sex harvesting are a consequence of their different effects on population equilibrium. If we explicitly account for both sexes in our model, assuming reproduction depends only on females, the equilibrium condition (eq. 4a) becomes

$$1 = \sum_{a=1}^{a_{\max}} \sigma_f(a) b(a) f \left\{ R_e \sum_{a=1}^{a_{\max}} [\sigma_f(a) c_f(a) + \sigma_m(a) c_m(a)] \right\}, \quad (29)$$

where the subscripts denote male and female. When only males are harvested, only male survival, $\sigma_m(a)$, is affected by the harvest; hence, the function $f(.)$ remains constant, and equilibrium recruitment, $R_e$, must increase as $\sigma_m(a)$ decreases. Recruitment increases, but the levels of density (i.e., effective population size), and hence density dependence, remain the same. Recruitment and harvest increase as the harvesting rate of males increases, but density remains high (Fig. 10). Male-only harvesting changes stability through its influence on age structure; it has no effect on $K$. If

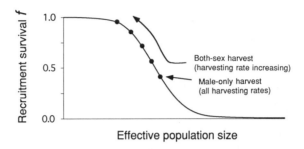

FIGURE 10. *Equilibrium levels of density-dependent recruitment survival as male-only and two-sex harvesting increases. The equilibrium level changes with harvest only with two-sex harvesting.*

the harvest is age-selective, it concentrates the age structure over a smaller range of ages and, from the stability results above, can make the population less stable (i.e., more likely to oscillate and more sensitive to environmental variability).

When both males and females are harvested, both female survival, $\sigma_f(a)$, and male survival, $\sigma_m(a)$, change, thus changing both reproduction per recruit ($\Phi_b$) and effective population size per recruit ($\Phi_c$). In this case, the value of $f(.)$ increases by an amount equal to the decrease in the first sum on the right-hand side of the equation (i.e., the effect of harvesting on reproduction); hence, the effective population size, $C_e$, decreases. As the equilibrium point moves up along the recruitment survival curve, $K$, the normalized slope of the recruitment survival function, declines, thus making the population more stable (Fig. 10). The population can decrease to the point where $f(.) = 1$ (and $\Phi_b = 1$), beyond which the population declines with further harvesting.

This decline in the value of the normalized slope with increased harvesting rate is the essential stabilizing effect of two-sex harvesting when density dependence is in recruitment. With single-sex harvesting, there is no control on density; hence, it is free to irrupt to high values, incur density dependence, return to low values, and then possibly increase again and continue in a cyclic fashion. In some populations in which a single sex is harvested, other effects are important. For example, for deer populations, the density dependence of adult mortality is significant.

Another, more subtle way in which the shape of influence functions can vary and thereby change equilibrium is through the density dependence of the growth rate of individuals. The models presented here do not include density-dependent growth rate, but we can get some idea of its influence on population behavior by examining the implications of different fixed growth rates. Influence functions (eq. 6) are the product of survival to each age and either the reproductive or density-dependent effect of recruitment at each age. Both of these can depend on individual size. If growth rate changes, the influence functions can change in such a way that equilibrium recruitment changes (eqs. 4). An increase in $\Phi_b$ tends to reduce $R_e$; an increase in $\Phi_c$ tends to decrease $R_e$.

This mechanism was proposed to explain the observation that several populations with known density-dependent recruitment (cannibalism in all three cases) had been harvested to low levels but had not returned to former high levels when harvesting ceased or declined (Botsford 1981). These populations were the Pacific sardine off California, the central California Dungeness crab, and the Eurasian perch in Lake Windermere in England. In all of these populations, individual growth rate had increased after the population declined to lower density. The increase in growth rate may shift the density-dependent weighting function $c_a$ to lower ages, thus increasing $\Phi_c$ and thereby lowering equilibrium recruitment (Botsford 1981). Any shift in reproduction has a lesser effect either because the population is essentially an open population or because the effects of the change in $\Phi_b$ on equilibrium are much less than the effects of changes in $\Phi_c$. Biologically, after a population is harvested to a low density, growth rate increases, such that recruits reach cannibalistic size faster and thus endure lower mortality. This increases the effective cannibalism per recruit (i.e., $\Phi_c$) enough to lock the population into a low level (Fig. 11). Figure 11$a$ compares favorably to the central California catch record in Figure 4$d$. A similar effect is being evaluated in the recent decline of haddock stocks in the eastern United States.

## Nonlinear Aspects of Behavior

In spite of a general awareness of the importance of understanding the nonlinear behavior of models with density-dependent recruitment and age structure, there has been little work on this aspect. It is important to know the complete range of behavior to expect from these models and the life-history characteristics that lead to them. Few population time series resemble smooth cycles; most fluctuate, either because of random environmental buffeting or,

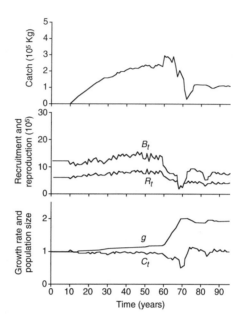

FIGURE 11. *Development of a fishery with gradually increasing harvesting rate on a size-structured population with density-dependent growth rate in addition to density-dependent recruitment (Botsford 1981). Catch increases as harvesting rate increases until* $\Phi_b$ *(i.e., $R_0$) is less than 1.0, at which time the population declines (the catch, reproduction, effective population size, and recruitment all decline). As the population density decreases, growth rate (g) increases, changing the influence functions, which changes the equilibrium condition and locks the population into a lower equilibrium, where it can remain even when the harvesting rate is reduced. (Redrawn from Botsford 1981.)*

possibly, because of inherent nonlinearity. We need to know the time scales and the amplitudes of fluctuations possible from deterministic models as a baseline for the study of the influence of random environments.

The initial cause for concern regarding the potential for exotic kinds of behavior from age structure and density-dependent recruitment was the chaotic behavior of semelparous models with density-dependent recruitment (May & Oster 1976). Early studies showed that chaotic behavior is also possible for models with two age classes, but that the amount of nonlinearity required to produce such behavior is greater than in the semelparous case (see,

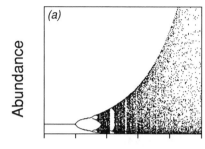 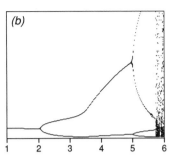

FIGURE 12. *The stabilizing effect of adding a small amount of age structure to the semelparous discrete logistic (Ricker function, Fig. 1). Comparison of* (a) *a bifurcation diagram of the semelparous case with* (b) *a bifurcation diagram for a minimally iteroparous population in which only 5 percent of adult individuals are allowed to survive each year. (Redrawn from Botsford 1992a.*

e.g., Guckenheimer et al. 1977). This is dramatically demonstrated for the delayed-recruitment model with a Ricker stock-recruitment model (Fig. 12; Botsford 1992*a*). Adding an adult survival of just 0.05 to a semelparous model shifts the value of the slope at which chaotic behavior occurs from greater than 3.0 to almost 6.0.

Simulations of this same delayed-recruitment model for typical values of harvested fish populations (i.e., $p = 0.6$ and $T = 3$ in eq. 9) indicate that as the slope increases, the population initially cycles with a period approximately twice the mean age of reproduction $\{2T + [2/(1 - p)]\}$, but then the period increases by 1 in either periodic or quasiperiodic cycles (Botsford 1992*a*; cf. Levin 1981). When the Ricker function is modified to make it more realistic by establishing a maximum value on recruitment, the increase in period with slope is no longer present. When the age structure is modified to remove the infinitely long tail by truncating it after age 7, the model displays behavior graphically similar to period-doubling (i.e., as the slope becomes steeper, every other cycle becomes depressed, thereby doubling the period (Botsford 1992*a*; note that Figs. 6 and 8 are reversed in that paper).

In a more exhaustive simulation study of the behavior of the delayed-recruitment model with a Ricker function, K. Higgins and his colleagues have shown that the expectations based on linearized analysis break down for low values of adult survival (pers. comm.).

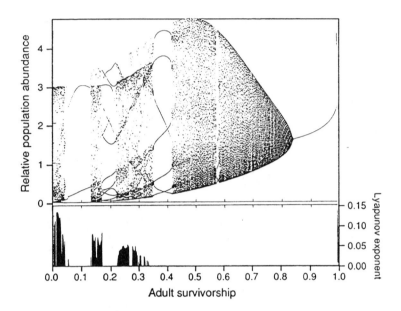

FIGURE 13. *A bifurcation diagram for a delayed-recruitment model using a Ricker stock-recruitment function with a* = 3.25 *and age of maturation of* $T = 3$, *as adult survival is varied. Positive values of the largest Lyapunov exponent indicate chaotic behavior. (Figure provided by K. Higgins.)*

For example, for Ricker parameter $a = 3.25$ and maturation age 3, nonperiodic behavior occurs for some values of adult survival below 0.33, parts of which are chaotic (Fig. 13).

Tuljapurkar (1987) explored the question of when limit cycles would occur in continuous-time models and what their period would be. He developed a method for essentially computing the value of $K$ at which nonlinear cycles occur, computing the period of the cycles in terms of the period of the cycles implied by the linearized analysis, and determining whether the nonlinear cycle was itself stable to small perturbations. Tuljapurkar et al. (1994) examined how the magnitude and period of cycles produced in a discrete-time, age-structured model with density-dependent recruitment depend on life-history parameters. They derived expressions for cycles of moderate amplitude, but they also found more-complex, possibly chaotic behavior.

Another analysis of explicitly nonlinear aspects of age-structured models with density-dependent recruitment is the exploration of the dynamics of a population with a rectangular-shaped recruitment function (i.e., an infinite negative slope at high density; see Fig. 2c). Nisbet and Bence (1989) modeled recruitment to populations of macroalgae (*Macrocystis* spp.) as being at a fixed level when adult density is below a certain density and at zero when adult density is above the threshold. On the basis of earlier results by an der Heiden and Mackey (1982), they showed that the population can be either at a stable equilibrium if mortality rates are high enough or at an unstable cyclic equilibrium. In the former case, the population is essentially an open population that never reaches the density-dependent threshold. In the latter, the period of the cycles is greater than twice the maturation delay (for details, see Nisbet & Bence 1989).

## 4 Discussion

Over the past 20 years we have made substantial progress in the development of a consistent theory of the dynamic behavior of age-structured populations with density-dependent recruitment. If presented with a new problem involving such a population, we have some idea of what to expect on the basis of the age structure and shape of density dependence in recruitment. Dramatic density dependence and narrow age structures at older ages tend to be less stable.

It is important to realize that the theory is not absolutely general in the strict mathematical sense; it specifies typical behavior rather than behavior that occurs in every instance. The class of populations with density-dependent recruitment and age structure may be simply too broad to allow strictly general statements regarding the behavior of all such populations. General statements are possible mathematically, of course, but they are possible only for specific models that do not necessarily cover all cases.

Elements of this theory have developed in many different areas—fisheries, general ecology, and human demography—with the various developers often not aware of each other's results and potentially valuable commonalities. Clearly, progress would be more rapid if investigators took a more synthetic view. For example, researchers working on space-limited recruitment to marine populations are working on a problem with many similarities to cannibalism in *Tribolium*. The shapes of the recruitment survival functions are slightly different, but some results are similar, such as the loss of stability as the rate of input of propagules increases ($B_c$ here; $s$ in Roughgarden et al. 1985; $b$ in Hastings & Costantino 1987). Although descriptions of equilibria and stability in terms of problem-specific quantities such as the lifetime amount of occupied space or total cannibalism over age are necessary, communication among a larger group of investigators would be facilitated if these were translated into a common language such as the lifetime influence on density dependence. Influence functions reflect the lagged effect of past recruitment on current recruitment regardless of the exact mechanism underlying the density dependence.

The development of this theory provides an awareness of a potential problem with conclusions regarding whether certain structural changes are stabilizing or destabilizing. It may not be possible to make simple statements (e.g., increased individual growth rate is always stabilizing) for all population variables. Rather, a critical part of our task is to determine which combination of population variables controls stability (i.e., essentially what it is about populations that makes them stable or unstable). Also, whether a specified change in a parameter is stabilizing or destabilizing depends on which other variables are being held constant. The difference between Table 1 and Table 2 is an example.

While we have focused on age-structured populations here, some investigators have explored the implications of including the size (or stage) of individuals in similar models. It is important to realize that size structure will not produce behavior different from that produced by age structure unless there is *not* a fixed relationship between size and age. Using the theory underlying a state-variable representation, one can show that the additional complexity of a size-structured model is not required unless growth rate varies with time (i.e., with density or environment; Caswell et al. 1972; Botsford 1992*b*). Several investigators have explored the influence of growth rate on the shape of influence functions in age-structured models (Roughgarden et al. 1985; Bence & Nisbet 1989). If the influence of individuals on density dependence in recruitment depends on individual size, then an increase in growth rate can have two effects on an influence function: (1) the influence function could shift closer to the origin (i.e., lag is reduced), which would be stabilizing, and (2) the lifetime effect of each individual could be greater because more survive to large size, which could change equilibrium and be destabilizing (see Bence & Nisbet 1989). Pascual and Caswell (1991) have explored the influence on stability of different growth rates as reflected in the transition rates of a stage-structured model. Abbiati et al. (1992) formulated a similar model for the analysis of harvested populations of red coral.

Another aspect of this type of population not touched on here is the influence of random environment. Several investigators have explored the effects on the variability in harvesting of adding a small level of randomness to recruitment in similar models of harvested populations (Horwood & Shepherd 1981; Horwood 1983, 1984; Reed 1983; Botsford 1986). Reed (1983) pointed out that harvesting has two effects, one on age structure and the other on equilibrium. As noted here, the latter does not occur with single-sex harvesting. Stable populations that are close to being unstable can be selectively sensitive to variability in the environment on time scales of twice the generation time (for a discussion of the consequences for temporal scales of variability in marine fisheries, see Botsford 1986; Botsford et al. 1989).

Another aspect of age-structured models with density-dependent recruitment not addressed here is their distribution over space. The fact that we have approximated some coastal marine populations by modeling them as open populations is a hint that we may be omitting dynamically important aspects. There is a growing awareness of the importance of spatial distribution to population dynamics in general (Roughgarden & Iwasa 1986; Gilpin & Hanski 1991; Hastings & Higgins 1994). Working out the combined effects of spatial and age structure on population dynamics will not be a simple process (Hastings 1992). Attempts to explain the dynamics of age-structured metapopulations in terms of the dynamics of local subpopulations have met with some success; however, there have been surprises. Hastings and Higgins (1994) demonstrated the existence of extremely long transients in the behavior of metapopulations of Ricker stock-recruitment relationships distributed along a coast, linked by larval dispersal. Not only does this provide a possible mechanism for episodic shifts in population behavior, but it also renders analysis of asymptotic behavior less useful. For age-structured populations, connecting a number of age-structured populations together with larval dispersal can lead to unexpected spatial patterns in cyclic behavior (Botsford et al. 1994).

Since this is a tutorial exposition aimed at the investigators of the future, I will emphasize some of the open problems touched on here. One area in need of better understanding is the dynamic consequence of there being several kinds of density dependence present. We need further investigation of combinations of density-dependent growth, recruitment, and survival. Another area is further investigation of where the intuition gained from analyses of linearized models breaks down (see Higgins et al., in press). Results regarding models with an explicit account of more than one species indicate similarities with results obtained here, but much more remains to be done to gain a complete understanding (see Hastings 1984). The addition of spatial structure will also be a fruitful avenue. The effects on populations of environmental forcing on interannual and longer time scales are also attracting interest. Our understanding of these populations will also be advanced by further efforts comparing modeling results with laboratory and field observations.

## Acknowledgments

I thank A. Hastings and K. Higgins for comments on a draft of this chapter. I thank K. Higgins for Figure 13 and J. Brittnacher for preparing the other figures.

## Literature Cited

Abbiati, M., G. Buffoni, G. Caforio, G. Di Col, and G. Santangelo. 1992. Harvesting, predation and competition effects on a red coral population. *Netherlands Journal of Sea Research* 30: 219–228.

Allen, R. L., and P. Basasibwaki. 1974. Properties of age structure models for fish populations. *Journal of the Fisheries Research Board of Canada* 31: 1119–1125.

an der Heiden, U., and M. C. Mackey. 1982. Dynamics of destruction and renewal. *Journal of Mathematical Biology* 16: 75–101.

Beddington, J. R. 1974. Age distribution and the stability of simple discrete time population models. *Journal of Theoretical Biology* 47: 65–74.

Bence, J. R., and R. M. Nisbet. 1989. Space-limited recruitment in open systems: The importance of time delays. *Ecology* 70: 1434–1441.

Bergh, M. O., and W. M. Getz. 1988. Stability of discrete age-structured and aggregated delay-difference population models. *Journal of Mathematical Biology* 26: 551-581.

Botsford, L. W. 1981. The effects of increased individual growth rates on depressed population size. *American Naturalist* 117: 38–63.

———. 1984. Effect of individual growth rates on expected behavior of the northern California Dungeness crab (*Cancer magister*) fishery. *Canadian Journal of Fisheries & Aquatic Science* 41: 99–107.

———. 1986. Effects of environmental forcing on age-structured populations: Northern California Dungeness crab (*Cancer magister*) as an example. *Canadian Journal of Fisheries & Aquatic Science* 43: 2345–2352.

———. 1991. Crustacean egg production and fisheries management. Pp. 379–394 *in* A. M. Wenner, ed., *Crustacean Egg Production*. A. A. Balkema, Rotterdam.

———. 1992*a*. Further analysis of Clark's delayed-recruitment model. *Bulletin of Mathematical Biology* 54: 275–293.

———. 1992*b*. Individual state structure in population models. Pp. 213–236 *in* D. L. DeAngelis and L. J. Gross, eds., *Individual-Based Models and Approaches in Ecology*. Chapman & Hall, New York.

Botsford, L. W., and R. C. Hobbs. 1995. Recent advances in the understanding of cyclic behavior of Dungeness crab (*Cancer magister*) populations. *International Council for the Exploration of the Sea Marine Sciences Symposium* 199: 157–196.

Botsford, L. W., and D. E. Wickham. 1978. Behavior of age-specific, density-dependent models and the northern California Dungeness crab (*Cancer magister*) fishery. *Journal of the Fisheries Research Board of Canada* 35: 833–843.

Botsford, L. W., R. D. Methot, and W. E. Johnston. 1983. Effort dynamics of the northern California Dungeness crab (*Cancer magister*) fishery. *Canadian Journal of Fisheries & Aquatic Science* 40: 337–346.

Botsford, L. W., D. A. Armstrong, and J. Shenker. 1989. Oceanographic influences on the dynamics of commercially fished populations. Pp. 511–565 *in* M. R. Landry and B. M. Hickey, eds., *Coastal Oceanography of Washington and Oregon*. Elsevier, Amsterdam.

Botsford, L. W., C. L. Moloney, A. Hastings, J. L. Largier, T. M. Powell, K. Higgins, and J. F. Quinn. 1994. The influence of spatially and temporally varying oceanographic conditions on meroplanktonic metapopulations. *Deep-Sea Research* II 41: 107–145.

Caswell, H., H. E. Koenig, J. A. Resh, and Q. E. Ross. 1972. An introduction to systems science for ecologists. Pp. 3–78 *in* B. C. Patten, ed., *Systems Analysis and Simulation in Ecology*. Academic Press, New York.

Caughley, G. 1970. Eruption of ungulate populations with emphasis on Himalayan Thar in New Zealand. *Ecology* 51: 53–72.

Clark, C. W. 1976. A delayed-recruitment model of population dynamics, with an application to baleen whale populations. *Journal of Mathematical Biology* 3: 381–391.

Diekmann, O., R. M. Nisbet, W. S. C. Gurney, and F. van den Bosch. 1986. Simple mathematical models for cannibalism: A critique and a new approach. *Mathematical Biosciences* 78: 21–46.

Dong, Q., and G. A. Polis. 1992. The dynamics of cannibalistic populations: A foraging perspective. Pp. 13–37 *in* M. A. Elgar and B. Crespi, eds., *Cannibalism: Ecology and Evolution among Diverse Taxa*. Oxford University Press.

Fox, L. R. 1975. Cannibalism in natural populations. *Annual Review of Ecology & Systematics* 6: 87–106.

Frauenthal, J. C. 1975. A dynamic model for human population growth. *Theoretical Population Biology* 8: 64–73.

Gabriel, W. 1985. Overcoming food limitation by cannibalism: A model study on cyclopoids. *Archiv für Hydrobiologie* 21: 373–381.

Gaines, S., and J. Roughgarden. 1985. Larval settlement rate: A leading determinant of structure in an ecological community of the marine intertidal zone. *Proceedings of the National Academy of Sciences (USA)* 82: 3707–3711.

Gilpin, M., and I. Hanski. 1991. *Metapopulation Dynamics: Empirical and Theoretical Investigations.* Academic Press, London.

Guckenheimer, J., G. Oster, and A. Ipaktchi. 1977. The dynamics of density-dependent population models. *Journal of Mathematical Biology* 4: 101–147.

Gurney, W. S. C., R. M. Nisbet, and J. H. Lawton. 1983. The systematic formulation of tractable single-species population models incorporating age structure. *Journal of Animal Ecology* 52: 479–495.

Hastings, A. 1987. Cycles in cannibalistic egg-larval interactions. *Journal of Mathematical Biology* 24: 651–666.

———. 1992. Age-dependent dispersal is not a simple process: Density dependence, stability and chaos. *Theoretical Population Biology* 41: 388–400.

Hastings, A., and R. F. Costantino. 1987. Cannibalistic egg-larva interactions in *Tribolium*: An explanation for the oscillations in population numbers. *American Naturalist* 130: 36–52.

———. 1991. Oscillations in population numbers: Age-dependent cannibalism. *Journal of Animal Ecology* 60: 471–482.

Hastings, A., and K. Higgins. 1994. Persistence of transients in spatially structured ecological models. *Science (Washington, D.C.)* 263: 1133–1136.

Higgins, K., A. Hastings, and L. W. Botsford. In press. Adult survivorship in the delayed-recruitment model: Influence on dynamics. *American Naturalist*.

Hobbs, R. C., and L. W. Botsford. 1989. Dynamics of an age-structured prey with density- and predation-dependent recruitment: The Dungeness crab and a nemertean egg predator worm. *Theoretical Population Biology* 36: 1–22.

Horwood, J. W. 1983. A general linear theory for the variance of yield from fish stocks. *Mathematical Biosciences* 64: 203–225.

———. 1984. The variance and response of biological systems to variability in births and survivals. *(Institute of Mathematics and Its Applications) Journal of Mathematics Applied in Medicine & Biology* 1: 309–323.

Horwood, J. W., and J. A. Shepherd. 1981. The sensitivity of age-structured populations to environmental variability. *Mathematical Biosciences* 57: 59–82.

Keyfitz, N. 1972. Population waves. Pp. 1–38 *in* T. N. E. Greville, ed., *Population Dynamics*. Academic Press, New York.

Lee, R. 1974. The formal dynamics of controlled populations and the echo, the boom and the bust. *Demography* 11: 563–585.

Leopold, A., L. K. Sowls, and D. L. Spencer. 1947. A survey of over-populated deer ranges in the United States. *Journal of Wildlife Management* 11: 162–177.

Levin, S. A. 1981. Age structure and stability in multiple-age spawning populations. Pp. 21–45 *in* T. L. Vincent and J. M. Skowronski, eds., *Renewable Resource Management*. Springer-Verlag, New York.

Levin, S. A., and C. P. Goodyear. 1980. Analysis of an age-structured fishery model. *Journal of Mathematical Biology* 9: 245–274.

May, R., and G. F. Oster. 1976. Bifurcations and dynamic complexity in simple ecological models. *American Naturalist* 110: 573–599.

McKelvey, R., D. Hankin, K. Yanosko, and C. Snygg. 1980. Stable cycles in multistage recruitment models: An application to the northern California Dungeness crab (*Cancer magister*) fishery. *Canadian Journal of Fisheries & Aquatic Science* 37: 2323–2345.

Menshutkin, V. V. 1964. Population dynamics studied by representing the population as a cybernetic system. *Voprosy Ikthiologii* 1: 23–33.

Murdoch, W. W., and E. McCauley. 1985. Three distinct types of dynamic behavior shown by a single planktonic system. *Nature* 316: 628–630.

Nisbet, R. M., and J. R. Bence. 1989. Alternative dynamic regimes for canopy-forming kelp: A variant on density-vague population regulation. *American Naturalist* 134: 377–408.

Pascual, M., and H. Caswell. 1991. The dynamics of a size-classified benthic population with reproductive subsidy. *Theoretical Population Biology* 39: 129–147.

Pennycuick, C. J., R. M. Compton, and L. Beckingham. 1968. A computer model for simulating the growth of a population of two interacting populations. *Journal of Theoretical Biology* 18: 316–329.

Polis, G. 1981. The evolution and dynamics of intraspecific predation. *Annual Review of Ecology & Systematics* 12: 125–151.

Possingham, H., S. Tuljapurkar, J. Roughgarden, and M. Wilks. 1994. Population cycling in space-limited organisms subject to density-dependent predation. *American Naturalist* 143: 563–582.

Reed, W. J. 1983. Recruitment variability and age structure in harvested animal populations. *Mathematical Biosciences* 65: 239–268.

Ricker, W. E. 1954. Stock and recruitment. *Journal of the Fisheries Research Board of Canada* 11: 559–623.

Rorres, C. 1976. Stability of an age-specific population with density-dependent fertility. *Theoretical Population Biology* 10: 26–46.

Roughgarden, J., and Y. Iwasa. 1986. Dynamics of a metapopulation with space-limited subpopulations. *Theoretical Population Biology* 29: 235–261.

Roughgarden, J., Y. Iwasa, and C. Baxter. 1985. Demographic theory for an open marine population with space-limited recruitment. *Ecology* 66: 54–67.

Strong, D. R. 1986. Density-vagueness: Abiding the variance in the demography of real populations. Pp. 257–268 *in* J. Diamond and T. J. Case, eds., *Community Ecology*. Harper & Row, New York.

Tuljapurkar, S. 1987. Cycles in nonlinear age-structured models. I. Renewal equations. *Theoretical Population Biology* 32: 26–41.

Tuljapurkar, S., C. Boe, and K. W. Wachter. 1994. Nonlinear feedback dynamics in fisheries: Analysis of the Deriso-Schnute model. *Canadian Journal of Fisheries & Aquatic Science* 51: 1462–1473.

Usher, M. B. 1972. Developments in the Leslie matrix model. Pp. 29–60 *in* J. N. R. Jeffers, ed., *Mathematical Models in Ecology*. British Ecological Society Symposium 12. Blackwell Scientific, Oxford.

# CHAPTER 13

# Models for Marine Ecosystems

## *Eileen E. Hofmann*

The first mathematical models of marine planktonic ecosystems (Riley 1946, 1947) were formulated with no distinction between phytoplankton or zooplankton. Rather, the models were for homogeneous populations, and the rates applied to these components were assumed to be representative of average conditions. The structure of these first models was in large part determined by the available data sets, which, in general, were not of sufficient resolution to warrant complex models or complex formulations for the model components. As understanding and measurement capability of marine ecosystems have advanced, ecosystem models have undergone an evolution from the initial bulk-approach models to models that include structure within the various components. This has made the models more realistic and has allowed marine-plankton models to remain current with the level of understanding of how marine systems function.

By adding structure based on size or animal stage, distinct features of a population, such as feeding or migration behavior, can be isolated in a model. The disadvantage of adding increasing levels of complexity within ecosystem components is that more knowledge is needed to set the appropriate parameters for the model processes. For example, physiological rates, behavioral characteristics, and growth forms for the different sizes or stages need to be measured. Often, whether a model is size- or stage-structured is determined more by the available measurements than by the research questions being posed. For example, measurements of marine zooplankton are typically made on the basis of animal stage. Measurements of benthic invertebrates or marine phytoplankton are typically done on the basis of size.

This chapter describes several models developed for investigating processes that regulate the distribution and abundance of the lower trophic levels of marine food webs. The models chosen for inclusion represent a variety of approaches that range from simple bulk formulations for the biological components to complex stage- or size-structured formulations. These models also include time- and space-dependent frameworks. The latter is important for marine food webs since much of the observed structure in marine plankton populations results from the effects of the circulation pattern. In discussing the models, emphasis is placed on what has been learned from the different approaches, the shortcomings and strengths of the approaches, and how future models can be improved. (For more-detailed reviews of models constructed for marine ecosystems, see Wroblewski & Hofmann 1989; Fransz et al. 1991; Evans & Fasham 1993; Hofmann 1993; Hofmann & Lascara, in press.)

## 1 Time-Dependent Models

*Bulk Food and Predator*

The first models to be developed for marine systems were attempts to explain the seasonal variation in the standing stocks of phytoplankton and zooplankton observed on Georges Bank (Riley 1946, 1947). These models were based on time-dependent equations that described increases and decreases in the phytoplankton and zooplankton populations. The models were simple in design, which reflects the limited data on which they were based and the level of understanding at the time.

The governing equation describing the time-dependent changes in the standing stock of phytoplankton ($P$) was of the form

$$\frac{dP}{dt} = P(P_h - R - G),\qquad(1)$$

where $P_h$ is the photosynthetic rate per unit of population, $R$ is the phytoplankton respiration rate, and $G$ is the rate of grazing by zooplankton. Similarly, changes in the standing stock of herbivorous zooplankton ($H$) were assumed to be described by an equation of the form

$$\frac{dH}{dt} = H(A - R - E - D),\qquad(2)$$

where $A$ is the rate of food assimilation by the herbivores, $R$ is the herbivore respiration rate, $E$ is the rate at which the herbivores

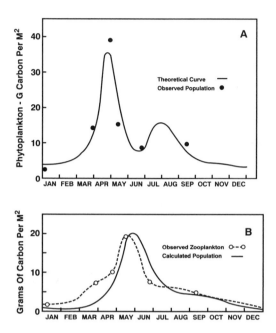

FIGURE 1. *Observed and simulated seasonal cycle of* (a) *phytoplankton and* (b) *zooplankton on Georges Bank. (Adapted from Riley 1946, 1947.)*

are consumed by their predators, and $D$ is the herbivore natural death rate.

These formulations treat phytoplankton and zooplankton as bulk quantities. No distinction is made between species, stage, or size. Thus, the rates included in the model represent population or community averages and, as such, integrate large ranges of possible values. The model, therefore, can provide only an average distribution, which may or may not be consistent with the measurements available for calibration and verification.

The simulated phytoplankton distribution (Fig. 1a) shows that the model reproduces the observed spring peak in abundance although with a slightly lower magnitude. In addition, the decay time of the simulated spring bloom is slightly longer than that in the observed distribution. The model predicts a second peak in phytoplankton in late summer or early fall; however, observations to support the existence of this feature were not available.

The simulated zooplankton distributions (Fig. 1b) reproduce the magnitude of the spring increase in zooplankton abundance. Yet, the timing of the onset and decay of this event in the simulated distributions is delayed relative to the observed timing. The primary mismatch between simulated and observed zooplankton distributions occurs in the winter and early spring, for which the model significantly underestimates observed abundances.

Given the limited data base and simple nature of the phytoplankton and zooplankton models, it is encouraging that the simulated distributions are as consistent with observations. However, reproducing the timing and magnitude of events, such as spring blooms, is a basic result that should come from any model.

*Size-Structured Food and Bulk Predator*

About 45 years after Riley's initial modeling studies, Fasham et al. (1990) developed a model that focused on simulating seasonal changes in phytoplankton standing stock and primary production off Bermuda. Their model uses more biological detail, consisting of equations for time-dependent changes in nitrate, ammonium, dissolved organic nitrogen, phytoplankton, bacteria, zooplankton, and detritus. In an attempt to provide more realism, their model allows for selective feeding by zooplankton on a range of prey types. The equation describing time-dependent changes in zooplankton biomass is of the form

$$\frac{dZ}{dt} = \beta_1 \, G_1 + \beta_2 \, G_2 + \beta_3 \, G_3 \; - \; \text{mortality} \; - \; \text{loss}, \qquad (3)$$

where $\beta_1$, $\beta_2$, $\beta_3$ and $G_1$, $G_2$, $G_3$ represent assimilation efficiencies and grazing functions for feeding on phytoplankton, bacteria, and detritus, respectively. This approach provides a size-structured food distribution to the zooplankton predator. However, a bulk formulation is used to represent the dynamics within each prey (food) component. Thus, the model rates are assumed to represent averages and as such do not undergo seasonal variations.

The simulated time-dependent distribution of primary production (Fig. 2a) obtained from the Fasham et al. (1990) model shows an increase in the spring, low levels during the summer, and a slight increase in the fall. These features in general match those seen in observations made at Bermuda. However, upon closer comparison, the simulated distribution misses the timing and amplitude of the spring bloom for both rates (10 md$^{-1}$ and 1 md$^{-1}$ of mixed-layer shallowing. The simulated fall increase in primary production also differs from the observed (Fig. 2a, b).

Even with the increased realism afforded by the additional bio-

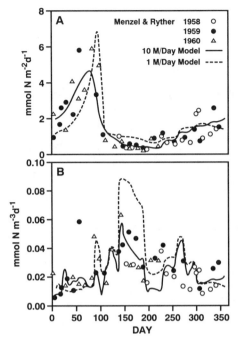

FIGURE 2. *Comparison of observed data from Menzel and Ryther (Bermuda Biological Station, 1960) with simulated distributions of primary production over the annual cycle (a) by area and (b) by volume. The Bermuda data were recalculated to reflect mixed-layer values only. (From Fasham et al.1990.)*

logical components and associated dynamics, the simulated seasonal plankton cycle obtained with the Fasham et al. (1990) model still has many of the same problems noted for the Georges Bank plankton simulations (Riley 1946, 1947). Fasham et al. (1990) attributed the mismatches between their simulated distributions and observations to insufficient resolution in the formulation of the zooplankton predator and too simple a representation of the mixed-layer dynamics.

## Size-Structured Food and Predator

One argument frequently made is that more detail in the prey and predator components will improve the realism of the simulated distributions for marine ecosystems. Moloney and Field (1991) developed a size-based model for investigating nitrogen and car-

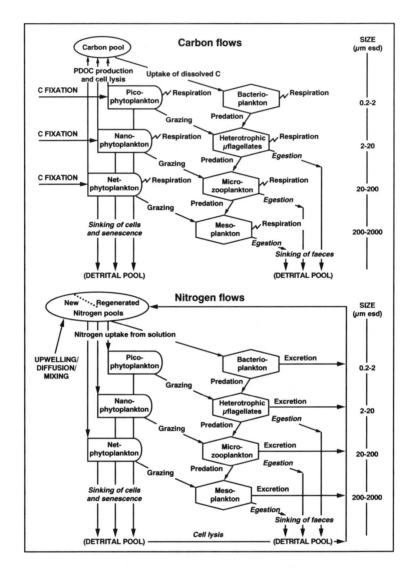

FIGURE 3. *Schematic of the size-based plankton model and the flows of carbon and nitrogen. (From Moloney & Field 1991.)*

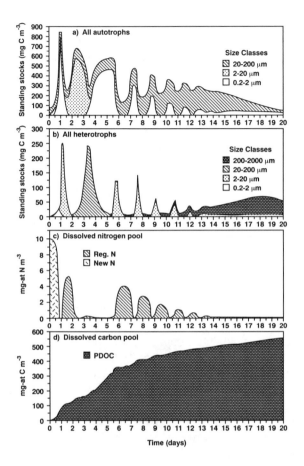

FIGURE 4. *Simulated time-dependent distribution of the standing stocks of four groups. (Adapted from Moloney & Field 1991.)*

bon cycling in marine plankton communities. The structure of this model is determined from general ecological principles, and as a result, it is not designed to represent a specific system. The model includes three autotroph (food) and four heterotroph (predator) size groups (Fig. 3). The other model components—dissolved and particulate carbon; dissolved, regenerated, and new nitrogen; and detrital material—have no explicit size dependence. The model structure and transfer processes are based on size, and all model parameters are formulated in terms of body size.

Governing equations are developed for each of the size-dependent components of the model ecosystem. For example, the equations describing the time-dependent changes in heterotrophic carbon $(HC)$ for any size class, $j$, are of the basic form

$$\frac{d}{dt}(HC_j) = \tag{4}$$

$$\sum_{i=1}^{n} \text{ingestion}_{i,j} - \text{egestion}_j - \text{respiration}_j - \text{predation}_j \,,$$

where $i$ represents the size class of food that is ingested. The structure of processes, such as ingestion, within each size class may differ as a function of size. The model given by equation (4) allows the concentration of carbon and nitrogen to change within a size class, but individuals do not transfer between size classes.

The simulated time-dependent distribution of the model components expressed in terms of size classes is shown in Figure 4. The simulated autotroph community (Fig. 4a) shows large initial blooms, which decay in amplitude over time and eventually level off to a steady value. The simulated heterotroph community (Fig. 4b) also shows fluctuations that are out of phase with those in the autotroph distributions, which is in agreement with the behavior expected from predator-prey systems. Moloney and Field (1991) attributed the fluctuations in the simulated communities to the interaction of size-dependent grazing pressure and nutrient limitation. However, they indicated that the frequency and magnitude of these fluctuations may not be realistic and that this behavior can be minimized by assigning different parameter values.

The size-based model developed by Moloney and Field (1991) is a start toward the development of a generic model for the dynamics of marine plankton communities. The structure of the model makes it straightforward to add other components, and because the model interactions are based on size, the necessary linkages between the different model components are already in place. Moreover, the model can be applied to other environments by changing parameter values and the input biomass spectrum, as illustrated by Moloney et al. (1991). However, migration of the size-based model into other ecosystems requires a considerable data base on plankton size distribution, trophic interactions, and the size dependence of rates and processes.

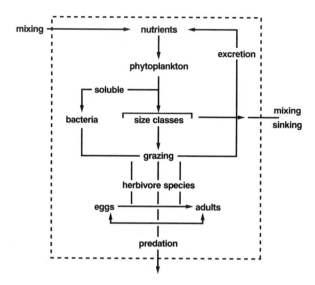

FIGURE 5. *Schematic of the processes and interactions in the Steele &*
*Frost model. Processes outside of the dashed line represent external forc-*
*ing and losses from the model ecosystem. (From Steele & Frost 1977.)*

## Size-Structured Food and Stage-Structured Predator

One of the objectives in modeling marine ecosystems is to predict the population structure of plankton communities. For many
reasons, the population structure tells more about how marine food
webs operate than about the total biomass at each trophic level.
The first model designed to simulate the structure of a marine
plankton community was developed by Steele and Frost (1977).
This model was designed to simulate the observed seasonal changes
in the size distribution of phytoplankton and zooplankton in Loch
Striven, Scotland. The available observations show certain patterns
in the relative composition of the phytoplankton and zooplankton
communities, suggesting that these patterns might be predictable.
Furthermore, the interactions of the herbivores with their phytoplankton food appear to depend on the size composition of the
organisms at both trophic levels.

In order to investigate community interactions, a model was constructed that includes size-structured phytoplankton, which are
grazed upon by two herbivores (*Pseudocalanus* and *Calanus*) with

stage-structured life histories (Fig. 5). An additional model component is an equation for nitrogen (nitrate and ammonium), which is removed by the phytoplankton and supplied by mixing and recycling from the herbivores.

The equation describing the time-dependent changes in phytoplankton ($P$) community structure is of the form

$$\frac{dP_i}{dt} = \text{growth}_i - \sum_{j=1}^{n} \text{grazing by zooplankton}_{i,j}$$
$$- \text{sinking}_i - \text{mixing} , \tag{5}$$

where $i$ represents the phytoplankton size class and $j$ represents grazing by herbivorous stages, which do not necessarily have to be of the same species. The phytoplankton biomass spectrum is partitioned into 20 size classes. Processes within each size class are formulated on the basis of cell size. For example, phytoplankton growth includes formulations for nutrient uptake, photosynthetic rate, respiration, and cell sinking, which are based on cell size.

The equation describing the time-dependent changes in the size classes of the different herbivorous stages of different species is of the form

$$\frac{dZ_j}{dt} = \text{recruitment from } Z_{j-1}$$
$$- \text{loss to } Z_{j+1} - \text{predation}_j . \tag{6}$$

Transfers between stages for a given herbivore are determined by growth processes, which in turn are determined by the rates of ingestion, assimilation, and metabolic losses. As with the phytoplankton processes, all herbivore processes are formulated in terms of animal size (biomass). In addition, the herbivore ingestion rates depend on phytoplankton size, with the smaller copepod ingesting primarily the smaller cells and the larger copepod effectively ingesting only the larger cells. Equation (6) differs from equation (4) in that carbon (or nitrogen) can transition between size classes; that is, animals grow to larger size over time.

The time-dependent changes in total phytoplankton and herbivore biomass (Fig. 6) show a bloom in phytoplankton, which is then followed by blooms in the *Calanus* and *Pseudocalanus* populations. Nitrate decreases over the time of the simulations. The simulated distributions shown in this figure provide gross features of the changes in total biomass at each trophic level; however, this

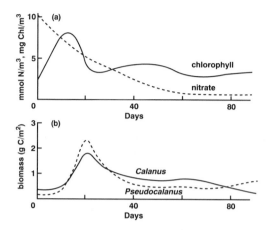

FIGURE 6. *Simulated time-dependent changes in* (a) *nitrate and chloro-phyll and* (b) Calanus *and* Pseudocalanus *obtained from the basic model described by Steele & Frost. All phytoplankton and zooplankton size classes are combined. (From Steele & Frost 1977.)*

is not very instructive in determining interactions among and be-tween trophic levels.

When the simulated phytoplankton and herbivore distributions are viewed in terms of cell diameter and animal biomass (size), the interactions between the trophic levels become more appar-ent (Fig. 7). The initial domination of the phytoplankton commu-nity by large cells corresponds to a spring diatom bloom. Over the 90-day simulation, the large cells decrease in abundance and the phytoplankton community becomes dominated by small cells. The herbivore community changes from one dominated by the larger *Calanus* to one dominated by the smaller *Pseudocalanus* during the course of the simulation. In fact, by the end of the 90-day period, the *Calanus* is essentially extinct.

The growth of the herbivore components is to a large extent determined by the phytoplankton size distribution. After the initial phytoplankton bloom, sufficient food exists to stimulate growth of the larger copepod. However, this animal ingests the larger cells more rapidly than they are replenished. Following the decrease in large cells, there is a decrease in the growth rate of the larger stages of both copepods. The lack of grazing pressure by the herbivores

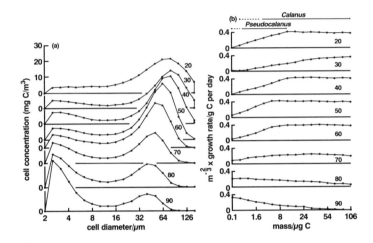

FIGURE 7. *Simulated changes in the* (a) *phytoplankton size composition and* (b) *growth rate of each size class of herbivore from day 20 to 90 obtained from the basic model run. Herbivore growth rate is expressed in terms of animal biomass (m). (From Steele & Frost 1977.)*

and reduced competition for nutrients from the large cells allows the smaller phytoplankton cells to bloom. This then provides a food source for *Pseudocalanus* and the smaller stages of *Calanus*. Thus, this simulation shows that the interactions and feedbacks between the herbivore and its food source are critical in structuring the community at both trophic levels. Additional simulations show this interaction to be an integral part of the co-occurrence of different zooplankton species.

## 2 Spatially Dependent Models

*Complex Circulation with Bulk Food and Predator*

The model developed by Wroblewski (1977) for investigating processes controlling primary production in the Oregon upwelling region was one of the first to attempt to combine a marine food-web model with a theoretical circulation model. The food-web model was simple, consisting of five components representing phytoplankton, zooplankton, nitrate, ammonium, and detritus. No attempt was made to distinguish among phytoplankton of different sizes or among zooplankton of different stages or species. The circulation model provided simulations of the horizontal and vertical velocities in an across-shelf vertical plane. No alongshore variation in the circulation field was included. The governing equations for this model are of the form

$$\frac{\partial B}{\partial t} + u \frac{\partial B}{\partial x} + w \frac{\partial B}{\partial z} - K_x \frac{\partial^2 B}{\partial x^2} - K_z \frac{\partial^2 B}{\partial z^2}$$
$$= \text{biological sources/sinks}\,, \tag{7}$$

where the terms on the left represent the changes in the biological variable, $B$, as the result of local time-dependent processes, horizontal and vertical advective effects, and horizontal and vertical diffusive effects, respectively. The circulation effects are contained in separately specified distributions of the across-shelf ($u$) and vertical ($w$) velocities. Values for the advective velocities are obtained from a theoretical circulation model that is developed for the Oregon upwelling region (Thompson 1974). The circulation model includes the effects of variable wind forcing, bottom topography, incident solar radiation, and surface, interfacial, and bottom stresses. The horizontal and vertical diffusion coefficients are given by $K_x$ and $K_z$, respectively. The terms on the right side of equation (7) represent the biological processes that control the growth or loss in specific biological variables. Thus, the model constructed by Wroblewski (1977) consists of a system of five coupled equations that provide space and time distributions for each of its food-web components.

The simulated circulation field (Fig. 8a) shows the structure that is expected in an upwelling region. Surface waters are transported offshore by the prevailing wind stress, and onshore flow occurs at depth. Between 10 and 15 km offshore, an upwelling front forms. Maximum velocities occur at the surface and bottom and along the upwelling front. Flow velocities at intermediate depths are less than those at the surface and bottom.

The simulated phytoplankton field (Fig. 8b,c) shows the highest concentration in the inshore region. This maximum in concentration results from the enhanced nutrient input to the lighted portion of the water column by the upwelling vertical velocity. Offshore, the phytoplankton field forms a coherent plume that extends downward in the region where the upwelling front is found. Overall, the structure and magnitude of the simulated phytoplankton fields agree with observations from the Oregon upwelling region (Wroblewski 1977).

Although this model uses a simple bulk approach for repre-

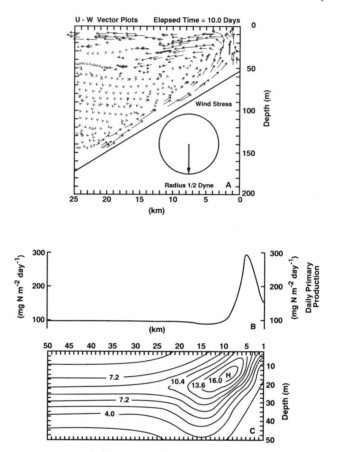

FIGURE 8. *Simulated distributions for the Oregon upwelling after ten days of elapsed time: (a) circulation in a transverse plane normal to the coast, the bottom topography, and the wind stress; (b) daily gross primary production of the water column; (c) distribution of phytoplankton with a contour interval of 1.6 µg at $Nl^{-1}$. (From Wroblewski 1977.)*

senting the lower trophic levels and lacks one spatial dimension (alongshore), it provides considerable insight into the interactions of physical and biological processes in upwelling regions. For example, the structure of the biological fields is, to a large extent, determined by the structure of the upwelling circulation. In addition, the maximum in phytoplankton concentration occurs after the period of most intense upwelling, which has implications for designing field measurements. However, the neglect of one spatial dimension and portions of the ecosystem, and the simple structure used for the marine food web, limits the space and time scales over which this model is valid.

*Complex Circulation and Stage-Structured Zooplankton without and with Behavior*

Two additional models were developed for the Oregon upwelling region in an attempt to explain the observed distributions of the zooplankton *Acartia clausi* and *Calanus marshallae* (Wroblewski 1980, 1982). Each of these models includes a stage-structured framework to describe the life history of the copepods. For *A. clausi*, the life history is represented by four components: eggs, nauplii, copepodids, and adults. The life history of *C. marshallae* is subdivided as eggs, nauplii, copepodid stages I to III, copepodid stages IV and V, and adults. The subdivision of the *C. marshallae* copepodid stages allows the vertical migration of stages IV and V and of adults to be explicitly included in the model; the other stages do not migrate. An implicit assumption in each of these models is that the copepods are not limited by food availability; hence, no equations for the dynamics of the copepod food source are included.

The governing equations for these models are similar to equation (7) except that the terms on the left side of the equation represent the growth or loss processes of the different copepod life stages as

$$\frac{\partial Z_j}{\partial t} + u\,\frac{\partial Z_j}{\partial x} + w\,\frac{\partial Z_j}{\partial z} - K_x\,\frac{\partial^2 Z_j}{\partial x^2} - K_z\,\frac{\partial^2 Z_j}{\partial z^2} =$$
$$\text{growth}_{j-1} - \text{growth}_{j+1} - \text{mortality}_j\,, \tag{8}$$

where growth represents the development from one life stage to the next and $j$ designates the particular life stage. The mortality term represents processes such as predation by higher trophic levels and natural mortality, but the many sources of mortality are not differentiated. Equation (8) was modified for the egg and adult components. (For details of the parameters used to model the copepod growth and mortality processes, see Wroblewski 1980, 1982.)

The horizontal and vertical advective velocities used in the two copepod models are obtained from the same circulation model that is used by Wroblewski (1977) for simulating the distribution of phytoplankton in the Oregon upwelling zone. For the *C. marshallae* model, however, the vertical velocity, $w$, is modified:

$$w = w_{\rm a} + w_{\rm b}\,, \tag{9}$$

where $w_a$ represents the advective vertical velocity and $w_b$ represents the biological vertical velocity due to migration behavior. For *C. marshallae*, $w_b$ is assumed to have a diel (24-hour) sinusoidal dependence of the form

$$w_b = w_s \sin(2\pi t), \tag{10}$$

where $w_s$ is the maximum vertical migration speed. The diel migration is set such that the late copepodids and adults are at the surface at night and at depth during the day. Hence, this model includes the effects of ontogenetic migration as well as diel migration on the distribution of the copepod life stages.

The *A. clausi* model is used to explore the question of how high concentrations of this animal persist in the inshore regions during periods of offshore transport that occur during coastal upwelling. Simulations with continuous wind forcing of the upwelling show that the extent of *A. clausi* would be maximal (extending 25 km or more offshore) during strong upwelling and minimal during relaxation periods (Wroblewski 1980). However, field observations show that these simulated distributions overestimate the offshore range of the copepod by about 50 percent. Additional simulations with realistic wind forcing, in which upwelling favorable winds are intermittent, result in a reduced offshore transport of *A. clausi* adults. Increasing the mortality rate for the adults results in even more reduced offshore transport. Hence, Wroblewski (1980) concluded that realistic winds are necessary to simulate the distribution of plankton populations in coastal upwelling regions. The intermittency in periods of upwelling (offshore transport) and relaxation (onshore transport) results in the animals being retained in coastal waters closer to shore. Moreover, the simulated distributions suggest that all stages of *A. clausi* must undergo high rates of mortality; otherwise, the animals would be transported out of the coastal zone during their potential 60-day life span. Hence, these simulations illustrate the importance of the circulation pattern, as well as the importance of stage-related mortality, in determining the structure of the copepod distribution in Oregon coastal waters.

The *C. marshallae* model uses the same simulated circulation field, but with the modification for the vertical velocity described above. The hypothesis tested with this model is that the distribution of this copepod is governed by the combined effects of upwelling circulation and vertical migration. The adults of this animal release their eggs in the food-rich nearshore waters; however, the offshore surface circulation from the upwelling carries the nonmi-

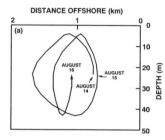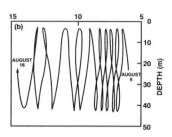

FIGURE 9. *Simulated zonal-plane trajectories for a copepod that undergoes a diel migration between 3 and 44 m: (a) two-day trajectory for a copepod that started 1 km from the coast; (b) ten-day trajectory for a copepod that started 6 km from the coast. (From Wroblewski 1982.)*

grating eggs and younger stages offshore. Diel vertical migration by the older copepodids and adults moves these animals into the deep-water onshore flow, which returns them to the nearshore regions.

The simulated distributions show clearly that the original distance from shore is important, and vertical migration greatly reduces the offshore transport of the copepod (Fig. 9). Migration by adults in nearshore waters is sufficient to prevent their offshore transport. The amount of offshore transport is determined by how close the animal is to the depth at which the return flow is strong. As the animal is moved offshore, the region of return flow is deeper; hence, migration must occur over a larger vertical distance to take advantage of the subsurface onshore flow. This is especially important for the older copepodid stages, which are transported the farthest offshore. The nonmigrating stages can return closer to shore during periods in which the upwelling circulation relaxes. Hence, the distribution of this animal is controlled by the existing circulation pattern and stage-dependent migration behavior.

*Complex Circulation with Size-Structured Food and Predators*

The Coastal Transition Zone (CTZ) off the coast of California is influenced by large filaments that develop along the outer portion of the California Current. These filaments are believed to be important in the transport of physical quantities, such as heat and salt, as well as providing a mechanism for the offshore transport of nitrogen, organic carbon, and animal populations. To investigate the

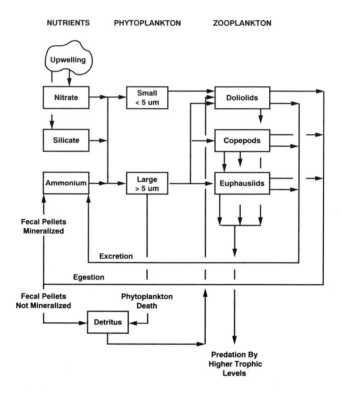

FIGURE 10. *Schematic of the state variables and interactions in the marine food-web model developed for the Coastal Transition Zone. (From Moisan & Hofmann, in press.)*

role of the California Current filaments in regulating the structure of plankton populations and the flux of biological properties in the CTZ, a model of the lower trophic levels is constructed (Moisan & Hofmann, in press; Moisan et al., in press). This model is designed to provide three-dimensional, time-dependent simulations for biological distributions in the CTZ.

The CTZ model consists of equations for three nutrients (nitrate, ammonium, and silicate), two phytoplankton size fractions (> 5 and < 5 $\mu$m), three zooplankton species (*Eucalanus californicus*, *Euphausia pacifica*, and *Dolioletta gegenbauri*), and detritus (Fig. 10). The three zooplankton species account for about 75 percent of the observed zooplankton biomass in the CTZ. These species have different rates of grazing on the two phytoplankton sizes. The governing equations for the CTZ model are of the form

$$\frac{\partial B}{\partial t} + u\,\frac{\partial B}{\partial x} + v\,\frac{\partial B}{\partial x} + w\,\frac{\partial B}{\partial z} - K_x\,\frac{\partial^2 B}{\partial x^2} - K_y\,\frac{\partial^2 B}{\partial y^2} - K_z\,\frac{\partial^2 B}{\partial z^2}$$
$$= \text{biological sources/sinks}\,. \tag{11}$$

Equation (11) is similar to equation (7) except that the along-shore dimension $(y)$ has been included. The advective velocity and diffusion effects in this direction are given by $v$ and $K_y$, respectively. The advective velocities in equation (11) are obtained from a theoretical circulation model developed for the CTZ (Haidvogel et al. 1991). In addition, the biological model is coupled with a spectrally dependent, bio-optical model that allows the two phytoplankton size fractions to absorb light at different wavelengths in accordance with their absorption spectra. Thus, the bio-optical model provides time-varying simulations of the spectral characteristics of the underwater light field.

The simulated phytoplankton distributions obtained with the CTZ model adequately reproduce the development and maintenance of the subsurface chlorophyll maximum. Analysis of the terms in the model equations shows that in situ phytoplankton growth is primarily responsible for the creation and maintenance of the subsurface chlorophyll maximum (Moisan & Hofmann, in press). Moreover, the simulated bio-optical fields show that the e-folding scale of the subsurface light field is influenced by zooplankton grazing. Removal of the large phytoplankton cells by the larger zooplankton allows light to penetrate deeper into the water column. This allows the smaller phytoplankton cells, which are deeper in the water column, to bloom and provide a food source for the smaller zooplankton. Since the smaller phytoplankton cells are a suboptimal food source for the larger zooplankton, these animals decrease in abundance over time. Hence, zooplankton grazing alters the phytoplankton distribution, which in turn alters the spectral characteristics of the underwater light field. The modified light field then results in the growth and dominance of a different zooplankton species.

The three-dimensional simulated nutrient and plankton fields obtained from the CTZ model show that the patterns of carbon flux near the surface are similar in shape to those of the meander structure of the filaments formed from the California Current. Moreover, the across-shore area-integrated carbon flux is predominantly offshore and is maximum in the region where the filaments form.

## 3 Discussion and Summary

The models described here provide some examples of the theoretical frameworks that have been developed to understand the processes controlling the distribution and abundance of marine plankton communities. The first of these time- and space-dependent models uses bulk formulations to represent the lower trophic levels of marine food webs. Although these models are relatively simple in design, they provide insight into the processes controlling general patterns in marine plankton abundance. These models also provide a theoretical basis for the development of more-complex marine food-web models as understanding, laboratory technique, and field-data collection advance.

The addition of size and stage structure adds realism to the models and allows the explicit inclusion of important phenomena that are missing from the bulk models. Moreover, the inclusion of size- or stage-dependent population dynamics allows different processes to affect different sizes and stages at different times. This point is well illustrated by the simulations from Steele and Frost (1977), which show that the relative size structure of the herbivores and their food is more important than the total biomass of each trophic level in determining transfers between the phytoplankton and herbivores. Furthermore, in the case of the CTZ simulations, grazing by the zooplankton on the phytoplankton modifies the spectral characteristics of the underwater light field, which in turn modifies the composition of the phytoplankton community and ultimately the zooplankton community composition. These subtle and nonlinear interactions are possible only in models that include multiple types of predators and prey. Prediction of the community structure of marine food webs requires models that include realistic ecosystem structures as well as realistic representations of the physical environment.

Interactions among herbivores or with their food may occur through only a subset of the different life-history stages. Steele and Frost (1977) showed that grazing by certain copepod stages alters the size spectrum of food that is available. An indirect effect of grazing, they noted, is that selective grazing by zooplankton affects the sequestering of nutrients by phytoplankton. In their model, smaller cells have higher specific rates of nutrient uptake, and depletion of the larger cells by grazing allows the nutrients to be used by smaller cells. Hence, when the grazing pressure on the larger cells is relaxed, a bloom does not necessarily occur because of reduced nutrient concentrations. The conclusion from these simulations is that the presence of the smaller copepod, which controls the smaller phytoplankton, is needed to ensure the existence of the larger copepod. Again, this type of subtle interaction is revealed only in models that allow size- and stage-structured population dynamics.

Prediction of plankton community structure and changes in community structure is an important aspect of forecasting the effects of natural or man-induced climatic changes, and it is also an important component of developing management strategies for commercial fisheries. For these applications, where the prediction of the population structure of the lower trophic levels in marine systems is of practical importance, models of marine food webs must include details of the life history or size composition of the plankton components. Stage or size structure can play a potentially crucial role in species interactions and, ultimately, in the relative abundance of species.

The existing models for marine plankton communities contain only those species and life stages that have been observed in a particular region, such as was done for the CTZ model (Moisan & Hofmann, in press; Moisan et al., in press). However, models constructed for investigating changes in species composition that may arise through climatic change will require inclusion not only of species found in a particular region but of those that are not. The model ecosystem must then be comprehensive enough to describe the dynamics of several different species when the environmental and biological interactions vary. The need for increased realism and detail in marine-plankton models, especially marine-zooplankton models, has been highlighted as an area for development as part of the International and U.S. Global Ocean Ecosystem Dynamics (GLOBEC) program (GLOBEC 1993, 1994, 1995), which focuses on understanding the marine population variability in response to physical forcing.

The models developed by Wroblewski (1980, 1982) for the distribution of copepods in the Oregon upwelling region represent a class of models that have not been fully exploited in studies of biological oceanography. The simulated distributions from these models can be portrayed as Lagrangian trajectories, such as those shown in Figure 9. These are constructed by tracking the movement of a particle (e.g., phytoplankton or zooplankton) over space and time. The resulting trajectory includes the effect of circulation as well as the contribution from biological processes such as ontogenetic and diel migration. Thus, the simulated trajectories obtained from Lagrangian models are useful for determining residence times and general transport patterns of marine plankton populations. This approach can be used for different life stages of zooplankton and the larval forms of fish or benthic animals to investigate processes controlling recruitment to these populations.

As a final comment, a critical aspect of improving the structure of marine population models is the availability of data on which to base the models. Without additional measurements, the understanding of how marine systems function will be limited, and as a result, the models designed to investigate these systems will be limited. Development of integrated sampling systems, the concurrent measurement of physical and biological properties, the development of data-assimilation techniques for biological models (see, e.g., Lawson et al. 1995), and advances in model construction are all necessary components for the improvement of marine-plankton models. In particular, experimentalists and modelers must work together in order to allow for information transfer between the two communities. These tasks will require input from numerous individuals, coordinated research programs, and considerable cooperation at national and international levels.

## Acknowledgments

Support for completion of this paper was provided by the Office of Naval Research and the National Aeronautics and Space Administration. Computer resources and facilities were provided by the Center for Coastal Physical Oceanography at Old Dominion University.

## Literature Cited

Bermuda Biological Station. 1960. *The Plankton Ecology, Related Chemistry and Hydrography of the Sargasso Sea.* Final report, Part 3, U.S. AEC Contract AT(30-1)-2078.

Evans, G. T., and M. J. R. Fasham, eds. 1993. *Towards a Model of Ocean Biogeochemical Processes.* NATO ASI Ser. I, Vol. 10. Springer-Verlag, Berlin.

Fasham, M. J. R., H. W. Ducklow, and S. M. McKelvie. 1990. A nitrogen-based model of plankton dynamics in the oceanic mixed layer. *Journal of Marine Research* 48: 591–639.

Fransz, H. G., J. P. Mommaerts, and G. Radach. 1991. Ecological modelling of the North Sea. *Netherlands Journal of Sea Research* 28: 67–140.

GLOBEC. 1993. *Population Dynamics and Physical Variability.* International GLOBEC Report 2. GLOBEC – International Executive Secretary, University of Massachusetts Dartmouth, North Dartmouth, Mass.

———. 1994. *Numerical Modeling.* International GLOBEC Report 6. GLOBEC – International Executive Secretary, University of Massachusetts Dartmouth, North Dartmouth, Mass.

———. 1995. *Secondary Production Modeling Workshop Report.* U.S. GLOBEC Report 13. U.S. GLOBEC Office, University of California, Berkeley, Calif.

Haidvogel, D. B., A. Beckman, and K. S. Hedström. 1991. Dynamical simulations of filament formation and evolution in the Coastal Transition Zone. *Journal of Geophysical Research* 96: 15017–15040.

Hofmann, E. E. 1993. Coupling of circulation and marine ecosystem models. Pp. 136–161 *in* S. A. Levin, T. M. Powell, and J. H. Steele, eds., *Patch Dynamics.* Lecture Notes in Biomathematics 96. Springer-Verlag, Berlin.

Hofmann, E. E., and C. M. Lascara. In press. A review of predictive modeling for coastal marine ecosystems. *In* C. N. K. Mooers, ed., *Coastal Ocean Prediction.* Coastal and Estuarine Sciences, American Geophysical Union, Washington, D.C.

Lawson, L. M., Y. H. Spitz, E. E. Hofmann and R. B. Long. 1995. A data assimilation technique applied to a predator-prey model. *Bulletin of Mathematical Biology* 57: 593–617.

Moisan, J. R., and E. E. Hofmann. In press. Modeling nutrient and plankton processes in the California Coastal Transition Zone 1. A time- and depth-dependent model. *Journal of Geophysical Research.*

Moisan, J. R., E. E. Hofmann, and D. B. Haidvogel. In press. Modeling nutrient and plankton processes in the California Coastal Transition Zone 2. A three-dimensional physical-bio-optical model. *Journal of Geophysical Research.*

Moloney, C. L., and J. G. Field. 1991. The size-based dynamics of plankton food webs. I. A simulation model of carbon and nitrogen flows. *Journal of Plankton Research* 13: 1003–1038.

Moloney, C. L., J. G. Field, and M. I. Lucas. 1991. The size-based dynamics of plankton food webs. II. Simulations of three contrasting southern Benguela food webs. *Journal of Plankton Research* 13: 1039–1092.

Riley, G. A. 1946. Factors controlling phytoplankton populations on Georges Bank. *Journal of Marine Research* 6: 54–73.

————. 1947. A theoretical analysis of the zooplankton population of Georges Bank. *Journal of Marine Research* 6: 104–113.

Steele, J. H., and B. W. Frost. 1977. The structure of plankton communities. *Philosophical Transactions of the Royal Society of London* 280: 485–534.

Thompson, J. D. 1974. The coastal upwelling cycle on a beta-plane: Hydrodynamics and thermodynamics. Ph.D. diss. Florida State University, Tallahassee.

Wroblewski, J. S. 1977. A model of phytoplankton plume formation during variable Oregon upwelling. *Journal of Marine Research* 35: 357–394.

————. 1980. A simulation of the distribution of *Acartia clausi* during Oregon upwelling, August 1973. *Journal of Plankton Research* 2: 43–68.

————. 1982. Interaction of currents and vertical migration in maintaining *Calanus marshallae* in the Oregon upwelling zone–a simulation. *Deep-Sea Research* 29: 665–686.

Wroblewski, J. S., and E. E. Hofmann. 1989. U.S. interdisciplinary modeling studies of coastal-offshore exchange processes: Past and future. *Progress in Oceanography* 23: 65–99.

CHAPTER 14

# Frequency Response of a Simple Food-Chain Model with Time-Delayed Recruitment: Implications for Abiotic-Biotic Coupling

*Bruce C. Monger, Janet M. Fischer,*
*Brian A. Grantham, Vicki Medland, Bing Cai,*
*and Kevin Higgins*

Understanding the complex interactions between the physical environment and biological dynamics is a fundamental goal in ecology (E. Odum 1971; Ricklefs 1979; McIntosh 1985). It has been proposed that the strength of coupling between a fluctuating environment and associated biota depends on the extent of overlap of the time scales of the biotic and abiotic components (Haury et al. 1978; Powell 1989; Steele 1989, 1991). Despite the recognition that the overlap in scale plays an important role in determining coupling, few studies have directly examined specific scale-dependent linkages between biotic and abiotic components.

Mathematical modeling is a useful tool for investigating the linkages between the abiotic and biotic components of ecosystems. Several studies (Nisbet & Gurney 1976, 1982; Steele & Henderson 1984) have addressed the issue of coupling between the environment and biota by using population models of a single trophic level and time-dependent environmental forcing. Nisbet and Gurney (1976, 1982) studied the response of the delayed logistic model to forced sinusoidal variations in the carrying capacity. Steele and Henderson (1984) used a population model that incorporated logistic growth and predator-mediated mortality. They examined the model's frequency response to forced variations in predation.

Models of a single trophic level, such as the two described above, do not always contain adequate details to properly define system dynamics (Carpenter & Kitchell 1984; Hastings & Powell 1991). As a natural extension of the studies by Nisbet and Gurney (1976, 1982) and by Steele and Henderson (1984) on single-trophic-level models, we explore here the dependence on scale of physical-biological coupling using a model of a three-level food chain described previously by Hastings and Powell (1991). This model, although simplistic, crudely approximates a functioning ecosystem, thus providing more realism than single-trophic-level models.

We modify the model to include age structure in the form of a time-delayed development at the second trophic level. The original unforced and undelayed model has been shown to exhibit chaotic dynamics over a broad range of parameter space (Hastings & Powell 1991). The time-delayed model appears to have retained this chaotic behavior, but it also exhibits instability that is expressed by the extinction of the third trophic level when relatively small changes are made in parameter-space values.

The main objective of our effort is to explore the response of the delayed Hastings and Powell model to different frequencies of environmental forcing. The model can be forced from the bottom in a manner similar to that employed by Nisbet and Gurney (1982), by periodically varying the carrying capacity of the lowest trophic level. A direct physical analogy of this type of forcing is the periodic input of nitrate or phosphate into aquatic systems. Alternatively, the model could be forced from the top by periodically varying the specific mortality rate of the top consumer population. This type of model perturbation is similar to that used by Steele and Henderson (1984), who modeled the effects of fluctuations in fishing pressure or environmental factors affecting the survival of top consumers. In our present analysis, we follow Nisbet and Gurney (1982) and examine model response to periodic fluctuations in the carrying capacity of the lowest trophic level. Model parameters are adjusted such that when the carrying capacity is periodically varied, the model either remains well within the stable region of parameter space or crosses a boundary separating stable and unstable regions (referred to as fully and marginally stable models, respectively). We present results of the frequency response of the model for these two cases and discuss the findings within the context of the modeling efforts of Nisbet and Gurney (1976, 1982) and

Steele and Henderson (1984). Furthermore, we discuss the implications of our findings for understanding abiotic-biotic coupling in natural systems.

## 1 Model Background

The model of a three-level food chain, as developed by Hastings and Powell (1991), incorporates logistic growth at the lowest trophic level and type-II (saturating) functional feeding responses for the two consumer levels in the model. Following Hastings and Powell's (1991) model notation, $X$, $Y$, and $Z$ represent, respectively, the species at the first, second, and third trophic level. The coupled differential equations describing their simple food chain are given by the following expressions:

$$\frac{dX}{dT} = R_0 X \left(1 - \frac{X}{K_0}\right) - C_1 F_1(X) Y, \tag{1}$$

$$\frac{dY}{dT} = F_1(X) Y - F_2(Y) Z - D_1 Y, \tag{2}$$

$$\frac{dZ}{dT} = C_2 F_2(Y) Z - D_2 Z, \tag{3}$$

$$F_i(U) = \frac{A_i U}{B_i + U}, \tag{4}$$

where $R_0$ and $K_0$ are the intrinsic growth rate and carrying capacity of the lowest trophic level, respectively; $C_1^{-1}$ and $C_2$ are gross growth efficiencies; and $A_i, B_i,$ and $D_i$ $(i = 1, 2)$ are, respectively, maximal ingestion rates, half-saturation constants, and specific mortality rates for the two consumer populations. Following Hastings and Powell (1991), with $R_0$ and $K_0$ as natural time and biomass scales, respectively, equations (1)–(4) may be rewritten using dimensionless variables, as

$$\frac{dx}{dt} = x(1 - x) - f_1(x) y, \tag{5}$$

$$\frac{dy}{dt} = f_1(x) y - f_2(y) z - d_1 y, \tag{6}$$

$$\frac{dz}{dt} = f_2(y) z - d_2 z, \tag{7}$$

$$f_i(u) = \frac{a_i u}{1 + b_i u}. \tag{8}$$

The relationships between dimensionless parameters ($a_i$, $b_i$, $d_i$, $x$, $y$, $z$, $t$) and dimensional parameters are given in Table 1. Unless

TABLE 1. *Default Values of Dimensionless Constants (Where Applicable) and the Relationship Between Dimensionless and Dimensional Parameters*

| Dimensionless Parameter | Dimensional Relationship | Model Value |
|:---:|:---:|:---:|
| $a_1$ | $K_0 A_1 / R_0 B_1$ | 5.0 |
| $a_2$ | $C_2 A_2 K_0 / C_1 R_0 B_2$ | 0.1 |
| $b_1$ | $K_0 / B_1$ | variable |
| $b_2$ | $K_0 / C_1 B_2$ | 2.0 |
| $d_1$ | $D_1 / R_0$ | 0.4 |
| $d_2$ | $D_2 / R_0$ | 0.01 |
| $m$ | $M / R$ | 0.0 |
| $t$ | $R_0 T$ | variable |
| $\omega$ | $F / R_0$ | variable |
| hz | $F / 2\pi R_0$ | variable |
| $x$ | $X / K_0$ | variable |
| $y$ | $C_1 Y / K_0$ | variable |
| $z$ | $C_1 Z / C_2 K_0$ | variable |

NOTE.—$T$ is dimensional time; $F$, dimensional frequency; $M$, dimensional specific mortality rate for juveniles. See the text for all other dimensional constants.

otherwise stated, the value of the dimensionless parameters used in the present analysis are the same as those given by Hastings and Powell (1991, Table 1).

In order to incorporate stage structure in the model, we use a time delay in adult recruitment to create an invulnerable juvenile class at the second trophic level. This approach is directly analogous to the Nicholson blowfly model of Gurney et al. (1980). Time-dependent carrying capacity, $K(t) = K_0[1 - A\sin(2\pi\omega t)]$, was incorporated according to the model of Nisbet and Gurney (1982). The resulting periodically forced delay-differential expressions used in the present structured model are given by

$$\frac{dx(t)}{dt} = x(t)\left(1 - \frac{x(t)}{1 - A\sin(2\pi\omega t)}\right) - f_1[x(t)]y(t), \qquad (9)$$

$$\frac{dy(t)}{dt} = f_1[x(t-\tau)]y(t-\tau)e^{-m\tau} - f_2[y(t)]z(t) - d_1 y(t), \quad (10)$$

$$\frac{dz(t)}{dt} = f_2[y(t)]z(t) - d_2 z(t) \,, \tag{11}$$

$$f_1[u(t)] = a_i u(t)/[1 + b_i u(t)]. \tag{12}$$

The variables $\tau$ and $m$ represent dimensionless juvenile developmental period and specific mortality rate, respectively (Table 1). The term expressing the recruitment rate of adult $y$'s,

$$f_1[x(t - \tau)]y(t - \tau)e^{m\tau},$$

depends on the abundance of adults and prey at a time equal to $t - \tau$. For the purposes of the present analysis, we set $m$ equal to 0. We vary the carrying capacity sinusoidally at a fixed amplitude ($A$) of 0.1 and prescribed frequency ($\omega$). For clarity, we present variations in carrying capacity in terms of the ratio $K(t)/K_0 = [1 - A\sin(2\pi\omega t)]$. Consequently, $K(t)/K_0$ varies sinusoidally between 0.9 and 1.1. Finally, we drop the explicit expression of the time dependence of $K(t)/K_0$ and use $K/K_0$ with the implicit assumption, unless otherwise stated, of time dependence.

## 2 Model Analysis

Model solutions were obtained using the program SOLVER (Gurney & Tobia 1995). The program uses a fourth-order Runge-Kutta integration method with adaptive time stepping. Integrations were carried out for 10,000 time steps to eliminate transient behavior. Maximum and minimum time steps were set at 0.004 and 0.04, respectively. Tolerable error was set at 0.0001. Every fifth point in the time series was entered into a data file for subsequent spectral analysis and graphing. Initial $x, y$, and $z$ values were set, respectively, at 0.74, 0.17, and 9.8 to be consistent with Hastings and Powell (1991). The initial-history variable (number of juvenile $y$'s) was set at zero at the beginning of each model run.

In order to determine proper parameter configuration to establish fully and marginally stable model configurations, we explore the effects of the values of $b_1$, $K/K_0$, and time delay ($\tau$) on model stability. We accomplish this by evaluating model behavior for a series of model runs in which parameters are varied systematically between successive runs. Model runs in which the top predator persists for 10,000 iterations (i.e., does not become extinct) are considered stable.

We arbitrarily choose an intermediate time-delay value and ex-

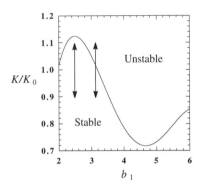

FIGURE 1. *Stability of the unforced time-delayed Hastings and Powell model (eqs. 9–12) expressed as a function of $b_1$ and $K/K_0$ for a fixed time delay of 1.0. The arrows depict paths in parameter space traversed by the forced time-delayed Hastings and Powell model. Stability is functionally defined in terms of the top consumer population persisting for 10,000 time units. See text and Table 1 for explanations of $b_1$, $K/K_0$, and time delay.*

amine the effect of changes in $b_1$ and $K/K_0$ on model stability. To examine the combined effects of changes in the value of $b_1$ and $K/K_0$, we use a factorial approach in which $\tau$ is set at 1.0, $b_1$ is systematically varied from 2.0 to 6.0 by 0.3 increments, and $K/K_0$ ranges from 0.55 to 1.20 by 0.05 increments. Model stability as a function of $b_1$ and $K/K_0$ exhibits a small degree of variability about a general trend. Because we are more interested in finding general regions of model stability and instability, a fourth-order polynomial is fit by least-squares regression through the individual stability points (Fig. 1). On the basis of an analysis of this stability diagram, we fix $b_1$ at 2.5 and 3.1 and vary $K/K_0$ by $\pm 10$ percent to produce fully and marginally stable models, respectively. The arrows in Figure 1 show the parameter-space paths for the two model configurations. To illustrate unforced model behavior for the range of parameter values used in model simulations, a series of model runs is performed for $b_1$ equal to 2.5 and 3.1, where $K/K_0$ equals 0.9, 1.0, and 1.1.

For the fully stable case ($b_1 = 2.5$), the model is run at constant $K/K_0$, with $K/K_0$ varying sinusoidally by $\pm 10$ percent at the following frequencies: 0.1, 0.01, and 0.001 hz (hz = dimensionless frequency; Table 1). For the marginally stable case ($b_1 = 3.1$), the model is run at constant $K/K_0$, with $K/K_0$ varying sinusoidally by $\pm 10$ percent at frequencies of 0.1, 0.05, 0.01, and 0.001 hz.

Output from the fully stable model is analyzed using spectral techniques. Nisbet and Gurney (1982) pointed out that spectral analysis is a powerful tool for describing the behavior of chaotic systems because the positions of dominant spectral peaks are independent of initial conditions and map smoothly across bifurcations. We transformed model output into frequency space using the Fast Fourier Transform utility provided with the SOLVER software package, with smoothing and mean-suppression options activated. We present the power spectra of the time series for only the top predator because its dynamics are qualitatively similar to that of the other trophic levels.

## 3  Results

### Unforced-Model Behavior

The dynamics of the unforced model at fixed values of $K/K_0$ (0.9, 1.0, and 1.1) are illustrated in Figures 2 and 3 for the fully and marginally stable models, respectively. The model remains stable for all values of $K/K_0$ when $b_1$ is equal to 2.5 (Fig. 2) but becomes unstable when $K/K_0 = 1.1$ when $b_1 = 3.1$ (Fig. 3).

### Frequency Response of the Fully Stable Model

The frequency of environmental forcing influences the dynamics of food-chain components in the fully stable model (Fig. 4). Similar dynamics, in terms of the position of the dominant spectral peak, are observed in the unforced system ($\omega = 0.0$ hz) and in the system with high-frequency environmental forcing ($\omega = 0.1$ hz). At an intermediate frequency of environmental forcing ($\omega = 0.01$ hz), the dynamics are more erratic and exhibit large-amplitude fluctuations about the mean. When the forcing frequency is low relative to the natural frequency of the unforced system ($\omega = 0.001$ hz), mean abundance tracks environmental forcing, and moderate fluctuations of a higher frequency are observed about this oscillating mean.

The power spectra of the time series reflect quantitative differ-

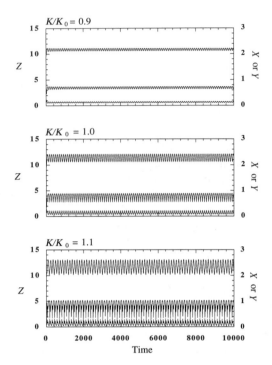

FIGURE 2. *Time series for the unforced fully stable model at equilibrium and at high and low values of $K/K_0$.*

ences in model behavior at the three frequencies of environmental forcing used in this study (Fig. 5). The frequency of the natural (unforced) oscillations in the abundance of the top predator when $K/K_0 = 0.9$, 1.0, and 1.1 are presented in the top panel. There is a moderate peak at 0.0095 hz, a small peak around 0.0098 hz, and a large peak around 0.0077 hz for $K/K_0$ fixed at, respectively, 1.0, 0.9, and 1.1.

Periodically forcing the system at high frequency relative to the natural frequency of oscillation produces a moderate peak at 0.0095 hz (Fig. 5; $\omega = 0.1$ hz). The position of this peak is essentially identical to the peak obtained in the unforced model simulation, where $K/K_0$ was fixed at 1.0 ($\omega = 0.0$ hz). However, the amplitude of this peak is diminished somewhat in comparison with the peak of the unforced case.

When the system is forced near its natural frequency ($\omega = 0.01$ hz), it exhibits a moderate peak at 0.01 hz (a frequency slightly

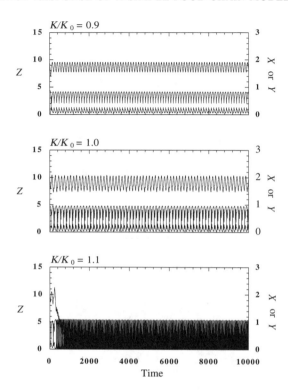

FIGURE 3. *Time series for the unforced marginally stable model at equilibrium and at high and low values of $K/K_0$. Left and right scales correspond to $Z$ and $X$ or $Y$ populations, respectively.*

higher than its natural frequency). A large peak at 0.005 hz and two intermediate peaks at 0.0017 and 0.0034 hz also occur with intermediate forcing frequency.

At a low forcing frequency ($\omega = 0.001$ hz), the model exhibits a moderate peak at the forcing frequency, and two peaks at approximately the same frequencies that occur in the unforced model when $K/K_0$ is fixed at 1.1 and 1.0.

*Frequency Response of the Marginally Stable Model*

The frequency of environmental forcing dramatically influences the long-term behavior of food-chain components in the marginally stable model (Fig. 6). At high forcing frequencies ($\omega = 0.1$ hz), the dynamics are essentially identical to the result obtained for the un-

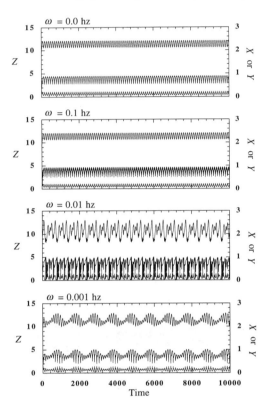

FIGURE 4. *Time series of the fully stable model when unforced* (top) *and when forced sinusoidally at high, intermediate, and low frequencies* (lower 3 panels).

forced model when $K/K_0$ is set at 1.0. The dynamics become much more erratic when the forcing frequency is approximately 5 times the natural frequency ($\omega = 0.05$ hz). At a forcing frequency near or below the natural frequency of oscillation ($\omega = 0.01$ and $0.001$ hz), the model becomes unstable and the top predator becomes extinct before completion of the model run.

## 4 Discussion

*Fully Stable Model*

The effects of environmental forcing on trophic dynamics are undetectable at high forcing frequency. We observe similar dynamics in

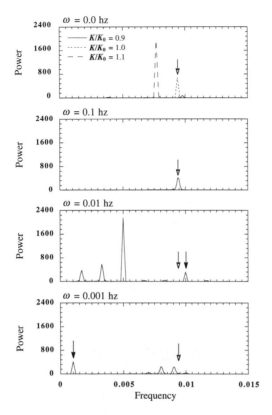

FIGURE 5. *Three separate spectra, overlaid into a single panel* (top), *correspond to an unforced, fully stable model time series in which $K/K_0$ is set at high, low, and intermediate values. Spectra for the fully stable model forced at high, intermediate, and low frequencies are depicted, respectively, in the lower three panels.* Open arrows, *The natural frequency of oscillation (i.e., unforced model with $K/K_0 = 1.0$);* closed arrows, *the forcing frequency. Note that the highest forcing frequency ($\omega = 0.1$ hz) is off the scale.*

the unforced system and in the system forced at approximately 10 times its natural frequency. This result is consistent with previous population-level studies. For example, Nisbet and Gurney (1982) found no detectable response in population abundance when the carrying capacity of a delay-logistic model was varied periodically on a time scale that was short relative to the return time of the population.

The trophic dynamics of the model forced at an intermediate fre-

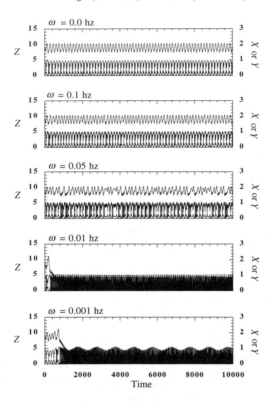

FIGURE 6. *Time series of the marginally stable model forced at high and progressively lower frequencies. Note that 0.01 hz corresponds, approximately, to the frequency of population oscillations in the unforced model (not shown in this figure).*

quency are qualitatively different from those of the unforced model. In the unforced model, the top predator oscillates at 0.095 hz. In model simulations forced at 0.01 hz, however, the top predator population oscillates at the higher forcing frequency (0.01 hz). We believe the underlying cause of this behavior is the synchronization of the unforced population oscillations with the forced oscillations in carrying capacity. Nisbet and Gurney (1982) explained this synchronization with special regard to the ecological consequences.

The forced population also exhibits strong synchronization at a subharmonic ($^1/_5$) of the forcing frequency. The presence of subharmonic oscillations in nature may impede our ability to link observed population fluctuation to underlying environmental fluctuations (Nisbet & Gurney 1976).

At low forcing frequencies, the system tracks environmental fluctuations. As the environment varies, mean abundance oscillates at all three trophic levels. Although a dominant spectral peak is observed at the forcing frequency, we prefer to refer to the behavior of the top trophic level as tracking rather than synchronization. We reserve the term synchronization for situations in which the behavior of the forced system is qualitatively different from the behavior of the unforced model. We choose to refer to the behavior of the low-frequency forcing model as tracking because the top predator continues to oscillate at frequencies near the natural (i.e., unforced) frequencies (as noted by the spectral peaks at 0.008 and 0.009 hz in Fig. 5, $\omega = 0.001$ hz). Nisbet and Gurney (1976) and Steele and Henderson (1984) observed similar results in their single-trophic-level models. Specifically, when the environment is varied slowly, relative to the return of the population, the mean population density tracks the forced environmental variable.

We conclude that the behavior of the fully stable model depends strongly on forcing frequency. Our analysis suggests that the effects of environmental fluctuations are easier to detect when they occur on moderately slow time scales. In contrast, very rapid environmental fluctuations may not greatly influence system behavior.

*Marginally Stable Model*

In the marginally stable model, long-term behavior is influenced dramatically by forcing frequency. Specifically, at high forcing frequencies the model remains stable; at intermediate to low frequencies, the model becomes unstable. The top predator persists at high forcing frequencies because the model passes in and out of the unstable regions of parameter space rapidly. At lower forcing frequencies, however, the length of time spent in the unstable region is sufficient to cause the model to "lock onto" the unstable solution, in which the top predator becomes extinct.

Somewhat similar model behavior has been shown to occur with the model of Steele and Henderson (1984). Their model population is forced by top-down control in predation, but an analogous result would have been obtained if they had employed bottom-up forcing in logistic growth. Consequently, their results can be compared, indirectly, with ours. In the absence of forcing, the Steele and Henderson model exhibits two fundamentally different solutions that depend on model parameter values. The population density in their model drops abruptly from a high to a low level when predation exceeds a critical level. This behavior is analogous to our model's response when $K/K_0$ increases past a critical value. When their model is forced with low-amplitude fluctuations of high frequency about one of the unforced solutions, the population is observed to remain near the unforced solution. Our model is stable at high frequency because the average model solution corresponds to the solution for the stable unforced model when $K/K_0 = 1.0$ (Figs. 3, 6).

Similarities in the behavior of our model and that of Steele and Henderson (1984) also occur at low-frequency forcing. As the forcing frequency is lowered below a critical value, their model abruptly begins to lock alternately onto the two unforced solutions. Their results resemble ours, in the sense that their model exhibits an abrupt shift in behavior at forcing frequencies far below the characteristic time scale of the system. The critical difference in their model is that the population density for large predators is finite, giving their model population an opportunity to recover as predation is subsequently reduced. In our model, the population at the third trophic level becomes extinct and, therefore, cannot exhibit recovery behavior when conditions return to a favorable state. We conclude that forcing frequency determines the long-term behavior of a model in which extinction is one of two alternate solutions. In our model, the top predator persists only when environmental fluctuations are rapid. These findings may have important implications for the conservation of species in fluctuating environments. In addition to identifying environmental conditions that are stressful to threatened species, conservation biologists should consider both the time scale of the environmental fluctuations and the maximum time a population can persist under stressful conditions. It should be emphasized that our model results pertain to a closed system. An open system may be expected to exhibit considerably different dynamics because of recruitment from an influx of individuals from outside the system.

## *Implications for Abiotic-Biotic Linkages in Natural Systems*

We show here that a system's response to environmental fluctuations depends strongly on the frequency of environmental changes. Our findings have important implications for understanding the linkages between environmental fluctuations and the resulting fluctuations in population abundance.

Recent reviews have contrasted abiotic-biotic linkages in disparate ecosystems and explored the extent of overlap in the temporal scales of environmental fluctuations and biological dynamics in marine and terrestrial systems (Steele 1989, 1991). Specifically, Steele argued that the time scales of physical and biological cycles in marine systems are similar, whereas in terrestrial systems the physical processes occur on a shorter time scale than that of biological processes. Thus, he concluded that at a gross level, abiotic and biotic processes are more tightly coupled in marine systems than on land.

As an example of the importance of temporal scale in determining biological responses to abiotic fluctuations, we contrast the responses of primary producers to seasonal environmental fluctuations in terrestrial and marine ecosystems. In forest ecosystems, the life span of the dominant primary producer is long—often a hundred years or more. Consequently, the effects of short-term environmental fluctuations like seasonal temperature changes or fluctuations in nutrient availability are seldom detected in the dynamics of tree populations (H. Odum 1984). In contrast, the dominant primary producers in most aquatic systems have generation times on the order of a couple of days. Indeed, phytoplankton communities often exhibit characteristic patterns that have been related to rapid changes in temperature and in nutrient and light availability (Sverdrup 1953; Denman & Powell 1984; Evans & Parslow 1985).

We manipulate the overlap between abiotic and biotic scales in our model by changing the frequency of environmental forcing while holding all biological parameters constant. Thus, our results are relevant to contrasting abiotic-biotic linkages in marine and terrestrial systems. For the purposes of extending our findings to natural systems, we accept Steele's arguments and consider model simulations with intermediate forcing frequency to be representative of marine environments and high-frequency simulations to be more characteristic of terrestrial systems. We observe synchronization of biological cycling to environmental forcing in the simulation with intermediate forcing frequency. One might expect, therefore, to find evidence of environmental fluctuations in biological time series from marine systems, though subharmonic synchronization may obscure this linkage.

Abiotic and biotic components may be more loosely coupled in terrestrial ecosystems. Our results suggest that when environmental fluctuations occur more rapidly than biological processes, the effects are undetectable in population dynamics. Thus, high-frequency environmental fluctuations may be undetectable in biological time series from terrestrial ecosystems.

These model results suggest that the tendency for marine models to include strong coupling between physical and biological processes, while terrestrial models tend to focus more on biological interactions (Steele 1989, 1991), has a firm empirical and increasingly theoretical grounding. If Steele's arguments hold, one would not expect to improve model fit to field data by incorporating a detailed physical component into a terrestrial model. For marine models, by contrast, one would expect qualitative changes in model behavior upon the inclusion of time-dependent physical forcing. Such changes have been shown to improve the fit between model and field data. For example, by including mechanistic representations of marine circulation into their models of barnacle populations, Roughgarden et al. (1988, 1994) and Possingham and Roughgarden (1990) produced models that provide qualitatively better descriptions of population dynamics than earlier models that did not incorporate the physical environment.

Another consequence of variation in the degree of abiotic-biotic coupling in disparate ecosystems is that ecological generalities will be harder to come by. As suggested by Roughgarden et al. (1994, p. 79), generalization in ecology may consist of simply examining a variety of different systems in order to accumulate "a small collection of big particulars." Since the time scales of the physical parameters, as well as the biological components of many ecosystems, can be expected to vary widely within, as well as across, ecosystems, it may be difficult to produce general models that adequately describe population and community processes in all but particular cases.

## 5 Conclusions

This study lays the groundwork for further investigation of abiotic-biotic coupling in moderately complex food-chain models. Future modeling efforts may expand on our analysis by exploring the frequency response of the model in more detail and by further examining the effects of time delay on model behavior. Abiotic-biotic coupling in disparate environments could be addressed more explicitly by incorporating more-realistic parameter values for biological components in the model, forcing the model using variance spectra of red or white noise rather than sine waves (e.g., Steele 1985), or using time series from real physical data to force the model. Moreover, analysis of abiotic and biotic time series from disparate ecosystems may help elucidate complex interactions between the physical environment and biological dynamics.

The importance of identifying relevant "natural" population time scales is clear in this study. A knowledge of these time scales, as well as data on the time scales of environmental phenomena, is important for predicting possible couplings between population and environmental changes. Studies aimed at understanding how trophic and population structures influence natural time scales may also provide new insights into abiotic-biotic coupling.

To our knowledge, these results represent the first explicit documentation of synchronization of a forced food-chain model. In our model of three trophic levels, we periodically force the lowest trophic level and assess the frequency response of the highest trophic level. Consequently, the effects of environmental forcing must pass through two trophic levels, with associated chaotic dynamics, relatively unaltered. It may be concluded, therefore, that the addition of trophic complexity does not necessarily lead to greater intractability of the linkage between environmental fluctuations and population dynamics. As a parting caveat, we note that linking observed population fluctuations with environmental variability will be problematic if the strong subharmonic synchronization observed in our model response to periodic environmental forcing is prevalent in natural systems.

## Literature Cited

Carpenter, S. R., and J. F. Kitchell. 1984. Plankton community structure and limnetic primary production. *American Naturalist* 124: 159–172.

Denman, K., and T. Powell. 1984. Effects of the physical processes on planktonic ecosystems in the coastal ocean. *Oceanography & Marine Biology Annual Review* 22: 125–168.

Evans, G. T., and J. S. Parslow. 1985. A model of plankton cycles. *Biological Oceanography* 3: 327–347.

Gurney, W. S. C., and S. Tobia. 1995. SOLVER A program template for initial value problems expressible as sets of ordinary or delay differential equations. STAMS, University of Strathclyde, Glasgow.

Gurney, W. S. C., S. P. Blythe, and R. M. Nisbet. 1980. Nicholson's blowflies revisited. *Nature* 287: 17–21.

Hastings, A., and T. Powell. 1991. Chaos in a three-species food chain model. *Ecology* 72: 896–903.

Haury, L. R., J. A. McGowan, and P. H. Wiebe. 1978. Patterns and processes in the time-space scales of plankton distributions. Pp. 277–327 *in* J. H. Steele, ed., *Spatial Patterns in Plankton Communities*. Plenum, New York.

McIntosh, R. P. 1985. *The Background of Ecology: Concepts and Theory.* Cambridge University Press.

Nisbet, R. M., and W. S. C. Gurney. 1976. Population dynamics in a periodically varying environment. *Journal of Theoretical Biology* 56: 459–475.

———. 1982. *Modeling Fluctuating Populations.* Wiley, New York.

Odum, E. 1971. *Fundamentals of Ecology.* Saunders, Philadelphia.

Odum, H. T. 1984. Summary: Cypress swamps and their regional role. Pp. 416–443 *in* K. C. Ewel and H. T. Odum, eds., *Cypress Swamps.* University of Florida Press, Gainesville.

Possingham, H. P., and J. Roughgarden. 1990. Spatial population dynamics of a marine organism with a complex life cycle. *Ecology* 71: 973–985.

Powell, T. 1989. Physical and biological scales of variability in lakes, estuaries, and the coastal ocean. Pp. 157–176 *in* J. Roughgarden, R. M. May, and S. A. Levin, eds., *Perspectives in Ecological Theory.* Princeton University Press, Princeton, N.J.

Ricklefs, R. E. 1979. *Ecology.* Chiron, New York.

Roughgarden, J., S. Gaines, and H. P. Possingham. 1988. Recruitment dynamics in complex life cycles. *Science* 241: 1460–1466.

Roughgarden, J., T. Pennington, and S. Alexander. 1994. Dynamics of the rocky intertidal zone with remarks on generalization in ecology. *Philosophical Transactions of the Royal Society of London B, Biological Sciences* 343: 79–85.

Steele, J. H. 1985. Comparison of terrestrial and marine ecological systems. *Nature* 313: 355–358.

———. 1989. The ocean landscape. *Landscape Ecology* 3: 185–192.

———. 1991. Can ecological theory cross the land-sea boundary? *Journal of Theoretical Biology* 153: 425–436.

Steele, J. H., and E. W. Henderson. 1984. Modeling long-term fluctuations in fish stocks. *Science* 224: 985–987.

Sverdrup, H. U. 1953. On the conditions for the vernal blooming of phytoplankton. *Journal du Conseil Conseil International pour l'Exploration de la Mer* 18: 287–295.

# CHAPTER 15

# Stochastic Demography for Conservation Biology

## C. S. Nations and M. S. Boyce

"The use of deterministic rather than stochastic models can only be justified by mathematical convenience."—Maynard Smith 1974

Models used in conservation biology often include demographic structure, but usually these models are deterministic. Stochastic models receive less attention perhaps because of data limitations or the intimidating appearance of the computations. Nevertheless, we believe that stochastic demography needs to be given more consideration by conservation biologists both because stochastic processes are fundamental to extinction and because population projections from stochastic models may differ substantially from those of deterministic models.

It is our objective to provide a brief review of stochastic demography and to consider possible applications. We focus on simple matrix models of exponential growth in which vital rates vary randomly. Some analytical results exist for these models, and understanding these may be helpful in comprehending the behavior of more-realistic but complex models. These models include some important properties. First, in stochastic environments, population size assumes a skewed distribution in which very large populations occur only rarely. As a consequence of this skewness, mean population size tends to overestimate most populations. Second, stochastic growth rate is lower than deterministic growth rate based on average vital rates. Ignoring temporal variation in estimates of demographic parameters may lead to serious overestimates of actual growth rate. Third, increased variation in the vital rates produces a lower growth rate. Fourth, variation in vital rates with high sensitivity has greater impact on stochastic growth rate; that is, deterministic sensitivity analysis has implications for stochastic dynamics. Finally, and not surprisingly, lower population growth rates translate into increased extinction probabilities. Yet, a finite probability of extinction exists even when the growth rate is positive. Furthermore, increased variation in growth rate is associated with a higher extinction probability and shorter persistence times for the population.

## 1 Deterministic Demography

*Matrix Structure*

Matrix models incorporate details of life history that may be fundamental to understanding population dynamics. In particular, matrix models can be used to examine the effects of variation in the demographic parameters of survival and fecundity on population-level parameters such as growth rate and extinction probability. This has relevance to management if the demographic parameters in question are amenable to manipulation. We refer to the general stage-based projection matrix, in which any element may be nonzero, representing a stage-specific vital rate, such as growth, survival, or fecundity. The age-based Leslie (1945) matrix is a special case of the stage-based matrix.

Deterministic population growth can be described by a discrete-time recursive equation,

$$\mathbf{n}_{t+1} = \mathbf{X}\mathbf{n}_t, \tag{1}$$

where $\mathbf{X}$ is the matrix composed of stage-specific transition probabilities, and $\mathbf{n}_t$ is a column vector representing the number or proportion of individuals in each stage class at time $t$. Discrete time may not be justifiable for some biological populations, although in the case of birth-pulse populations (Caughley 1977), it may be a better representation than simple continuous-time models. Equation (1) leads to asymptotically exponential growth at a rate $\lambda$, with certain extinction if $\lambda < 1$.

Previous conservation applications of deterministic demography have suggested that when field estimates of vital rates yield $\lambda < 1$, more demographic information should be gathered or more effort should be expended on conservation (Anderson & Burnham 1992). The projection of stationary or increasing growth is thought to be more reassuring (except when dealing with an undesired species).

*Sensitivity Analysis*

Sensitivity analysis in the context of demography involves estimating the change in population growth rate resulting from changes in vital rates (Caswell 1978). Because sensitivities depend on the magnitude of vital rates, these are often normalized to create "elasticities" (de Kroon et al. 1986).

To manage for positive growth in populations that are rare or have economic value, it seems wise to invest most effort in demographic parameters that have the largest potential impact on the growth rate (Goodman 1980). For example, the growth rate of loggerhead sea turtles is more sensitive to the survival of juveniles than to the survival of eggs or hatchlings (Crouse et al. 1987). Thus, minimizing juvenile mortality by installing devices to exclude turtles from fishing trawls may be more effective than massive efforts aimed at protecting eggs on beaches. Similarly, adult survival is more important for many long-lived species with low reproductive rates, such as spotted owls (Lande 1988*b*) and grizzly bears (Shaffer 1981). For these species, an investment of time and resources in maximizing reproductive output may not be the most effective strategy. The practical application of sensitivity analysis must also be tempered by a consideration of which demographic parameters can be feasibly manipulated (Goodman 1980).

The sensitivity or elasticity of a vital rate may not reflect the variable's contribution to population fluctuations, simply because any particular vital rate may not exhibit much variation. The contribution of a demographic parameter to variation in $\lambda$ is characterized by its variance times the sensitivity squared, adjusted for covariances among vital rates (Brault & Caswell 1993, eq. 15).

## 2 Stochastic Demography

The assumption of constant vital rates is usually inappropriate. There is variation in the real world, and much of it is sufficiently unpredictable to be modeled as random or stochastic behavior. The inclusion of stochasticity complicates the analysis of population growth and persistence but does not render it intractable.

Shaffer (1981) categorized four types of stochasticity that might affect population dynamics: genetic, demographic, environmental, and catastrophic. Although this scheme is often cited, the distinctions among the four types are murky (Simberloff 1988; Dennis et al. 1991). A nearly formal classification excluding genetics is based on the relationship between the mean and variance of population size: demographic stochasticity is important for small populations, but its effects decrease in magnitude with increasing size; population variability induced by environmental stochasticity is independent of population size; and with catastrophes, the coefficient of variation of population size actually increases as size increases (Dennis et al. 1991). Genetic stochasticity is expected to be most important in small populations, so its effects may aggravate demographic stochasticity. Furthermore, the ultimate effect of genetic stochasticity is demographic in character; that is, a decline in fitness due to inbreeding or loss of heterozygosity may be measured by declines in fecundity and survival (Goodman 1987a; Lande 1988a). Note, however, that genetic viability and demographic viability are fundamentally distinct; results from one analysis are not directly transferable to the other (Ewens et al. 1987).

We focus on so-called environmental stochasticity. This approach is partly justified both by simulation and by analytical results (Simberloff 1988), indicating that environmental stochasticity is likely to have greater impact than demographic stochasticity except in extremely small populations. In addition, observations of natural populations indicate that demographic stochasticity is insufficient to account for rates of extinction (Leigh 1981). With environmental stochasticity, it can be assumed that all individuals in the population are subject to the same variation impinging directly on stage-specific vital rates. Then, $\mathbf{X}$ in equation (1) becomes $\mathbf{X}_t$, effectively a new matrix of transition probabilities at each time interval.

## Lognormal Distribution of Population Size

For stochastic exponential-growth models, population size follows a lognormal distribution (Lewontin & Cohen 1969; Tuljapurkar & Orzack 1980). As time progresses, skewness increases. The increasing skewness is produced because favorable, albeit rare, sequences of environments lead to populations much larger than those that result from more-probable environmental sequences. Consequently, mean population size may not offer an adequate description of the distribution—most populations are smaller than the mean. The importance of this phenomenon in determining growth rate and extinction probability is not always fully appreciated (see, e.g., Burgman & Lamont 1992).

## Stochastic Growth Rate

For unstructured populations, the one-period growth rate is $L_t = N_{t+1}/N_t$, where $N_t$ is the population size at time $t$, and the growth rate of the mean population is $\bar{L} = \Sigma L_t/t$. The most probable growth rate is the geometric mean of observed one-period growth rates, $(\Pi L_t)^{1/t}$, which is always less than the arithmetic mean (Lewontin & Cohen 1969).

For structured populations, the most probable growth is also less than the growth rate of expected population size (Tuljapurkar 1989). However, because of the multiplicative properties of matrices (Furstenberg & Kesten 1960), it is no longer true that the geometric mean adequately describes growth rate. Tuljapurkar and Orzack (1980) described a measure for long-run growth rate:

$$a = \lim_{t \to \infty} \left[ \mathrm{E}(\ln N_t) \right]/t \tag{2}$$

(Cohen 1979 used the notation $\ln \lambda$ rather than $a$). The exact solution of $a$ is difficult even for simple matrices, and the computations are not general.

Tuljapurkar (1982$a,b$) developed a "small-noise" approximation of $a$. When variation in the matrix elements is relatively small and there is no serial correlation in the environment,

$$a \approx \ln \lambda_0 - \tau^2/2\lambda_0^2, \tag{3}$$

$$\sigma_a^2 \approx \tau^2/\lambda_0^2, \tag{4}$$

$$\tau^2 = \sum_{i,j} \left( \partial \lambda_0 \partial x_{i,j} \right)^2 \mathrm{var}(x_{i,j})$$
$$+ \sum_{(i,j),(k,\ell)} \frac{\partial \lambda_0}{\partial x_{i,j}} \frac{\partial \lambda_0}{\partial x_{k,\ell}} \mathrm{cov}(x_{i,j}, x_{k,\ell}), \tag{5}$$

where $\lambda_0$ is the dominant eigenvalue of the average matrix $\bar{\mathbf{X}}$ and the $\partial \lambda_0/\partial x_{i,j}$ are sensitivities (Caswell 1978). This set of equations has a straightforward interpretation. It can be seen that $a$ is reduced relative to $\lambda_0$ by (1) the greater sensitivity of $\lambda_0$ to the average vital rates and (2) greater variances and covariances among the vital rates. Furthermore, factors that reduce $a$ tend to increase $\sigma_a^2$. Lande and Orzack (1988) found that the small-noise approximation works well with the coefficient of variation (the ratio of standard deviation to mean) in vital rates as large as 38 percent.

Given an empirical or simulated time series of population counts
($N_t$) generated by a stochastic process, one can estimate long-run
growth rate (Cohen et al. 1983; Heyde & Cohen 1985). For a period
of $T$ years,

$$\hat{a} = \frac{\ln N_T - \ln N_1}{T - 1}. \tag{6}$$

Heyde and Cohen (1985) provided a variance estimator that is
cumbersome but not particularly difficult. A simpler estimator for
var($\hat{a}_t$) (Cohen et al. 1983) is based on the collection of one-period
growth rates: $\hat{a}_t = \ln N_{t+1} - \ln N_t$, for $t = 1, \dots, T$. Both $\hat{a}$ and $\sigma_a^2$
are functions of total population size and time; neither estimator
uses the vital rates or the projection matrices. Dennis et al. (1991)
used this formulation to compute similar maximum-likelihood estimates based on population counts. Heyde and Cohen (1985) and
Dennis et al. (1991) applied the method to fairly short time series (9- to 50-sample intervals). For an endangered species, even as
much as 50 years of data is extraordinary.

Using estimates of growth rate and its variance, Heyde and Cohen (1985) projected population trends. Even for short-term projections, confidence intervals are extremely wide, partly because of
limited data and partly because the estimator for $\sigma$ may be in-
efficient, according to Heyde and Cohen. Short-term projections
and confidence intervals were also addressed by Gerrodette et al.
(1985).

Cohen et al. (1983) considered another possible measure for long-
run growth rate:

$$\mu = \lim_{t \to \infty} \left[ \ln E(N_t) \right] / t. \tag{7}$$

When successive environments are independent, $\mu$ is the natural
logarithm of the dominant eigenvalue of the average projection
matrix; that is, $\mu = \ln \lambda_0$, with $\lambda_0$ from equations (3) and (4).
Furthermore, it should be clear from (3) and (4) that $\mu \geq a$. That
is, in stochastic environments, the most probable growth rate, $a$,
is always less than the growth rate of the mean population size,
$\mu$. This inequality follows from the logarithmic relationship and
Jensen's rule (Karlin & Taylor 1975).

It might appear that $a$ and $\mu$ are similar to the arithmetic and geometric means for non-structured populations. But the similarity
is superficial, because neither $\mu$ nor $a$ has any direct correspondence with the series of eigenvalues of observed matrices. Both the
arithmetic and geometric means of a series of eigenvalues are unre-

liable measures; Cohen (1979) has shown that the former may be either less than or greater than $a$.

## Comparative Statics

We are often tempted to seek a simple explanation for the reduced long-term growth trajectory for stochastic populations. The concavity in the deterministic growth rate, ln $\lambda$, as a function of vital rates, might account for the decline in stochastic long-term growth rates as the variance of vital rates increases (Boyce 1977). But this interpretation applies only where a vital rate is sampled over a large number of populations at one point in time. When age-structured populations are projected through time, however, ln $\lambda$ is not a useful measure of long-term population growth (Cohen 1979), as we have seen.

A connection between these "statics" and dynamics was formalized by Tuljapurkar (1982a). Numerical examples may help to illustrate. Consider a Leslie matrix model with five age classes for the great tit (sensu Meyer & Boyce 1994). Let adult survival $(s_a)$ = 0.38 and fecundity $(f)$ = 0.64. The deterministic growth rate, $\lambda$, is only slightly less sensitive to survival $(\partial\lambda/\partial s_a = 0.96)$ than to fecundity $(\partial\lambda/\partial f = 1.02)$. In considering sensitivities alone, it appears from equations (3) and (5) that the impact on $a$ would be greater for a given magnitude of variation in $f$ than in $s_a$. Assume that there is no environmental autocorrelation, so that $\mu = \ln \lambda_0$, and var$f$ = var$s_a$ = 0.03. Then, the small-noise approximation shows that $a$ is smaller when variation is imposed on $f$ $(a = 0.0016)$ than when imposed on $s_a$ $(a = 0.0026)$.

Extinction probability also depends on which vital rate is varying. Assume an initial population of 200 and an extinction threshold of 10. When var $f$ = 0.03, the probability of ultimate extinction is 0.64. The same variance of $s_a$ results in a lower extinction probability, 0.46. When $f$ varies, most populations become extinct more rapidly, but the distribution of extinction times is more skewed, reflecting larger $\sigma_a^2$: the mode is 135 years, and the mean is 1,824 years. Variation in $s_a$ produces a mode of 146 and a mean of 1,157.

It is also helpful to compare different life histories. Van Sickle (1990) examined the effects of random variation in juvenile survival on the long-run growth rate of large African ungulates. For two hypothetical life histories with similar vital rates—one with a life span of 10 years, and one of 20 years—the impact on $a$ is greater in the shorter-lived animals.

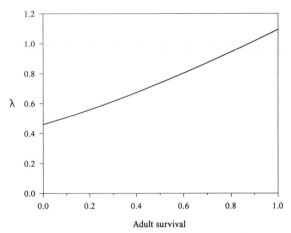

FIGURE 1. *Deterministic growth rate, $\lambda$, as a convex function of adult survival for a two-stage class model for the Everglades kite. Nonlinearities in $\lambda$ can complicate the outcome of population projections. A relative scale for survival is created by dividing values used in simulations by the mean survival observed in a field study (Nichols et al. 1980), thereby scaling the observed value to 1.*

These observations confirm a more general result by Tuljapurkar (1982$a$) that the effect of variation in a vital rate depends on mean generation length, $T$. By definition, $T = \langle \mathbf{w}, \mathbf{v} \rangle$, the dot product of the right ($\mathbf{w}$) and left ($\mathbf{v}$) eigenvectors (Leslie 1966). This quantity appears in the growth-rate sensitivities:

$$\frac{\partial \lambda}{\partial x_{i,j}} = \frac{v_i w_j}{\langle \mathbf{w}, \mathbf{v} \rangle} . \tag{8}$$

Decreases in mean generation length tend to increase sensitivities. The effect may be more pronounced for some vital rates, such as juvenile survival in this case. The general relationship between generation length and sensitivity is also consistent with the observation that sensitivity to vital rates tends to be greater in age-based models than in similar stage-based models without senescence (Meyer & Boyce 1994). Therefore, when vital rates vary randomly, age-based models assuming finite life spans should predict higher extinction probabilities and shorter persistence times than similar stage-based models.

A final example serves to illustrate the potential difficulties in drawing conclusions from comparative statics. In the Everglades kite, $\lambda$ is more sensitive to adult survival than to other vital rates

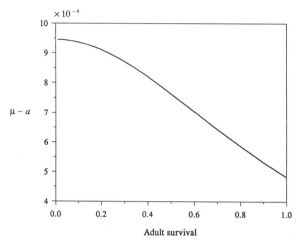

FIGURE 2. *Impact on growth rate, $\mu - a$, plotted against adult survival for the Everglades kite. Variance in survival is constant (SD = 0.04). Note increase in $\mu - a$ at low survival, contrary to expectation from sensitivity shown in Figure 1.*

(Nichols et al. 1980), as is true for many long-lived species. A stage-based model might be constructed as follows:

$$\begin{bmatrix} 0 & f \\ s_0 & s_a \end{bmatrix}, \tag{9}$$

where $s_0$ is juvenile survival. In this case, $\lambda$ is a convex function of $s_a$ (Fig. 1). Convexity implies declining sensitivity with declining vital rate. For example, observed values of $s_a$ yield $\partial\lambda/\partial s_a = 0.85$. If $s_a$ decreases to 20 percent of its normal value, $\partial\lambda/\partial s_a = 0.60$. One might expect this decreased sensitivity to result in a smaller impact on $a$. One way to view the effect on $a$, especially over a wide range of vital rates, is to examine the difference, $\mu - a$. This difference should decrease in magnitude when a constant variance is imposed on progressively smaller values of $s_a$.

This is not the case, however, because $\mu - a$ is inversely related to $s_a$ (Fig. 2). The reason for this behavior may be discerned by reexamining equation (3). The value of $a$ relative to $\mu$ is affected not only by the sensitivities and variances summarized in $\tau^2$, but by $\lambda_0^2$. As expected, $\tau^2$ does decrease with decreasing $s_a$, but $\lambda_0^2$ decreases to a greater degree (Fig. 3), ensuring that $\tau^2/\lambda_0^2$ increases. The quotient $\tau^2/\lambda_0^2$ may be an even more complex function of vital rates, affected by the nonlinearities in both components. This

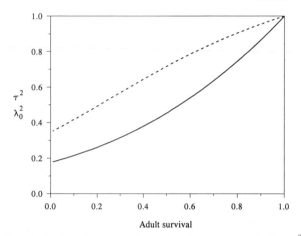

FIGURE 3. *Effect of stochastic variation in adult kite survival on* $\lambda_0^2$ *(solid line) and* $\tau^2$ *(dashed line). Both parameters normalized by their value at baseline survival.* $\lambda_0^2$ *declines more rapidly than* $\tau^2$ *with decreases in survival, accounting for the relationship shown in Figure 2. (Same calculations used to generate both figures.)*

should serve as a caution against predicting the effects of stochastic variation only on the basis of the concavity or convexity of $\lambda$ as a function of vital rate.

The crucial distinction here lies in the comparisons one wishes to make. The relative sensitivities may have importance in two situations: in comparing different vital rates for a single life history, and in comparing the same vital rates in different life histories (Tuljapurkar 1982a). The first situation corresponds to the great tit example above, the second to Van Sickle's comparison of the effects of different life spans on juvenile survival in ungulates. It does not follow that comparing the effects of different values of the same vital rate will be as meaningful.

## 3 Applications

*Sensitivities*

In stochastic environments, effective management depends not only on sensitivity and ease of manipulation, but also on natural variability exhibited by different vital rates. Goodman (1987a,b) suggested that controlling the variance of growth rate should be effective in reducing extinctions (we wish we could think of an example!). Equations (3) to (5), here, indicate that the variance of vital rates is a critical component of growth-rate variance. Thus, controlling vital-rate variance may be as important as management based on sensitivities. Indeed, it is the product of the variance and the squared sensitivity that determines the contribution of a particular vital rate to the growth rate, $a$, and $\sigma_a^2$, parallel to the deterministic case discussed above. Given that various extinction parameters in turn depend on growth rate and its variance (see below), population management should focus on those vital rates for which this product is largest.

*Extinction*

> "They must die out, because Time fights against them."
> —Buffon (1707–1788)

Stochastic variation in population growth obviously poses a threat to the persistence of a population; relatively large declines can quickly eliminate all members of a population or reduce the population to a level from which it cannot recover. The relationship between long-run growth rate and extinction probability is more complex, however. Using a diffusion approximation and assuming a stationary process, Tuljapurkar and Orzack (1980) showed that the probability of ultimate extinction is

$$\pi = \begin{cases} 1 & \text{for } a \leq 0, \\ \exp(-2ax_{\mathrm{d}}/\sigma_a^2) & \text{for } a > 0, \end{cases} \qquad (10)$$

where $x_{\mathrm{d}}$ is the difference between the initial population size ($N_0$) and extinction threshold size ($N_{\mathrm{c}}$), on a log scale (i.e., $x_{\mathrm{d}} = \ln N_0 - \ln N_{\mathrm{c}}$; Lande & Orzack 1988). Extinction may occur even when $a$ is positive (eq. 10). Just as important, increases in growth-rate variance increase the extinction probability in direct proportionality.

When $a > 0$, for those populations that are driven to extinction, the time to extinction can be described by the conditional distribution of extinction times (Lande & Orzack 1988). This inverse Gaussian distribution is strongly right-skewed, with the mean greater than the mode, and is thus qualitatively similar to the lognormal distribution of population size. Simulation studies typically show right-skewed distributions for population persistence times (see, e.g., Shaffer & Samson 1985; Stacey & Taper 1992). Characteristics of the distribution were described in detail by Lande and

Orzack (1988) and Dennis et al. (1991). Of particular interest, the mean time to extinction is

$$\bar{t} = x_d / |a| \, ; \tag{11}$$

the mode (i.e., the most probable time of extinction) is given by

$$t^* = \frac{x_d}{|a|} \left\{ \left[ 1 + \frac{9}{4} \left( \frac{\sigma_a^2}{x_d |a|} \right)^2 \right]^{1/2} - \frac{3\sigma_a^2}{2 x_d |a|} \right\}, \tag{12}$$

when $a \neq 0$. Both the mean and the mode decrease with increases in the absolute value of $a$. This appears odd because it means that large positive growth rates are associated with rapid extinctions. Although few population trajectories end in extinction, those few tend to become extinct early, while population sizes are small. Note in equation (12) that $t^*$ is a decreasing function of $\sigma_a^2$ (though the strength of this effect diminishes as $|a|$ increases). Thus, increases in growth-rate variance not only increase the probability of extinction but also decrease the probable lifetime of populations that do become extinct. Using a birth-and-death model, Goodman (1987a) also found that population persistence time is strongly dependent on growth-rate variance. For the models considered here, analytical approximations and simulation results produce similar values for the conditional probability of extinction (Lande & Orzack 1988), obviating the need for time-consuming simulations.

Calculations become messy when taking into account autocorrelation in the environment. For annual breeders, environmental autocorrelation may be relatively unimportant because weather conditions typically have low autocorrelation over periods of a year or more (Lande 1987). Furthermore, autocorrelation generally has a small effect on $a$ (Tuljapurkar 1982b). To some extent, this effect is balanced because stage structure induces its own serial correlation in growth rate. Nevertheless, a positive serial correlation may have a large effect on $\sigma_a^2$ (Tuljapurkar 1982b; Tuljapurkar & Orzack 1980; see also "Comparative Statics," above). In simulations, a large positive autocorrelation increases the probability of extinction and decreases the extinction time (Kalisz & McPeek 1993).

## Harvest Policy

Harvest policies designed under the assumption of a deterministic environment may have undesired consequences when conditions are variable (Goodman 1981). Both nonselective harvests and adult harvests reduce adult survival. If there are compensatory increases in fecundity and juvenile survival, a population is especially sensitive to stochastic fluctuations in the environment, since juvenile survival and fecundity tend to be more closely tied to environmental variation than is adult survival (Charnov 1986). Indeed, harvested populations are more vulnerable even without such compensatory processes. The reduction in mean generation length by itself lowers $a$ relative to $\lambda_0$ (Tuljapurkar 1982$a$), so there is the potential for stochastic and deterministic analyses to yield different lessons for management.

*Limitations*

The effects on growth rate associated with vital-rate sensitivities may be small, especially in comparison with growth-rate variance. Furthermore, sampling variance may be overwhelming (Daley 1979), resulting in a low power of analysis (Taylor & Gerrodette 1993). For certain life histories in some environments, the reduction in $a$ relative to $\mu$ may seem inconsequential. If $\mu$ is much greater than 0, then in many cases, $a$ is also large and positive. But when $\mu \approx 0$ or only slightly positive, $a < 0$, with high probability. In any event, the potential impacts of stochasticity on growth rate and extinction probability should not be ignored.

*Density dependence.*    An obvious shortcoming of the models discussed above is the absence of density-dependent population regulation. Most structured-population models addressing density dependence assume a deterministic environment. In discrete time, these models can show a range of behaviors from stability to chaos (see, e.g., Beddington 1974; Guckenheimer et al. 1977; Desharnais & Cohen 1986). Analysis is greatly complicated by the addition of stochasticity.

Density dependence has been modeled as a ceiling to population size in populations without age structure (Lande 1993; Foley 1994), and generalizations of extinction parameters to structured-population models may be possible. The more general formulation by Tuljapurkar and Semura (1979) based on a Lyapunov function may also hold promise for structured-population models. Simulation models are certainly the easiest alternative, a view long championed by Shaffer (1981; Shaffer & Samson 1985).

Generalizing from the results of simulations can be risky, but, reassuringly, recent reports confirm, in a broad sense, some analytical results of exponential models. For example, Emlen and Pikitch (1989) examined the sensitivities of mean population size and extinction probability to changes in vital rates for hypothetical life histories. For small animals with low survival and high fecundity, both population size and extinction probability are most sensitive to changes in juvenile survival. Conversely, in animals with high survival and low fecundity, adult survival is the most sensitive vital rate.

Yet, fundamental differences can exist in the structure of density-dependent and exponential-growth models. For example, stochastic variation in vital rates results in normally distributed population sizes in density-dependent populations (Pielou 1977), in contrast to lognormally distributed populations in exponential-growth models. Consequently, methods for characterizing the population trajectory for exponential-growth models (Heyde & Cohen 1985; Dennis et al. 1991) may be biased when applied to density-regulated populations.

Extinction dynamics can also be quite different, since density dependence dampens stochastic perturbations that can greatly reduce the risks of extinction (Ferson et al. 1989). Conversely, the functional form and strength of density dependence may aggravate extinction probabilities (Ginzburg et al. 1990); that is, deterministic instability may interact with random fluctuations. As in exponential models, mean size often provides an inadequate description of population trajectory (Ferson et al. 1989), and the time to extinction is skewed to the right (Akçakaya 1992).

*Spatial considerations.* Metapopulation structure adds an even greater complexity to the structured-population models dealt with here. Direct analysis is all but impossible. General results from simulations are harder to obtain, partly because of the increase in parameter number. Metapopulation models without stochasticity may exhibit multiple equilibria, thresholds, and breakpoints (Hanski 1991). Random variation in combination with such behavior can yield wondrously complex dynamics. In some cases, population persistence is more sensitive to various dispersal parameters than to fecundity or survival (see, e.g., Doak 1989). Some studies show that dispersal among subpopulations tends to increase stability, that is, to decrease the extinction probability and increase the persistence time of the entire population (Burkey 1989). However, dispersal may be destabilizing, depending both on the magnitude of dispersal and on which stage class is dispersing (Davis & Howe 1992).

## 4 Conclusion

Conservation for an endangered species may be as straightforward as preserving critical habitat. But in some cases, opportunities for habitat preservation may be limited by economics or politics. Or the population may be limited by factors other than available habitat, leading to the needs for both greater understanding of an organism's ecology and management that is more active. It is under these circumstances that stochastic demography may have useful applications.

Neglecting density dependence and spatial structure, as well as species interactions, may be unacceptable. But this loss of structure is accompanied by a gain in generality that facilitates our understanding of population processes. Examinations of relatively simple models is justifiable on several grounds.

First, in some instances, simple models may capture the dynamics of real populations for short periods (Cohen et al. 1983; Dennis et al. 1991). Exponential models cannot be expected to provide useful projections for long periods, but most applications to conservation biology should probably focus on short-term projections anyway (Burgman et al. 1993).

Second, these models have heuristic value by helping us to understand the mechanisms driving population dynamics. Consider again the interaction between vital-rate sensitivity and variance, and its contribution to stochastic growth rate. Such analyses indicate which demographic parameters are most important given particular conditions. Sensitivity analysis advanced by Caswell (1978) may not apply when, for instance, vital rates are nonlinear functions of density. Yet, information gained under the assumption of exponential growth may be useful in determining appropriate management strategies. Furthermore, depending on the form and strength of the density function, predictions about population behavior may be qualitatively similar to those based on exponential models. For example, extinction probabilities given by both types of model may be similarly sensitive to changes in certain vital rates.

Finally, simpler exponential models may provide a first step in the development of population viability analyses for adaptive management (Boyce 1992, 1993). As better information becomes available through management experiments and further study, model refinements could include ecological complications such as spatial structure or species interactions. Such refinements will almost certainly require simulation models that are specific to the species in question. Stage-structured population models are readily modified to handle these complications.

## Literature Cited

Akçakaya, H. R. 1992. Population viability analysis and risk assessment. Pp. 148–157 *in* D. R. McCullough and R. H. Barrett, eds., *Wildlife 2001: Populations*. Elsevier, New York.

Anderson, D. R., and K. P. Burnham. 1992. Demographic analysis of northern spotted owl populations. Pp. 319–328 *in* J. Bart et al., eds., *Recovery Plan for the Northern Spotted Owl*. Ms. U.S. Fish and Wildlife Service, Washington, D.C.

Beddington, J. 1974. Age distribution and stability of simple discrete-time population models. *Journal of Theoretical Biology* 47: 65–74.

Boyce, M. S. 1977. Population growth with stochastic fluctuations in the life table. *Theoretical Population Biology* 12: 366–373.

———. 1992. Population viability analysis. *Annual Review of Ecology and Systematics* 23: 481–506.

———. 1993. Population viability analysis: Adaptive management for threatened and endangered species. *Transactions of the North America Wildlife and Natural Resources Conference* 58: 520–527.

Brault, S., and H. Caswell. 1993. Pod-specific demography of killer whales (*Orcinus orca*). *Ecology* 74: 1444–1455.

Burgman, M. A., and B. B. Lamont. 1992. A stochastic model for the viability of *Banksia cuneata* populations: Environmental, demographic and genetic effects. *Journal of Applied Ecology* 29: 719–727.

Burgman, M. A., S. Ferson, and H. R. Akçakaya. 1993. *Risk Assessment in Conservation Biology*. Chapman & Hall, New York.

Burkey, T. V. 1989. Extinction in nature reserves: The effect of fragmentation and the importance of migration between reserve fragments. *Oikos* 55: 75–81.

Caswell, H. 1978. A general formula for the sensitivity of population growth rate to changes in life history parameters. *Theoretical Population Biology* 14: 215–230.

Caughley, G. 1977. *Analysis of Vertebrate Populations*. Wiley, New York.

Charnov, E. 1986. Life history evolution in a "recruitment population": Why are adult mortality rates constant? *Oikos* 47: 129–134.

Cohen, J. E. 1979. Comparative statics and stochastic dynamics of age-structured populations. *Theoretical Population Biology* 16: 159–171.

Cohen, J. E., S. W. Christensen, and C. P. Goodyear. 1983. A stochastic age-structured population model of striped bass (*Morone saxatilis*) in the Potomac River. *Canadian Journal of Fisheries and Aquatic Sciences* 40: 2170–2183.

Crouse, D. T., L. B. Crowder, and H. Caswell. 1987. A stage-based population model for loggerhead sea turtles and implications for conservation. *Ecology* 68: 1412–1423.

Daley, D. J. 1979. Bias in estimating the Malthusian parameter for Leslie matrices. *Theoretical Population Biology* 15: 257–263.

Davis, G. J., and R. W. Howe. 1992. Juvenile dispersal, limited breeding sites, and the dynamics of metapopulations. *Theoretical Population Biology* 41: 184–207.

de Kroon, H., A. Plaiser, J. van Groenendael, and H. Caswell. 1986. Elasticity: The relative contribution of demographic parameters to population growth rate. *Ecology* 67: 1427-1431.

Dennis, B., P. L. Munholland, and J. M. Scott. 1991. Estimation of growth and extinction parameters for endangered species. *Ecological Monographs* 61: 115–143.

Desharnais, R. A., and J. E. Cohen. 1986. Life not lived due to disequilibrium in heterogeneous age-structured populations. *Theoretical Population Biology* 29: 385–406.

Doak, D. 1989. Spotted owls and old growth logging in the Pacific Northwest. *Conservation Biology* 3: 389–396.

Emlen, J. M., and E. K. Pikitch. 1989. Animal population dynamics: Identification of critical components. *Ecological Modelling* 44: 253–273.

Ewens, W. J., P. J. Brockwell, J. M. Gani, and S. I. Resnick. 1987. Minimum viable population size in the presence of catastrophes. Pp. 59–68 *in* M. E. Soulé, ed., *Viable Populations for Conservation*. Cambridge University Press.

Ferson, S., L. Ginzburg, and A. Silvers. 1989. Extreme event analysis for age-structured populations. *Ecological Modelling* 47: 175–187.

Foley, P. 1994. Predicting extinction times from environmental stochasticity and carrying capacity. *Conservation Biology* 8: 124–137.

Furstenberg, H., and H. Kesten. 1960. Products of random matrices. *Annals of Mathematical Statistics* 31: 457–469.

Gerrodette, T., D. Goodman, and J. Barlow. 1985. Confidence limits for population projections when vital rates vary randomly. *Fishery Bulletin* 83: 207–217.

Ginzburg, L. R., S. Ferson, and H. R. Akçakaya. 1990. Reconstructibility of density dependence and the conservative assessment of extinction risks. *Conservation Biology* 4: 63–70.

Goodman, D. 1980. Demographic intervention for closely managed populations. Pp. 171–195 *in* M. E. Soulé and B. A. Wilcox, eds., *Conservation Biology*. Sinauer, Sunderland, Mass.

———. 1981. Life history analysis of large mammals. Pp. 415–436 *in* C. W. Fowler and T. D. Smith, eds., *Dynamics of Large Mammal Populations*. Wiley, New York.

———. 1987*a*. Consideration of stochastic demography in the design and management of biological reserves. *Natural Resource Modeling* 1: 205–234.

———. 1987*b*. The demography of chance extinction. Pp. 11–34 *in* M. E. Soulé, ed., *Viable Populations for Conservation*. Cambridge University Press.

Guckenheimer, J., G. Oster, and A. Ipaktchi. 1977. The dynamics of density dependent population models. *Journal of Mathematical Biology* 4: 101–147.

Hanski, I. 1991. Single-species metapopulation dynamics: Concepts, models and observations. *Biological Journal of the Linnean Society* 42: 17–38.

Heyde, C. C., and J. E. Cohen. 1985. Confidence intervals for demographic projections based on products of random matrices. *Theoretical Population Biology* 27: 120–153.

Kalisz, S., and M. A. McPeek. 1993. Extinction dynamics, population growth and seed banks. *Oecologia* 95: 314–320.

Karlin, S., and H. M. Taylor. 1975. *A First Course in Stochastic Processes*. Academic Press, New York.

Lande, R. 1987. Extinction thresholds in demographic models of territorial populations. *American Naturalist* 130: 624–635.

———. 1988*a*. Genetics and demography in conservation biology. *Science* 241: 1455–1460.

———. 1988*b*. Demographic models of the northern spotted owl (*Strix occidentalis caurina*). *Oecologia* 75: 601–607.

———. 1993. Risks of population extinction from demographic and environmental stochasticity and random catastrophes. *American Naturalist* 142: 911–927.

Lande, R., and S. H. Orzack. 1988. Extinction dynamics of age-structured populations in a fluctuating environment. *Proceedings of the National Academy of Sciences (USA)* 85: 7418–7421.

Leigh, E. G. 1981. The average lifetime of a population in a varying environment. *Journal of Theoretical Biology* 90: 213–239.

Leslie, P. H. 1945. On the use of matrices in certain population mathematics. *Biometrika* 33: 183–212.

———. 1966. The intrinsic rate of increase and the overlap of successive generations in a population of guillemots (*Uria aalge* Pont.). *Journal of Animal Ecology* 35: 291–301.

Lewontin, R. C., and D. Cohen. 1969. On population growth in a randomly varying environment. *Proceedings of the National Academy of Sciences (USA)* 62: 1056–1060.

Maynard Smith, J. 1974. *Models in Ecology*. Cambridge University Press.

Meyer, J. S., and M. S. Boyce. 1994. Life historical consequences of pesticides and other insults to vital rates. Pp. 349–363 *in* R. J. Kendall and T. E. Lacher, eds., *Wildlife Toxicology and Population Modeling.* Lewis, Washington, D.C.

Nichols, J. D., G. L. Hensler, and P. W. Sykes. 1980. Demography of the Everglades kite: Implications for population management. *Ecological Modelling* 9: 215–232.

Pielou, E. C. 1977. *Mathematical Ecology.* Wiley, New York.

Shaffer, M. L. 1981. Minimum population sizes for species conservation. *BioScience* 31: 131–134.

Shaffer, M. L., and F. B. Samson. 1985. Population size and extinction: A note on determining critical population sizes. *American Naturalist* 125: 144–152.

Simberloff, D. 1988. The contribution of population and community biology to conservation science. *Annual Review of Ecology and Systematics* 19: 473–511.

Stacey, P. B., and M. Taper. 1992. Environmental variation and the persistence of small populations. *Ecological Applications* 2: 18–29.

Taylor, B. L., and T. Gerrodette. 1993. The uses of statistical power in conservation biology: The vaquita and northern spotted owl. *Conservation Biology* 7: 489–500.

Tuljapurkar, S. D. 1982a. Population dynamics in variable environments. II. Correlated environments, sensitivity analysis and dynamics. *Theoretical Population Biology* 21: 114–140.

———. 1982b. Population dynamics in variable environments. III. Evolutionary dynamics or r-selection. *Theoretical Population Biology* 21: 141–165.

———. 1989. An uncertain life: Demography in random environments. *Theoretical Population Biology* 35: 227–294.

Tuljapurkar, S. D., and S. H. Orzack. 1980. Population dynamics in variable environments. I. Long-run growth rates and extinction. *Theoretical Population Biology* 18: 314–342.

Tuljapurkar, S. D., and J. S. Semura. 1979. Stochastic instability and Liapunov stability. *Journal of Mathematical Biology* 8: 133–145.

Van Sickle, J. 1990. Dynamics of African ungulate populations with fluctuating density-dependent calf survival. *Theoretical Population Biology* 37: 424–437.

CHAPTER 16

# Sensitivity Analysis of Structured-Population Models for Management and Conservation

*Philip Dixon, Nancy Friday, Put Ang, Selina Heppell, and Mrigesh Kshatriya*

Population management can be broadly defined by three goals: conservation, harvest, and control. Conservation management focuses on increasing population sizes or reducing the probability of extinction for rare species (see, e.g., Crouse et al. 1987; Slooten & Lad 1991; Crowder et al. 1994; Heppell et al. 1994). Management for the harvesting of fish and wildlife seeks to maintain population size or optimize the yield from some harvest population (see, e.g., Mendoza & Setyarso 1986; Cattan & Glade 1989; Eberhardt 1990; Basson & Beddington 1991). Pest and disease control attempts to reduce population size and prevent outbreaks of harmful or nuisance organisms (see, e.g., Mount & Haile 1987). Population models can provide insight into the causes of population dynamics as well as qualitative predictions of future trends.

The types of models that are used for population management depend on the questions being asked and the available data. Predictive models attempt to make quantitative predictions of population dynamics, whereas projection models focus on qualitative trends. Both types of models are used in the three management groups. Fisheries models, which are used to predict population trends, generally use density-dependent difference equations based on estimated recruitment curves and age-length measurements taken from commercial catches. Many conservation managers use age- or stage-based matrix models to make long-term projections for their populations. Often data-rich systems, such as harvested ungulate populations and insect pests, lead managers to complex simulation models to make predictions of population size through time.

Model development and analysis can guide research and management efforts by elucidating the driving forces behind population dynamics. In particular, sensitivity analysis may be used to identify which vital rates have the greatest effect on population growth. While detailed experiments may be the most thorough means to test a series of management alternatives, often managers are faced with limited time and funding for large-scale manipulations. Such manipulations may not be ethically sound, such as the release of a known pest to simulate an outbreak, or demographically feasible, such as manipulating endangered populations. Sensitivity analysis of simulation or analytical models allows us to make predictions about the relative effects of various management alternatives before field manipulations or conservation strategies are implemented.

Population transition matrices, although certainly not the best models for all management problems, provide a simple, analytical way to look at the sensitivity of a model population to perturbations that affect vital rates. In deterministic matrix models, the population multiplication rate is given by the dominant eigenvalue, $\lambda$. The sensitivity of $\lambda$ to small changes in the $ij$th element of the transition matrix, $\mathbf{A}$, is

$$S_{ij} = \delta\lambda/\delta a_{ij} = V_i U_j / \lambda \mathbf{V}'\mathbf{U},$$

where $\mathbf{V}$ and $\mathbf{U}$ are the left and right eigenvectors of the matrix $\mathbf{A}$. Proportional sensitivity, or elasticity, is used when perturbations to the matrix are expressed as percent changes. Elasticity is given by

$$E_{ij} = \delta \ln \lambda / \delta \ln a_{ij} = a_{ij} S_{ij} / \lambda.$$

We use sensitivity and elasticity to analyze three applications of structured-population models for managed populations. Each example illustrates a different aspect of modeling: the importance of population structure, the appropriate model complexity, and the incorporation of environmental variability.

We discuss the importance of structure in our first example, which adds age structure to a density-dependent, stochastic simulation model for elk and wolf in Yellowstone Park (Boyce 1992). Boyce's original model used regression relationships of observed growth in elk populations at different levels of density and winter severity. Our matrix model uses life-table data to estimate age- and stage-specific survival, growth, and fecundity. Adding structure to the elk model allows us to ask more-detailed questions about the effects of human harvest and wolf predation on elk populations; for instance, what are the consequences if wolves prey only on younger animals? Managers may rely on this type of information to set harvest rates for proper maintenance of both elk and wolf populations.

Model complexity is the topic of our second example. Managers attempting to control disease-vector tick species have collected detailed information about the effects of temperature, density, and host quality on tick population dynamics. An elaborate simulation model has been developed to predict tick dynamics through the year (Mount et al. 1991). We argue that increasing the number of parameters in a model may add "realism," but at the cost of a loss in interpretability. Although analytical models are often oversimplified, they can provide information about the underlying mechanisms of population structure. We develop a transition-matrix model for *Boophilus* ticks by using a continuous-time simulation model to estimate the coefficients in a series of weekly transition matrices. Environmental variability is incorporated into the tick model by estimating separate transition matrices for each week. In other cases, stochastic transition-matrix models can be used to describe the influence of random environmental variation on population dynamics.

In our third example, we evaluate the influence of environmental variation on the estimates of elasticity. We compare the stochastic equivalent of elasticity to the traditional elasticity computed from the average transition matrix. We use two examples: a multi-year study of the population dynamics of a rare plant, and a study of the influence of burning on the dynamics of a savanna grass.

## 1 Modeling and Managing Elk Populations

Elk (*Cervus elaphus*) are the most abundant ungulate and dominant herbivore in the Yellowstone National Park (Houston 1982). From 1955 to 1968, hunting reduced the northern herd from around 12,000 individuals to 4,000. A moratorium on reduction subsequently went into effect, and the number increased again to 12,000. There are two conflicting views on defining appropriate management for natural areas. Some people maintain that ecological processes, including plant succession and animal regulation, should be allowed to unfold without any human interference. Others believe that the "pristine relationship" among organisms in the park has been so altered by man that interference is required to improve or to restore the vegetation-herbivore equilibria (Houston 1982). Indeed, an example of human alteration of this ecosystem is the large-scale reduction of the gray wolf (*Canis lupus*), the primary natural predator of elk, carried out by the U.S. government during 1914–1926 (Boyce 1992). Both points of view on park management have been considered valid, and both approaches have been tried with this northern herd of elk. Attempts were made to reconcile these two points of view. An extensive study of elk biology was carried out from 1970 to 1979 to determine whether population reduction is necessary in the park. This study evaluated the roles of density dependence and environmental forcing in determining changes in population size (Houston 1982).

In recent years, public attitude toward the wolf has changed. With an increasing awareness of wolf as an integral part of the park ecosystem, the U.S. Fish and Wildlife Service approved a recovery plan for the northern Rocky Mountain wolf (USFWS 1987). A logistic predator-prey model with stochastic function was developed by Boyce (1992) in a computer program called WOLF5 to predict the probable consequences of wolf introduction or recovery on the northern elk population in the park. Simulation results from this model indicate that wolf predation should dampen the fluctuation of the elk population size due to environmental variation and that wolves should reduce ungulate abundance by 10–30 percent.

This model includes several species of ungulates as alternative prey for the wolves. All these prey populations are unstructured; that is, all individuals within the population are treated as equal. In many natural populations, different age or stage classes of individuals may have different vital rates. It is therefore of interest to examine whether projections from structured-population models differ from those from unstructured models and to consider the management implications of those differences.

In this section we develop a structured-population model for elks and wolves in Yellowstone National Park. This model combines a matrix model that incorporates age and stage structures of the elk population with the density-dependence, environmental-forcing, and wolf-predation functions from the WOLF5 model (Boyce 1992).

### The WOLF5 *Unstructured Model*

The WOLF5 model estimates the effects of winter severity and summer production on per capita population growth rates by regression:

$$r(t) = r_0 - \beta_1 n(t) - \beta_2 W(t) + \beta_3 P(t),$$

where $r_0$ is the maximum population growth rate, $\beta_i$'s are regression coefficients, $n(t)$ is the number of elk, $W(t)$ is winter severity, and $P(t)$ is green herbaceous phytomass during summer (Boyce 1992). This is incorporated into a difference equation for predicting the dynamics of the elk population number:

$$n(t+1) = n(t) \exp\left[r_0 - \beta_1 n(t) - \beta_2 W(t) + \beta_3 P(t)\right],$$

where $W(t)$ and $P(t)$ are independent random normal variables with mean 0 and standard deviations of 6.5 and 309, respectively (Boyce & Merrill 1991). Analysis of the climatic variation among years found that it showed no significant first-year lag or autocorrelation (Boyce 1992).

The carrying capacity, $K$, for elk is determined by solving for $r(t) = 0$. By using the mean values of climate and herbaceous phytomass in equation (1), $K = r_0/\beta_1$. The difference equation can be written as

$$n(t+1) = n(t) \exp\left[r_0 - r_0/Kn(t) - \beta_2 W(t) + \beta_3 P(t)\right].$$

The elk population may be affected by wolf predation. The per capita rate at which prey are captured is a function of the number of prey available to the predator, and can be modeled as a Type-III functional response (Holling 1959). The annual number of elk killed by wolves is given by

$$f(n) = Rn^2/(1 + RT_h n^2),$$

where $n$ is the number of elk, $R$ is the per capita attack rate, and $T_h$ is the handling time. The coefficients for the functional response are taken from WOLF5 (Boyce 1992) and are modified for predation on a single species. In the WOLF5 model, wolf functional response depends both on the number of elk and on the number of other ungulates present in the park (Boyce 1992). In our simulation, we remove the other ungulates, such that the wolf population is responding only to the number of elk. We therefore modified wolf functional response by increasing the attack rate to $R = 1.2 \times 10^{-6}$, a value ten times Boyce's (1992) original estimate; $T_h$ is set at 0.04, corresponding to a maximum rate of 25 elk killed per wolf per year.

The growth rate of the wolf population depends on the number of wolves and the rate at which captured prey are converted into predator offspring. The wolf population size at time $t+1$ is given by

$$w(t+1) = w(t) \exp\left\{0.075 f[n(t)] - 0.6/150w(t) - 0.63\right\},$$

where $f[n(t)]$ is the wolf functional response. Depending on the abundance of elk, the wolves have a maximum potential growth rate of 1.4.

## Formulation of a Structured Model

We construct a matrix model based on the life table of elk given by Houston (1982). Males and females have an equal sex ratio in utero, so only females are considered in the model. Because information on pregnancy rate is available only for groups of age classes, we use a $7 \times 7$ stage-structured model with age classes comprising the following ages: <1 year (calves), 1 year (yearling), 2 years, 3–7 years, 8–15 years, 16–20 years, and >20 years. Fecundity for an individual in stage $i$, $F_i$, was calculated from pregnancy rates using the formula for a pre-breeding birth pulse (Caswell 1989$b$), $F_i = 0.677\, m_i$, where $m_i$ is the pregnancy rate, and 0.677 is the probability of surviving from birth to one year.

Elk yearlings and 2-year-olds do not always reproduce. Their pregnancy rate depends on density as well as on winter condition ($W$). Houston (1982) fit regression models to estimate the pregnancy rate for yearlings, $F_2 = 0.195 - 0.000059\, n(t) - 0.0536\, W$, and for 2-year-olds, $F_3 = 2.12 - 0.000063\, n(t) - 0.0375\, W$, where $n(t)$ is the total population size. The winter index, $W$, ranges from $-9$ to 13 (mean $= 1.5$, SD $= 6.38$; Houston 1982). Pregnancy rates of the other stage classes are density-independent.

In order to set up the basic matrix, we set the fecundity values of yearlings and 2-year-olds to be the pregnancy rate when $n(t) = 0$ (i.e., with no density dependence). The resulting matrix was iterated with an initial $\lambda = 1.0$ (Caswell 1989$b$). A constant $\lambda$ of 1.19 was attained in two iterations. The final transition matrix is

$$\mathbf{A} = \begin{bmatrix} 0 & 0.066 & 0.289 & 0.322 & 0.305 & 0.118 & 0.112 \\ 0.981 & 0 & 0 & 0 & 0 & 0 & 0 \\ 0 & 0.997 & 0 & 0 & 0 & 0 & 0 \\ 0 & 0 & 0.997 & 0.808 & 0 & 0 & 0 \\ 0 & 0 & 0 & 0.185 & 0.916 & 0 & 0 \\ 0 & 0 & 0 & 0 & 0.054 & 0.706 & 0 \\ 0 & 0 & 0 & 0 & 0 & 0.063 & 0.703 \end{bmatrix}.$$

The first three classes are age classes, and the last four are stage classes (here, stage classes are lumped age classes, as described earlier). The diagonal entries of the matrix give the probability of staying in the same stage class in each time step of one year. According to this matrix, the population of elk grows at an exponential rate of about 20 percent a year under density-independent conditions and without any environmental forcing. The elasticity matrix (de Kroon et al. 1986) is

$$
\begin{bmatrix}
0 & 0.0067 & 0.024 & 0.070 & 0.044 & 0.0019 & 0.0002 \\
0.148 & 0 & 0 & 0 & 0 & 0 & 0 \\
0 & 0.141 & 0 & 0 & 0 & 0 & 0 \\
0 & 0 & 0.116 & 0.243 & 0 & 0 & 0 \\
0 & 0 & 0 & 0.046 & 0.152 & 0 & 0 \\
0 & 0 & 0 & 0 & 0.002 & 0.0031 & 0 \\
0 & 0 & 0 & 0 & 0 & 0.0002 & 0.0003
\end{bmatrix}.
$$

The largest elasticities are associated with the survival, growth, and reproduction of stage 4 (3–7-year-old) individuals.

The final detail in the structured model is to specify how wolf kills are distributed across the different age or stage classes. Wolf kill is modeled in three ways: equally distributed through all the age or stage classes, proportionally distributed through the age or stage classes, and restricted to only the first four age or stage classes. The third resembles wolf kill restricted to younger elks.

## Model Simulation

We compare projections from the structured and unstructured models by running the two models simultaneously, with the same sequence of random environmental forcing. The projection equation for the structured model is

$$
\mathbf{N}_{t+1} = \exp\left[ -\frac{r_0}{K} \sum \mathbf{N}_t - \beta_2 W_t + \beta_3 P_t \right] \mathbf{A}\,\mathbf{N}_t. \tag{2}
$$

The quantity $\sum \mathbf{N}_t$ is the total number of individuals in the structured population. The unstructured projection equation is given in equation (1). The quantity $\mathbf{A}\,\mathbf{N}_t$ in the structured model is equivalent to $n(t)\,\exp(r_0)$ in the unstructured model. To compare the dynamics of the two models, we set $r_0$ for the unstructured model equal to $\ln(\max \lambda_\mathbf{A})$. Using Boyce's (1992) value of $r_0$ does not change the qualitative dynamics, but it does change the projected population size.

The structured and unstructured models use identical parameters for density dependence, $r_0/Kn(t)$, winter forcing, $\beta_2 W(t)$, and phytomass forcing, $\beta_3 P(t)$. Hence, any differences in the responses reflect the effects of adding structure to the elk population. All simulations were run for 100 years.

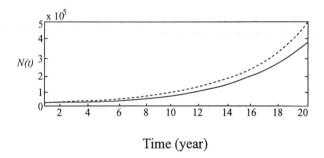

Time (year)

FIGURE 1. *Elk population increase for first 20 years, from simulations of the unstructured model* (dashes) *and structured model* (solid line), *showing exponential growth.*

*Simulation Results*

*Elk population dynamics without wolves.* In the absence of density dependence and environmental forcing, and given an initial condition of 7,000 individuals (1,000 individuals per age or stage class), the population size projected by the unstructured model is higher than that projected by the structured model (Fig. 1). This is largely an effect of the transient behavior of the structured matrix model. Given an initial condition, the population modeled by the unstructured model immediately grows at the constant rate defined by $r_0$, whereas the population modeled by the matrix model grows at that rate only after a stable distribution is attained. The time to reach a stable distribution from the initial condition in the structured model determines the length of the time before populations projected by the two models grow at the same exponential rate. If the population at a stable distribution is given as the initial condition, then the two models yield identical population projections.

In both models, the population size converges to the carrying capacity in the presence of density dependence. Given an initial population size larger than the carrying capacity, the population projected by the unstructured model converges to the carrying capacity at a constant rate. Population projected by the structured model decreases initially below the carrying capacity before increasing again to reach the carrying capacity (Fig. 2). This is again an effect of the transient behavior of the structured model as the population converges to the stable age structure.

In both models, winter forcing alone can produce an irregular fluctuation in growth. This is also true when food is limited, although the fluctuation in population size is less variable.

When density dependence is imposed on top of environmental

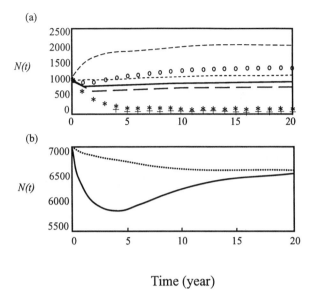

Time (year)

FIGURE 2. *Elk age- or stage-class distribution for first 20 years from simulations of the unstructured and structured models:* (a) *class distribution of individuals;* (b) *distribution of total population size for unstructured model* (dotted line) *and structured model* (solid line). (a) *Classes:* dots, *calves;* solid line, *yearlings;* long dashes, *2-year-olds;* short dashes, *3–7-year-olds;* circles, *8–15-year-olds;* asterisks, *16–20-year-olds;* plus signs, *>20-year-olds. Carrying capacity, K, of the population is 6,428.*

forcing, the population size can overshoot the carrying capacity and the mean population size become higher than the carrying capacity (Fig. 3). Populations projected from the two models do not behave differently with environmental forcing. The stochastic effect of winter forcing causes the population size to fluctuate, but the stable distribution of the stage classes is maintained even though the sizes of each stage can fluctuate.

*Elk population dynamics with modified matrix elements.* Given that there is no significant or apparent difference in the behavior of the population projected by either the structured or the unstructured models, we manipulated the structured model by changing some elements of the matrix. This approach has the advantage of determining the effect of any specific change in any particular stage or age class(es) on the population growth.

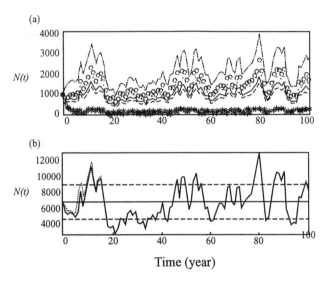

Time (year)

FIGURE 3. *Elk sample runs of simulations of the unstructured and struc-*
*tured models with density-dependent survivorship, winter-forcing func-*
*tion, and food limitation imposed on the total population.* (a) *Age- or*
*stage-class distribution of individuals:* dash-dot line, *3–7-year-olds; other*
*symbols as in Figure 2. Individuals >20 years old decrease from an initial*
*population of about 700 to zero by year 10.* (b) *Fluctuations in popula-*
*tion size from unstructured model* (dotted line) *and structured model*
(solid line). *Mean population size is 6,830.*

When the fecundity rates of stage (age) classes 2 and 3 in the
structured model are made density-dependent and subjected to
a winter-forcing function, the resulting mean population size is
higher than the carrying capacity calculated from an unmodified
matrix (Fig. 4). However, the transient behavior remains similar to
that observed when density dependence is imposed on the whole
population (cf. Figs. 2, 4).

Elasticity analysis of the structured model indicates that stage
4 has contributed most to the population growth rate. When the
fecundity values for stages 2–4 are reduced to 0 and the growth
from stage 4 to stage 5 is reduced to a small probability (0.01
percent), thus creating a bottleneck in the population flow, the
population experiences a net decrease in growth rate. When the
effect of density dependence is doubled in stage 4, the population
size is much smaller (Fig. 5).

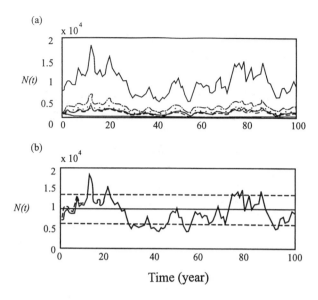

FIGURE 4. *Elk sample runs, with additional density dependence, of simulations of unstructured and structured models. In addition to conditions imposed in Figure 3, the fecundity rates of stage (age) classes 2 and 3 in the structured model are made density-dependent and are subjected to a winter-forcing function. (a) Class distribution relative to population size; symbols as in Figure 3. (b) Mean population size is 8,770.*

*Elk population dynamics with wolves.* The effect on the elk population size of adding structure to the population model depends on the choice of predation model. With nonselective predation of wolves on elk, the structured model projects a larger elk population than does the unstructured model (Fig. 6). In this case, a structured population appears to be able to withstand higher predation and random fluctuation without crashing. With selective predation on younger age classes, the structured model projects a smaller population size than does the unstructured model. However, both model projections are qualitatively similar. The peaks in the population fluctuation coincide; only their magnitudes differ (Fig. 7).

## *Discussion*

Some of the models developed for the management of species have been unsuccessful partly because they fail to take into account the

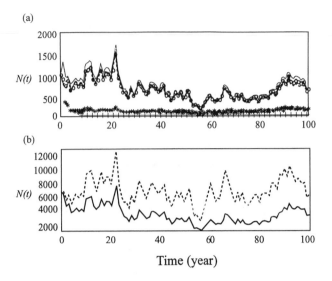

FIGURE 5. *Elk sample runs, with reduced survivorship, of simulations of the unstructured and structured models. In addition to the conditions in Figure 4, the density-dependent survivorship is further reduced by 50 percent in the structured model. Dots, calves; otherwise, panels and symbols as in Figure 3. Mean population size is 6,830.*

dynamics of the population structure. There are many examples of this in both wildlife and fisheries literature. One advantage of the structured model lies in its ability to assess the relative impact of environmental forcing, density dependence, or predation on a selected group of individuals within the population. This may be particularly significant if individuals within the population are not equal. Alternatively, an unstructured model may be able to capture the essential dynamics of the population when the vital rates are similar among individuals, irrespective of age or size or other demographic parameters. Because elk mortality and fecundity do not vary much with age or stage class, the classes can be treated almost alike. This explains the similar responses of the structured and the unstructured models. Winter severity imposes a significant density-independent mortality on the population; thus, predation alone may not be sufficient to alter the pattern imposed by winter forcing.

Neither the unstructured model originally proposed by Boyce (1992) nor the structured model developed here claims to com-

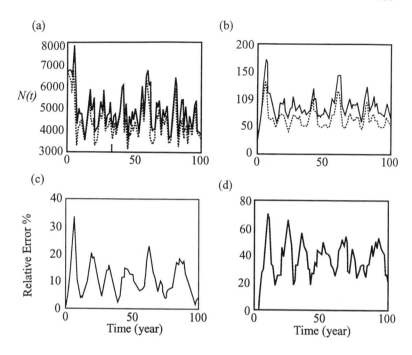

FIGURE 6. *Population sizes, with nonselective predation, for elk* (a) *and wolf* (b) *from the unstructured model* (dotted line) *and the structured model* (solid line); *attack rate* (eq. 4) *for wolf functional response is multiplied by 10. Relative error for elk* (c) *and wolf* (d) *is the percentage discrepancy between the population sizes from the structured model and the unstructured model.*

pletely reflect the dynamics of both the elk and the wolf populations. However, changes in the functional response of the predator result in more crashes of elk and wolf populations in the unstructured model. This suggests the inherent resilience of the structured model to perturbation; that is, the presence of individuals in different age or stage groups allows some of them to escape the perturbation, hence forming the basis for population recovery. This may be a more realistic depiction of what is actually happening in nature.

Some of the problems associated with the management of species are related to knowing whether the population is in a stable state. This concern applies when dealing with the introduction of any new individuals, the reintroduction of a predator into the population, or the restarting of a new population in a depleted area. Transient

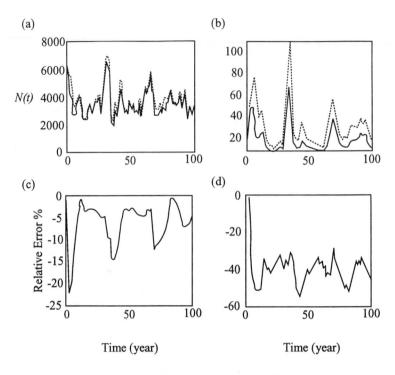

FIGURE 7. *Population sizes, with selective predation for elk* (a) *and wolf* (b) *from the unstructured model* (dotted line) *and the structured model* (solid line), *with wolf predation restricted to elk in stage (age) classes 1–4. Relative error for elk* (c) *and wolf* (d) *is the percentage discrepancy between the population sizes from the structured model and the unstructured model.*

behavior exhibited by the population may be more appropriately depicted by a structured model than by an unstructured model.

It is important to assess the dynamics of the structure of the population before any conservation measure for any endangered species can be addressed. If a certain stage or age group of the population makes a disproportionately higher contribution to the overall population growth, conserving that particular group may be the best strategy for ensuring the survival of the species. Failing to recognize this can entail unnecessary expenditure of resources with only a limited chance of success for the conservation effort.

## 2 Models and Control for Tick Populations

Frequently, successful "management" of a population is the reduction or eradication of the organism, at least on a local scale. We devote significant amounts of time and resources to the control of numerous pest organisms, including agents of disease (bacteria, viruses, protozoans, and helminths), disease vectors (mosquitoes, ticks), crop pests (red scale, mealy bugs), and annoyance pests (nonbiting midges). Pest managers have often used population models to explore various control strategies before initiating elaborate field programs (Hassell 1980; Gutierrez et al. 1988).

Examples of models used in this role, which range from the simple to the extremely complex, include both structured (e.g., life-history stage-specific) and unstructured systems. The structured-population models of Anderson and May have been used to explore the dynamics of epidemics and disease-control strategies (Anderson 1982, 1988; Anderson & May 1989, 1991). Population biologists have tended to follow one of two approaches to modeling pest and natural-enemy interactions (Godfray & Waage 1991): simple, analytical models (Hassell 1978; Taylor 1984; Hassell & May 1990) and detailed simulation models (Gutierrez et al. 1984; Carter 1985; Mount & Haile 1987; Sutherst et al. 1988; Mount et al. 1991). While complex simulation models contain more explicit biological functions and usually fewer assumptions, the large number of parameters in these models makes formal analyses difficult. However, because complete data are rarely available for a pest system and parameter estimates often have large confidence intervals, a sensitivity analysis is crucial to understanding how model results, such as population growth rate, are affected by changes in survival, growth, and fecundity.

Here, we simplify a complex simulation model designed to assess control methods for cattle ticks (*Boophilus* spp.; Mount et al. 1991). By modifying the simulation model to a series of stage-structured matrix models, we are able to apply some formal analytical techniques to examine the sensitivities and elasticities of the stage transitions involved. This analysis of how population dynamics are affected by changes in model parameters should help pest managers focus their control efforts on critical life stages.

This study is based on a particular example from Mount et al.'s (1991) simulation model of *Boophilus* cattle ticks. We restrict our efforts to *Boophilus microplus* because of the detailed descriptions

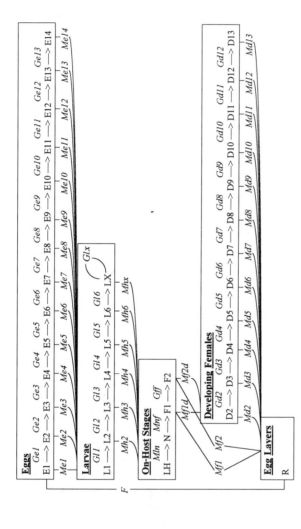

FIGURE 8. *Boophilus microplus life cycle. Italics indicate life-cycle transitions; other abbreviations indicate life-cycle stages. Numbers indicate weeks. Egg stages include young eggs (E1–E4) and old eggs (E5–E14). Ge transitions are the survival of eggs from one week to the next, Ei to E(i+1). Me: maturation from Ei to L1. Off-host larval stages: young larvae (L1–L6), larvae >6 weeks old (LX). Gl: survival of Li to L(i+1). Mh: maturation from Li to LH. On-host stages: larvae (LH), nymphs (N), females (F1, F2). Mln, maturation from LH to N; Mnf, maturation of N to F1; Gff, survival from F1 to F2. Mf1d and Mf2d: maturation of F1 and F2 to R. Developing female (engorged, free-living) stages: young females (D2–D4) and old females (D5–D13). Gd: survival of Di to D(i+1). Md: maturation from Di to R, the egg-laying class; F is fecundity. Males are removed from the model in the transition from nymphs to on-host females.*

of the functions used to calculate the parameters provided for this species and the availability of the original references used for parameter estimation. However, the methods we describe here are applicable to many other discrete-time simulation models.

## Model Description

Mount et al. (1991) used a weekly time step in their system, which sufficiently matches many of the time intervals required for tick development. However, the weekly time step may have been longer than some of the feeding periods, as we discuss below. Using the functional forms from the original model whenever possible and few simplifications, we recast the model into a series of weekly transition matrices. This puts the model into the framework of a periodic matrix model, for which analytical techniques have been developed (Caswell 1989$a$,$b$; Caswell & Trevisan 1994; Tuljapurkar 1985, 1990). Each weekly transition matrix ($\mathbf{B}_w$) projects the population forward in time from week $w$ to week ($w + 1$). These weekly transition matrices can be multiplied together to calculate a yearly transition matrix ($\mathbf{A}$), which projects the population forward from year $n$ to year ($n + 1$): $\mathbf{A} = \mathbf{B}_{52}\,\mathbf{B}_{51}\ldots\mathbf{B}_2\,\mathbf{B}_1$. This formulation allows the retention of temporal change in the parameters but still allows the model to be evaluated using analytical techniques.

In the original model (Mount et al. 1991), each simulation required a set of basic parameters, such as species, location, cattle type and density, and weather. To illustrate how analytical techniques for matrix models can be used for tick systems, we choose one specific set of parameter values: *Boophilus microplus*, from Willowbank, Australia (near Brisbane), on hybrid cattle *Bos taurus* × *Bos indicus* at a density of 0.6 per hectare; temperatures calculated from a 30-year average of mean daily temperature for each month for Brisbane (Elsevier 1971). We attempt to retain as much of the life cycle as possible, categorizing the ticks into age and stage classes (Fig. 8).

Temperature significantly affects many of the transitions in the matrix and is approximated with a cosine function, $T = 20 + 5\cos[(2\pi/52)(\text{week} - 4)]$, which is fit to data on monthly mean daily temperatures for Brisbane (Elsevier 1971). This function is used to estimate the temperature value for each week of the year. These 52 weekly temperature values are used to calculate survival and maturation rates using the functions given by Mount et al. (1991), as described below.

TABLE 1. *Effects of Temperature, Precipitation, Saturation Deficit, and Pasture Type on Survival*

| | | |
|---|---|---|
| Temperature ($T_e$) | | |
| Young | $T_e = -0.00097\,T^2 + 0.031\,T + 0.75$ | |
| Old | $T_e = -0.00115\,T^2 + 0.036\,T + 0.71$ | |
| | | |
| Simplified Effects | | |
| | Young | Old |
| Precipitation ($P_e$) | 0.93 | 0.92 |
| Saturation deficit ($D_e$) | 1.00 | 1.00 |
| Pasture type ($V_e$)–Eggs | 0.92 | 0.77 |
| Pasture type ($V_e$)–Larvae | 0.92 | 0.88 |
| Pasture type ($V_e$) | | |
| – Engorged Females | 0.93 | 0.77 |

NOTE.—Constants are given where the model is simplified because temporal variability is low. In the equations for temperature effect, $T$ is the predicted weekly average temperature in degrees Celsius.

*Survival rates of free-living stages.*    Survival rates of the free-living stages vary with temperature, precipitation, saturation deficit, and pasture type (Mount et al. 1991; Teel 1984). There are separate functions for the young and the old classes of eggs, larvae, and developing, engorged females. Eggs 1–4 weeks old are young eggs; eggs more than 4 weeks old are old eggs. Larvae 1–6 weeks old are young larvae; the larval accumulation class (LX) includes larvae older than 6 weeks. Free-living, developing females 1–4 weeks old are young females; developing females more than 4 weeks old are old females. Survival rates ($S$) for each age group are estimated from the product of temperature, precipitation, saturation deficit, and pasture-type effects (Mount et al. 1991): $S = T_e P_e D_e V_e$. We estimate each of these individual effects from Mount et al.'s equations (Table 1). Because the temporal variability in precipitation and saturation-deficit effects for this region is small, we simplify the model by setting these effects to constants. Pasture-type effects are calculated for an area of 60 percent unimproved light and 40 percent improved light.

*Temperature-dependent developmental rates.*    Following Mount et al. (1991), degree-week accumulation thresholds are used for the development of both eggs to larvae and engorged females to egg layers. A thermal constant for cumulative degree-weeks of 36 for eggs and 30 for females and a minimum developmental threshold temperature ($T_d$) of 15 for both eggs and females are used (Hitchcock 1955).

Degree-week accumulation is a process whereby eggs, or females, accumulate degree-weeks ($T_i - T_d$) until they reach their cumulative degree-week constant, at which time they mature to larvae, or to egg layers (Fig. 8). For example, eggs born in week one (E1), when the temperature is 24.7°C, accumulate 9.7 degree-weeks (24.7−15). This is below their cumulative degree-week constant of 36, so they remain as eggs and, subject to egg survival, move to the E2 egg class. In week two, they accumulate an additional 9.9 degree-weeks. Their total degree-weeks are now 19.6, and they move to the E3 egg class. In week three, they accumulate 10 degree-weeks for a total of 29.6 and move to the E4 egg class. In week four, they accumulate 10 degree-weeks for a total of 39.6 degree-weeks. Therefore, in week four they reach their cumulative degree-week constant of 36 (39.6 > 36), and they mature to the first free-living larval stage (L1).

We determine the number of weekly stage classes for eggs and engorged females by calculating when the survival through the last stage drops below 5 percent. Ticks that do not mature out of the last class die.

*Host seeking by larvae, on-host survival and maturity rates, and fecundity rates.* The probability that a larva finds a host ($H$) is a function of host density, temperature, and tick density (Mount et al. 1991). It is difficult to incorporate density dependence into this type of analysis because nonlinearities usually make the partial differential equations of the sensitivity analysis unsolvable. Therefore, density dependence is not incorporated into our matrix version of the model, and the effect of tick density is set at one. The temperature effect is calculated using the equation above (Mount et al. 1991) and the weekly temperature values from the cosine curve. Host density is chosen to be 0.6 cattle per hectare. The resulting probability of finding a host is $H = 0.20 \times 0.6^{0.51} \times -0.008\,T^2 + 0.4\,T - 4$.

In the original model, density dependence was included in the survival of the on-host stages and defined as cattle resistance developed from their exposure to ticks. We use an average of the maximum and minimum on-host survival rates given by Mount et al. (1991) for the tick exposure of the *B. taurus* × *B. indicus* hybrid. The survival from larvae to nymphs is 0.25. The survival from nymphs to adults is 0.75. Males are removed from the model at this point; therefore, the nymph survival is multiplied by 0.68, the percentage of females in the population. The survival of the adult females is 0.65. The time spent on the host by the females is split, with half remaining for one week and the second half remaining for two weeks (Fig. 8). Therefore, the transition from the first female class to the second is 0.65 (adult survival) × 0.5 (with 50 percent remaining on the host).

We add a class of "egg layers" to the life cycle. The egg-laying class ensures that females are removed from the population after laying. To balance the extra time step this requires, on-host females develop directly after feeding is completed either into the egg-laying class or into two-week-old developing, engorged females (Fig. 8). Fecundity is temperature-dependent with temperature calculated from the cosine curve $F = -2540 + 553\,T - 10.5\,T^2$.

*Matrix Analysis*

We use the methods of Caswell and Trevisan (1994) to analyze these matrices. The matrix product of the weekly matrices gives the projection matrix over the whole interval, specific to each starting time:

$$\mathbf{n}_{t+w} = [\mathbf{B}_w \mathbf{B}_{w-1} \ldots \mathbf{B}_1]\, \mathbf{n}_t = \mathbf{A}_1 \mathbf{n}_t \ .$$

For a year cycle composed of 52 weeks ($w = 52$), $\mathbf{B}_1$ projects the population from week 1 to week 2, and $\mathbf{A}_1$ projects the population from week 1 in year $n$ to week 1 in year $(n+1)$. Changing the order of multiplication of the $\mathbf{B}$ matrices defines a new $\mathbf{A}_i$ matrix, which projects the population from week $i$ to week $i$ in a subsequent year. The asymptotic growth rate of the population is given by the largest eigenvalue $\lambda_i$ of $\mathbf{A}_i$. The $\lambda$'s of all $\mathbf{A}$'s are identical and equal to $\lambda$. The matrix of sensitivities of $\lambda$ to the elements of the $\mathbf{B}_i$'s is given by

$$\mathbf{S}(\mathbf{B}_i) = [\mathbf{B}_{i-1} \mathbf{B}_{i-2} \ldots \mathbf{B}_1 \mathbf{B}_w \mathbf{B}_{w-1} \ldots \mathbf{B}_{i+1}]\, \mathbf{S}(\mathbf{A}_i) \ ,$$

where $\mathbf{S}(\mathbf{A}_i)$ is the sensitivity matrix of $\mathbf{A}_i$ (Caswell & Trevisan 1994). The elasticity of the $jk$th transition in the $i$th $\mathbf{B}$ matrix is given by

$$E_{jk}(\mathbf{B}_i) = (\mathbf{B}_i)_{jk}\, [\mathbf{S}(\mathbf{B}_i)]_{jk} / \lambda \ .$$

The population growth rate, $\lambda$, is a function of all the parameters of the matrix, but some parameters may be more "important" in determining $\lambda$ (Caswell 1989$b$). Sensitivities provide a measure of the changes in $\lambda$ due to changes in the matrix elements. Calculating sensitivities allows researchers to evaluate the effect of errors in their parameter estimation and to evaluate alternative management strategies. Elasticities translate sensitivities into proportional changes in $\lambda$ due to proportional changes in the matrix elements, such that parameters with different scales can be compared (Caswell 1989$b$). All matrix manipulations are conducted using MATLAB.

*Results and Discussion*

For ticks, the long-term population growth rate, $\lambda$, is high (6.45). This is not surprising for a population with no density dependence and favorable environmental conditions. The elasticities show some interesting patterns. Figures 9–16 show the parameter values and the elasticities across the year for each transition element (i.e., survival, maturation, or fecundity). For many of these parameters (e.g., in Figs. 9, 10, 14, 15), the elasticity peaks during periods of change in the parameter value. The elasticities for the survival rates of free-living larvae (Fig. 11) are erratic.

The highest elasticity observed is for the survival of old (>6 weeks), free-living larvae at the coldest time of the year (week 30 is the coldest week). Larvae that survive this cold period are ready to find a host as soon as the temperature increases, since the probability of finding a host depends on temperature. They then move through the on-host stages, which are time-dependent, and go on to develop into engorged females, which takes 3–4 weeks. This timing allows them to develop as the temperature increases, putting them into the egg-laying class when the temperature is warm. Since fecundity depends on temperature, the timing of this developmental sequence can have a strong impact on $\lambda$.

The elasticities of the on-host survival rate and maturation rate are high (Fig. 13), especially at certain times. This indicates that these variables are important to overall population growth and that host resistance should be explored further, as Mount et al. (1991) concluded. In our analysis, however, the elasticities of fecundity equal the on-host elasticities (cf. Figs. 13, 16). This is worth exploring further, since fecundity may be affected by many forces other than temperature.

Because the elasticity of $\lambda$ is the response to a proportional change, elasticities can be calculated only for nonzero matrix entries.

Egg Survival

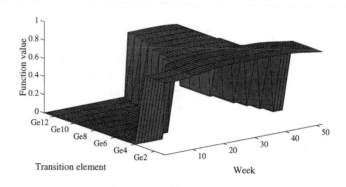

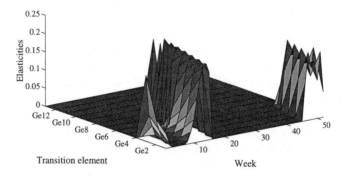

FIGURE 9. *Rates and elasticities for the survival of each egg class (Ge1–Ge13) for each week of the year.*

Looking at the sensitivities is also useful, since it can indicate the effect of change in a transition probability that is zero in our matrices but might be biologically possible (Fig. 17). For example, in our model, larvae cannot move onto a host until they are in the second age class, but it may be biologically possible for larvae to find a host during their first week (panel *a*). The sensitivity of $\lambda$ to this element of the matrices is high in the spring and fall. Therefore, if conditions allow larvae to find a host during their first week, the population growth rate can be dramatically altered.

Likewise, $\lambda$ is sensitive to any transitions implying faster feeding times (larvae to on-host females in one time step, see panel *b*; nymphs to developing females in one step, panel *c*). In this model, as in that of Mount et al. (1991), it is assumed that these feed-

Egg Maturation to Larva

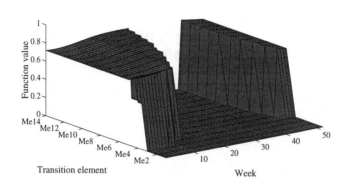

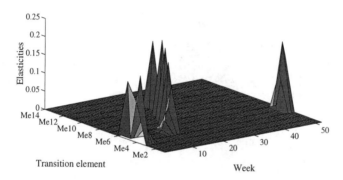

FIGURE 10. *Rates and elasticities for the maturation of each egg class (Me1–Me14) to the first free-living larval class* (L1) *for each week of the year.*

ing periods take at least one week. Experimental data (Hitchcock 1955), however, indicate that feeding periods may be as short as 2–3 days at high temperatures. Thus, the structure of the model may force slower development than might actually occur.

A more extensive analysis of the dynamics of *B. microplus* can be conducted using matrix models. Many of the complexities of the ecology of *B. microplus* omitted here could be added. For example, precipitation effects, saturation-deficit effects, and the choice of other cattle types and locations can be included in this matrix structure. Density dependence can also be included in the matrix structure but may limit the extent of the analytical techniques that can be used on the model.

Off–Host Larval Survival

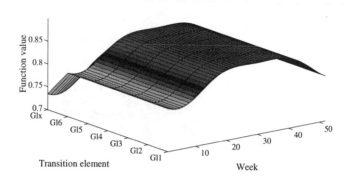

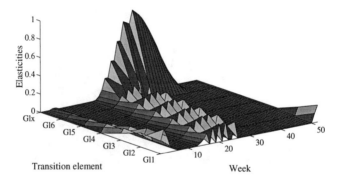

FIGURE 11. *Rates and elasticities for the survival of each larval class (Gl1–Gl6 and Glx) for each week of the year. Note that the scale on the elasticity panel is four times that of the elasticity scale in all other figures. This is entirely due to the elasticity for the Glx transition, which has a maximum of 0.92. The maximum elasticity for the transitions from Gl1 to Gl6 is 0.26.*

## Conclusions

Even with the simplified model we use, we are able to identify some of the same sensitive points identified by Mount et al. (1991), in particular, the survivals and maturations of the on-host stages. We also identify the fecundity and off-host survival of old larvae as potentially important contributors to the population growth rate. In addition, transitions corresponding to shorter developmental times also affect the long-term population growth rate, indicating that

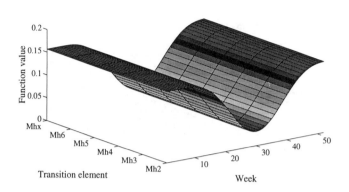

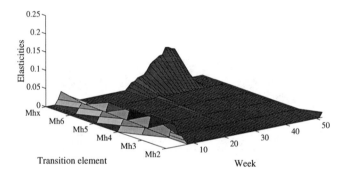

FIGURE 12. *Rates and elasticities for the maturation of each larval class* *(Mh2–Mhx) to the on-host larval class* (LH) *for each week of the year.*

an exploration of different time steps might be revealing.

This approach is a powerful addition to simulation models and can be used to focus attention on the most sensitive points of a more complex simulation model. In addition, by examining elasticities or sensitivities across the year, we can highlight when the population is most sensitive to changes in the parameters. With a complex simulation model, many runs would be needed to explore the parameter space for even one element of the matrix. This type of matrix analysis can provide a useful technique for evaluating the effect of errors in parameter estimation and lead to alternative management strategies, adding to the pest manager's arsenal of control models.

On–Host Survival and Maturation

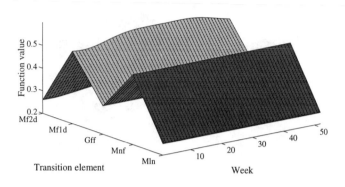

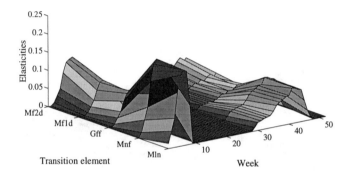

FIGURE 13. *Rates and elasticities for the survival and maturation of the on-host classes (Mln, Mnf, Gff, Mf1d, Mf2d) for each week of the year. Mln is the maturation from on-host larva to on-host nymph. Mnf is the maturation from nymph to on-host female (F1). Gff is the survival of an on-host female to remain on the host for a second week. Mf1d and Mf2d are the maturations of the on-host female classes to the engorged, free-living female class.*

## 3 Elasticity under Environmental Variation

Elasticity analysis, or proportional-sensitivity analysis, has led to a number of management recommendations for conservation (Crouse et al. 1987; Crowder et al. 1990; Dixon & Cook 1990; Heppell et al. 1994). In general, sensitivity has been calculated from deterministic models in order to identify life stages whose vital rates exert large effects on population dynamics. Occasionally, these analyses

Female Survival (Off Host)

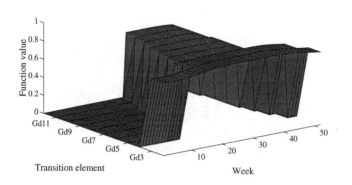

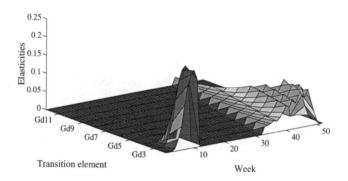

FIGURE 14. *Rates and elasticities for the survival of each free-living, engorged female class (Gd2–Gd12) for each week of the year.*

provide critical insight; Crouse et al. (1987) showed that management efforts for the recovery of loggerhead sea turtles in the southeastern United States should be refocused from the egg or hatchling stage to the large-juvenile stage. This type of information can be incorporated into recovery plans for endangered species, harvesting regimens, and control efforts for pest species.

But how reliable are these analyses in the presence of spatial or temporal variability? Deterministic matrix models are highly simplified, usually consisting of pooled or averaged rates from several years or locations; such pooling masks the variability among years or locations. Stochastic matrix models explicitly include environmental variation, but they are more complicated to construct and

Female Development to Egg Layers

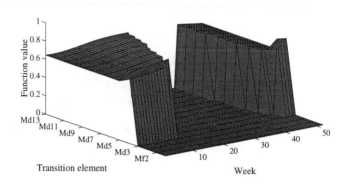

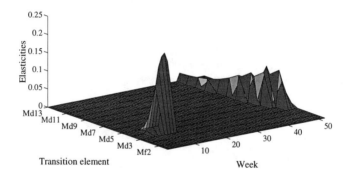

FIGURE 15. *Rates and elasticities for the maturation of the two on-host female classes (Mf1 and Mf2) and each free-living, engorged female class (Md2–Md13) to the egg-laying class* (R) *for each week of the year.*

interpret. In this section, we compare the deterministic elasticity, computed from the average transition matrix, with the stochastic elasticity.

Before deterministic and stochastic results can be compared, it is necessary to define analogous quantities. For a deterministic model, the sensitivity is $\delta\lambda/\delta x(ij)$ and the elasticity is $\delta\log\lambda/\delta\log x(ij)$. The stochastic population growth rate is given by $a = \log\lambda$. Tuljapurkar's (1990) method for computing stochastic sensitivity evaluates $\delta a/\delta x(ij)$. Hence, stochastic sensitivity is not directly comparable to deterministic sensitivity. The quantity $x(ij)\,\delta a/\delta x(ij) = \delta a/\delta\log x(ij)$, which we define as the stochastic elasticity, can be considered analogous to the deterministic elasticity.

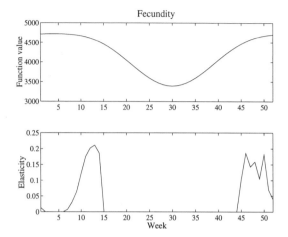

FIGURE 16. *Rates and elasticities for the fecundity (F) of the egg-laying class* (R) *for each week of the year.*

We compare deterministic and stochastic elasticities in two cases. The first is from a five-year study of the population dynamics of a rare plant with moderate year-to-year variation in transition probabilities. We investigate the effects of environmental autocorrelation on the stochastic growth rate and elasticities. The second case is a tropical savanna grass in burned and unburned environments (Silva et al. 1991). In this case, the transition-matrix elements differ considerably between the two environments.

The notation used in this section is the same as that used by Tuljapurkar (1990), in which much of the theory of stochastic matrix models is presented.

### Methods

The annual transition matrices observed in multi-year studies usually show considerable variability. One way to model the effects of this environmental variability is to consider each annual matrix as representative of a particular environment, and then to generate a stochastic sequence of environments. One possible stochastic sequence is a first-order Markov chain, specified by a specific environmental transition matrix, **B**. The transition matrix determines the frequency and temporal autocorrelation of each environment. If all the probabilities in the transition matrix are the same, the environments occur equally frequently and are not cor-

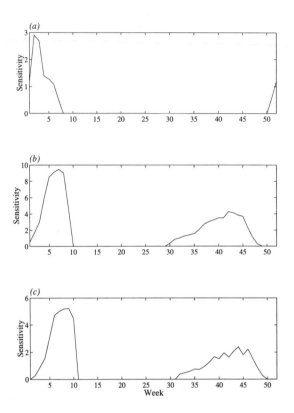

FIGURE 17. *Sensitivities when developmental time decreases from two weeks to one week, for three transitions:* (a) *from the free-living larval class to the on-host larval class in one week;* (b) *from the on-host larval class to the on-host female class in one week;* (c) *from the on-host nymph class to the free-living, engorged female class in one week.*

related. If the elements on the diagonal of the transition matrix are larger than the off-diagonal elements, the environments are autocorrelated. If larger elements are on the subdiagonal, the environments are negatively correlated or semi-periodic. That is, environment 1 is more likely to be followed by environment 2, which is more likely to be followed by environment 3, and so on. Examples of transition matrices with different levels of autocorrelation are given in the Appendix.

Given the environmental transition matrix, $\mathbf{B}$, and the set of population transition matrices $\{\mathbf{X}_i\}$, then the stochastic growth rate, $a$, and the stochastic sensitivity matrix, $S$, can be calculated by simulation (Tuljapurkar 1990). A sequence of environments is generated as a realization of the environmental Markov chain. At each time, $t$, the environment specifies a population transition matrix, $\mathbf{X}_t$. Starting from some initial vector of population structure, $\mathbf{U}_0$, the population growth rate, $\lambda_t$, and the population structure, $\mathbf{U}_t$, at each time can be calculated from the recursions

$$\lambda_{t+1} = \sum \mathbf{X}_{t+1}\mathbf{U}_t \; ,$$

$$\mathbf{U}_{t+1} = \mathbf{X}_{t+1}\mathbf{U}_t/\lambda_{t+1} \; .$$

Reproductive values can be calculated by projecting backward from some final reproductive value, $\mathbf{V}_{t_{\max}}$, using the recursion

$$\mathbf{V}_{t-1} = \mathbf{X}_t'\mathbf{V}_t/\lambda_t \; .$$

In our implementation of this algorithm, we choose $\mathbf{U}_0$ and $\mathbf{V}_{t_{\max}}$ as the deterministic stable age distribution and reproductive-value vector for the population transition matrix averaged across environments. These are projected for 100 iterations forward (for $\mathbf{U}$) and 100 iterations backward (for $\mathbf{V}$) to allow these vectors to converge to their stochastic distributions. The population is then projected for 1,000 iterations.

The stochastic growth rate, $a$, has been defined as follows (Tuljapurkar 1990, eq. 12.1.5):

$$a = \lim_{t \to \infty} \frac{1}{t} \, \mathrm{E} \, \log \left( \mathbf{V}'\mathbf{X}_t\mathbf{X}_{t-1} \ldots \mathbf{X}_2\mathbf{X}_1\mathbf{U} \right) , \tag{3}$$

where $\mathbf{X}_i$ is the population transition matrix at time $i$, $\mathbf{U}$ is an initial population stage vector, $\mathbf{V}$ is an initial reproductive-value vector, and $\mathrm{E}$ is the expectation operator. The stochastic population growth rate and its standard deviation are estimated from the simulated sequence $\{\lambda_t\}$ as

$$\hat{a} = \frac{1}{N} \sum_{t=1}^{N} \log \lambda_t \, ,$$

$$\hat{\sigma}_a = \left[ \frac{1}{N-1} \sum_{t=1}^{N} (\log \lambda_t - \hat{a})^2 \right]^{1/2} .$$

If the sequence of environments is not correlated, the small-noise approximation to the stochastic growth rate can be calculated without simulation (Tuljapurkar 1990, eq. 12.2.3) as

$$a \approx \log \lambda - \frac{1}{2\lambda^2} \sum_{ij,k\ell} \frac{\delta\lambda}{\delta_{ij}} \frac{\delta\lambda}{\delta a_{k\ell}} \, \mathrm{cov} \, a_{ij}a_{k\ell} . \tag{4}$$

The sensitivity of $a$ to a sequence of random perturbations $\{C_t\}$ to the population transition matrix is

$$\mathbf{S}_{ij} = \mathrm{E}\left(\frac{\mathbf{V}_t' C_t \mathbf{U}_{t-1}}{\lambda_t \mathbf{V}_t' \mathbf{U}_t}\right).$$

Many different sensitivities can be defined by the appropriate choice of $C_t$. One that is closely analogous to the deterministic sensitivity is to perturb the $ij$th element of every population transition matrix by some small constant. In this case, the sensitivity of $a$ to a small change in the $ij$th element of each transition matrix is

$$\mathbf{S}_{ij} = \mathrm{E}\left(\frac{V_{t,i}\, U_{(t-1),j}}{\lambda_t \mathbf{V}_t' \mathbf{U}_t}\right) \tag{5}$$

(Tuljapurkar 1990, eq. 11.2.8), where $\mathbf{U}_t$, $\mathbf{V}_t$, and $\lambda_t$ are the random sequences of population structure, reproductive value, and population growth rate. $\mathbf{S}_{ij}$ is calculated by approximating the expectation in equation (5) by the average over the 1,000 iterations. All calculations are performed in MATLAB.

The stochastic elasticities are calculated from the stochastic sensitivities by

$$\mathbf{E}_{ij} = \bar{\mathbf{X}}_{ij} \times \mathbf{S}_{ij}, \tag{6}$$

where $\bar{\mathbf{X}}$ is the average demographic transition matrix from $k$ environments that occur with probabilities $p_i$: $\bar{\mathbf{X}} = \sum_{i=1}^{k} p_i \mathbf{X}_i$. The stochastic elasticities describe the change in growth rate in response to a proportional change in one parameter of the transition matrix. Equation (6) is not the only possible definition of stochastic elasticities. Another possible definition is to incorporate $\mathbf{X}_{ij}$ inside the expectation of equation (5), giving

$$\mathbf{E}_{ij}^* = \mathrm{E}\left(\frac{V_{t,i} X_{t,ij} U_{(t-1),j}}{\lambda_t \mathbf{V}_t' \mathbf{U}_t}\right).$$

*Examples*

*Northern monkshood.* Northern monkshood (*Aconitum noveboracense*) is an herbaceous perennial plant found in cool, damp, semi-shaded habitats in New York, Ohio, Wisconsin, and Iowa. In 1978 it was listed as Threatened under the U.S. Endangered Species Act. The demography of 14 populations in New York State was studied from 1985 to 1990 (Dixon & Cook 1990). Each year all individuals in a population were marked, measured, and classified into one of five size classes: seedlings, juveniles (2 or 3 leaves, but no stem),

TABLE 2. *Monkshood Growth Rates for Four Environmental Transition Matrices*

| Environmental Transition Matrix | Deterministic $\lambda$ ($\log \lambda$) | Stochastic $a$ | SD |
|---|---|---|---|
| Not correlated ($\mathbf{B}_1$) | 0.94 ($-0.063$) | $-0.071$ | 0.10 |
| Periodic ($\mathbf{B}_2$), $\rho = -0.6$ | 0.94 ($-0.063$) | $-0.076$ | 0.11 |
| Autocorr. ($\mathbf{B}_3$), $\rho = 0.6$ | 0.94 ($-0.063$) | $-0.058$ | 0.10 |
| Autocorr. ($\mathbf{B}_4$), $\rho = 0.9$ | 0.94 ($-0.063$) | 0.037 | 0.09 |

NOTE.—Approximate stochastic growth rate, from eq. (4): $\hat{a} = -0.065$.

small (stem diameter < 2 mm), medium (2 mm $\geq$ stem diameter <4 mm), and large (stem diameter $\geq$4 mm). Like many herbaceous perennial plants, individuals often shrink in size; hence, there are nonzero transition probabilities from large plants to smaller-size classes.

The demography of the New York populations is driven by clonal reproduction. Although plants produce abundant viable seeds, few seedlings germinate and almost none survive until the first census date. Most new individuals result from clonal propagation. Large stems frequently produce two or more smaller stems.

A stage-based deterministic matrix model was used to calculate growth rates and sensitivities (Dixon & Cook 1990). Growth rates are calculated for each population in each year, along with sensitivities, for a single pooled transition matrix, pooled across all years and sites. Because sample sizes are small, for this analysis we construct year-specific transition matrices by pooling sites. Variation between years in the transition-matrix elements is moderately high. Coefficients of variation range from 13 percent to 180 percent. Yearly population growth rates range from a low of $\lambda = 0.83$ in 1987–88 to a high of $\lambda = 1.05$ in 1988–89. Elasticity analysis of the pooled transition matrix indicates that the largest elasticities are associated with the survival rates of the juvenile, small-sized, and medium-sized classes, that is, the (2 2), (3 3), and (4 4) elements of the transition matrix.

Stochastic growth rates and sensitivities are computed for four environmental transition matrices (see Table 2): an uncorrelated transition matrix ($\mathbf{B}_1$), a periodic matrix with moderate negative autocorrelation ($\rho = -0.6$, $\mathbf{B}_2$), a transition matrix with a moderate autocorrelation ($\rho = 0.6$, $\mathbf{B}_3$), and a transition matrix with

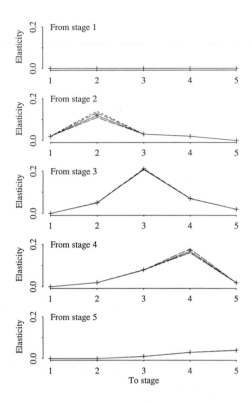

FIGURE 18. *Elasticities for each transition of the northern monkshood:* +, *mean deterministic matrix;* solid lines, *stochastic uncorrelated environments;* short dashes, *stochastic periodic environments;* medium dashes, *stochastic autocorrelated environments;* long dashes, *stochastic highly autocorrelated environments.*

a high autocorrelation ($\rho = 0.9$, $\mathbf{B}_4$). Incorporating environmental variability reduces slightly the estimated growth rate in the uncorrelated and periodic environments and increases slightly the growth rate in the two autocorrelated sequences (Table 2). However, these differences are of little practical consequence because they are all much less than the uncertainty in the growth rate.

The stochastic elasticities for each of the four environmental transition matrices are almost identical to those from the deterministic analysis of the average matrix (Fig. 18). The highest elasticities are associated with the survival probabilities of juvenile, small-sized, and medium-sized plants, whereas the fecundities have

TABLE 3. *Summary of Elasticities for Monkshood*

| Data Set | Environmental Transition | Largest Elasticity | Matrix Elements with Large $E_{ij}$ |
|---|---|---|---|
| Monkshood | Deterministic | 0.221 | (3 3), (4 4), (2 2) |
| | Uncorrelated | 0.226 | (3 3), (4 4), (2 2) |
| | Periodic | 0.230 | (3 3), (4 4), (2 2) |
| | Positively correlated | 0.222 | (3 3), (4 4), (2 2) |
| | Highly correlated | 0.218 | (3 3), (4 4), (2 2) |

low elasticities (Table 3). The magnitude of the stochastic elasticities is essentially the same as that of the deterministic elasticities. The choice of environmental autocorrelation changes the numerical values of the elasticities, but not which elements have large elasticities. If conservation or management decisions are made by identifying transitions with large elasticities, the same decisions would be made in the deterministic case and all the stochastic cases.

*Tropical savanna grass.*   South American savannas are subject to stochastic fire regimes, which affect the vital rates of the grasses and other species that have evolved there. The perennial grass *Andropogon semiberbis* depends on fire; without it, the population can decline rapidly to extinction ($\lambda = 0.28, r = -1.29$) (Silva et al. 1991). Burned populations, in contrast, may increase quickly because of higher fecundity and growth rates ($\lambda = 1.25$, $r = 0.23$). Silva et al. (1991) calculated population growth rates for different fire frequencies in a stochastic model with three levels of autocorrelation ($\rho = -0.5$, 0.0, and +0.5). They found that fire must occur in 85 percent of the years in order to maintain the population; they observed little effect of autocorrelation on the population growth rate.

Values for the deterministic elasticity in the burned and unburned models are qualitatively different, although peak elasticities occur in the growth from stage 1 to 2 and in the survival of stage 2 for both matrices. We ran a series of stochastic simulations to calculate elasticities for five fire regimes (two burn levels, three autocorrelation levels; Table 4). Autocorrelation exerts negligible effects on the elasticities, although the stochastic growth rate does decrease with increasing values of $\rho$. The stochastic-elasticity

TABLE 4. *Savanna Grass Population Growth Rates for Two Environmental Transition Matrices (Burned and Unburned)*

| Environmental Transition Matrix | $\lambda$ $(\log \lambda)$ | $a$ | SD | Elasticity $(1 \to 2)$ | Elasticity $(2 \to 2)$ |
|---|---|---|---|---|---|
| Burned matrix | 1.25 (0.22) | — | — | 0.20 | 0.26 |
| Unburned matrix | 0.28 ($-1.3$) | — | — | 0.14 | 0.35 |
| 50% burn, mean matrix | 0.78 ($-0.24$) | — | — | 0.19 | 0.26 |
| 50% burn, $\rho = 0.0$ | — | $-0.48$ | 0.65 | 0.45 | 0.51 |
| 50% burn, $\rho = -0.5$ | — | $-0.46$ | 0.54 | 0.41 | 0.53 |
| 50% burn, $\rho = +0.5$ | — | $-0.52$ | 0.71 | 0.48 | 0.52 |
| 85% burn, mean matrix | 1.11 (0.11) | — | — | 0.20 | 0.26 |
| 85% burn, $\rho = 0.0$ | — | $-0.023$ | 0.46 | 0.29 | 0.22 |
| 85% burn, $\rho = +0.5$ | — | $-0.038$ | 0.54 | 0.19 | 0.24 |

NOTE.—Percent burn refers to number of years of fire; $\rho$ is the level of autocorrelation. Approximate stochastic growth rate, from eq. 4, is $\hat{a} = -3.2$.

values are quantitatively different from the mean matrix elasticities, but there are no qualitative changes in the most critical matrix parameters (Fig. 19*a,b*). As in the northern monkshood example, management recommendations suggested by the deterministic elasticities for the mean matrices and the stochastic matrices are the same: maintain the health of stage-2 individuals through frequent burning.

## *Discussion*

In both case studies, deterministic elasticity is a good indicator of stochastic elasticity. Other responses—for example, the stochastic population growth rate—are not well predicted by a deterministic analysis of the mean matrix. In the grass study, the stochastic growth rate is quite different from either the approximate stochastic growth rate (eq. 4) or the deterministic growth rate from the

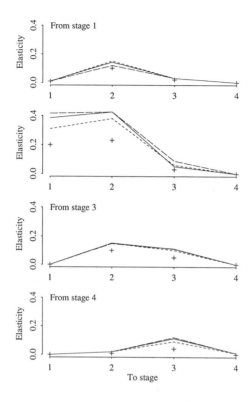

FIGURE 19. *Elasticities for each transition of* Andropogon: +, *mean deterministic matrix;* solid lines, *stochastic uncorrelated environments;* short dashes, *stochastic periodic environments,* $\rho = -0.5$; long dashes, *stochastic autocorrelated environments,* $\rho = 0.5$.

mean matrix (Table 4). In both cases, the difference can be attributed to the large differences between the burned and unburned transition matrices.

Our results indicate that elasticities from mean deterministic matrices are qualitatively invariant, even in radically different environments. In both case studies, stochastic-elasticity analysis does not change the management recommendations based on a much simpler deterministic analysis. This does not prove that stochastic elasticities are always equivalent to the deterministic elasticities. The two case studies include one with relatively little variation among years and one with large variation between environments, but they do not cover all the possible ways in which transition matrices can vary. However, the qualitative similarity between the deterministic and stochastic analyses is reassuring.

## 4 Conclusions

A common theme in this chapter has been the estimation of parameter sensitivity in structured models for population management. The choice of model structure and parameter values, while always dependent on the availability of data, is also dependent on the question to be answered; in these three cases, the models attempt to address how management should be focused on particular life stages. In the analysis of elk-and-wolf interaction, we evaluate the effect of adding age structure to a predator-prey model; although most of the simulations do not change qualitatively, we are able to analyze the potential impacts of management or wolf predation that specifically affect calf and yearling survival. Thus, the addition of age structure allows predictions for more-detailed management. In the tick section, we recast a simulation model as a matrix-transition model to evaluate analytically the sensitivity of population growth rate to parameter changes. Our elasticity analysis enables us to pinpoint the most critical time of year and life stages for tick population growth, thus providing a way to focus management efforts for population control. Finally, our comparison of stage-specific elasticities in deterministic and stochastic population models shows that in most cases, elasticity estimates do not change qualitatively in variable environments. This type of analysis is thus seen to be qualitatively invariant for populations with moderate amounts of temporal variability in the vital rates.

These sorts of calculations have two different implications for population managers. As tools for model assessment, sensitivity calculations indicate critical components of a model. Particular effort should be made to ensure that parameters with high sensitivity are accurately known and that critical parts of the model structure are appropriately specified. In this role, sensitivity and elasticity analyses are tools for model validation. As tools for management, sensitivities indicate where management actions to change vital rates have large effects on the population growth rate. In this role, sensitivity and elasticity analyses are one component among many that contribute to effective and scientifically justifiable decision making.

## Acknowledgments

Cynthia Lord was the sixth member of the group that developed the ideas in this chapter. Unfortunately, she was not able to be listed as an author because we could not locate her two years after the summer school. We would like to especially thank her for her work on the section "Models and Control for Tick Populations." We thank Mark Boyce for allowing us to use WOLF5 and making the source code available to us. He helped us greatly in transforming WOLF5 into a structured-population model, and the source code allowed us to translate his Pascal code into MATLAB code. We thank the Summer School staff and the Applied Mathematics Department at Cornell University for providing us with the environment and resources to learn and to develop this project. In particular, we thank Dr. Guckenheimer and the student assistants for allowing us access and for helping us with their computer facilities. Finally, we would like to thank Shripad Tuljapurkar and Hal Caswell for their ongoing support and their initial vision of the Summer School course. It's been a privilege to work with them and the diverse group of people they assembled.

The manuscript preparation was partially supported by contract DE-AC09-76-SROO-819 between the U.S. Department of Energy and the Savannah River Ecology Laboratory of the University of Georgia.

## Appendix

*Transition Matrices for Northern Monkshood, Pooled across Sites*

$$
\begin{array}{cc}
1985\text{--}1986 & 1986\text{--}1987 \\[4pt]
\begin{bmatrix}
0 & 0 & 0 & 0 & 0 \\
0 & 0.422 & 0.065 & 0.174 & 0.071 \\
0 & 0.244 & 0.455 & 0.174 & 0.286 \\
0 & 0.022 & 0.351 & 0.628 & 0.286 \\
0 & 0 & 0 & 0.047 & 0.571
\end{bmatrix}
&
\begin{bmatrix}
0 & 0 & 0.001 & 0.013 & 0.041 \\
0.500 & 0.808 & 0.166 & 0.095 & 0.103 \\
0 & 0.239 & 0.621 & 0.175 & 0.138 \\
0 & 0.047 & 0.187 & 0.574 & 0.207 \\
0 & 0 & 0.006 & 0.052 & 0.586
\end{bmatrix}
\end{array}
$$

*(continued)*

1987–1988                                 1988–1989

$$\begin{bmatrix} 0 & 0 & 0.001 & 0.011 & 0.033 \\ 0.455 & 0.481 & 0.167 & 0.100 & 0.206 \\ 0 & 0.176 & 0.583 & 0.345 & 0.294 \\ 0 & 0.030 & 0.119 & 0.437 & 0.559 \\ 0 & 0 & 0.005 & 0.026 & 0.353 \end{bmatrix} \quad \begin{bmatrix} 0 & 0 & 0.013 & 0.286 & 0.869 \\ 0.750 & 0.588 & 0.101 & 0.107 & 0.048 \\ 0 & 0.320 & 0.592 & 0.266 & 0.048 \\ 0 & 0.039 & 0.223 & 0.469 & 0.286 \\ 0 & 0 & 0.018 & 0.170 & 0.571 \end{bmatrix}$$

Average matrix

$$\begin{bmatrix} 0 & 0 & 0.004 & 0.078 & 0.235 \\ 0.426 & 0.575 & 0.125 & 0.119 & 0.107 \\ 0 & 0.245 & 0.563 & 0.240 & 0.191 \\ 0 & 0.034 & 0.220 & 0.527 & 0.334 \\ 0 & 0 & 0.007 & 0.074 & 0.521 \end{bmatrix}$$

*Environmental Transition Matrices for Stochastic Analysis of Northern Monkshood*

Uncorrelated                              Negatively Correlated,
Environments                          Semi-Periodic Environments
$(\rho = 0)$                                   $(\rho = -0.6)$

$$\mathbf{B}_1 = \begin{bmatrix} 0.25 & 0.25 & 0.25 & 0.25 \\ 0.25 & 0.25 & 0.25 & 0.25 \\ 0.25 & 0.25 & 0.25 & 0.25 \\ 0.25 & 0.25 & 0.25 & 0.25 \end{bmatrix} \quad \mathbf{B}_2 = \begin{bmatrix} 0.1 & 0.1 & 0.1 & 0.7 \\ 0.7 & 0.1 & 0.1 & 0.1 \\ 0.1 & 0.7 & 0.1 & 0.1 \\ 0.1 & 0.1 & 0.7 & 0.1 \end{bmatrix}$$

Positively Correlated                    Positively Correlated
Environments                             Environments
$(\rho = 0.6)$                                  $(\rho = 0.9)$

$$\mathbf{B}_3 = \begin{bmatrix} 0.7 & 0.1 & 0.1 & 0.1 \\ 0.1 & 0.7 & 0.1 & 0.1 \\ 0.1 & 0.1 & 0.7 & 0.1 \\ 0.1 & 0.1 & 0.1 & 0.7 \end{bmatrix} \quad \mathbf{B}_4 = \begin{bmatrix} 0.925 & 0.025 & 0.025 & 0.025 \\ 0.025 & 0.925 & 0.025 & 0.025 \\ 0.025 & 0.025 & 0.925 & 0.025 \\ 0.025 & 0.025 & 0.025 & 0.925 \end{bmatrix}$$

## Literature Cited

Anderson, R. M. 1982. *Population Dynamics of Infectious Diseases: Theory and Applications.* Chapman & Hall, New York.
———. 1988. The role of models in the study of HIV transmission and the epidemiology of AIDS. *Journal of AIDS* 1: 241–256.

Anderson, R. M., and R. M. May. 1989. The transmission dynamics of human immunodeficiency virus (HIV). *Philosophical Transactions of the Royal Society of London Biological Sciences* 321: 565–607.

———. 1991. *Infectious Diseases of Humans.* Oxford Science Publications, Oxford.

Basson, M., J. R. Beddington, and R. M. May. 1991. An assessment of the maximum sustainable yield of ivory from African elephant populations. *Mathematical Biosciences* 104: 73–95.

Boyce, M. S. 1992. Wolf recovery for Yellowstone National Park: A simulation model. Pp. 123–138 *in* D. R. McCullough and R. H. Barrett, eds., *Wildlife 2001: Populations.* Elsevier Applied Science, London.

Boyce, M. S., and Z. Merrill. 1991. Effects of the 1988 fires on ungulates in Yellowstone National Park. *Proceedings of the Tall Timbers Fire Ecology Conference* 17: 121–132.

Carter, N. 1985. Simulation modeling of the population dynamics of cereal aphids. *BioSystems* 18: 111–119.

Caswell, H. 1989a. Analysis of life table response experiments I. Decomposition of effects on population growth rate. *Ecological Modeling* 46: 221–237.

———. 1989b. *Matrix Population Modeling.* Sinauer, Sunderland, Mass.

Caswell, H., and M. C. Trevisan. 1994. Sensitivity analysis of periodic matrix models. *Ecology* 75: 1299–1303.

Cattan, P. E., and A. A. Glade. 1989. Management of the vicuna (*Vicuna vicuna*) in Chile: Use of a matrix model to assess harvest rates. *Biological Conservation* 49: 131–140.

Crouse, D. T., L. B. Crowder, and H. Caswell. 1987. A stage-based population model for loggerhead sea turtles and implications for conservation. *Ecology* 68: 1412–1423.

Crowder, L. B., D. T. Crouse, S. S. Heppell, and T. H. Martin. 1994. Predicting the impact of turtle excluder devices on loggerhead turtle populations. *Ecological Applications* 4: 437–445.

de Kroon, H., A. Plaisier, J. van Groenendael, and H. Caswell. 1986. Elasticity: The relative contribution of demographic parameters to population growth rate. *Ecology* 67: 1427–1431.

Dixon, P. M., and R. E. Cook. 1990. *Life History and Demography of Northern Monkshood ('Aconitum noveboracense') in New York State.* Cornell Plantations, Ithaca, N.Y.

Eberhardt, L. L. 1990. A fur seal population model based on age structure data. *Canadian Journal of Fisheries & Aquatic Science* 47: 122–127.

Elsevier. 1971. *World Survey of Climatology, Volume 13: Climates of Australia and New Zealand.* Elsevier, New York.

Godfrey, H. C. J., and J. K. Waage. 1991. Predictive modeling in biological control: The mango mealy bug (*Rastrococcus invadens*) and its parasitoids. *Journal of Applied Ecology* 28: 434–453.

Gutierrez, A. P., J. V. Baumgaertner, and C. G. Summers. 1984. Multitrophic models of predator prey energetics. II. A realistic model of plant-herbivore-parasitoid-predator interactions. *Canadian Entomologist* 116: 933–949.

Gutierrez, A. P., P. Neuenschwander, F. Schulthess, H. R. Herren, J. V. Baumgaertner, B. Wermelinger, B. Lohr, and C. K. Ellis. 1988. Analysis of a biological control of cassava pests in Africa. II. Cassava mealybug *Phenacoccus manihoti*. *Journal of Applied Ecology* 25: 921–940.

Hassell, M. P. 1978. *The Dynamics of Arthropod Predator-Prey Systems*. Princeton University Press, Princeton, N.J.

———. 1980. Foraging strategies, population models and biocontrol: A case study. *Journal of Animal Ecology* 49: 603–628.

Hassell, M. P., and R. M. May. 1990. *Population, Regulation and Dynamics*. Royal Society, London.

Heppell, S. S., J. R. Walters, and L. B. Crowder. 1994. Evaluating management alternatives for red-cockaded woodpecker: A modeling approach. *Journal of Wildlife Management* 58: 479–487.

Hitchcock, L. F. 1955. Studies of the non-parasitic stages of the cattle tick, *Boophilus microplus* (Canestrini) (Acarina: Ixodidae). *Australian Journal of Zoology* 3: 295–311.

Hollings, C. S. 1959. Some characteristics of some simple types of predation and parasitism. *Entomologist* 91: 385–398.

Houston, D. B. 1982. *The Northern Yellowstone Elk: Ecology and Management*. Macmillan, New York.

Mendoza, G. A., and A. Setyarso. 1986. A transition matrix forest growth model for evaluating alternative harvesting schemes in Indonesia. *Forest Ecology & Management* 15: 219–228.

Mount, G. A., and D. G. Haile. 1987. Computer simulation of population dynamics of the Lone Star tick, *Amblyomma americanum* (Acari: Ixodidae). *Journal of Medical Entomology* 24: 356–369.

Mount, G. A., D. G. Haile, R. B. Davey, and L. M. Cooksey. 1991. Computer simulation of *Boophilus* cattle tick (Acari: Ixodidae) population dynamics. *Journal of Medical Entomology* 28(3): 223–240.

Silva, J. F., J. Raventos, H. Caswell, and M. C. Trevisan. 1991. Population responses to fire in a tropical savanna grass, *Andropogon semiberbis*: A matrix model approach. *Journal of Ecology* 79: 345–356.

Slooten, E., and F. Lad. 1991. Population biology and conservation of Hector's dolphin. *Canadian Journal of Zoology* 69: 1701–1707.

Sutherst, R. W., G. F. Maywald, A. S. Bourne, I. D. Sutherland, and D. A. Stegeman. 1988. Ecology of the cattle tick (*Boophilus microplus*) in subtropical Australia. II: Resistance of different breeds of cattle. *Australian Journal of Agricultural Research* 39: 299–308.

Taylor, R. J. 1984. *Predation*. Chapman & Hall, New York.

Teel, P. D. 1984. Effect of saturation deficit on eggs of *Boophilus annulatus* and *B. microplus* (Acari: Ixodidae). *Annals of the Entomological Society of America* 77: 65–68.

Tuljapurkar, S. 1985. Population dynamics in variable environments. VI: Cyclical environments. *Theoretical Population Biology* 28: 1–17.

———. 1990. *Population Dynamics in Variable Environments.* Springer-Verlag, New York.

USFWS. 1987. *Northern Rocky Mountain Wolf Recovery Plan.* U.S. Fish and Wildlife Service, Denver, Colo.

# CHAPTER 17

# Nonlinear Ergodic Theorems and Symmetric versus Asymmetric Competition

## Kathleen M. Crowe

From the earliest discrete linear models of age-structured populations through the most recent work in nonlinear size- and stage-structured models, the use of theorems concerning the existence of a stable normalized class distribution has been a key facet of the study of structured populations. Not only do the fundamental theorem of demography and later ergodic theorems provide knowledge about the behavior of the population's distribution among its classes, they also allow the analysis of these models to be greatly simplified by providing a limiting equation for total population size.

One main necessary assumption in these ergodic theorems is that the interactions between individuals be symmetric; that is, the per unit effects caused by an individual are the same regardless of the class to which the individual belongs. However, in many ecological examples the interactions between species are asymmetric (see Neill 1975, 1988; Werner & Gilliam 1984). In such cases, one or more classes are responsible for a "greater than average" per unit effect on some other class or classes. Few models of this sort of interaction exist, but of the ones that have been analyzed, many of the populations fail to exhibit a stable normalized class distribution.

There are two main goals of this chapter. The first is to consider several ergodic theorems, from the simplest linear case to more-complicated versions incorporating density dependence. These theorems are applied to a model of a size-structured population with symmetric competition. Then, two simple examples are used to test models of asymmetric competition between two size-structured populations. The final section is a discussion of open questions in the areas of nonlinear ergodic theorems and the use of limiting equations in the analysis of structured-population models.

## 1 Single-Species Models

The system of difference equations is given by

$$\mathbf{x}(t+1) = \mathbf{A}\left[\mathbf{x}(t)\right]\mathbf{x}(t),$$

where the $\mathbf{x}(t) \in \mathbf{R}^{m+}$ is the size-class distribution vector, and the (nonnegative) $m \times m$ projection matrix $\mathbf{A}$ may be written as the sum of a fertility matrix and a transition matrix; that is, $\mathbf{A}(\mathbf{x}) = \mathbf{F}(\mathbf{x}) + \mathbf{T}(\mathbf{x})$. Corresponding to the distribution vector at time $t$ is the total population size at time $t$, which we denote by $P(t)$. To arrive at $P(t)$, let $\mathbf{w}$ be any nonnegative vector. Then, the inner product

$$P(t) = \mathbf{w}^{\mathrm{T}}\mathbf{x}(t) = \sum_{i=1}^{m} w_i x_i(t)$$

provides a measure of total population size.

In the simplest case, that of a constant environment, the matrix $\mathbf{A}(\mathbf{x})$ is identically equal to $\mathbf{A}$, a constant matrix. If we assume that $\mathbf{A}$ is nonnegative, irreducible, and primitive, then a theorem of Frobenius states that $\mathbf{A}$ has a dominant eigenvalue $\lambda$, with corresponding right eigenvector $\mathbf{v}$. As $t$ increases, the class distribution vector $\boldsymbol{\eta}(t) = \mathbf{x}(t)/P(t)$ converges to $\mathbf{v}$, and the dynamics of total population size approach $P(t+1) = \lambda P(t)$. Also, as shown in Chapter 3, by Tuljapurkar, note that the dynamics of the class-distribution vector are governed by the equation

$$\boldsymbol{\eta}(t+1) = \frac{\mathbf{A}\boldsymbol{\eta}(t)}{|\mathbf{A}\boldsymbol{\eta}(t)|},$$

where $|\cdot|$ indicates the sum of elements in the vector enclosed by the vertical lines.

### *Density-Dependent Survival*

To incorporate density dependence into these models, consider the kinds of nonlinear matrix models that still have a stable size distribution. First is the case in which survival in all classes is reduced by a common fraction $h\,[\mathbf{x}(t)]$, a function of the class distribution at time $t$. The system of difference equations to be studied may then be written as

$$\mathbf{x}\,(t+1) = h\,[\mathbf{x}(t)]\,\mathbf{A}\mathbf{x}(t)\,,$$
$$\mathbf{x}\,(0) = \mathbf{x}_0 > \mathbf{0}\,, \tag{1}$$

where it is again assumed that $\mathbf{A} \geq 0$, is irreducible, and is primitive and, in addition, that $h(\mathbf{0}) = 1$ (i.e., there is no reduction in survival at low population levels).

The normalized distribution vector must satisfy the equation

$$\boldsymbol{\eta}(t+1) = \frac{h\,[\mathbf{x}(t)]\,\mathbf{A}\mathbf{x}(t)}{|h\,[\mathbf{x}(t)]\,\mathbf{A}\mathbf{x}(t)|} = \frac{\mathbf{A}\mathbf{x}(t)}{|\mathbf{A}\mathbf{x}(t)|}\,,$$

and, just as in the linear case, $\boldsymbol{\eta}(t)$ satisfies equations (1) with the condition that $|\boldsymbol{\eta}(t)| = 1$. Thus, the normalized distribution vector $\boldsymbol{\eta}(t)$ again approaches $\mathbf{v}$, the (unit) eigenvector of $\mathbf{A}$ corresponding to the maximal eigenvalue $\lambda$, as $t \to +\infty$. This constitutes the nonlinear ergodic result shown by Cushing (1989) and again allows the derivation of an autonomous limiting equation for total population size.

In this case, the expression for total population size may be written as

$$P(t+1) = h\,[\mathbf{x}(t)]\,\mathbf{w}^{\mathrm{T}}\mathbf{A}\mathbf{x}(t) = h\,[P(t)\boldsymbol{\eta}(t)]\,\mathbf{w}^{\mathrm{T}}\mathbf{A}\boldsymbol{\eta}(t)P(t)\,.$$

Again, this is a scalar nonautonomous difference equation, but since $\boldsymbol{\eta}(t)$ converges to $\mathbf{v}$, the equation for $P(t)$ converges asymptotically to the autonomous limiting equation

$$P(t+1) = \lambda h\,[P(t)\mathbf{v}]\,P(t)\,.$$

## Density-Dependent Growth and Birthrates

Next, consider a size-structured model in which population density affects both the transition probabilities and the birthrates of the species. Furthermore, assume that all births occur into the first size class, that no individual may grow more than one size class in one unit of time, that no individual may shrink, and that the survival probability per unit of time (represented by $\pi$) is constant over all classes. For simplicity, also assume that the amount of resource available at each time unit is constant, although the results hold if the resource is dynamically modeled. (For a more complete derivation of the model, see Crowe 1991; Cushing 1991.)

Under these assumptions, some fraction $f_j$ of the surviving individuals in class $j$ advances to class $j+1$, with the fraction $1 - f_j$ remaining in class $j$. In addition, at each time unit the first size class gains $b_j$ offspring of surviving $j$-class individuals. It is assumed that population density affects the transition probabilities and fertility, through the $f_j$ and $b_j$, as follows.

A competition, or density-effects, term is denoted by $c(P)$, with the assumptions that $c(0) = 1$, $c'(P) \leq 0$, and $c(P) \leq 1$ for $P > 0$. The density effects are thus negligible at low population levels, increasing monotonically with increasing population levels. This density term may be modeled in various ways ranging from $c(P) \equiv 1$, representing purely exploitative competition, to stronger forms of competition such as $c(P) = 1/(1 + dP)$ and $c(P) = e^{-dP}$. It is this last form that will be used for the examples below.

Given this competition term, the transition probabilities may be expressed as $f_j = \beta_j c(P)$, and the births from each class as $b_j = \gamma_j c(P)$, where the growth and birth coefficients, $\beta_j$ and $\gamma_j$, respectively, are derived from such individual physiological parameters as body density and fraction of resource allocated to growth, etc. The following expressions then describe the transition and fertility matrices:

$$
\mathbf{T} = \pi
\begin{bmatrix}
1 - \beta_1 c(P) & 0 & 0 & \cdots & 0 \\
\beta_1 c(P) & 1 - \beta_2 c(P) & 0 & \cdots & 0 \\
0 & \beta_2 c(P) & 0 & \cdots & 0 \\
0 & 0 & \ddots & & \vdots \\
\vdots & \vdots & & \ddots & \ddots & \vdots \\
0 & 0 & & \cdots & \beta_{m-1} c(P) & 1
\end{bmatrix},
$$

$$
\mathbf{F} = \pi c(P)
\begin{bmatrix}
\gamma_1 & \gamma_2 & 0 & \gamma_m \\
0 & 0 & \cdots & 0 \\
\vdots & \vdots & \ddots & \vdots \\
0 & 0 & \cdots & 0
\end{bmatrix}.
$$

In contrast to the previous case, the nonlinearity, $c(P)$, cannot be simply factored out of the projection matrix $\mathbf{A}$, since it occurs as a scalar multiple on $\mathbf{F}$ but not as a scalar multiple on $\mathbf{T}$. Thus, we need another ergodic theorem.

**Theorem 1** *Consider the system of difference equations given by*

$$
\mathbf{x}(t+1) = \mathbf{A}(t)\mathbf{x}(t),
$$

$$
\mathbf{A}(t) = a_0(t)\mathbf{I} + a_1(t)\mathbf{L} + a_2(t)\mathbf{L}^2 + \cdots + a_r(t)\mathbf{L}^r, \tag{2}
$$

*where* $\mathbf{I}$ *is the* $m \times m$ *identity matrix and* $\mathbf{L}$ *is a constant non-negative* $m \times m$ *matrix. Assume that the following two hypotheses hold.*

**H1** *There exist constants* $a_0$ *and* $b_0$ *such that* $0 \le a_0(t) \le a_0$ *and* $0 < b_0 \le a_i(t)$, $i = 1, \dots, r$ *for all* $t = 0, 1, 2, \dots$ .

**H2** $\mathbf{L}$ *has a strictly dominant, positive, simple eigenvalue* $\lambda^+ > 0$, *with an associated positive eigenvector* $\mathbf{v}^+$, *normalized such that* $\mathbf{w}^{\mathrm{T}}\mathbf{v}^+ = 1$.

*Also assume that* $\mathbf{x}(t) \ge \mathbf{0}$ *is a solution of equations (2), such that* $\mathbf{x}(0) \ge \mathbf{0}$, $\mathbf{x}(0) \ne \mathbf{0}$, *and* $P(t) > 0$ *for* $t \ge 0$. *Then,* $\boldsymbol{\eta}(t) \to \mathbf{v}^+$ *as* $t \to +\infty$.

At first glance, this theorem seems unrelated to our example, but if $\beta_q$ denotes the maximum of all the $\beta_j$, then the projection matrix $\mathbf{A}$ may be rewritten as

$$\mathbf{A} = \pi \left[ 1 - \beta_q c(P) \right] \mathbf{I} + \pi c(P) \mathbf{L},$$

where the nonnegative matrix

$$\mathbf{L} = \begin{bmatrix} \beta_q - \beta_1 + \gamma_1 & \gamma_2 & \cdots & \gamma_q & \cdots & \gamma_{m-1} & \gamma_m \\ \beta_1 & \beta_q - \beta_2 & \cdots & 0 & \cdots & 0 & 0 \\ \vdots & \vdots & \ddots & \vdots & & \vdots & \vdots \\ 0 & 0 & \cdots & 0 & \cdots & 0 & 0 \\ \vdots & \vdots & & \vdots & \ddots & \vdots & \vdots \\ 0 & 0 & \cdots & 0 & \cdots & \beta_q - \beta_{m-1} & 0 \\ 0 & 0 & \cdots & 0 & \cdots & \beta_{m-1} & \beta_q \end{bmatrix}.$$

To satisfy the first assumption requires using Frobenius' theorem; to satisfy the second, it is sufficient to know that all solutions of equations (2) are bounded for $t \ge 0$. (For details, see Crowe 1991.)

This theorem once again yields a stable, normalized size distribution for this example, and as before, there is an autonomous limiting equation for total population size. The inner product of equations (2) with the vector $\mathbf{w}$ yields an equation for total population size: $P(t+1) = \left[ a_0(t)\mathbf{w}^{\mathrm{T}}\mathbf{I} + a_1(t)\mathbf{w}^{\mathrm{T}}\mathbf{L} \right] \mathbf{x}(t)$, which may be rewritten as

$$P(t+1) = \pi \left\{ \left[ 1 - \beta_q c(P) \right] \mathbf{w}^{\mathrm{T}}\mathbf{I} + c(P)\mathbf{w}^{\mathrm{T}}\mathbf{L} \right\} \mathbf{x}(t).$$

Because $\mathbf{w}^{\mathrm{T}}\mathbf{L}\mathbf{x}(t) = \mathbf{w}^{\mathrm{T}}\mathbf{L}\boldsymbol{\eta}(t)P(t)$, the equation for $P(t)$ may be rewritten further as the scalar, asymptotically autonomous equation

$$P(t+1) = \pi \left[ 1 - \beta_q c(P) + c(P)\mathbf{w}^{\mathrm{T}}\mathbf{L}\boldsymbol{\eta}(t) \right] P(t).$$

Because, by Theorem 1, $\mathbf{w}^{\mathrm{T}}\mathbf{L}\mathbf{x}(t) \to \mathbf{w}^{\mathrm{T}}\mathbf{L}\mathbf{v} = \lambda\mathbf{w}^{\mathrm{T}}\mathbf{v} = \lambda$, the total population size has the limiting equation

$$P(t+1) = \pi\left[1 + \theta c(P)\right]P(t),$$

where $\theta = \lambda - \beta_q$ is the eigenvalue of the matrix $\mathbf{L} - \beta_q\mathbf{I}$ with the largest real part.

In all of the preceding cases, the normalized class distribution stabilizes, regardless of the dynamics of the total population size, $P(t)$. Figure 1 illustrates this for the third example, that of the effects of density on the birth and growth rates. In these simulations, the competition function is taken to be $c(P) = e^{-0.1dP}$. In the bar graphs, the normalized class distributions approach a constant vector while the total population numbers equilibrate (views $a,b$) or oscillate (views $c,d$).

## 2 Interspecific Competition

Competition between two species may take various forms, may rely on exploitation or interference, and may be symmetric (across all classes) or asymmetric (allowing some classes to affect others selectively). Models of such interactions can be of many types, with varying levels of complexity. To keep things as simple as possible and to allow a comparison of the dynamics of populations under various assumptions, the model used here includes only two species and two size classes. A further simplification adds the assumption that only the larger of the two classes in each species is fertile (i.e., $\gamma_1 = 0$ for each species). Hence, it will be convenient to think of (and refer to) the first, or smaller, size class as the juvenile class and the second, or larger-size, class as the adult class. The first model, of symmetric competition, is followed by models of two simple kinds of asymmetric interactions; the dynamics resulting from the two kinds of models differ.

*Symmetric Competition*

Label the two species $A$ and $B$. For species $A$, let $\mathbf{x}(t) = [x_1(t), x_2(t)]^{\mathrm{T}}$ denote the vector of the size-class distribution, let $s_1 < s_2$ represent the sizes of individuals in the two classes of $\mathbf{x}$, and denote the total population size at time $t$ by $X(t) = s_1x_1(t) + s_2x_2(t)$. The corresponding quantities for species $B$ are $\mathbf{y}(t) = [y_1(t), y_2(t)]^{\mathrm{T}}$, $\sigma_1 < \sigma_2$, and $Y(t) = \sigma_1y_1(t) + \sigma_2y_2(t)$. Let $P(t) = X(t) + Y(t)$ denote the total population size of both species at time $t$. As in the

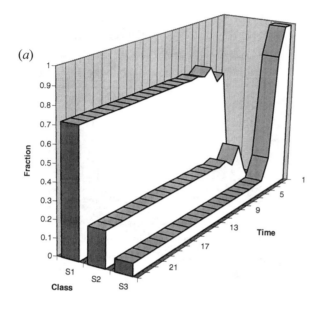

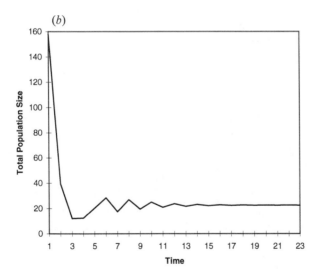

FIGURE 1. (Above and overleaf) *Three size-class simulations, with* $\beta_1 =$ 0.99, $\beta_2 = 0.95$, $\gamma_2 = 1000$: (a,b) *for equilibrating population;* (c,d) *for oscillating population.* (a,c) *The normalized class distribution;* (b,d) *the corresponding time series of the total population size* $P$.

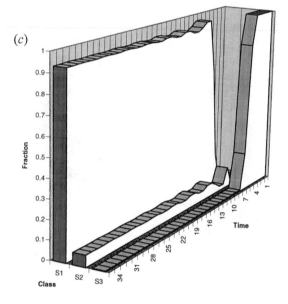

(c)

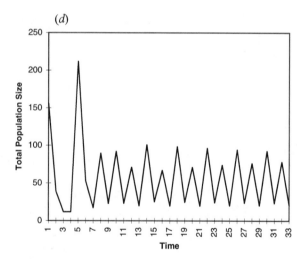

(d)

FIGURE 1. (Continued)

single-species model, let $1 - c(P)$ denote the (symmetric) competitive effects caused by $P$ units of competitors ($x_i$ and $y_i$). Then, the symmetric-competition model may be written as

$$\begin{bmatrix} x_1(t+1) \\ x_2(t+1) \end{bmatrix} = \pi_x \begin{bmatrix} 1 - \beta_x c(P) & \gamma_x c(P) \\ \beta_x c(P) & 1 \end{bmatrix} \begin{bmatrix} x_1(t) \\ x_2(t) \end{bmatrix},$$

$$\begin{bmatrix} y_1(t+1) \\ y_2(t+1) \end{bmatrix} = \pi_y \begin{bmatrix} 1 - \beta_y c(P) & \gamma_y c(P) \\ \beta_y c(P) & 1 \end{bmatrix} \begin{bmatrix} y_1(t) \\ y_2(t) \end{bmatrix}.$$

(For more about models of this form, see Crowe 1991; Cushing 1991.) Theorem 1 applies to each matrix difference equation to give stable, normalized class distributions, and hence limiting equations, for the total population size of each species:

$$X(t+1) = \pi_x \left[ 1 + \theta_x c(P) \right] X(t),$$
$$Y(t+1) = \pi_y \left[ 1 + \theta_y c(P) \right] Y(t).$$

The competitive outcome depends on the relative values of $\theta_x$ and $\theta_y$ and hence depends in a complicated way on the individual parameters in the model. A key feature of the competition is that, if the species are "similar" (i.e., the physiological parameters are close), then at equilibrium the species with larger adults will have driven the other species to extinction. An example of this is shown in Figure 2.

## Asymmetric Competition

The following models differ from the preceding one by introducing asymmetries into the interaction between the species. The juvenile bottlenecks documented by Neill (1975, 1988) and others provide a biological basis for these models, in which the adults of a smaller species negatively affect the juveniles of a larger species. Three assumptions about the symmetric model are introduced. First, the parameters in the symmetric model are such that equilibrium dynamics occur. Second, the juveniles of both species are the same size ($s_1 = \sigma_1$), but adults of $A$ are smaller than adults of $B$ ($s_2 < \sigma_2$). Finally, species $B$ outcompetes species $A$; that is, the equilibrium $(0, 0, y_{1e}, y_{2e})$ is locally asymptotically stable, whereas the equilibrium $(x_{1e}, x_{2e}, 0, 0)$ is unstable.

Next, the symmetric model is modified to incorporate asymmetries whereby the adults of species $A$ negatively affect the juveniles of species $B$. In one type of asymmetry, the benefits to species $A$

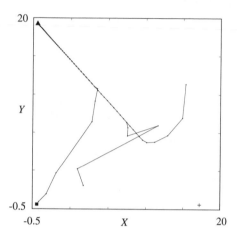

FIGURE 2. *Competitive exclusion of species $A$ under symmetric compe-*
*tition. Here, $\pi_x = \beta_x = \pi_y = 0.5$, $\beta_y = 0.7$, $\gamma_x = 50$, $\gamma_y = 75$. The sizes*
*are given by $s_1 = \sigma_1 = 1$, $s_2 = 4$, $\sigma_2 = 9$. +, Saddle; □, source; △, sink.*

are indirect (through a decrease in the survival rate of $B$ juveniles);
in a second type, these benefits are direct (through predation upon
the juveniles of species $B$).

*Indirect benefits.*   Assume that the presence of adults of species
$A$ decreases the survival of juveniles of species $B$ over one unit of
time by a fraction $1 - e^{-\delta s_2 x_2(t)}$, leading to the following system of
difference equations:

$$\begin{bmatrix} x_1(t+1) \\ x_2(t+1) \end{bmatrix} = \pi_x \begin{bmatrix} 1 - \beta_x c\,(P) & \gamma_x c\,(P) \\ \beta_x c\,(P) & 1 \end{bmatrix} \begin{bmatrix} x_1(t) \\ x_2(t) \end{bmatrix},$$

$$\begin{bmatrix} y_1(t+1) \\ y_2(t+1) \end{bmatrix} = \pi_y \begin{bmatrix} [1 - \beta_y c\,(P)]\,e^{-\delta s_2 x_2(t)} & \gamma_y c\,(P) \\ \beta_y c\,(P)\,e^{-\delta s_2 x_2(t)} & 1 \end{bmatrix} \begin{bmatrix} y_1(t) \\ y_2(t) \end{bmatrix}.$$

Note that species $A$ receives no direct benefits from the reduction
in survival of the species $B$ juveniles: there is no change in the
equations for species $A$. This has the immediate consequence that
Theorem 1 may be applied to the equations for $A$, even though it
may not be applied to the equations for $B$. The ergodic theorem
implies that the normalized distribution vector corresponding to
$\mathbf{x}$ approaches a positive vector, $\mathbf{v} = (v_1, v_2)^{\mathrm{T}}$; the limiting system

thus consists of three autonomous difference equations:

$$X(t+1) = \pi_x \left[1 + \theta_x c(P)\right] X(t),$$

$$y_1(t+1) = \pi_y \left\{ \left[1 - \beta_y c(X,\mathbf{y})\right] e^{-\delta s_2 v_2 X} y_1(t) + \gamma_y c(X,\mathbf{y}) y_2(t) \right\},$$

$$y_2(t+1) = \pi_y \left[ \beta_y c(X,\mathbf{y}) e^{-\delta s_2 v_2 X} y_1(t) + y_2(t) \right],$$

$$(3)$$

where $v_2 X = x_2$, the number of adults of species $A$.

A study of the dynamics of this system begins with the equilibrium states. If $\mathbf{F}(X,\mathbf{y},\delta)$ denotes the right-hand sides of equations (3), then the equilibria of this system are characterized by the equation $(X,\mathbf{y}) = \mathbf{F}(X,\mathbf{y},\delta)$. The dependence of $\mathbf{F}$ on $\delta$ is important, since $\delta$ is a bifurcation parameter in the analysis below.

Clearly, there is a trivial $(0,0)$ equilibrium; in addition, there are two "exclusion equilibria," in which only one species is present. These two are given by the equations $\mathbf{F}(X_e,\mathbf{0},\delta) = (X_e,\mathbf{0})$ (equilibrium A) and $\mathbf{F}(0,\mathbf{y}_e,\delta) = (0,\mathbf{y}_e)$ (equilibrium B), each of which has a unique solution for each value of $\delta > 0$ (since these are simply the single-species equilibria, which exist by assumption).

Denote by $D\mathbf{F}$ the Jacobian of $\mathbf{F}$. For equilibrium B, the eigenvalues of $D\mathbf{F}(0,\mathbf{y}_e,\delta)$ are independent of $\delta$. Then, because of the assumption that this equilibrium is locally asymptotically stable when $\delta = 0$, the equilibrium B is locally asymptotically stable under this asymmetric competition as well.

The Jacobian at equilibrium A is $D\mathbf{F}(X_e,\mathbf{0},\delta)$. Some rearranging of terms shows that there exists a unique $\delta_{\mathrm{cr}} > 0$ such that

$$D\mathbf{F}(X_e,\mathbf{0},\delta_{\mathrm{cr}}) = (X_e,\mathbf{0}).$$

In other words, there is a unique value of $\delta$, the parameter governing the strength of the asymmetry in competition, at which the linearized system $\mathbf{z} = D\mathbf{F}(X_e,\mathbf{0},\delta)\mathbf{z}$ has an eigenvalue of 1. Furthermore, it may be shown that for $\delta < \delta_{\mathrm{cr}}$, no eigenvalue of the matrix $D\mathbf{F}(X_e,\mathbf{0},\delta)\mathbf{z}$ has a magnitude of 1. Thus, $\delta = \delta_{cr}$ is a bifurcation value for equilibrium A. The direction of bifurcation and the stability of the bifurcating equilibrium need to be determined (see Crowe 1991; Cushing, Chapter 4), which can be done by finding the sign of the product $\mathbf{u}_{\mathrm{L}}^{\mathrm{T}} D\mathbf{F}_\delta(X_e,\mathbf{0},\delta_{\mathrm{cr}})\mathbf{u}_{\mathrm{R}}$, where $\mathbf{u}_{\mathrm{L}}$ and $\mathbf{u}_{\mathrm{R}}$ are the left and right eigenvectors of the Jacobian $D\mathbf{F}(X_e,\mathbf{0},\delta_{\mathrm{cr}})$ corresponding to the eigenvalue 1. Here $D\mathbf{F}_\delta$ is the (element-by-element) derivative of $D\mathbf{F}$ with respect to $\delta$. A calculation shows that the product above is less than zero; thus, there is a nontrivial $(X \neq X_e,\ \mathbf{y} \neq \mathbf{0})$ equilibrium locally about $(X_e,\mathbf{0})$, with an eigen-

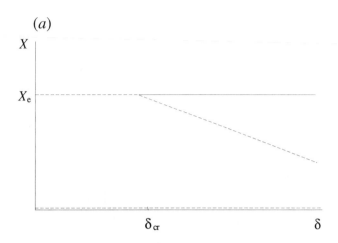

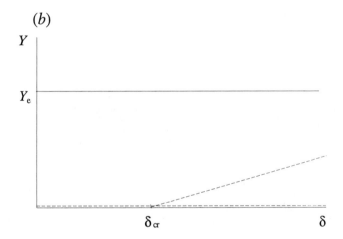

FIGURE 3. *Bifurcation diagrams for X and Y with bifurcation parameter*
$\delta$. (a) $Y = 0$; (b) $X = 0$. Solid line, *Stable equilibrium;* dashed lines,
*unstable equilibria.*

value $\mu(\delta)$ such that $\mu(\delta_{cr}) = 1$ and $\mu(\delta) > 1$ for $\delta \gtrsim \delta_{cr}$. Thus,
an unstable equilibrium $(X^*, \mathbf{y}^*)$ bifurcates from the equilibrium
$(X_e, \mathbf{0})$ into the interior of the positive cone, and the equilibrium A
gains (local) stability. This is illustrated schematically in Figure 3.

In terms of the competitive outcome, it is important that at the
bifurcation point, not only does a positive equilibrium appear, but
the formerly unstable equilibrium $(X_e, \mathbf{0})$ becomes asymptotically

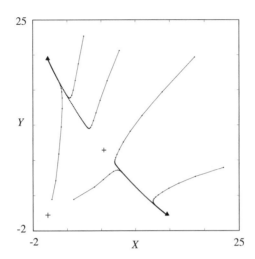

FIGURE 4. *Phase portrait in the population-size variables $X$ and $Y$ for asymmetric competition with indirect benefits.* $\beta_x = \beta_y = 0.3$, $\pi_x = \pi_y = 0.75$, $\gamma_x = 10$, $\gamma_y = 22.5$, $d = 0.1$, $\delta = 0.1$. +, *Saddles.*

stable as well. This implies that there exists some set of positive initial conditions under which species $B$, in spite of its larger adults, is driven to extinction by species $A$. This is illustrated in Figure 4, which shows the (unstable) saddle in the interior of the positive cone and various trajectories, some approaching equilibrium A and others approaching equilibrium B. Thus, the incorporation of a competitive asymmetry into the model allows (under some circumstances) a previously inferior competitor to survive and drive to extinction the competing species.

*Direct benefits.* Now consider another form of competitive asymmetry, in which the adults of species $A$ receive direct benefits. Specifically, assume that at low resource levels (i.e., high population levels $P$), the adults of species $A$ prey upon the juveniles of species $B$. We model this predation as follows.

Let $\bar{\Psi}(P)$ denote the "inherent" ($x_2$ and $y_1$ low) probability that a juvenile of species $B$ is eaten (by an adult of species $A$). Assume that $\bar{\Psi}(0) = 0$, $\bar{\Psi}(+\infty) = k$, and for all $P$, $0 \leq \bar{\Psi}(P) \leq 1$ and $\bar{\Psi}'(P) \geq 0$. Rewrite $\bar{\Psi}$ as $k\Psi$ to isolate the constant $k$, which will be used as a bifurcation parameter in the analysis of this model. Next, let $\phi(y_1)$ denote the fractional change in $\Psi$ due to the presence of $y_1$

units of juveniles of species $B$, and assume that $\phi(0) = 1$, $\phi(+\infty) > 0$, and for all $y_1$, $0 \leq \phi(y_1) \leq 1$, $\phi'(y_1) \leq 0$, and $d[y_1\phi(y_1)]/dy_1 > 0$. Finally, denote by $\xi(x_2)$ the fractional change in $\Psi$ due to the presence of $x_2$ units of adults of species $A$, and assume that $\xi(0) = 0$, $\xi(+\infty) = 1$, and for all $x_2$, $0 \leq \xi(x_2) \leq 1$ and $\xi'(x_2) \geq 0$.

With these definitions, we see that the probability that a juvenile of species $B$ survives one unit of time is

$$\pi_y \left[1 - k\Psi(P)\xi(x_2)\phi(y_1)\right],$$

and the number surviving is

$$\pi_y \left[1 - k\Psi(P)\xi(x_2)\phi(y_1)\right] y_1 .$$

This level of consumption of juveniles of species $B$ by adults of species $A$ results in an additional $\pi_x\gamma_x\zeta k\Psi(P)\xi(x_2)\phi(y_1)y_1$ offspring of species $A$, where $\zeta$ is a factor converting $y_1$ to resource units. The following system of four difference equations is then obtained for this asymmetric model:

$$
\begin{aligned}
x_1(t+1) &= \pi_x \left[(1 - \beta_x c)x_1(t) + \gamma_x cx_2(t) + \gamma_x\zeta k\Psi\xi\phi y_1(t)\right], \\
x_2(t+1) &= \pi_x \left[\beta_x cx_1(t) + x_2(t)\right], \\
y_1(t+1) &= \pi_y \left[(1 - k\Psi\xi\phi)(1 - \beta_y c)y_1(t) + \gamma_y cy_2(t)\right], \\
y_2(t+1) &= \pi_y \left[\beta_y(1 - k\Psi\xi\phi)cy_1(t) + y_2(t)\right].
\end{aligned}
\tag{1}
$$

In this case, none of the ergodic theorems available seems to apply, and the full four-dimensional system must be analyzed without using a limiting equation to reduce the number of equations under consideration. The study of this model begins with the exclusion equilibria. In addition to the trivial equilibrium, there are two unique exclusion equilibria, $X_e = (x_{1e}, x_{2e}, 0, 0)$ and $Y_e = (0, 0, y_{1e}, y_{2e})$. Let $\mathbf{F}(\mathbf{x}, \mathbf{y}, k)$ denote the right-hand side of equations (4). To study the stability of the equilibrium, $X_e$, write the Jacobian of $\mathbf{F}$ at $X_e$ as $D\mathbf{F}(X_e)$; two of its eigenvalues depend on $k$, the bifurcation parameter, while (by assumption) the other two have modulus not equal to 1. Calculation shows that as $k$ increases from 0, a single (real) eigenvalue crosses out of the unit circle through $+1$ when $k$ reaches a critical value $k_x$. Employing the bifurcation theorem stated by Cushing (Chapter 4) shows that (just as in the indirect-benefits case), as $k$ increases through $k_x$, the exclusion equilibrium, $X_e$, gains stability and a saddle bifurcates into the interior of the positive cone.

Turning to the equilibrium $Y_e$ again, two of the four eigenvalues of the Jacobian $D\mathbf{F}(Y_e)$ depend upon $k$, and as $k$ increases from 0,

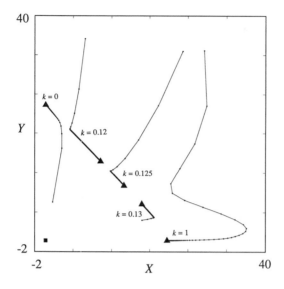

FIGURE 5. *Phase portrait for asymmetric competition with direct benefits. Parameter values are* $\zeta = 50, \gamma_x = 25, \beta_x = \beta_y = 0.5, \pi_x = \pi_y = 0.7, \gamma_y = 50$. *Asymmetric-competition functions are given by* $\phi(y_1) = 1/(1 + 5y_1)$, $\xi(x_2) = x_2/(10 + x_2)$, $\Psi(P) = P/(5 + P)$.

the first eigenvalue crosses out of the unit circle when $k$ reaches a critical value $k_y$. In this case, as $k$ increases through $k_y$, the equilibrium $Y_e$ loses stability and an asymptotically stable equilibrium bifurcates into the interior of the positive cone. Figure 5 shows an example of this situation for several values of $k$.

Thus, in this case of asymmetric competition with direct benefit to the smaller species, we see that either competitive exclusion (of either species) or competitive coexistence may occur. Which outcome occurs depends largely on the relative values of $k_x$ and $k_y$, since as $k$ increases from 0 (i.e., as the strength of the asymmetry increases), the effect of the asymmetry is determined by which critical value of $k$ is reached first.

Although this model does not meet the hypotheses of the available ergodic theorems, in numerical studies it appears that the normalized class distribution of species $B$ approaches a constant vector, so the model may possess ergodic properties.

The results in this section concern only equilibria; however, the same asymmetry also allows an inferior competitor to coexist in an oscillatory setting, as is illustrated in Figure 6.

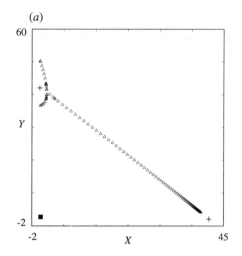

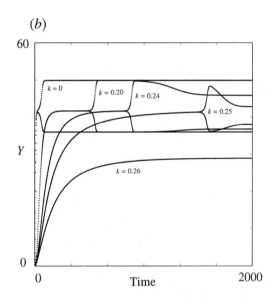

FIGURE 6. *Oscillatory trajectories for asymmetric competition with direct benefits.* (a) *Phase portrait for* $k$ *ranging from* 0 *to* 1. +, *Saddles;* □, *source;* △, *sinks.* (b) *Time series for five values of* $k$. *Parameter values are* $\zeta = 50$, $\gamma_x = 100$, $\beta_x = \pi_y = 0.5$, $\pi_x = \beta_y = 0.7$, $\gamma_y = 6000$, $d = 0.1$. *The functions* $\phi$, $\xi$, *and* $\Psi$ *are as in Figure 5.*

## 3 Concluding Remarks

Several ergodic theorems are applicable to models incorporating nonlinearities, for both single species and multispecies competition. Using the stable size distribution given by these theorems allows the derivation of a scalar limiting equation for total population size, greatly simplifying the analysis of these systems. However, there remain open mathematical questions concerning the relationship between the limiting behaviors of the asymptotically autonomous equation and its limit equation (see Crowe 1993, where persistence and limit set properties are studied). In cases that have been studied for almost all parameter values, the two equations do exhibit qualitatively similar behavior.

For some size-structured models of asymmetric competition, existing ergodic theorems cannot be applied. Such asymmetries are exhibited by many species and provide a possible explanation for the observed coexistence of similar species. Models of interactions of this type are necessarily more complicated, and hence more difficult to analyze, than are symmetric-competition models, in part because the species may not have a stable, normalized class distribution or do not satisfy the conditions of the known ergodic theorems. Clearly, many types of asymmetries remain to be modeled, and possibly less restrictive theorems concerning their normalized distributions remain to be developed. The results here suggest a connection between the "degree" of asymmetry in the competition and the ergodicity shown; precise knowledge of such a relationship would greatly increase our understanding of these types of interactions.

## Acknowledgments

This work was supported at Cornell University by the U.S. Army Research Office through the Mathematical Sciences Institute, contract DAAL03-91-C-0027.

## Literature Cited

Crowe, K. M. 1991. A discrete size-structured competition model. Ph.D. diss. University of Arizona, Tucson.
———. 1993. Persistence in asymptotically autonomous systems of difference equations. Fields Institute Technical Report.

————. 1994. A nonlinear ergodic theorem for discrete systems. *Journal of Mathematical Biology* 32: 179–191.

Cushing, J. M. 1989. A strong ergodic theorem for some nonlinear matrix models for structured population growth. *Natural Resource Modeling* 3(3): 331–357.

————. 1991. Competing size-structured species. Pp. 27–44 in O. Arino, D. Axelrod, and M. Kimmel, eds., *Mathematical Population Dynamics*. Marcel Dekker, New York.

Gantmacher, F. R. 1959. *Applications of the Theory of Matrices*. Interscience, New York.

Neill, W. E. 1975. Experimental studies of microcrustacean competition, community composition and efficiency in resource utilization. *Ecology* 56: 809–826.

————. 1988. Community responses to experimental nutrient perturbations in oligotrophic lakes: The importance of bottlenecks in size-structured populations. Pp. 236–255 *in* B. Ebenman and L. Persson, eds., *Size-Structured Populations: Ecology and Evolution*. Springer-Verlag, Berlin.

Werner, E., and J. F. Gilliam. 1984. The ontogenetic niche and species interactions in size-structured populations. *Annual Review of Ecology & Systematics* 15: 393–425.

# CHAPTER 18

# The Evolution of Age-Structured Marriage Functions: It Takes Two to Tango

## *Carlos Castillo-Chavez and Shu-Fang Hsu Schmitz*

Two-sex marriages in socially structured populations can be characterized as multiplicative perturbations of heterosexually random, or proportionate, mixing (Castillo-Chavez & Busenberg 1991). Such perturbations are expressed in terms of preferences, or affinities, of males for females, and vice versa. Male and female preferences are obviously not independent, since they depend on the availability of male and female behavioral classes. Knowledge of the preferences of one gender can characterize the preferences of both genders in socially structured populations; in other words, "it takes two to tango." This is the basic content of the $T^3$ Theorem. Different sets of preferences, that is, distinct behavioral classes, may give rise to identical mating probabilities, the determinants of the behavioral "phenotypes" (Hsu Schmitz 1994; Hsu Schmitz et al. 1993). Hence, different sets of individual decisions can lead to identical social dynamics, a fact well established in genetics. The importance of the incorporation of mating systems at the population level is a neglected but central area in evolutionary biology.

Marriage functions are solutions to this two-sex pairing problem. Despite their importance in areas such as population genetics (mating functions), demography (population projection), cultural anthropology (preservation and dissemination of cultural traits), and evolutionary biology (life history), their application has been quite limited. Most researchers have addressed theoretical issues in these areas through the use of single-sex models or highly simplified two-sex models. The difficulties involved are evident from the pioneering work of Kendall (1949), Keyfitz (1949), Fredrickson (1971), McFarland (1972), Parlett (1972), Pollard (1973), and Caswell and Weeks (1986).

We have developed an axiomatic framework for a systematic study of marriage functions (Busenberg & Castillo-Chavez 1989, 1991; Castillo-Chavez & Busenberg 1991; Hsu Schmitz 1994; Hsu Schmitz et al. 1993; Blythe et al. 1991). Our work has been applied in areas as diverse as cultural anthropology (Lubkin & Castillo-Chavez, in press), demography (Castillo-Chavez, Fridman & Luo, in press), epidemiology and food-web dynamics (Castillo-Chavez et al. 1995), and parameter estimation (Castillo-Chavez et al. 1992; Hsu Schmitz & Castillo-Chavez 1994, in press). We provide a summary of our characterization of marriage functions for populations defined through fixed characteristics such as race, language, biological species, religion, level of education, and socioeconomic level. We then provide a detailed characterization of age-structured pairing functions. The discrete framework described in this chapter can be incorporated into deterministic or stochastic models with finite dimension, while the continuous framework is easily incorporated into age-structured models.

A useful interpretation of our work is to think of preferences as a method for assigning parameters to a family of conditional probability distributions. Then, our main result establishes that setting the parameters of the male distribution completely determines its associated female distribution, and vice versa. In fact, the $T^3$ Theorem makes it possible to generate new parametric families of distributions (marriage functions) systematically. One-, two-, or $n$-parameter mixing distributions can be easily constructed to model a pre-specified set of affinities associated with a two-sex mating system. For example, the two-parameter family of functions given by $\phi(a, a') = \nu \exp[-\kappa(a - a')^2]$, which would model like-with-like or preferred affinities between males of age $a$ and females of age $a'$, can be easily incorporated into dynamic two-sex models via our two-parameter age-structured pairing function with like-with-like preferences given by $\phi(a, a')$. Structured-population models that *explicitly* incorporate mating functions of this type at the population level have never been studied. Hence, the role of selection in mating systems at the population level has not been properly explored.

Section 1 introduces a framework for two-sex pairing in a population defined in terms of fixed characteristics based on earlier work (Castillo-Chavez & Busenberg 1991; Rubin et al. 1992); Section 2 introduces flexible parametric families of pairing solutions that make connections to data possible; Section 3 discusses the relationship between male and female preferences through the $T^3$ Theorem; and Section 4 establishes analogous results in a continuous framework.

## 1 Framework for Discrete Two-Sex Mixing

Consider a population with $L$ types of males and $N$ types of females. Let $T_i^m(t)$ denote the number of males of type $i$ ($i = 1, \ldots, L$) at time $t$, and $T_j^f(t)$ denote the number of females of type $j$ ($j = 1, \ldots, N$) at time $t$. Let $C_i$ ($i = 1, \ldots, L$) and $B_j$ ($j = 1, \ldots, N$) denote the per capita pairing rates for males of type $i$ and for females of type $j$, respectively. These rates must in general be assumed to be functions of the state variables $\mathbf{T}^m = (T_1^m, \ldots, T_L^m)^T$ and $\mathbf{T}^f = (T_1^f, \ldots, T_N^f)^T$. This is an important assumption since the absence of either males or females makes mating impossible; that is, if one sex is not present, these rates must be zero.

We characterize two-sex pairing functions using two matrices: $\mathbf{P}(\cdot, t) = \{p_{ij}(\cdot, t)\}$ and $\mathbf{Q}(\cdot, t) = \{q_{ij}(\cdot, t)\}$. Here, $p_{ij}(\cdot, t)$ denotes the conditional probability that a male of type $i$ pairs with a female of type $j$, given that he has formed a heterosexual partnership at time $t$; $q_{ji}(\cdot, t)$ denotes the conditional probability that a female of type $j$ pairs with a male of type $i$, given that she has formed a heterosexual partnership at time $t$. These matrices are functions of average weighted pairing rates of the various groups; that is, they are functions of the "abundance" of partners and their affinities for particular types. The pair $(\mathbf{P}(\cdot, t), \mathbf{Q}(\cdot, t))$ is called a marriage, a mixing matrix, or a pair-formation matrix if and only if it satisfies the following properties at all times (for more-general properties, which include those below, see Castillo-Chavez, Huang & Li, in press):

(A1)    $p_{ij}(\cdot, t) \geq 0$  and  $q_{ji}(\cdot, t) \geq 0$
         for  $i = 1, \ldots, L, \ j = 1, \ldots, N$ ;

(A2)    $\displaystyle\sum_{j=1}^{N} p_{ij}(\cdot, t) = 1$  for  $i = 1, \ldots, L$ ;

         $\displaystyle\sum_{i=1}^{L} q_{ji}(\cdot, t) = 1$  for  $j = 1, \ldots, N$ ;

(A3)  $\quad C_i(\mathbf{T}^{\mathrm{m}}, \mathbf{T}^{\mathrm{f}}) T_i^{\mathrm{m}}(t) p_{ij}(\cdot, t)$

$\qquad = B_j(\mathbf{T}^{\mathrm{m}}, \mathbf{T}^{\mathrm{f}}) T_j^{\mathrm{f}}(t) q_{ji}(\cdot, t)$  for  $i=1,\dots,L, j=1,\dots,N$ .

Note that (A2) and (A3) imply that

$$\sum_{i=1}^{L} C_i(\mathbf{T}^{\mathrm{m}}, \mathbf{T}^{\mathrm{f}})\, T_i^{\mathrm{m}}(t) = \sum_{j=1}^{N} B_j(\mathbf{T}^{\mathrm{m}}, \mathbf{T}^{\mathrm{f}})\, T_j^{\mathrm{f}}(t), \qquad (1)$$

which in fact provides a necessary and sufficient condition to guarantee the existence of a solution. The only separable solution (Castillo-Chavez & Busenberg 1991) to axioms (A1)–(A3) is the Ross solution: $p_{ij} = \bar{p}_j$ and $q_{ji} = \bar{q}_i$, where

$$\bar{p}_j = \frac{B_j T_j^{\mathrm{f}}}{\sum_{j=1}^{N} B_j T_j^{\mathrm{f}}} \quad \text{and} \quad \bar{q}_j = \frac{C_i T_i^{\mathrm{m}}}{\sum_{i=1}^{L} C_i T_i^{\mathrm{m}}} . \qquad (2)$$

All solutions to axioms (A1)–(A3) can be characterized as multiplicative perturbations of the Ross solution. These perturbations are defined in terms of two matrices: $\mathbf{\Phi}^{\mathrm{m}} = \{\phi_{ij}^{\mathrm{m}}\}$ and $\mathbf{\Phi}^{\mathrm{f}} = \{\phi_{ji}^{\mathrm{f}}\}$, the male and female preference matrices, which are a measure of the mating preferences and/or affinities of individuals of each gender for the opposite gender. These preferences may be time-dependent or may change as a result of changes in the frequency or abundance of mixing types. To explicitly state the Castillo-Chavez and Busenberg (1991) characterization theorem of two-sex marriage functions requires some definitions:

$$\ell_i^{\mathrm{m}} \equiv \sum_{j=1}^{N} \bar{p}_j \phi_{ij}^{\mathrm{m}}, \qquad R_i^{\mathrm{m}} \equiv 1 - \ell_i^{\mathrm{m}}, \qquad V^{\mathrm{m}} \equiv \sum_{i=1}^{L} \bar{q}_i R_i^{\mathrm{m}},$$

$$\ell_j^{\mathrm{f}} \equiv \sum_{i=1}^{L} \bar{q}_i \phi_{ji}^{\mathrm{f}}, \qquad R_j^{\mathrm{f}} \equiv 1 - \ell_j^{\mathrm{f}}, \qquad V^{\mathrm{f}} \equiv \sum_{j=1}^{N} \bar{p}_j R_j^{\mathrm{f}}. \qquad (3)$$

**Theorem 1**  *For each marriage function* $(\mathbf{P}, \mathbf{Q})$, *matrices* $\mathbf{\Phi}^{\mathrm{m}}$ *and* $\mathbf{\Phi}^{\mathrm{f}}$ *can be found such that*

$$p_{ij} \equiv \bar{p}_j \left( \frac{R_j^{\mathrm{f}} R_i^{\mathrm{m}}}{V^{\mathrm{f}}} + \phi_{ij}^{\mathrm{m}} \right),$$

$$q_{ji} \equiv \bar{q}_i \left( \frac{R_i^{\mathrm{m}} R_j^{\mathrm{f}}}{V^{\mathrm{m}}} + \phi_{ji}^{\mathrm{f}} \right), \qquad (4)$$

*where* $0 \le R_i^{\mathrm{m}} \le 1$ $(i = 1, \dots, L)$, $0 \le R_j^{\mathrm{f}} \le 1$ $(j = 1, \dots, N)$,

$\sum_{i=1}^{L} \ell_i^{\mathrm{m}} \bar{q}_i < 1$, and $\sum_{j=1}^{N} \ell_j^{\mathrm{f}} \bar{p}_j < 1$  *if and only if*

$$\phi_{ij}^{\mathrm{m}} = \phi_{ji}^{\mathrm{f}} + R_i^{\mathrm{m}} R_j^{\mathrm{f}} \left( \frac{1}{V^{\mathrm{m}}} - \frac{1}{V^{\mathrm{f}}} \right). \tag{5}$$

Condition (5) shows the implicit frequency and time-dependent relationship forced by the property (A3) between the elements of $\mathbf{\Phi}^{\mathrm{m}}$ and $\mathbf{\Phi}^{\mathrm{f}}$. Letting

$$\bar{\mathbf{p}} = \begin{pmatrix} \bar{p}_1 \\ \vdots \\ \bar{p}_N \end{pmatrix} \quad \text{and} \quad \bar{\mathbf{q}} = \begin{pmatrix} \bar{q}_1 \\ \vdots \\ \bar{q}_L \end{pmatrix}$$

and using matrix notation, we can summarize the constraints imposed by (5) in an implicit nonlinear relationship

$$\mathbf{\Phi}^{\mathrm{m}} = \mathbf{\Psi} \left( \bar{\mathbf{p}}, \ \bar{\mathbf{q}}, \ \mathbf{\Phi}^{\mathrm{m}}, \ \mathbf{\Phi}^{\mathrm{m}} \right), \tag{6}$$

where the components of $\mathbf{\Psi}$ are defined by (5). The representation theorem does not explicitly determine the class of matrices $\mathbf{\Phi}^{\mathrm{m}}$ and $\mathbf{\Phi}^{\mathrm{f}}$ that may result naturally from the mixing constraints imposed by heterosexually mixing populations. The natural reduction in the number of degrees of freedom available to a heterosexually mixing population is not completely obvious and, in fact, is the content of the T$^3$ Theorem.

The roles of the affinity matrices are not entirely clear from the above result. A naive interpretation, based on an assumption that both are matrices of constants, may lead to the wrong interpretation that all we have managed to do is to set parameters for a model (see Altman & Morris 1994). A clearer understanding of this model can be gained directly from Blythe et al. (1995; this paper was used by Altman and Morris as the basis for their paper and never cited!). A preliminary result (Castillo-Chavez & Busenberg 1991) gives an insight into the role of $\mathbf{\Phi}^{\mathrm{m}}$ and $\mathbf{\Phi}^{\mathrm{f}}$.

**Theorem 2**  *If either $\phi_{ij}^{\mathrm{m}} = \alpha$ $(0 \leq \alpha < 1$ for all $i, j)$ or $\phi_{ji}^{\mathrm{f}} = \beta$ $(0 \leq \beta < 1$ for all $j, i)$, where $\alpha$ and $\beta$ are constants, then $p_{ij} = \bar{p}_j$ and $q_{ji} = \bar{q}_i$. That is, equation (4) reduces to the unique separable Ross solution in equation (2).*

That is, in a mating system in which either males or females have identical affinities, random (proportionate) two-sex mixing is the only possibility. Other models are pursued in the next section.

## 2  Parameters for Preference Matrices

Equation (4) encapsulates all possible mixing patterns in terms of two preference matrices. It may be argued, to some degree correctly, that this representation transfers the difficulties from one set of matrices, $(\mathbf{P}, \mathbf{Q})$, to another, $(\mathbf{\Phi}^m, \mathbf{\Phi}^f)$. However, as shown below, the use of preference matrices $(\mathbf{\Phi}^m, \mathbf{\Phi}^f)$ increases our understanding of the social structure of a population. Specific preference matrices facilitate the modeling of nontrivial parametric mixing patterns (again, see Blythe et al. 1995). Indeed, it is possible to construct easily a large class of marriage functions that go well beyond those found in the literature, for example, by allowing the clear modeling of asymmetrical age-structured interactions. Asymmetrical age-structured interactions (e.g., females dating older males), although common in human populations, have never been incorporated in ordinary age-structured pairing models because of the lack of an appropriate model. From the results here, it is obvious how to construct mating systems that incorporate realistic affinities with few parameters.

Most of the prior theoretical work on mating systems was based on random mating, or on specific types of assortative mating with some minor variations. This is particularly clear from the literature in population genetics (Crow & Kimura 1970). Modelers interested in mating systems at the population level began to move away from random mating through the use of special mixing matrices including like-with-like, preferred mixing, and biased mixing (see Sattenspiel & Castillo-Chavez 1990). Other forms of mixing—such as asymmetrical heterosexual mixing, whereby females prefer to mix with older males and males prefer to mix with younger females—have been avoided because of the lack of a clear modeling framework. The resulting limitations are evident from the literature on marriage functions, demography, and mathematical demography. Furthermore, a glance at the literature on mathematical ecology and epidemiology shows that efforts to understand disease dynamics and demographic effects in gender-specific populations have been conducted using mostly unrealistic mating structures. We must therefore ask the obvious question: to what extent is our current theoretical understanding of mating systems or of population dynamics solely dependent on our use of specialized forms of mixing? The use of our preference matrices $(\mathbf{\Phi}^m, \mathbf{\Phi}^f)$ makes the construction of flexible and parameter-poor mating systems possible. Future applications of these models will measure the success and utility of this approach.

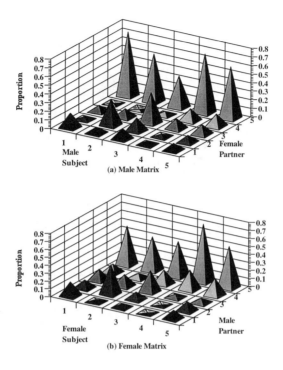

FIGURE 1. *Mixing matrix estimated from data:* (a) *for males,* $\hat{\mathbf{P}}$; (b) *for females,* $\hat{\mathbf{Q}}$.

A simple result allows for the explicit construction of flexible-pairing parameter-poor structures.

**Theorem 3** $V^{\mathrm{f}} = V^{\mathrm{m}}$ *if and only if* $\boldsymbol{\Phi}^{\mathrm{m}} = (\boldsymbol{\Phi}^{\mathrm{f}})^{\mathrm{T}}$.

The only solutions to axioms (A1)–(A3) with frequency-independent $\boldsymbol{\Phi}^{\mathrm{m}}$ and $\boldsymbol{\Phi}^{\mathrm{f}}$ are those for which $\boldsymbol{\Phi}^{\mathrm{m}} = (\boldsymbol{\Phi}^{\mathrm{f}})^{\mathrm{T}}$. This simple case corresponds to the situation where males and females have matching and fixed preferences (preferences that do not change with the dynamics of $T_i^{\mathrm{m}}(t)$ and $T_j^{\mathrm{f}}(t)$). The class of solutions for which $\boldsymbol{\Phi}^{\mathrm{m}} = (\boldsymbol{\Phi}^{\mathrm{f}})^{\mathrm{T}}$ is restrictive, yet they help define parametrically the pairing structures available in the literature with a limited number of parameters, since many of the entries can be (and should be, if the model is to be of *practical* use) filled with zeros. The use of preference matrices $\boldsymbol{\Phi}^{\mathrm{m}}$ and $\boldsymbol{\Phi}^{\mathrm{f}}$ with constant entries provides a class of parametric mixing distributions that are rich, flexible, and connectable to data. In Figure 1, we illustrate the use of these ma-

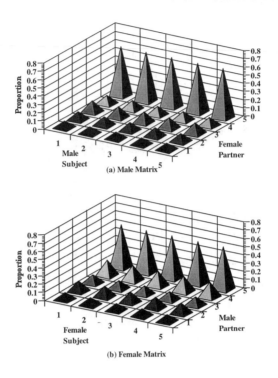

FIGURE 2. *Random-mixing matrix for our data:* (a) *for males,* $\bar{\mathbf{P}}$*;* (b) *for females,* $\bar{\mathbf{Q}}$.

trices with data derived elsewhere (Rubin et al. 1992; Hsu Schmitz 1994; Hsu Schmitz & Castillo-Chavez 1994).

Figure 1 illustrates a mixing matrix exhibiting like-with-like mixing (individuals prefer to mix with those of the same class or age) coupled with an additional trend, whereby females tend to pair with older males and males tend to pair with younger females. Thus, the use of constant preference matrices that satisfy the relationship $\mathbf{\Phi}^m = (\mathbf{\Phi}^f)^T$ provides a reasonable model that captures this mating structure with only "a couple" of parameters, regardless of the complexity of the social structure of the population in terms of the number of classes or categories. Here, the number of classes is fixed. Classifying a population by age introduces dynamic categories.

Figure 2 shows the corresponding random-mixing pattern associated with the same parameters. To put emphasis on the flexibility of this approach, observe that we could fit the data of Figure 1 with a one-parameter class of mixing matrices. One example (from Hsu Schmitz 1994) is the following $\mathbf{\Phi}^{\mathrm{f}}$:

$$\mathbf{\Phi}^{\mathrm{f}} = \begin{bmatrix} 1 & d & d & d & d \\ 0 & 1 & d & d & d \\ 0 & 0 & 1 & d & d \\ 0 & 0 & 0 & 1 & d \\ 0 & 0 & 0 & 0 & 1 \end{bmatrix}.$$

If the preference matrix $\mathbf{\Phi}^{\mathrm{m}} = (\mathbf{\Phi}^{\mathrm{f}})^{\mathrm{T}}$ is fixed, that is, if all its elements are constant, then relationship (5) is always satisfied. Therefore, once we have computed an estimate $\hat{d}$ of $d$ from data, we can predict $p_{ij}(\hat{d}, \cdot)$ and $q_{ji}(\hat{d}, \cdot)$ for all future times using equation (4) (or eq. 21) and a model, deterministic or stochastic, for the dynamics of the populations $T_i^{\mathrm{m}}(t)$ and $T_j^{\mathrm{f}}(t)$. Mixing matrices given parameters in this fashion are flexible enough to capture the qualitative features observed in the data with a single parameter, $d$. Hence, the assumption that $\mathbf{\Phi}^{\mathrm{m}} = (\mathbf{\Phi}^{\mathrm{f}})^{\mathrm{T}}$ makes it possible to construct pairing functions modeled by few parameters. Further applications and extensions are possible, including stochastic and deterministic demographic and epidemiological models that incorporate the contact structure just described (Castillo-Chavez, Fridman & Luo, in press). We now proceed to discuss the key modeling result of this chapter.

## 3 The Two-Body Problem in a Discrete Framework

In Section 2, the preferences of males for females and vice versa satisfy in general a complex relationship, namely,

$$\mathbf{\Phi}^{\mathrm{m}} = \mathbf{\Psi}\left(\bar{\mathbf{p}}, \bar{\mathbf{q}}, \mathbf{\Phi}^{\mathrm{f}}, \mathbf{\Phi}^{\mathrm{m}}\right). \tag{7}$$

Common sense dictates that if one set of preferences (e.g., $\mathbf{\Phi}^{\mathrm{f}}$) is known, then so must be the other (e.g., $\mathbf{\Phi}^{\mathrm{m}}$). Consequently, it should be possible to solve (5) in terms of a single affinity matrix. Therefore, we expect that a functional $\mathbf{\Psi}$ can be found such that

$$\mathbf{\Phi}^{\mathrm{m}} = \mathbf{\Psi}\left(\bar{\mathbf{p}}, \bar{\mathbf{q}}, \mathbf{\Phi}^{\mathrm{f}}\right).$$

This result is the main content of our $\mathrm{T}^3$ Theorem.

To find a solution of $\mathbf{\Phi}^{\mathrm{m}}$ in terms of $\bar{p}_j$, $\bar{q}_i$, and $\phi^{\mathrm{f}}_{ji}$, we multiply $\bar{p}_j$ on both sides of equation (5) and sum over $j$. The resulting relationships are

$$\sum_{j=1}^{N} \bar{p}_j \phi^{\mathrm{m}}_{ij} = \sum_{j=1}^{N} \bar{p}_j \phi^{\mathrm{f}}_{ji} + \sum_{j=1}^{N} \bar{p}_j R^{\mathrm{m}}_i R^{\mathrm{f}}_j \left( \frac{1}{V^{\mathrm{m}}} - \frac{1}{V^{\mathrm{f}}} \right)$$

$$\Leftrightarrow \; 1 - R^{\mathrm{m}}_i = \sum_{j=1}^{N} \bar{p}_j \phi^{\mathrm{f}}_{ji} + R^{\mathrm{m}}_i V^{\mathrm{f}} \left( \frac{1}{V^{\mathrm{m}}} - \frac{1}{Vf} \right)$$

$$\Leftrightarrow \; 1 - \sum_{j=1}^{N} \bar{p}_j \phi^{\mathrm{f}}_{ji} = R^{\mathrm{m}}_i \frac{V^{\mathrm{f}}}{V^{\mathrm{m}}} \,. \tag{8}$$

Let

$$U^{\mathrm{f}}_i \equiv \sum_{j=1}^{N} \bar{p}_j \phi^{\mathrm{f}}_{ji} \,; \tag{9}$$

then, from equation (8),

$$(1 - U^{\mathrm{f}}_i)/V^{\mathrm{f}} = \frac{R^{\mathrm{m}}_i}{V^{\mathrm{m}}} \,, \tag{10}$$

which demonstrates that male preferences can be obtained from female preference if we can solve (10). If, by definition,

$$\beta^{\mathrm{f}}_i \equiv \frac{1 - U^{\mathrm{f}}_i}{V^{\mathrm{f}}} \,, \tag{11}$$

then the system that must be solved becomes

$$R^{\mathrm{m}}_i - \beta^{\mathrm{f}}_i V^{\mathrm{m}} = 0 \,, \tag{12a}$$

or

$$R^{\mathrm{m}}_i - \beta^{\mathrm{f}}_i \sum_{k=1}^{L} \bar{q}_k R^{\mathrm{m}}_k = 0 \,. \tag{12b}$$

It can also be formulated in matrix notation as follows:

$$\left( \mathbf{I} - \boldsymbol{\beta}^{\mathrm{f}} \bar{\mathbf{q}}^{\mathrm{T}} \right) \mathbf{R}^{\mathrm{m}} = \mathbf{0} \,, \tag{12c}$$

where

$$\boldsymbol{\beta}^{\mathrm{f}} = \begin{pmatrix} \beta^{\mathrm{f}}_1 \\ \vdots \\ \beta^{\mathrm{f}}_L \end{pmatrix} \quad \text{and} \quad \mathbf{R}^{\mathrm{m}} = \begin{pmatrix} R^{\mathrm{m}}_1 \\ \vdots \\ R^{\mathrm{m}}_L \end{pmatrix} .$$

Let $\mathbf{B} = \mathbf{I} - \boldsymbol{\beta}^{\mathrm{f}}\,\bar{\mathbf{p}}^{\mathrm{T}}$; then, $\mathbf{B}$, an $L \times L$ matrix, is a rank-one perturbation of the identity. Furthermore, a simple computation shows that

$$\det \mathbf{B} = 1 - \sum_{i=1}^{L} \bar{q}_i \beta_i^{\mathrm{f}} = 0,$$

and hence, all solutions are given by

$$\mathbf{R}^{\mathrm{m}} = \gamma\,\boldsymbol{\beta}^{\mathrm{f}}, \tag{13}$$

where $\gamma$ is an arbitrary "constant" for each time $t$. In other words, the null space of $\mathbf{I} - \boldsymbol{\beta}^{\mathrm{f}}\,\bar{\mathbf{q}}^{\mathrm{T}}$ is equal to span $\{\boldsymbol{\beta}^{\mathrm{f}}\}$. Substituting solution (13) into (5) gives the relationship

$$\phi_{ij}^{\mathrm{m}} = \phi_{ji}^{\mathrm{f}} + \gamma\beta_i^{\mathrm{f}}R_j^{\mathrm{f}}\left(\frac{1}{\gamma} - \frac{1}{V^{\mathrm{f}}}\right) = \phi_{ji}^{\mathrm{f}} + \beta_i^{\mathrm{f}}\,R_j^{\mathrm{f}}\left(1 - \frac{1}{V^{\mathrm{f}}}\right). \tag{14}$$

The condition $\sum_{j=1}^{N} \ell_j^{\mathrm{f}}\bar{p}_j < 1$ in Theorem 1 implies that $\sum_{i=1}^{L} \bar{q}_i U_i^{\mathrm{f}} < 1$ because

$$\sum_{j=1}^{N} \ell_j^{\mathrm{f}}\bar{p}_j = \sum_{j=1}^{N}\sum_{i=1}^{L} \bar{q}_i\,\phi_{ji}^{\mathrm{f}}\bar{p}_j = \sum_{i=1}^{L} \bar{q}_i U_i^{\mathrm{f}}.$$

Further constrain $\phi_{ji}^{\mathrm{f}}$ by requiring that $U_i^{\mathrm{f}} = \Sigma_{j=1}^{N}\bar{p}_j\phi_{ji}^{\mathrm{f}} \le 1$; then, $\beta_i^{\mathrm{f}} \ge 0$ for all $i$ (by its own definition), since negative values of $\beta_i^{\mathrm{f}}$ would imply that $U_i^{\mathrm{f}} > 1$. Finally, in order to have $0 \le R_i^{\mathrm{m}} \le 1$, $\gamma$ must be chosen such that

$$0 \le \gamma \le \frac{1}{\max_i \beta_i^{\mathrm{f}}}. \tag{15}$$

Not all $\beta_i^{\mathrm{f}}$ can be zero, or not all $U_i^{\mathrm{f}}$ can be one; otherwise $V^{\mathrm{f}} = 0$. The parameter $\gamma$ gives an extra degree of freedom in the choice of $R_i^{\mathrm{m}}$ and $\phi_{ij}^{\mathrm{m}}$. To simplify the expression, rescale the free parameter $\gamma$, and let

$$\Gamma \equiv 1 - \frac{\gamma}{V^{\mathrm{f}}};$$

thus,

$$\gamma = V^{\mathrm{f}}(1 - \Gamma). \tag{16}$$

Hence, equation (14) becomes

$$\phi_{ij}^{m} = \phi_{ji}^{f} + \beta_i^{f} R_j^{f} \left[ 1 - \frac{V^{f}(1-\Gamma)}{V^{f}} \right] = \phi_{ji}^{f} + \Gamma \beta_i^{f} R_j^{f}$$
$$= \phi_{ji}^{f} + \frac{\Gamma(1 - U_i^{f}) R_j^{f}}{V^{f}} .$$

Plugging equation (16) into inequality (15) results in

$$0 \leq V^{f}(1-\Gamma) \leq \frac{1}{\max_i \beta_i^{f}}$$

$$\Leftrightarrow 1 - \frac{1}{V^{f} \max_i \beta_i^{f}} \leq \Gamma \leq 1 \qquad (17)$$

$$\Leftrightarrow -\min_i \frac{U_i^{f}}{1 - \min_i U_i^{f}} \leq \Gamma \leq 1 .$$

These computations allow us to state our main result, the T³ Theorem.

**Theorem 4 (Discrete T³ Theorem)** *The preference matrices at all times obey the explicit relation*

$$\phi_{ij}^{m} = \phi_{ji}^{f} + \frac{\Gamma (1 - U_i^{f}) R_j^{f}}{V^{f}} , \qquad (18)$$

*where $\Gamma$ is an implicitly time-dependent arbitrary "constant" satisfying*

$$\frac{-\min_i U_i^{f}}{1 - \min_i U_i^{f}} \leq \Gamma \leq 1 \quad \text{and} \quad U_i^{f} \leq 1;$$

*conversely,*

$$\phi_{ji}^{f} = \phi_{ij}^{m} + \Delta (1 - U_j^{m}) R_i^{m}/V^{m} , \qquad (19)$$

*where $\Gamma$ is an implicitly time-dependent arbitrary "constant" satisfying*

$$U_j^{m} = \sum_{i=1}^{L} \bar{q}_i \phi_{ij}^{m} \leq 1, \qquad (20)$$

*and $\Delta$ is an implicitly time-dependent arbitrary "constant" satisfying*

$$-\min_j U_j^{m}/(1 - \min_j U_j^{m}) \leq \Delta \leq 1. \qquad (21)$$

The function $\boldsymbol{\Psi}$ given by equation (7) is thus defined explicitly by equation (18). Note that the above result shows that the original model has an unnecessarily large number of parameters (and assumes constant affinities). However, the explicit relationship between female and male affinities allows the construction of parametric families of mating functions with few parameters, which capture a rich variety of situations never found in the literature.

If $\phi_{ji}^{\mathrm{f}} = \alpha$ (constant, and $0 \leq \alpha < 1$) for all $i$ and $j$, then $0 < R_j^{\mathrm{f}} = 1 - \alpha \leq 1$ for all $j$, $0 < V^{\mathrm{f}} = 1 - \alpha \leq 1$ by (3), and $0 \leq U_i^{\mathrm{f}} = \alpha < 1$ for all $i$ by (9). Thus, $\beta_i^{\mathrm{f}} = (1 - \alpha)/(1 - \alpha) = 1$ by (11) and $0 \leq R_i^{\mathrm{m}} = \gamma \leq 1$ by (13) for all $i$. Hence, from (14),

$$\phi_{ij}^{\mathrm{m}} = \alpha + (1 - \alpha)\left(1 - \frac{\gamma}{1 - \alpha}\right) = 1 - \gamma = \text{constant}$$

for all $i$ and $j$, and $0 \leq 1 - \gamma < 1$. Hence, lack of selectivity (preference) in one sex implies lack of selectivity (preference) in the other (Theorem 2). In this case, $p_{ij} = \bar{p}_j$ and $q_{ji} = \bar{q}_i$; that is, the population mixes at random (Ross solutions; for simple versions used by Ross in his malaria work, see Blythe et al. 1992). If $\Gamma = 0$ or $\Delta = 0$ for all times, then $\mathbf{\Phi}^{\mathrm{m}} = (\mathbf{\Psi}^{\mathrm{f}})^{\mathrm{T}}$, and we recover the frequency-independent mixing matrices of Theorem 3.

Using the mixing solution given by (13) and (14), the general mixing matrix in (4) can be rewritten as follows:

$$\begin{aligned} p_{ij} &= \bar{p}_j \left(\beta_i^{\mathrm{f}} R_j^{\mathrm{f}} + \phi_{ji}^{\mathrm{f}}\right), \\ q_{ji} &= \bar{q}_i \left(\beta_i^{\mathrm{f}} R_j^{\mathrm{f}} + \phi_{ji}^{\mathrm{f}}\right). \end{aligned} \tag{22}$$

If we visualize the preference matrices $\mathbf{\Phi}^{\mathrm{m}}$ and $\mathbf{\Psi}^{\mathrm{f}}$ as behavioral classes, then equations (22), which express a mating system as a function of behavioral classes, is independent of $\gamma$. Therefore, behavioral $\gamma$ classes can give rise to identical behavioral "phenotypes" or, equivalently, the same set of mixing probabilities. In other words, the matching relations set by axioms (A1)–(A3) show that (22) gives a consistent resetting of the parameters of the initial model. That the same set of parameters gives rise to identical mating systems can be interpreted as the result of having a model with more parameters than necessary (an issue addressed above). However, the possibility that the different sets of biological parameters give rise to the same observed behaviors (pairing systems) is common in biology and, in some sense, the essence behind the definition of phenotype (an equivalence class that expresses the observable characteristics of distinct phenotypes). Section 4 extends these results to age-structured populations.

## 4 Two-Sex Mixing in Age-Structured Populations

To describe the two-sex pairing framework in an age-structured population, use the following notation:

$M(a,t)$     number of males aged $a$ to $a + \Delta a$ at time $t$,
$F(a',t)$     number of females aged $a'$ to $a' + \Delta a$ at time $t$,
$C(a,\cdot)$     per capita pairing rate for females of age $a$,
$B(a',\cdot)$     per capita pairing rate for females of age $a'$,
$p(a,a',t,\cdot)$     probability that a male of age $a$ pairs with a female of age $a'$ given that he pairs at time $t$,
$q(a',a,t,\cdot)$     probability that a female of age $a'$ pairs with a male of age $a$ given that she pairs at time $t$,

where $a$, $a'$, $t \in [0,\infty)$, and $C(a,\cdot)$ and $B(a',\cdot)$ are assumed to be functions of $T^{\mathrm{m}}(t) \equiv \int_0^\infty M(a,t)da$ and $T^{\mathrm{f}}(t) \equiv \int_0^\infty F(a',t)da'$. Two-sex pairing functions $(p,q)$ must satisfy the following properties at all times:

(B1)         $p(a,a',t,\cdot) \geq 0$ and $q(a',a,t,\cdot) \geq 0$
                 for all $a$ and $a'$ ;

(B2)         $\displaystyle\int_0^\infty p(a,a',t,\cdot)da' = 1$ for all $a$, and

                 $\displaystyle\int_0^\infty q(a',a,t,\cdot)da = 1$ for all $a'$ ;

(B3)         $C(a,\cdot)M(a,t)p(a,a',t,\cdot)$
                 $= B(a',\cdot)F(a',t)q(a',a,t,\cdot)$ for all $a$ and $a'$

(Castillo-Chavez et al. 1991). These properties should be stated as holding almost everywhere. (However, these technicalities are avoided here.) Note that (B2) and (B3) imply that

$$\int_0^\infty C(a,\cdot)M(a,t)da = \int_0^\infty B(a',\cdot)F(a',t)da' , \qquad (23)$$

a condition that guarantees the existence of age-structured two-sex pairing functions (Castillo-Chavez, Huang & Li, in press). The only separable two-sex pairing function satisfying properties (B1)–(B3), the so-called Ross solution, is given by

$$
\bar{p}(a',t,\cdot) = B(a',\cdot)F(a',t) \bigg/ \int_0^\infty B(u',\cdot)F(u',t)du',
$$
$$
\bar{q}(a,t,\cdot) = C(a,\cdot)M(a,t) \bigg/ \int_0^\infty C(u,\cdot)M(u,t)du.
$$
(24)

This separable function models a heterosexually age-structured, randomly (proportionate) mixing, two-sex population. All two-sex pairing functions for a population with distinct subgroups can be characterized as multiplicative perturbations of the corresponding separable solution (Castillo-Chavez & Busenberg 1991). Similarly, for an age-structured population, all solutions to axioms (B1)–(B3) can also be characterized as multiplicative perturbations of the separable solution (24). To describe this formulation, let $\phi^m\,(a,a')$ denote the preference or affinity of males of age $a$ for females of age $a'$, and $\phi^f(a',a)$ denote the preference or affinity of females of age $a'$ for males of age $a$. The affinities can be independent of time, or they can change with time and/or group size. (We omit this level of detail below to simplify the notation.) The following definitions are needed:

$R^m(a) \equiv 1 - \int_0^\infty \bar{p}(a')\phi^m(a,a')da'$  average disaffinity of males of age $a$,

$R^f(a') \equiv 1 - \int_0^\infty \bar{q}(a)\phi^f(a',a)da$  average disaffinity of females of age $a'$,

$V^m \equiv \int_0^\infty \bar{q}(a)R^m(a)da$  weighted average disaffinity of males,

$V^f \equiv \int_0^\infty \bar{p}(a')R^f(a')da'$  weighted average disaffinity of females.

In principle, the functions $\phi^m$ and $\phi^f$ can take any value, positive or negative, whereby $R^m \geq 0$, $R^f \geq 0$, $V^m > 0$, and $V^f > 0$ at all times. With the additional restrictions that $0 \leq R^m \leq 1$ and $0 \leq R^f \leq 1$, these functions can take only nonnegative values (Hsu Schmitz 1994). Now to express the representation theorem of the two-sex pairing function in an age-structured population.

**Theorem 5** *For each two-sex pairing function (p, q) at any time, nonnegative affinity functions $\phi^{\mathrm{m}}(a, a')$ and $\phi^{\mathrm{f}}(a', a)$ can be found such that, for $a \in [0, \infty)$ and $a' \in [0, \infty)$,*

$$
\begin{aligned}
p(a, a') &= \bar{p}(a') \left[ \frac{R^{\mathrm{f}}(a') R^{\mathrm{m}}(a)}{V^{\mathrm{f}}} + \phi^{\mathrm{m}}(a, a') \right], \\
q(a', a) &= \bar{q}(a) \left[ \frac{R^{\mathrm{m}}(a) R^{\mathrm{f}}(a')}{V^{\mathrm{m}}} + \phi^{\mathrm{f}}(a', a) \right],
\end{aligned}
\tag{25}
$$

*if and only if*

$$
\phi^{\mathrm{m}}(a, a') = \phi^{\mathrm{f}}(a', a) + R^{\mathrm{m}}(a) R^{\mathrm{f}}(a') \left( \frac{1}{V^{\mathrm{m}}} - \frac{1}{V^{\mathrm{f}}} \right), \tag{26}
$$

*where $0 \leq R^{\mathrm{m}}(a) \leq 1$ for all $a$, $0 \leq R^{\mathrm{f}}(a') \leq 1$ for all $a'$, but not all $R^{\mathrm{m}}(a)$ and not all $R^{\mathrm{f}}(a')$ can be zeros.*

The incorporation of a preference structure is not intended to yield complex pairing patterns but rather to make the patterns richer and closer to reality. In fact, this view and the representation theorem bring population-biology models closer to those in the existing literature on mating functions. (Population geneticists have long recognized the important role of preferences in mating patterns; see, e.g., Wilson 1973; Wagener 1976; Karlin 1979*a–d*, 1980; Burley 1983; Gimelfarb 1988*a, b*.)

Condition (26) exhibits a relationship between male affinity $\phi^{\mathrm{m}}$ and female affinity $\phi^{\mathrm{f}}$. It might be possible to obtain $\phi^{\mathrm{m}}$ in terms of $\phi^{\mathrm{f}}$, $\bar{p}$, and $\bar{q}$, that is, to find the function $\boldsymbol{\Psi}$ such that $\phi^{\mathrm{m}} = \boldsymbol{\Psi}(\phi^{\mathrm{f}}, \bar{p}, \bar{q})$. Since we are dealing with a continuous variable, age, a direct solution of the nonlinear equation system for $\phi^{\mathrm{m}}$ may not be feasible. Nevertheless, we can still establish the same results as in the discrete case.

First, a frequency-independent function $\boldsymbol{\Psi}$ can be easily derived, as expressed in the following theorem.

**Theorem 6** $V^{\mathrm{f}} = V^{\mathrm{m}}$ *if and only if $\phi^{\mathrm{m}}(a, a') = \phi^{\mathrm{f}}(a', a)$ for $a \in [0, \infty)$ and $a' \in [0, \infty)$.*

(The simple proof is in Hsu Schmitz 1994; Hsu Schmitz et al. 1993.) This theorem defines a special function $\boldsymbol{\Psi}$:

$$
\phi^{\mathrm{m}}(a, a') = \boldsymbol{\Psi}[\phi^{\mathrm{f}}(a', a), \bar{p}, \bar{q}] = \phi^{\mathrm{f}}(a', a), \tag{27}
$$

which is independent of $\bar{p}$ and $\bar{q}$, that is, frequency-independent. Although the class of solutions for which $\phi^{\mathrm{m}}(a, a') = \phi^{\mathrm{f}}(a', a)$ is quite restrictive, this class extends considerably the age-dependent pairing structures available in the literature. Furthermore, the class of parametric pairing models becomes quite rich and flexible, as is clearly shown by the use of the preference function $\phi(a, a') = \nu \exp[-\kappa (a - a')^2]$.

To derive the general $\Psi$ function, results about integral equations are required (the Fredholm alternative; for details see Hsu Schmitz 1994; Hsu Schmitz et al. 1993). The final results are collected in the following theorem.

**Theorem 7 (Age-Structured $T^3$ Theorem)** *The nonnegative affinities of two genders, $\phi^m(a, a')$ and $\phi^f(a', a)$ for $a \in [0, \infty)$ and $a' \in [0, \infty)$, obey the following explicit relation at all times:*

$$\phi^m(a, a') = \phi^f(a', a) + \left(1 - \frac{\lambda}{V^f}\right) \frac{[1 - U^f(a)] R^f(a')}{V^f}, \qquad (28)$$

*where $U^f(a) \equiv \int_0^\infty \bar{p}(a')\phi^f(a', a)da' \leq 1$ for $a \in [0, \infty)$ but not all $U^f(a)$ can be 1, and $\lambda$ is an arbitrary constant within a frequency-dependent range:*

$$0 < \lambda \leq \frac{V^f}{1 - \min_a U^f(a)}. \qquad (29)$$

*Conversely,*

$$\phi^f(a', a) = \phi^m(a, a') + \left(1 - \frac{\gamma}{V^m}\right) \frac{[1 - U^m(a')] R^m(a)}{V^m}, \qquad (30)$$

*where $U^m(a) \equiv \int_0^\infty \bar{q}(a)\phi^m(a, a')da \leq 1$ for $a' \in [0, \infty)$ but not all $U^m(a)$ can be 1, and $\gamma$ is an arbitrary constant within a frequency-dependent range:*

$$0 < \gamma \leq \frac{V^m}{1 - \min_{a'} U^m(a')}. \qquad (31)$$

Expression (28) explicitly defines the general $\Psi$ function, and expression (30) defines the general function needed to obtain $\phi^f$ for given $\phi^m$, $\bar{p}$, and $\bar{q}$. However, for a given $\phi^f$ or $\phi^m$, several corresponding $\phi^m$ or $\phi^f$ are possible. For example, if $\lambda = V^f$ or $\gamma = V^m$, then $\phi^m(a, a') = \phi^f(a', a)$ for all $a$ and $a'$, and Theorem 6 is recovered.

If $\phi^f(a', a) = \alpha$, where $\alpha$ is a constant and $0 \leq \alpha < 1$, for all $a$ and $a'$—that is, females have no specific preferences for males of different ages—then $0 < R^f(a') = 1 - \alpha \leq 1$ for all $a'$, $0 \leq U^f(a) = \alpha < 1$ for all $a$, and $0 < V^f = 1 - \alpha \leq 1$. Hence, equation (28) becomes

$$\phi^m(a, a') = \alpha + \left(\frac{1 - \lambda}{1 - \alpha}\right)(1 - \alpha) = 1 - \lambda$$

for all $a$ and $a'$, where $1 - \lambda$ is a constant and $0 \leq 1 - \lambda < 1$ (males have no specific preferences for females of different ages, either). Furthermore, $0 < R^{\mathrm{m}}(a) = \lambda \leq 1$ for all $a$ and $0 < V^{\mathrm{m}} = \lambda \leq 1$. Thus, the pairing function in (25) reduces to the Ross solution, the function for heterosexually random pairing.

With $\phi^{\mathrm{m}}$ replaced by (28), the two-sex pairing function in (25) becomes

$$
\begin{aligned}
p(a, a') &= \bar{p}(a') \left\{ \frac{[1 - U^{\mathrm{f}}(a)]R^{\mathrm{f}}(a')}{V^{\mathrm{f}}} + \phi^{\mathrm{f}}(a', a) \right\}, \\
q(a', a) &= \bar{q}(a) \left\{ \frac{[1 - U^{\mathrm{f}}(a)]R^{\mathrm{f}}(a')}{V^{\mathrm{f}}} + \phi^{\mathrm{f}}(a', a) \right\}.
\end{aligned}
\tag{32}
$$

The free parameters $\lambda$ or $\gamma$ are absent in equations (32). This implies that for a given female affinity $\phi^{\mathrm{f}}$, the corresponding male affinity $\phi^{\mathrm{m}}$ can characterize several behavioral classes, represented by distinct $\lambda$, which all result in the same behavioral phenotype $(p, q)$ as expressed in (32). Equivalent results can be obtained using the male affinity $\phi^{\mathrm{m}}$, as given. Again, this finding agrees with Burley's (1983) statement that different preferences can result in identical mating patterns.

## 5 Conclusions

In this chapter we introduce a modeling framework that allows for the modeling of pairing density-dependent conditional probabilities in heterosexual populations. The incorporations of the probabilities in either stochastic or deterministic demographic or epidemiological models is straightforward (see Castillo-Chavez, Fridman & Luo, in press; Castillo-Chavez, Huang & Li, in press).

Several results have generated useful insights into the population dynamics and epidemiology of populations with mating systems. For example, for pair-formation models with an arbitrary number of types, the distribution of the average times spent in pairs plays a central role in determining the observed data on the distributions of pairs (Castillo-Chavez, Huang & Li, in press). If there is little variation—that is, if most types of individuals remain in pairings for approximately the same amount of time—then the observed data on the distribution of pairs may reflect the actual pairing system with high probability. If, by contrast, the resident time in partnerships is highly variable, then the observed data on the distributions of pair types (who pairs with whom) is not likely to reflect the mating system.

Further theoretical studies are needed to better understand the role of pairing systems in population biology. We hope that this chapter motivates further theoretical work in this area.

## Literature Cited

Altman, M., and M. Morris. 1994. A clarification of the $\phi$ mixing model. *Mathematical Biosciences* 124: 1–8.

Blythe, S. P., C. Castillo-Chavez, J. Palmer, and M. Cheng. 1991. Towards a unified theory of mixinging and pair formation. *Mathematical Biosciences* 107: 379–405.

Blythe, S. P., S. Busenberg, and C. Castillo-Chavez. 1995. Affinity in paired event probability. *Mathematical Biosciences* 128: 265–284.

Burley, N. 1983. The meaning of assortative mating. *Ethology & Sociobiology* 4: 191–203.

Busenberg, S., and C. Castillo-Chavez. 1989. Interaction, pair formation and force of infection terms in sexually-transmitted diseases. Pp. 289–300 *in* C. Castillo-Chavez, ed., *Mathematical and Statistical Approaches to AIDS Epidemiology.* Lecture Notes in Biomathematics 83. Springer-Verlag, Berlin.

————. 1991. A general solution of the problem of mixing sub-populations, and its application to risk- and age-structured epidemic models for the spread of AIDS. *IMA Journal of Mathematics Applied in Medicine & Biology* 8: 1–29.

Castillo-Chavez, C., and S. Busenberg. 1991. On the solution of the two-sex mixing problem. Pp. 80–98 *in* S. Busenberg and M. Martelli, eds., *Proceedings of the International Conference on Differential Equations and Applications to Biology and Population Dynamics.* Lecture Notes in Biomathematics 92. Springer-Verlag, Berlin.

Castillo-Chavez, C., S. Busenberg, and K. Gerow. 1991. Pair formation in structured populations. Pp. 47–65 *in* J. Goldstein, F. Kappel, and W. Schappacher, eds., *Differential Equations with Applications in Biology, Physics and Engineering.* Marcel Dekker, New York.

Castillo-Chavez, C., S.-F. Shyu, G. Rubin, and D. Umbach. 1992. On the estimation problem of mixing/pair formation matrices with applications to models for sexually-transmitted diseases. Pp. 384–402 *in* K. Dietz et al., eds., *AIDS Epidemiology: Methodology Issues.* Birkhäuser, Boston.

Castillo-Chavez, C., J. X. Velasco-Hernandez, and S. Fridman. 1995. *Modeling Contact Structures in Biology.* Lecture Notes in Biomathematics 100. Springer-Verlag, Berlin.

Castillo-Chavez, C., S. Fridman, and X. Luo. In press. Stochastic and deterministic models in epidemiology. *In Proceedings of the First World Congress of Non-Linear Analysts* (Tampa, Florida, August 19–26, 1992). Walter de Gruyter , Berlin.

Castillo-Chavez, C., W. Huang, and J. Li. In press. On the existence of stable pairing distributions. *Journal of Mathematical Biology.*

Caswell, H., and D. E. Weeks. 1986. Two-sex models: Chaos, extinction and other dynamic consequences of sex. *American Naturalist* 128: 707–735.

Crow, J. F., and M. Kimura. 1970. *An Introduction to Population Genetics Theory.* Harper & Row, New York.

Fredrickson, A. G. 1971. A mathematical theory of age structure in sexual populations: Random mating and monogamous marriage models. *Mathematical Biosciences* 10: 117–143.

Gimelfarb, A. 1988*a*. Processes of pair formation leading to assortative mating in biological populations: Encounter-mating model. *American Naturalist* 131: 865–884.

———. 1988*b*. Processes of pair formation leading to assortative mating in biological populations: Dynamic interaction model. *Theoretical Population Biology* 34: 1–23.

Hsu Schmitz, S.-F. 1994. Some theories, estimation methods and applications of marriage functions in demography and epidemiology. Ph.D. diss. Cornell University, Ithaca, N.Y.

Hsu Schmitz, S.-F., and C. Castillo-Chavez. 1994. Parameter estimation in non-closed social networks related to dynamics of sexually transmitted diseases. Pp. 533–559 *in* E. Kaplan and M. Brandeau, eds., *Modeling the AIDS Epidemic: Planning, Policy and Prediction.* Raven Press, New York.

———. In press. Completion of mixing matrices for non-closed social networks. *In Proceedings of the First World Congress of Non-Linear Analysts* (Tampa, Florida, August 19–26, 1992). Walter de Gruyter, Berlin.

Hsu Schmitz, S.-F., S. Busenberg, and C. Castillo-Chavez. 1993. On the evolution of marriage functions: It takes two to tango. Biometrics Unit Technical Report BU-1210-M. Cornell University, Ithaca, N.Y.

Karlin, S. 1979*a*. Models of multifactorial inheritance: I, Multivariate formulations and basic convergence results. *Theoretical Population Biology* 15: 308–355.

———. 1979*b*. Models of multifactorial inheritance: II, The covariance structure for a scalar phenotype under selective assortative mating and sex-dependent symmetric parental-transmission. *Theoretical Population Biology* 15: 356–393.

———. 1979*c*. Models of multifactorial inheritance: III, Calculation of covariance of relatives under selective assortative mating. *Theoretical Population Biology* 15: 394–423.

———. 1979*d*. Models of multifactorial inheritance: IV, Asymmetric transmission for a scalar phenotype. *Theoretical Population Biology* 15: 424–438.

————. 1980. Models of multifactorial inheritance: V, Linear assortative mating as against selective (non-linear) assortative mating. *Theoretical Population Biology* 17: 255–275.

Kendall, D. G. 1949. Stochastic processes and population growth. *Journal of the Royal Statistical Society B* 11: 230–264.

Keyfitz, N. 1949. The mathematics of sex and marriage. *Proceedings of the Sixth Berkeley Symposium on Mathematical Statistics & Probability* 4: 89–108.

Lubkin, S., and C. Castillo-Chavez. In press. A pair formation approach to modeling inheritance of social traits. *In Proceedings of the First World Congress of Non-Linear Analysts* (Tampa, Florida, August 19–26, 1992). Walter de Gruyter, Berlin.

McFarland, D. D. 1972. Comparison of alternative marriage models. Pp. 89–106 *in* T. N. E. Greville, ed., *Population Dynamics*. Academic Press, New York.

Parlett, B. 1972. Can there be a marriage function? Pp. 107–135 *in* T. N. E. Greville, ed., *Population Dynamics*. Academic Press, New York.

Pollard, J. H. 1973. *Mathematical Models for the Growth of Human Populations*. Cambridge University Press.

Rubin, G., D. Umbach, S.-F. Shyu, and C. Castillo-Chavez. 1992. Application of capture-recapture methodology to estimation of size of population at risk of AIDS and/or other sexually-transmitted diseases. *Statistics in Medicine* 11: 1533–1549.

Sattenspiel, L., and C. Castillo-Chavez. 1990. Environmental context, social interactions, and the spread of HIV. *American Journal of Human Biology* 2: 397–417.

Wagener, D. K. 1976. Preferential mating: Nonrandom mating of a continuous phenotype. *Theoretical Population Biology* 10: 185–204.

Wilson, S. R. 1973. The correlation between relatives under the multifactorial models with assortative mating: I, The multifactorial model with assortative mating. *Annals of Human Genetics* 37: 289–304.

CHAPTER 19

# Inverse Problems and Structured-Population Dynamics

## S. N. Wood

The "forward problem" in structured-population modeling is to work from observations and assumptions about the birth, death, and developmental rates of individuals within a population to predict population dynamics: how numbers of individuals within the population change through time. This is a well-behaved and respectable type of problem in that it is properly posed: the inputs to a model (information about births, deaths, and development) completely specify the outputs (population dynamics).

"Inverse problems" are more troublesome. The inverse problem for a structured population is to work back from observations about a population's dynamics to infer things about the birth, death, and developmental rates that must have caused those dynamics. The difficulty is that any particular set of observed dynamics can be caused by many different combinations of these rates. The inverse problem is ill posed: its inputs (population dynamics) do not uniquely specify its outputs (birth, death, and developmental rates).

The only way around this stumbling block of nonspecificity is to posit a model of the population and to fit it to the observed data from the structured population. The problems with doing so are twofold. Usually, if one knew what model to fit, one would not be trying to solve the inverse problem in the first place. Furthermore, it turns out that many rather obvious models do not produce stable estimates (Wood et al. 1989; Wood & Nisbet 1991); in particular, high death rates in one stage can often be compensated for by low death rates in the next, so that death-rate estimates can carry large uncertainties.

Additional problems afflict most published methods for solving the inverse problem. The "true model" for a population is unknown. Certainly, something happened to the population under observation, and there must be some set of equations that would accurately describe this, but there is no way of knowing what this true model might be. If this problem is ignored, and a simple model simply assumed, the population parameters estimated from the model fit will be in error because the model is wrong, and the uncertainties in these parameters will be underestimated, because they will be calculated without including the component of variability that results from not knowing the true model. If, alternatively, we specify a very general model, making almost no prior assumptions about model structure, then it is usually necessary to estimate almost as many parameters as there are data points; when data are noisy, this is a recipe for bad estimates with enormous uncertainties.

I discuss here conventional methods whereby a parameter-sparse model is fit to data and other methods that make only rather general assumptions about the underlying dynamics of the population. Only the general techniques by which the inverse problem may be solved are presented, rather than detailed discussions of particular published methods (for reviews, see Manly 1989; Wood & Nisbet 1991). Stochastic population dynamics are not discussed.

## 1 Models

Three formalisms are considered here: matrices, delay-differential equations, and partial differential equations. Matrix models, which are clearly most appropriate for discrete-time systems, assume that all individuals within a stage have an equal probability of progressing to any other stage and an equal probability of surviving. The general form of a matrix model is

$$\mathbf{n}(t_{j+1}) = \mathbf{A}(t_j)\,\mathbf{n}(t_j)\,, \tag{1}$$

where $\mathbf{n}(t)$ is a vector of stage populations at time $t$ and $\mathbf{A}(t_j)$ is a matrix determining the transition from populations at time $t_j$ to populations at time $t_{j+1}$. A particular example is the type introduced by Caswell (1989):

$$\mathbf{n}(t_{j+1}) = \begin{bmatrix} P_1 & 0 & . & . & \cdots & \beta \\ G_1 & P_2 & 0 & . & \cdots & 0 \\ 0 & G_2 & P_3 & 0 & \cdots & . \\ \vdots & \vdots & \vdots & \vdots & \ddots & \vdots \\ . & . & . & . & G_{m-1} & P_m \end{bmatrix} \mathbf{n}(t_j), \qquad (1)$$

where $P_i$ is the within-stage survival rate (individuals in stage $i$ survived but did not mature to the next stage), $G_i$ is the between-stage transition rate (the proportion in stage $i$ that survived and matured into stage $i + 1$), and $\beta$ is the birthrate. For forward modeling, these models clearly require a starting population vector. Inverse-problem methods using matrix models were presented by Caswell (1989) and Caswell and Twombly (1989).

The second model formalism of interest is the delay-differential equation. The key assumptions made here are that individuals are identical within a stage except for a developmental index (often just age, but possibly weight or something more complicated). These models are extremely flexible, but I consider only the simple case of a sequence of stages where $n_i(t)$ is the number in stage $i$, with

$$\frac{dn_i}{dt} = R_i(t) - M_i(t) - \mu_i(t)n_i(t), \qquad (2)$$

where

$$M_i(t) = R_i(t - \tau) \exp\left[ -\int_{t-\tau}^t \mu(x)dx \right] (1 - d\tau/dt),$$

where $R_i$ is the recruitment rate to stage $i$, $M_i$ is the maturation rate from stage $i$ to stage $i + 1$, $\mu$ is the per capita death rate, and $\tau_i$ is the duration of stage $i$. These models require an initial structure with continuous age (or development) within each stage as a boundary condition. In many ways, using delay-differential equations is the most natural, since in theory observations of a population could be taken as frequently as desired, but the population usually cannot be sampled at a finer resolution with respect to age or development than that dictated by the duration of stages. Unfortunately, the nonlinearity of this approach also makes model fitting more difficult than for models using matrices or partial differential equations. See Gurney et al. (1983, 1986) or Nisbet et al. (1983, 1985, 1989) for discussions of such models. Inverse methods using these models were given by, for example, Hay et al. (1988) and Parslow et al. (1979).

One detail worth noting about models like (2) using delay-differential equations is that their numerical solution usually entails storing past values of variables. This can be done efficiently in a computer program by using a "ring buffer," that is, an array of fixed length in which the position of "the present" is continuously updated. For example, the array might be of length 1,000, and the present might start at position 1. At each time step, the variable whose values must be stored are put in the location of the present and the present is moved forward by one. Past values can be worked out by counting back from the present to find the values lying on either side of the time for which a value is required and interpolating between these. When the present reaches the 1,000th element of the array, it moves back to the first element, overwriting the values from the preceding 1,000 time steps. Past-value calculation takes this wraparound into account. Through this approach, standard numerical integration routines can be used to integrate the differential equations (e.g., Runge-Kutta), although note that using linear interpolation in the calculation of lagged values reduces the order of accuracy of integration methods.

For models using partial differential equations (PDE), both the state of development and time are continuous variables. Because the models are general, it is possible to construct models that approach the "true" model of a population's dynamics, but this generality also means that a large amount of information is needed in order to predict anything. The simplest PDE model of an age-structured population is the McKendrick–von Foerster equation:

$$\frac{\partial n}{\partial t} + \frac{\partial n}{\partial a} + \mu(a,t)n(a,t) = 0\,. \tag{3}$$

Clearly, this model requires an initial age structure and birth process. If it is possible to map age to stage at any time in such a way that the population of stage $i$ at time $t$ consists of all individuals between the ages of $\alpha_{i-1}$ and $\alpha_i$, then

$$n_i(t) = \int_{\alpha_{i-1}}^{\alpha_i} n(x,t)dx\,. \tag{4}$$

(For fuller discussions of models of this type, see Nisbet & Gurney 1982; Metz & Diekmann 1986; Murray 1993. For inverse-problem methods based on these models, see Banks et al. (1991); Wood & Nisbet 1991; Wood 1994.)

## 2 Fitting Models to Data with Least Squares

This section reviews standard theory for fitting models to data largely independent of model details. For the present purposes, a model is a function for turning parameters into a series of stage population estimates. Usually the model takes the form of a function or subroutine in a computer program that solves or simulates the model and results in population estimates for each stage. See Gill et al. (1981) or Press et al. (1992) for fuller discussions of model fitting.

*Linear Least Squares*

In cases in which the output of a model can be written as a linear function of its parameters, least-squares fitting methods are straightforward. For example, if a particular set of model population estimates is written in a vector,

$$\mathbf{n}^T = [n_1(t_1), n_2(t_1), \ldots, n_m(t_1), n_1(t_2), \ldots, n_m(t_s)], \qquad (5)$$

and the predicted population is written as a linear combination of the model parameters, $\mathbf{p}$,

$$\mathbf{n} = \mathbf{W}\mathbf{p}, \qquad (6)$$

then the best-fit parameters are found by minimizing

$$
\begin{aligned}
F(\mathbf{p}) &= \sum_{j=1}^{s} \sum_{i=1}^{m} [y_i(t_j) - n_i(t_j)]^2 \\
&= (\mathbf{y} - \mathbf{n})^T(\mathbf{y} - \mathbf{n}) = (\mathbf{y} - \mathbf{W}\mathbf{p})^T(\mathbf{y} - \mathbf{W}\mathbf{p}) \\
&= \mathbf{p}^T\mathbf{W}^T\mathbf{W}\mathbf{p} - 2\mathbf{p}^T\mathbf{W}^T\mathbf{y} + \mathbf{y}^T\mathbf{y},
\end{aligned}
$$

where $y_i(t_j)$ is the observed population in stage $i$ at time $t_j$, and $\mathbf{y}$ is the vector of all observed population data ordered in the same way as $\mathbf{n}$.

The minimum of $F(\mathbf{p})$ is achieved when

$$\mathbf{p} = \left(\mathbf{W}^T\mathbf{W}\right)^{-1}\mathbf{W}^T\mathbf{y}.$$

From standard theory, if $\mathbf{Y} = \mathbf{A}\mathbf{X}$ and the covariance matrix of $\mathbf{X}$ is $\mathrm{cov}(\mathbf{X})$, then $\mathrm{cov}(\mathbf{Y}) = \mathbf{A}\,\mathrm{cov}(\mathbf{X})\mathbf{A}^T$. So if, as is usual, the errors of the observations are independent and all observations have equal variances, $\sigma^2$, then it is straightforward to obtain the result

$$\mathrm{cov}(\mathbf{p}) = \left(\mathbf{W}^T\mathbf{W}\right)^{-1}\sigma^2.$$

Unfortunately, it is fairly unusual to know $\sigma^2$ in advance, and it must be estimated. If $m_p$ is the number of parameters (i.e., the number of rows in $\mathbf{p}$,) then the estimate is

$$\hat{\sigma}^2 = (\mathbf{y} - \mathbf{W}\mathbf{p})^\mathrm{T}(\mathbf{y} - \mathbf{W}\mathbf{p})/(ms - m_p).$$

*Nonlinear Least Squares*

Most population models do not take the simple linear form of equation (6); rather, the model populations are complicated nonlinear functions of their parameters. Fitting such models uses the model as a vector function of its parameters; solution of the model with a given set of parameters yields a set of population estimates for the stages and times for which data are available. That is, if $\mathbf{n}$ is as defined in equation (5) and $\mathbf{p}$ is the vector of parameters, then the model is of the form

$$\mathbf{n} = \mathbf{f}(\mathbf{p}).$$

In practice, this $\mathbf{f}$ is usually a routine in a computer program that takes parameters $\mathbf{p}$ and returns population estimates $\mathbf{n}$. If observations of population density are written in a corresponding vector $\mathbf{y}$, the goal is to minimize

$$F(\mathbf{p}) = \sum_{j=1}^{s} \sum_{i=1}^{m} [y_i(t_j) - n_i(t_j)]^2$$
$$= (\mathbf{y} - \mathbf{n})^\mathrm{T}(\mathbf{y} - \mathbf{n}) = [\mathbf{y} - \mathbf{f}(\mathbf{p})]^\mathrm{T}[(\mathbf{y} - \mathbf{f}(\mathbf{p})].$$

Most of the sensible methods for solving this problem proceed iteratively, by expanding $\mathbf{f}(\mathbf{p})$ about the current value $\mathbf{f}(\mathbf{p}_0)$ to get

$$\mathbf{f}(\mathbf{p}_0 + \boldsymbol{\alpha}) \approx \mathbf{f}(\mathbf{p}_0) + \mathbf{J}\boldsymbol{\alpha}, \tag{7}$$

where $\boldsymbol{\alpha}$ is a vector of changes in $\mathbf{p}$, and $\mathbf{J}$ is a matrix of elements of the form $\partial f_i/\partial p_j$ in the $i$th row and $j$th column. It is now possible to write

$$F(\mathbf{p}) \approx \sum [y_i - f_i(\mathbf{p}_0) - (\mathbf{J}\boldsymbol{\alpha})_i]^2$$
$$= (\boldsymbol{\delta} - \mathbf{J}\boldsymbol{\alpha})^\mathrm{T}(\boldsymbol{\delta} - \mathbf{J}\boldsymbol{\alpha}) = \boldsymbol{\alpha}^\mathrm{T}\mathbf{J}^\mathrm{T}\mathbf{J}\boldsymbol{\alpha} - 2\boldsymbol{\alpha}^\mathrm{T}\mathbf{J}^\mathrm{T}\boldsymbol{\delta} + \boldsymbol{\delta}^\mathrm{T}\boldsymbol{\delta},$$

where $\boldsymbol{\delta} = \mathbf{y} - \mathbf{f}$; thus, minimization of $F(\mathbf{p})$ is close to

$$\boldsymbol{\alpha} = (\mathbf{J}^\mathrm{T}\mathbf{J})^{-1}\mathbf{J}^\mathrm{T}\boldsymbol{\delta}, \tag{8}$$

$\mathbf{p}_0$ is replaced by $\mathbf{p}_0 + \boldsymbol{\alpha}$, and the next iteration proceeds. (This derivation is actually a bit sloppy, since I expanded $\mathbf{f}$ to the first order but then left terms of the second order in the approximation of $F$. If one is rigorous and leaves in the second-order terms from the expansion of $\mathbf{f}$, they play an insignificant role as the minimum of $F$ is approached.)

The major problem with the approach outlined above is that sometimes the quadratic approximation to the least-squares objective is not very good, and by updating the parameters the fit becomes worse rather than better. In this case, it is a good idea to update $\mathbf{p}$ by simply taking a step "downhill," that is, a step in a direction in which $F$ is bound to get smaller if the step taken is small enough. From the expansion of $F(\mathbf{p})$ (eq. 7), it is clear that the vector whose $i$th element is $\partial F/\partial p_i$ is $-2\mathbf{J}^{\mathrm{T}}\boldsymbol{\delta}$, and the direction of steepest descent is thus $\mathbf{J}^{\mathrm{T}}\boldsymbol{\delta}$. (To help see this, consider the function $f(x, y) = x^2 + y^2$, which has a minimum at $(0, 0)$, and for which the direction of steepest descent is always $(-x, -y)$. In this case $\partial f/\partial x = 2x$ and $\partial f/\partial y = 2y$, which gives the up-hill direction. The downhill direction is just the opposite of this.) Therefore,

$$F(\mathbf{p} + k\mathbf{J}^{\mathrm{T}}\boldsymbol{\delta}) \; < \; F(\mathbf{p})$$

for some small constant $k$ provided that $\mathbf{p}$ is not already the vector minimizing $F$. In practice, it is best to combine the steepest-descent approach with the quadratic-approximation approach in the rather clever way suggested by Levenberg (1944) and Marquardt (1963) (see also Press et al. 1992). That is, replace equation (8) by

$$\boldsymbol{\alpha} = (\mathbf{J}^{\mathrm{T}}\mathbf{J} + \lambda \mathbf{I})^{-1}\mathbf{J}^{\mathrm{T}}\boldsymbol{\delta} \, ,$$

where $\lambda$ is a constant adjusted in the following way: if $F(\mathbf{p} + \boldsymbol{\alpha}) < F(\mathbf{p})$, then let $\lambda = \lambda/10$ and let $\mathbf{p} = \mathbf{p} + \boldsymbol{\alpha}$; if $F(\mathbf{p} + \boldsymbol{\alpha}) \geq F(\mathbf{p})$, then let $\lambda = \lambda \times 10$ and leave $\mathbf{p}$ unchanged.

In this way, if minimization is proceeding well, emphasis is given to the quadratic approximation to the least-squares objective (an approximation that should improve as the minimum is approached); but if the algorithm is failing to reduce $F$, then more and more influence is given to the steepest-descent method, at the same time the step length is reduced, so that eventually the algorithm must start to reduce $F$ by crawling in the local downhill direction.

The major problem with the methods suggested here, of course, is the need to obtain the terms $\partial f_i/\partial p_j$, which make up the matrix $\mathbf{J}$. Sometimes this can be economically performed as a result of model structure, but often finite differences have to be set up in a vector $\hat{\mathbf{p}} = [p_0, p_1, \ldots, p_i + \Delta p, \ldots, p_n]^{\mathrm{T}}$ and then an evaluation made of $[\mathbf{f}(\hat{\mathbf{p}}) - \mathbf{f}(\mathbf{p})]/\Delta p$, a vector whose $j$th element is approximately $\partial f_j/\partial p_i$. This process has to be repeated for each parameter in $\mathbf{p}$ separately (perturbing more than one element of $\mathbf{p}$ at a time does not produce *partial* derivatives). Fortunately, the method outlined here is quite impervious to the errors introduced by this finite-difference approximation to the partial derivatives— but it can be a numerically expensive process.

The final bit of theory needed from least-squares minimization is how to get error estimates for the parameter estimates. From the "law of propagation of errors," if a vector $\mathbf{X}$ of random variables has a covariance matrix $\mathrm{cov}(\mathbf{X})$, and if there is a matrix $\mathbf{A}$ such that $\mathbf{Y} = \mathbf{AX}$, then $\mathrm{cov}(\mathbf{Y}) = \mathbf{A}\,\mathrm{cov}(\mathbf{X})\mathbf{A}^{\mathrm{T}}$. At the minimum of the function using the least-squares objective,

$$\boldsymbol{\alpha} = (\mathbf{J}^{\mathrm{T}}\mathbf{J})^{-1}\mathbf{J}^{\mathrm{T}}(\mathbf{y} - \mathbf{f}) = (\mathbf{J}^{\mathrm{T}}\mathbf{J})^{-1}\mathbf{J}^{\mathrm{T}}\mathbf{e}\,,$$

where $\mathbf{e}$ is a vector of estimated errors in the observed data $\mathbf{y}$, and $\boldsymbol{\alpha}$ is now a vector of zeros. If, however, the $e_i$'s are observations of independent random variables, each with mean zero and variance $\sigma^2$, and the $\alpha_j$'s are observations of random variables representing the errors in the parameters (and also having expectation zero), $\mathrm{cov}(\mathbf{e}) = \mathbf{I}\sigma^2$, and hence,

$$\begin{aligned}\mathrm{cov}(\boldsymbol{\alpha}) = \mathrm{cov}(\mathbf{p}) &= (\mathbf{J}^{\mathrm{T}}\mathbf{J})^{-1}\mathbf{J}^{\mathrm{T}}\mathbf{J}\left[(\mathbf{J}^{\mathrm{T}}\mathbf{J})^{-1}\right]^{\mathrm{T}}\sigma^2 \\ &= (\mathbf{J}^{\mathrm{T}}\mathbf{J})^{-1}\sigma^2\end{aligned}$$

(which is exactly what is expected by analogy with linear least squares). Of course, more often than not, $\sigma^2$ is not known in advance, in which case it must be estimated:

$$\hat{\sigma}^2 = \mathbf{e}^{\mathrm{T}}\mathbf{e}/(ms - m_p)\,.$$

*Practicalities*

The solutions given to the least-squares problems in this section are not computationally optimal; in particular, for the solution of linear least-squares problems there are much more efficient algorithms than explicit formation of $(\mathbf{W}^{\mathrm{T}}\mathbf{W})^{-1}\mathbf{W}^{\mathrm{T}}$. See, for example, Gill et al. (1981), Watkins (1991), or Press et al. (1992).

## 3 Parametric Model Fitting

The most straightforward means of solving the inverse problem is to fit a parameter-sparse model to observed data. The difficulty with this approach is that the parameter estimates are reliable only if the model is a good description of the system to which it is being fit —and this is not something that can be ascertained simply by looking at the goodness of fit of the model. For every data set, many different models could fit equally well, but they are not equally valid descriptions of the system that generated the data. Furthermore, the family of models consistent with a given data set can have widely different values for basic quantities like mortality rates. This section presupposes, however, that a good model has somehow been revealed to the investigator and illustrates how fitting can be carried out in some simple examples.

*A Simple PDE Model of an Age-Structured Population*

Suppose that a particular age-structured population can be described by equation (3), and the stage populations are given by equation (4). The solution of (3) requires an initial age structure, $n(a, t_0)$, and an initial recruitment rate, $R(t) = n(0, t)$. Immediately,

$$n(a,t) = \begin{cases} R(t-a) \exp\left[-\int_0^a \mu(a-x, t-x)dx\right] \\ \quad \text{if } t - a > t_0, \\ n(a-t+t_0, t_0) \exp\left[-\int_0^{t-t_0} \mu(a-x, t_0+x)dx\right] \\ \quad \text{if } t - a \le t_0. \end{cases}$$

(9)

A suitable death-rate function might be the inclined plane

$$\mu(a, t) = p_1 + p_2 a + p_3 t,$$

and a model recruitment function might be the Gaussian

$$R(t) = \frac{p_4}{(2\pi)^{1/2} p_5} \exp\left[-\frac{(t - p_6)^2}{2 p_5^2}\right],$$

where $p_4$ is the total number of recruits to the population, $p_5$ is the characteristic width of the recruitment peak, and $p_6$ is the mean time of recruitment. The initial age structure might be the decaying exponential

$$n(a, t_0) = k \exp(-p_7 a),$$

where

$$k = \frac{p_4}{(2\pi)^{1/2}p_5} \exp\left[-\frac{(t_0 - p_6)^2}{2p_5^2}\right].$$

In this case, the model solution is simply

$$n(a,t) = \begin{cases} \dfrac{p_4}{(2\pi)^{-1/2}p_5} \exp\left[-\dfrac{(t - a - p_6)^2}{2p_5}\right] \\ \quad \times \exp\left[-(p_1 + p_2 a + p_3 t)a + \frac{1}{2}(p_2 + p_3)a^2\right], \\ \quad \text{if } t - a > t_0, \\[4pt] \frac{1}{2}k(t_0 - t + a)^2 \exp(-p_7 a) \\ \quad \times \exp\left[-(p_1 + p_2 a + p_3 t)(t_0 - t + a) + (p_2 + p_3)\right], \\ \quad \text{if } t - a \le t_0. \end{cases}$$

From this, it is straightforward to obtain any $\partial n/\partial p_i$ analytically, but to fit the model requires numerical evaluation of

$$n_i(t) = \int_{\alpha_{i-1}}^{\alpha_i} n(x,t)dx \quad \text{and} \quad \frac{\partial n_i}{\partial p_j} = \int_{\alpha_{i-1}}^{\alpha_i} \frac{\partial n(x,t)}{\partial p_j}\, dx\,.$$

These are the quantities, $\mathbf{f}(\mathbf{p})$ and $\mathbf{J}$, required for nonlinear least squares. In this example, much of the model calculation is performed analytically, which is rather unusual; normally the model must be solved numerically, which entails a numerical evaluation of the survival integrals (the terms in brackets in eqs. 9) in this case.

In this example, the age range for each stage at each sample time is assumed known. If the age boundaries of stages are being fit as free parameters, then it is worth noting that if $\alpha_{j-1}$ is the age of individuals entering the $j$th stage at time $t$, and hence $\alpha_j$ is the age of individuals leaving the stage, then

$$\frac{\partial n_i}{\partial \alpha_j} = \begin{cases} n(\alpha_i, t) & \text{if } i = j, \\ -n(\alpha_j, t) & \text{if } i = j + 1, \\ 0 & \text{if } i < j. \end{cases}$$

(Alternatively, these partial derivatives could be estimated by finite differencing.)

*Systems Identification*

"Systems identification" is the name given to the set of estimation techniques based on fitting differential-equation models of structured populations to observed data (the term was introduced in this context in Parslow et al. 1979). One model that seems quite appealing for inference from populations that display a single, but diffuse, cohort includes equations like (2) to model stage populations, giving each stage a separate mortality rate that does not vary through time and describing the total birthrate by a Gaussian:

$$\frac{dn_i}{dt} = R_i(t) - M_i(t) - \mu_i n_i(t), \tag{10}$$

where $M_i(t) = R_i(t - \tau_i) \exp(-\mu_i \tau_i)$ and

$$R_i(t) = \begin{cases} k \exp[-(t - p_2)^2 / 2p_1^2] & \text{if } i = 1, \\ m_{i-1}(t) & \text{if } i > 1, \end{cases}$$

where $k$ is total births, $\mu_i$ is the per capita mortality rate of the $i$th stage, and $p_1$ is the width parameter of the birth function. Given some initial populations, $n_i(t_0)$, these equations can be integrated numerically to obtain the vector of model population estimates at the sample times, $\mathbf{n}$. The partial derivatives of these estimates with respect to the model parameters can be estimated by using finite differences or by more-accurate means using a little algebra. As an illustration, consider obtaining the partial derivative of the population estimates for stage 1 with respect to the mortality rate of stage 1. Taking partial derivatives of both sides of equation (10) (letting $i = 1$) yields

$$\frac{d}{dt}\left(\frac{\partial n_1}{\partial \mu_1}\right) = \tau_1 R_1(t - \tau_1)\exp(-\mu_1\tau_1) - n_1 - \mu_1\left(\frac{\partial n_1}{\partial \mu_1}\right).$$

To solve this (usually numerically), use the fact that, since the initial population does not depend on $\mu_1$, the initial condition is $\partial n_1/\partial \mu_1 = 0$. The partial derivatives of the model population estimates with respect to any of the parameters of the model can be obtained in a similar manner.

Clearly, many other models may be estimated using the same ideas. The one presented above is actually problematic, because it yields estimated mortality rates, which tend to oscillate widely from one stage to the next in the face of anything except noise-free data (see Wood & Nisbet 1991, Chap. 5).

*A Matrix Model Example*

Consider fitting equation (1) to data, under the assumption that the parameters are not time-dependent. Defining $\mathbf{n}_0$ as the vector containing the initial stage populations allows writing the model approximations to the stage populations through time as

$$\begin{bmatrix} \mathbf{I} \\ \mathbf{A} \\ \mathbf{A}^2 \\ \mathbf{A}^3 \\ \vdots \\ \mathbf{A}^{s-1} \end{bmatrix} \mathbf{n}_0 \, .$$

The vector $\mathbf{n}_0$ has $m$ components, so the vector $\mathbf{n}$ above has $ms$ components that will be called $\tilde{n}_i$, in order. Let $y_i$ be the observed values of these components; we seek parameters that minimize $\Sigma(\tilde{n}_i - y_i)^2$.

There are $3m$ parameters to estimate (the entries of matrix $\mathbf{A}$), which we collect as the elements $p_i$ of a vector of parameters

$$\mathbf{p} = \begin{bmatrix} P_1, P_2, \ldots, P_m, G_1, \ldots, G_{m-1}, \beta, \mathbf{n}_0^{\mathrm{T}} \end{bmatrix} \, .$$

To use the method of nonlinear least squares requires the Jacobian matrix, $\mathbf{J}_\Sigma$, whose $ms \times 3m$ elements are $\partial \tilde{n}_i / \partial p_j$.

To compute the elements of $\mathbf{J}_\Sigma$, we must compute $\partial(\mathbf{A}^k \mathbf{n}_0)/\partial p_j$. Clearly,

$$\frac{\partial \mathbf{A}^k \mathbf{n}_0}{\partial \mathbf{n}_0} = \mathbf{A}^k \, .$$

To calculate the other derivatives, begin by computing, for every parameter $p_i$, the matrix of derivatives

$$\mathbf{Z}_{j,1} = \frac{\partial \mathbf{A}}{\partial p_j} \, .$$

Now, by repeated differentiation,

$$\frac{\partial \mathbf{A}^2}{\partial p_j} = \mathbf{A}\mathbf{Z}_{j,1} + \mathbf{Z}_{j,1}\mathbf{A} \, ,$$

$$\frac{\partial \mathbf{A}^3}{\partial p_j} = \mathbf{A}^2\mathbf{Z}_{j,1} + \mathbf{A}\mathbf{Z}_{j,1}\mathbf{A} + \mathbf{Z}_{j,1}\mathbf{A}^2 \, ,$$

and so forth. It is now apparent that

$$\mathbf{Z}_{j,k} = \frac{\partial \mathbf{A}^k}{\partial p_j} = \sum_{i=1}^{k} \mathbf{A}^{i-1} \mathbf{Z}_{j,1} \mathbf{A}^{k-i}$$

$$= \mathbf{Z}_{j,k-1} \mathbf{A} + \mathbf{A}^{k-1} \mathbf{Z}_{j,1} \, .$$

The matrix $\mathbf{J}_\Sigma$ is organized in blocks of $m$ rows: the first block of rows is

$$\mathbf{J}_1 = \left[ \frac{\partial \mathbf{n}_0}{\partial \mathbf{p}}, \frac{\partial \mathbf{n}_0}{\partial \mathbf{n}_0} \right] = [0, \, \mathbf{I}_m] \, ,$$

where $\mathbf{I}_m$ is the $m \times m$ identity matrix; the second block is

$$\mathbf{J}_2 = \left[ \frac{\partial \mathbf{A} \mathbf{n}_0}{\partial \mathbf{p}}, \frac{\partial \mathbf{A} \mathbf{n}_0}{\partial \mathbf{n}_0} \right]$$

$$= [\mathbf{Z}_{1,1} \mathbf{n}_0, \mathbf{Z}_{2,1} \mathbf{n}_0, \dots, \mathbf{Z}_{2m,1} \mathbf{n}_0, \mathbf{A}] \, ;$$

continuing in this way, the $k$th block is

$$\mathbf{J}_k = \left[ \frac{\partial \mathbf{A}^{k-1} \mathbf{n}_0}{\partial \mathbf{p}}, \frac{\partial \mathbf{A}^{k-1} \mathbf{n}_0}{\partial \mathbf{n}_0} \right] \, ,$$

$$= \left[ \mathbf{Z}_{1,k-1} \mathbf{n}_0, \mathbf{Z}_{2,k-1} \mathbf{n}_0, \dots, \mathbf{Z}_{2m,k-1} \mathbf{n}_0, \mathbf{A}^{k-1} \right] \, .$$

Finally,

$$\mathbf{J}_\Sigma = \begin{bmatrix} \mathbf{J}_1 \\ \mathbf{J}_2 \\ \vdots \\ \mathbf{J}_s \end{bmatrix} \, ,$$

which can now be used to fit the model to data using the Levenberg-Marquardt method (see Press et al. 1992) described above. (It is not difficult to extend this example to deal with time-varying parameters, although it helps to use the methods for imposing a smooth variation of parameters introduced in Section 5.)

## 4 Regression Methods

This section solves the inverse problem by a regression of the population estimates in one stage against those in another. There are statistical objections to taking this approach, but it does yield nice linear equations. The general idea is best illustrated with an example.

*A Matrix-Regression Method for Solving the Demographic Inverse Problem*

Consider a population described by equation (1). This can be rewritten as

$$\mathbf{n}(t_{i+1}) = \begin{bmatrix} n_1 & 0 & . & . & . & . & . & . & n_m \\ 0 & n_2 & 0 & . & n_1 & 0 & . & . & . \\ . & . & . & . & 0 & n_2 & 0 & . & . \\ . & . & . & . & . & . & . & . & . \\ . & . & . & n_m & 0 & . & . & n_{m-1} & 0 \end{bmatrix} \begin{bmatrix} P_1 \\ P_2 \\ \vdots \\ P_m \\ G_1 \\ G_2 \\ \vdots \\ G_{m-1} \\ \beta \end{bmatrix},$$

where the elements of the first matrix are evaluated at $t_i$. Denote this equation as $\mathbf{n}(t_{i+1}) = \mathbf{N}(t_i)\mathbf{p}$ and define $\mathbf{n}^-$ :

$$\mathbf{n}^- = \left[ \mathbf{n}^\mathrm{T}(t_2),\ \mathbf{n}^\mathrm{T}(t_3), \dots, \mathbf{n}^\mathrm{T}(t_s) \right]^\mathrm{T}.$$

Now,

$$\mathbf{n}^- = \begin{bmatrix} \mathbf{N}_1 \\ \mathbf{N}_2 \\ \vdots \\ \mathbf{N}_{s-1} \end{bmatrix} \mathbf{p},$$

which can be rewritten, by defining $\mathbf{N}_\Sigma$, as

$$\mathbf{n}^- = \mathbf{N}_\Sigma \mathbf{p}.$$

If $y_i(t_j)$ are the real data approximating the population $n_i(t_j)$ of stage $i$ at $t_j$, the data estimate of $\mathbf{N}_i$ can be defined as $\mathbf{Y}_i$. Then, the sum of the square of the differences between $\mathbf{Y}_\Sigma \mathbf{p}$ and $\mathbf{y}^-$,

$$(\mathbf{y}^- - \mathbf{Y}_\Sigma \mathbf{p})^\mathrm{T}(\mathbf{y}^- - \mathbf{Y}_\Sigma \mathbf{p}),$$

can be minimized to find the parameter vector $\mathbf{p}$, yielding the formal solution

$$\hat{\mathbf{p}} = (\mathbf{Y}_\Sigma^\mathrm{T} \mathbf{Y}_\Sigma)^{-1} \mathbf{Y}_\Sigma^\mathrm{T} \mathbf{y}^-.$$

There are serious problems with this approach. First, error estimates are difficult to obtain because the elements of the matrix $\mathbf{Y}_\Sigma$ contain the data, and therefore uncertainty, so the standard results for obtaining the covariance matrix of $\mathbf{p}$ do not hold. The second problem is that the population data are both the explanatory data and the response variables. Hence, the explanatory variables are uncertain, as well as the response variables, and the estimated regression coefficients ($\mathbf{p}$) are biased toward zero. Another problem with this sort of method is that it seeks the parameters giving the best projection of population from one time to the next, not necessarily the parameters that would best match the whole time series.

The final problem with using the model in this example is that if the model is a good one, the population tends to an equilibrium age structure (described by a single eigenvector) and increases by a fixed proportion (the dominant eigenvalue) at each time step. The dynamics rapidly become wholly described by $m + 1$ parameters, and from those dynamics we want to infer $2m$ parameters. In such cases, this method does not work even when there are long data runs.

## 5  Irritating Problems

*Unstable Parameter Estimates and Smoothness*

When fitting models to data to obtain information about underlying dynamics, one is unlikely to be certain about the correct form for the model. For example, it is unwise to assume that the mortality rate is the same for stages 1 and 2. The natural thing to do is to make the model rather general, so that it does not make too many prior assumptions. Doing this leads to trouble rather quickly. The parameters of structured-population models tend to be fairly colinear; that is, changing any of a number of different parameters can lead to similar changes in dynamics. For instance, an increase in recruitment to a stage can be counterbalanced by an increase in the mortality rate of that stage; and stage population is not altered. This example can be extended. Consider a sequence of stages for which we are trying to estimate birth and mortality rates using models of the type given by equation (2): an overestimation of the birthrate can be compensated for by overestimating the mortality rate in the first stage, but this leads to underestimating the recruitment rate to the second stage, which in turn is compensated for by underestimating the mortality rate in the second stage, leading to an overestimation of recruitment to the third stage, and so on. The errors actually grow (along with the organism) through the stages (Wood et al. 1989; Wood & Nisbet 1991). This tends to lead to

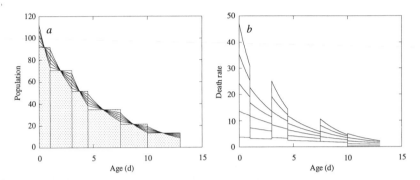

FIGURE 1. *Mortality estimation from structured-population data.* (a)
*Stage populations in each of six contiguous age classes (of different du-*
*rations). The area of each bar represents the population of a stage, and*
*the solid lines through the tops of the bars are continuous-age structures*
*perfectly consistent with the histogram of stage populations.* (b) *Deriva-*
*tives of the continuous-age structures yield death rates at age. Note the*
*wide range of death rates consistent with the stage populations.*

absurd parameter estimates as soon as the population data contain uncertainty. Figure 1 illustrates the problem. Suppose that the
stippled bars on the left are a histogram of populations in each of
six age classes. Assuming that the per capita death rate is constant
within each stage, then the family of solid lines through the top
of the histogram bars represents continuous age structures, consistent with the stage populations. The death rates in the stages
come from the first derivatives of the continuous age structures.
When these derivatives are calculated from the panel on the left,
the panel on the right results, where the solid lines represent part
of the family of death rates consistent with the stage populations.

One answer to this problem is to reformulate the model to include qualitative constraints on how quickly estimated quantities
can change with stage or age or time, as shown by the following
discussion.

The matrix method presented above as an example of a regression method suffers from instability problems when actually applied. One solution to this problem would be to simplify the model
to achieve stability, for example, by using a single $G$ or a single $P$
or by allowing the $G$'s and $P$'s to vary linearly with stage. A less
restrictive process is simply to require that the parameters vary
smoothly with stage. Imposing such a constraint requires a way

of measuring smoothness or wiggliness. A suitable measure of the wiggliness of an ordered set of parameters $a_1, a_2, \ldots, a_k$ is

$$\sum_{i=1}^{k-2}(a_i - 2a_{i+1} + a_{i+2})^2$$

$$= [a_1, a_2, \ldots, a_k] \begin{bmatrix} 1 & -2 & 1 & . & . & . & . \\ -2 & 5 & -4 & 1 & . & . & . \\ 1 & -4 & 6 & -4 & 1 & . & . \\ . & 1 & -4 & 6 & -4 & 1 & . \\ \vdots & \vdots & \vdots & \vdots & \vdots & \vdots & \vdots \\ . & . & . & . & 1 & -2 & 1 \end{bmatrix} \begin{bmatrix} a_1 \\ a_2 \\ \vdots \\ a_k \end{bmatrix}.$$

To measure the wiggliness of both the $G$'s and the $P$'s, form

$$[P_1, P_2, \ldots, P_m] \begin{bmatrix} 1 & -2 & 1 & . & . & . & . \\ -2 & 5 & -4 & 1 & . & . & . \\ 1 & -4 & 6 & -4 & 1 & . & . \\ . & 1 & -4 & 6 & -4 & 1 & . \\ \vdots & \vdots & \vdots & \vdots & \vdots & \vdots & \vdots \\ . & . & . & . & 1 & -2 & 1 \end{bmatrix} \begin{bmatrix} P_1 \\ P_2 \\ \vdots \\ P_m \end{bmatrix}$$

$$+ [G_1, G_2, \ldots, G_{m-1}] \begin{bmatrix} 1 & -2 & 1 & . & . & . & . \\ -2 & 5 & -4 & 1 & . & . & . \\ 1 & -4 & 6 & -4 & 1 & . & . \\ \vdots & \vdots & \vdots & \vdots & \vdots & \vdots & \vdots \\ . & . & . & . & 1 & -2 & 1 \end{bmatrix} \begin{bmatrix} G_1 \\ G_2 \\ \vdots \\ G_{m-1} \end{bmatrix}.$$

Next, convert this into an expression in terms of the parameter vector $\mathbf{p}$, namely, $\mathbf{p}^{\mathrm{T}} \mathbf{R}_\Sigma \mathbf{p}$.

The wiggliness measure is employed by adding it to the least-squares function that is minimized to fit the model, yielding

$$(\mathbf{y}^- - \mathbf{Y}_\Sigma \mathbf{p})^{\mathrm{T}}(\mathbf{y}^- - \mathbf{Y}_\Sigma \mathbf{p}) + \lambda \mathbf{p}^{\mathrm{T}} \mathbf{R}_\Sigma \mathbf{p},$$

which is itself minimized when

$$\mathbf{p} = \left(\mathbf{Y}_\Sigma^{\mathrm{T}} \mathbf{Y}_\Sigma + \lambda \mathbf{R}_\Sigma\right)^{-1} \mathbf{Y}_\Sigma^{\mathrm{T}} \mathbf{y}^-.$$

The fudge factor $\lambda$ controls the trade-off between smoothness of the solution and fidelity to the data.

In Section 7, relatively sophisticated methods for choosing the optimal smoothing parameter are presented. These methods are rather complicated to apply in the present case because of the data in the matrix $\mathbf{Y}_\Sigma$. Knowing in advance the variance of the error in the observations would allow the choice of a $\lambda$ that makes

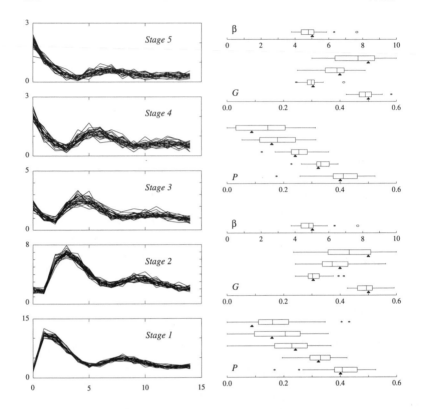

FIGURE 2. *Reconstruction of the parameters of a matrix model from noisy simulated data.* Left-hand curves, *Half of 50 replicate sets of noisy structured-population trajectories, using parameters estimated by two methods. Results are shown on the right.* Lower three box plots, *The distribution of estimated parameters when inequality constraints are applied in the estimation procedure but smoothness constraints are not.* Upper box plots, *A smoothness constraint is included on the P parameters of the matrix.* Triangles, *True parameter values. G, P, and β refer to the parameters of equation (1). The parameters for earlier stages are plotted below those for older stages in all plots. Note that the spread of estimates is lower when smoothness constraints are imposed. (If neither bound constraints nor smoothness constraints are applied, the distributions get very wide!)*

the residual sum of squares over its degrees of freedom match this value, but this is a rather crude method.

Figure 2 shows the results of applying the regression method presented here to simulated structured population data (although with the additional constraints on parameters discussed in the next subsection). A model of the form (1) is used to simulate a population with five stages, sampled assuming Poisson errors at 15 times; of the 50 replicates, 25 (selected randomly) are plotted on the left. The method is then applied to each of the 50 sets to obtain estimates of the parameters of the matrix model, with and without the smoothness constraint (see below). The smoothing parameter is chosen by a specially developed variant of cross-validation, which will be reported elsewhere. Note that the distributions of parameter estimates are narrower when smoothing constraints are imposed.

Smoothness constraints can be incorporated in the method suggested in Section 3 in much the same way as is done in this section and have the substantial advantage of increasing the goodness of the quadratic model used for nonlinear least-squares fitting. Details of this approach are omitted here. The optimal application of smoothness constraints is the subject of Section 7. (The method presented here can be extended to deal with time-dependent parameters.)

*Intrinsically Positive Parameters*

Death rate, population, birthrate, and developmental times are always positive. Unfortunately, the least-squares methods of Section 2 don't know that. In order to use such information about the parameters or processes in a model, constrained-fitting methods must be employed, or the problem must be transformed in such a way that parameters are forced to be positive. The transformation method simply involves replacing parameters like $G_i$ in equation (1) with $e^{g_i}$ or some other function that monotonically maps $(-\infty, \infty)$ on to $(0, \infty)$. The model is then fit by adjusting the new parameters, which can be transformed back to the old ones once the best fit has been found. As a more complicated example: to avoid a spontaneous generation of individuals or negative populations in model (1) requires that $0 \leq G_i + P_{i+1} \leq 1$, $0 \leq G_i \leq 1$, and $0 \leq P_{i+1} \leq 1$. This can be achieved using the function

$$\ell(x) = e^x/(1 + e^x)\,,$$

where $\ell^{-1}(x) = -\log_e\left([1/x] - 1\right)$. Using two new parameters, $\gamma_i$ and $\phi_i$, let $G_i = \ell(\gamma_i)\ell(\phi_i)$ and $P_{i+1} = \ell(\gamma_i)[1 - \ell(\phi_i)]$. The model fit would then proceed in terms of $\gamma$'s and $\phi$'s.

This approach has three disadvantages: (1) it makes any linear least-squares problem nonlinear, which is undesirable; (2) smoothness constraints no longer improve the quadratic approximation to the objective function used for nonlinear least squares; and (3) it can sometimes be rather difficult to impose the required constraints without changing the nature of the model. As an example of the third disadvantage: the necessary constraints required to ensure that a function is positive are seldom a simple set of bounds on the parameters of the constraints. It is always possible to transform the whole function so that it is always positive, but this changes its shape.

For linear problems it is better to use a quadratic programming algorithm, rather than attempting transformation. Quadratic programming is the name given to the process of solving the problem

$$\text{minimize } \mathbf{p}^{\mathrm{T}}\mathbf{G}\mathbf{p} + \mathbf{c}^{\mathrm{T}}\mathbf{p} \text{ with respect to } \mathbf{p}\,,$$
$$\text{subject to the constraint that } \mathbf{Ap} \geq \mathbf{b}\,, \tag{12}$$

where $\mathbf{A}$ is a matrix of coefficients for the constraints (which can have any number of rows), $\mathbf{b}$ and $\mathbf{c}$ are vectors of the coefficients, and $\mathbf{G}$ is a square matrix of coefficients that is assumed to be positive definite (or at least nonnegative definite). Algorithms for solving such problems are available in most major numerical-analysis libraries, and good explanations of how they work were given by Gill et al. (1981). (The author can also supply code to solve such problems.) For most of what follows, it is assumed that a problem written as a quadratic-programming problem can be solved.

For example, in the preceding matrix-regression example, the conditions on the $G_i$'s and $P_i$'s can be written

$$G_i + P_{i+1} \geq 0\,, \quad -G_i - P_{i+1} \geq -1\,, \quad G_i \geq 0\,,$$
$$-G_i \geq -1\,, \quad P_{i+1} \geq 0\,, \quad -P_{i+1} \geq -1\,.$$

The conditions for each $i$ can thus be stacked and written in one large matrix inequality of the form $\mathbf{Ap} \geq \mathbf{f}$, where $\mathbf{f}$ is a vector of zeros and ones. Furthermore, the quantity to be minimized in order to fit the model can be written in the form (12), where

$\mathbf{G} = \mathbf{Y}_\Sigma^{\mathrm{T}}\mathbf{Y}_\Sigma + \lambda\mathbf{R}_\Sigma$ and $\mathbf{c} = -2\mathbf{Y}_\Sigma^{\mathrm{T}}\mathbf{y}^-$, whereby the constrained-model fitting problem becomes a quadratic-programming problem and can therefore be solved.

A final piece of information relates to the way in which the quadratic program is minimized. Quadratic programming works by constraining the minimization of the objective function to a vector space (a subspace of the original parameter space) such that any constraints that would have been violated by an unconstrained minimization are either not violated or are satisfied as equality constraints. The algorithms for quadratic programming produce a matrix $\mathbf{Z}$ whose columns form an orthogonal basis for this space. This matrix is required in Section 7.

## 6 Model Uncertainty and a Way of Tackling It

As mentioned in the introduction, a major difficulty in solving the inverse problem is that the "true" population model is rarely, if ever, known a priori. Pretending that it is leads to biased estimates and an underestimation of the variance associated with these estimates. One way out of this difficulty is to define a population model flexibly, so that it is likely to be consistent with the true model, but then to use smoothness constraints of the type introduced in the preceding section to adjust the complexity of this model. Looking back to the example above, when $\lambda$ is set to a low value, each parameter of the model is treated independently, giving two parameters to estimate for every stage in addition to a birthrate parameter—a complicated model. Alternatively, if $\lambda$ is given a high value, then the parameters are constrained to vary linearly with stage, and a simple model results, with rather few degrees of freedom. Clearly, selecting appropriate $\lambda$'s can result in intermediate levels of model complexity; it is feasible to produce a whole family of models whose complexity is controlled by a single parameter. With such a family of models it is possible to address the question of model uncertainty, but seeing how this is done requires some additional bits of mathematical technology: a way of building smoothness into continuous models, and a way of choosing the "best" smoothing parameter, $\lambda$.

*Using Splines to Represent Unknown Smooth Functions*

Consider a set of points $(x_1, p_1), (x_2, p_2), \ldots, (x_n, p_n)$. The one-dimensional cubic spline is the function interpolating these points

TABLE 1. *Definitions for a Univariate Cubic Spline*

| Spline basis functions $(h_i = x_{i+1} - x_i)$ |
| --- |
| $\psi_{0i}(x) = (x_{i+1} - x)/h_i$ |
| $\zeta_{0i}(x) = \frac{1}{6}\left[(x_{i+1} - x)^3/h_i - h_i(x_{i+1} - x)\right]$ |
| $\psi_{1i}(x) = (x - x_i)/h_i$ |
| $\zeta_{1i}(x) = \frac{1}{6}\left[(x - x_i)^3/h_i - h_i(x - x_i)\right]$ |

Definitions of **H** and **B** $(h_i = x_{i+1} - x_i)$.
All elements are 0 except

| | | |
| --- | --- | --- |
| $H_{i,i} = h_i^{-1}$ | $H_{i,i+1} = -\left(h_i^{-1} + h_{i+1}^{-1}\right)$ | $1 \leq i \leq n - 2$ |
| $H_{i,i+2} = h_{i+1}^{-1}$ | $B_{i,i} = -\frac{1}{3}\left(h_i + h_{i+1}\right)$ | $1 \leq i \leq n - 2$ |
| $B_{i,i+1} = \frac{1}{6}h_{i+1}$ | $B_{i+1,i} = \frac{1}{6}h_{i+1}$ | $1 \leq i \leq n - 3$ |

NOTE.—The cubic spline interpolates at the set of points $\{(x_1, p_1), (x_2, p_2), \ldots, (x_n, p_n)\}$.

that minimizes

$$\int_{x_1}^{x_n} [f''(x)]^2 \, dx \,,$$

where $f''$ is the second derivative of $f$. This turns out to be a function made up of sections of cubic polynomial between each pair of points, joined so that continuity of the curve is maintained up to and including the second derivative. The curve can be expressed in terms of the $p_i$'s and the second derivatives, $d_i$, of the curve at the $x_i$'s:

$$f(x) = p_i\psi_{0i}(x) + p_{i+1}\psi_{1i}(x) + d_i\zeta_{0i}(x) + d_{i+1}\zeta_{1i}(x)$$
$$\text{for } x_i \leq x \leq x_{i+1} \,,$$

where

$$\mathbf{Bd} = \mathbf{Hp} \,.$$

**B** and **H** are coefficient matrices, and $\psi$ and $\zeta$ are cubic basis functions; all are defined in Table 1.

There are two convenient reasons for using splines in the methods discussed here. First, a continuous wiggliness measure is analogous to the measures used in the preceding section:

$$\int_{x_1}^{x_n} [f''(x)]^2 \, dx = \mathbf{p}^{\mathrm{T}}\mathbf{H}^{\mathrm{T}}\mathbf{B}^{-1}\mathbf{Hp} \,. \tag{13}$$

Second, since the spline at any point, $x$, is a linear function of its parameters $\mathbf{p}$, it can be written as

$$f(x) = \mathbf{s}^{\mathrm{T}}(x)\mathbf{p},$$

where $\mathbf{s}$ is a vector of coefficients that depend on $x$ but not on $\mathbf{p}$.

Splines can be used to represent unknown functions with any of the basic model formalisms used in this chapter. The next subsection gives an example.

*Using Splines to Represent a Continuous-Age, Continuous-Time Model*

Equation (3) defines a surface $n(a, t)$ that is the population per age interval at age $a$ and time $t$. This can be fit to observed data on different age classes within a population by representing $n(a, t)$ using cubic-spline functions. A method based on this approach has been described in detail (Wood 1994; for an application, see Ohman & Wood, in press). Here I describe a slightly more straightforward version of the approach, which is simpler and very nearly as effective.

To represent $n(a, t)$, a set of points in the age-time plane covers the age range of the pre-adult part of the population and the time interval of the samples. (Adults must be dealt with separately; see Ohman & Wood, in press.) Let these points be at time points $t_1, t_2, \ldots, t_s$ and at ages $a_1, a_2, \ldots, a_m$. These points are referred to as the "knots" of the surface (in accordance with spline terminology), each with an associated parameter $p_i$. The parameters determine the value of $n$ at the knots, and it is the parameters that are adjusted when the model is fit to data. So $\mathbf{p}$ can be interpreted as

$$\mathbf{p}^{\mathrm{T}} = [n(a_1, t_1), \ldots, n(a_m, t_1), n(a_1, t_2), \ldots, n(a_m, t_s)].$$

In this example there are as many knots as data points, but this assumption is not necessary. The values of $n$ between the knots is determined by interpolation with cubic splines, first in the age direction and then in the time direction. Starting with age, each set of parameters associated with a particular $t_i$ can be interpolated with a spline, leading to

$$n(a, t_i) = \mathbf{s}_a^{\mathrm{T}}\mathbf{p}_i, \tag{14}$$

where

$$\mathbf{p}_i^{\mathrm{T}} = [p_{im+1},\ p_{im+2},\ \ldots,\ p_{im+m}]$$
$$= [n(a_1, t_i), n(a_2, t_i), \ldots, n(a_m, t_i)]$$

and $\mathbf{s}$ is a vector of coefficients that map the parameters of the spline to its value at $a$. To find the value of the surface at any age and time, interpolate between values of $n(a, t_i)$, again using splines. If $\mathbf{s}_t$ is the vector that maps the coefficients of splines in the time direction to the value of the spline at $t$, then

$$n(a, t) = \mathbf{s}_t^{\mathrm{T}}\, \mathbf{q}(a)\,,$$

where $\mathbf{q}$ is a vector such that

$$q_i = n(a, t_i)\ \ [= \mathbf{s}_a^{\mathrm{T}} \mathbf{p}_i].$$

A routine multiplying of coefficients leads to a new vector $\mathbf{k}(a, t)$ such that

$$n(a, t) = \mathbf{k}^{\mathrm{T}}(a, t)\mathbf{p}\,,$$

thereby yielding the algorithm by which the population surface at any age and time can be obtained from its coefficients. So far, the representation is lacking some of the properties it needs, however. In order not to imply negative death rates (which are impossible no matter what the nice man from the cryogenics company tells you), it is essential that $n(a, t)$ decline along any line for which time is increasing and $da/dt = 1$. This restriction occurs because all individuals are born at age 0 and then age by one day per day until death; thus, the surface can only decline along these lines. An approximate version of this restriction can be imposed on the surface at any point by requiring that

$$[\mathbf{k}(a - \Delta/2,\ t - \Delta/2) - \mathbf{k}(a + \Delta/2,\ t + \Delta/2)]\ \mathbf{p}/\Delta \geq 0\,,$$

where $\Delta$ is a small positive time interval. The constraint can be applied approximately to the whole surface by requiring this condition to hold at a large number of points in the age-time plane. In a similar way, the surface must everywhere be positive, since populations cannot be negative (although, if the nice man from the cryogenics company were not talking nonsense, the contents of his freezer might qualify). This condition is simple to apply:

$$\mathbf{k}^{\mathrm{T}}(a, t)\, \mathbf{p} \geq 0\,.$$

If monotonicity conditions are used, it is sufficient to apply positivity constraints at the upper age boundary and upper time bound-

ary of the surface to ensure positivity everywhere else on the surface. Both sets of conditions can be assembled into a matrix inequality of the form $\mathbf{Ap} \geq 0$.

In order to relate the surface to some data, stage populations
at some time must be expressed as functions of the parameters $\mathbf{p}$.
Equation (4) provides the means for doing this. The population of
a stage at a time is just the integral of the population surface at
that time between the ages that the stage contains. From equation
(14) it is clear that the age and time dependence of the surface
are contained in the coefficient vector $\mathbf{k}$; to integrate across the
population surface, one simply integrates the elements of $\mathbf{k}$ to get
$n_i(t) = \mathbf{w}^{\mathrm{T}}\mathbf{p}$, where

$$\mathbf{w}^{\mathrm{T}} = \left[ \int_{\alpha_i}^{\alpha_{i+1}} k_1(a,t)da, \int_{\alpha_i}^{\alpha_{i+1}} k_2(a,t)da, \right.$$

$$\left. \ldots, \int_{\alpha_i}^{\alpha_{i+1}} k_{sm}(a,t)da \right];$$

hence, the stage populations, $\mathbf{n}$, corresponding to some set of observations, $\mathbf{y}$, can be obtained from an equation of the form

$$\mathbf{n} = \mathbf{Wp},$$

where $\mathbf{W}$ is made up of rows of the form $\mathbf{w}^{\mathrm{T}}$. A fittable model is
now fairly close. The deviation of the model from the data can be
measured by the sum of squares:

$$(\mathbf{y} - \mathbf{Wp})^{\mathrm{T}} (\mathbf{y} - \mathbf{Wp}).$$

The model could be fit as is, but it has a parameter for each data
point and cannot therefore be expected to yield very good or precise
estimates of anything. For this reason, one final part is added to
the formalism: a wiggliness measure for the surface. In keeping
with the way that the surface has been constructed, wiggliness is
measured as an average of the wiggliness measures for the splines
interpolating the original knots in the direction of age and time.
This yields a measure of the form

$$\frac{(t_s - t_1)^3}{sm} \sum_{i=1}^{m} \int_{t_1}^{t_s} [n_{tt}(a_i,t)]^2 dt + \frac{a_m^3}{sm} \sum_{i=1}^{s} \int_{0}^{a_m} [n_{aa}(a_i,t)]^2 da,$$

where $n_{tt} = \partial^2 n / \partial t^2$, and $n_{aa} = \partial^2 n / \partial a^2$.

The weighting factors in front of the summations are intended
to ensure that the same weight is given to wiggliness in age and
time, irrespective of the age range relative to the time interval. The

terms within the summations can be written as matrix expressions of the form (13), and thus the total wiggliness measure can be constructed in a manner similar to that used in equation (11) to yield an expression for the wiggliness of the surface of the form $\mathbf{p}^{\mathrm{T}}\mathbf{R}\mathbf{p}$.

To actually fit the model, it is necessary to minimize a term for the weighted sum of least squares plus a wiggliness measure, subject to the positivity and monotonicity constraints required for nonnegative populations and death rates. That is,

$$\text{minimize } (\mathbf{y} - \mathbf{W}\mathbf{p})^{\mathrm{T}} (\mathbf{y} - \mathbf{W}\mathbf{p}) + \lambda\mathbf{p}^{\mathrm{T}}\mathbf{R}\mathbf{p},$$

subject to the constraint that $\mathbf{A}\mathbf{p} \geq 0$.

Rearranging this leads again to a quadratic-programming problem. The smoothing parameter $\lambda$ controls the trade-off between smoothness of the model and fidelity to the data. High $\lambda$ gives a smooth and simple model that does not fit the data too closely; low $\lambda$ leads to a complicated model that gives a close fit to the data. Efficient means for choosing $\lambda$ are required and are the subject of the next section.

## 7 Choosing Model Complexity: Cross-Validation and Its Relatives

Cross-validation is a simple idea. If a model is to be a good model of the data to which it is fit, it should be good at predicting data points that were left out of the data set to which it was fit. This notion can be formalized. If $\mathbf{f}(\mathbf{p})$ is a model with parameter vector $\mathbf{p}$, which is to be fit to some data $\mathbf{y}$, then let $\mathbf{f}^{[i]}(\mathbf{p})$ be the model fit to all the data except $\mathbf{y}_i$. A measure of how good the model is at predicting missing data is given by

$$\sum_{i=1}^{n} \left( f_i^{[i]} - y_i \right)^2.$$

Minimizing this quantity provides a criterion for choosing between competing models. It is not hard to show that the above expression is the same as

$$\sum_{i=1}^{n} (f_i - y_i)^2 \bigg/ \left( 1 - \frac{\partial f_i}{\partial y_i} \right)^2 \tag{15}$$

(where $f$ is the model fit to all the data), minimization of which provides an efficient way of performing cross-validation. A more popular version of this approach is "generalized cross-validation," which minimizes

$$\sum_{i=1}^{n} (f_i - y_i)^2 \Bigg/ \sum_{i=1}^{n} \left(1 - \frac{\partial f_i}{\partial y_i}\right)^2 \qquad (16)$$

(for an explanation of its supposed advantages, see Wahba 1990). In the case of the models with smoothing constraints, like the model in Section 6, the smoothing parameter $\lambda$ can be chosen by cross-validation by minimizing equation (15) or (16). The calculation of the partial derivatives needed for calculating the cross-validation score requires some further effort. If the model is of the form

$$\mathbf{n} = \mathbf{W}\mathbf{p},$$

where $\mathbf{n}$ is a vector of population estimates to be fit to a vector of observation $\mathbf{y}$, and this model has been fit by quadratic programming (possibly with a smoothness constraint), then the matrix $\mathbf{V}$ for which $V_{i,j} = \partial p_i / \partial y_j$ is given by

$$\mathbf{V} = 2n^{-1}\lambda \mathbf{Z}(\mathbf{Z}^{\mathrm{T}}\mathbf{G}\mathbf{Z})^{-1}\mathbf{Z}^{\mathrm{T}}\mathbf{W}^{\mathrm{T}},$$

where $n$ is the number of data points, and $\lambda$ is replaced by 1 if there is no smoothness constraint in the model. Approximations for these partial derivatives can be obtained for nonlinear models. Forming $\mathbf{A} = \mathbf{W}\mathbf{V}$, the score from a generalized cross-validation now becomes

$$(\mathbf{n} - \mathbf{W}\mathbf{p})^{\mathrm{T}}(\mathbf{n} - \mathbf{W}\mathbf{p})[\mathrm{Tr}(\mathbf{w} - \mathbf{A})]^{-2}.$$

Here, Tr indicates the trace of a matrix.

For problems that include smoothness constraints and have their smoothing parameter chosen in this way, help comes from work pioneered by Wahba (1983) on incorporating model uncertainty into confidence-interval estimates for the model. Essentially, we assume that the true model is unknown but comes from the family of different models that results from choosing the full range of different smoothing parameters. It is then assumed that the fit model is a Bayes estimator of the true model; this implies a prior distribution for the models, which allows the calculation of a posterior covariance matrix for the fit model, which in turn makes allowance for this model uncertainty. This slightly dubious line of reasoning produces confidence intervals that appear to be effective when applied to simulated data. The covariance matrix for the model fit to the data is $\mathrm{cov}(\mathbf{f}) = \mathbf{A}\hat{\sigma}^2$, and for the parameters of the model, the

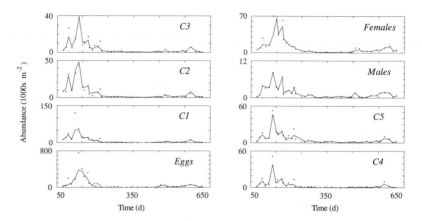

FIGURE 3. *Fit of a model* (solid line) *of the type described in Section 6 to data for* Pseudocalanus newmani (dots).

matrix is $\mathrm{cov}(\mathbf{p}) = \hat{\sigma}^2\, 2n^{-1}\lambda \mathbf{Z}(\mathbf{Z}^{\mathrm{T}}\mathbf{G}\mathbf{Z})^{-1}\mathbf{Z}^{\mathrm{T}}\mathbf{W}^{\mathrm{T}}$, where

$$\hat{\sigma}^2 = (\mathbf{W}\mathbf{p} - \mathbf{y})^{\mathrm{T}}(\mathbf{W}\mathbf{p} - \mathbf{y})[\mathrm{Tr}(\mathbf{w} - \mathbf{A})]^{-1};$$

the last term is the degrees of freedom associated with the residual sum of squares. For a more extensive discussion of splines, cross-validation, and confidence-interval calculations, see Wahba (1983, 1990), Craven and Wahba (1979), Villalobos and Wahba (1987), and Nychka (1990).

Figures 3–5 are the results of ongoing analysis (Ohman & Wood, in press) applying a method of the type described in Section 6 (with smoothness chosen by the methods of this section) to data from a population of the copepod *Pseudocalanus newmani* in a temperate fjord. Data were available for eggs and for 6 of the 12 copepod stages over a period of nearly two years. The analysis is complicated by the fact that the animals enter diapause over winter. Figure 3 shows the fit of the model to the data. Figure 4 shows the fit population and death-rate surfaces. Figure 5 shows the time-dependent, stage-specific death-rate estimates derived from the death-rate surface.

## 8 Conclusions

In conclusion, the message from this chapter is that the demographic inverse problem can be meaningfully solved only by using methods that recognize both measurement uncertainty and model uncertainty and that address both when estimating pop-

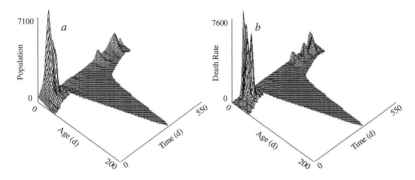

FIGURE 4. *Surfaces fit to the data from Figure 3:* (a) *population surface;* (b) *death-rate surface. The areas with no surface shown are age-time combinations for which there are no juvenile individuals.*

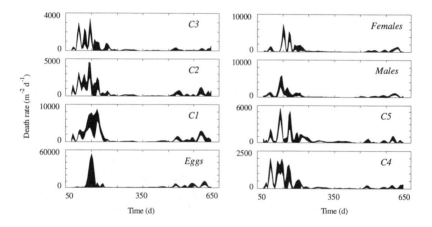

FIGURE 5. *For a population of* P. newmani, *95 percent confidence intervals for stage-specific total death rates. The rates are derived by integration of the death-rate surface of Figure 4.*

ulation rates and parameters. There remain many areas in which the technology for solving the demographic inverse problem could be improved. In this chapter, only least-squares fitting methods are used; for non-Gaussian errors, another form of maximum-likelihood estimation would probably be preferable. Models with multiple smoothing parameters remain problematic; there is a clear need for user-friendly software implementing this general framework, without imposing specific population models. As computer power increases, it may be preferable to examine computer-intensive resampling methods as the means for dealing with model uncertainty; the different candidate models could be fit to resampled population data, with the best-fit model being chosen each time. If enough realizations are performed, approximate distributions for the estimated population parameters could be obtained, without the sophisticated mathematical technology presented in this chapter (although it is not clear how best to generate the resampled data sets required by such an approach).

Furthermore, when developing a method, it is important to be aware of some basic instabilities that plague this particular estimation problem and (more helpfully) of the reality that many population rates are intrinsically positive quantities. The recipe suggested here is that a model formalism should be chosen, and a model constructed, in such a way that parts of the model for which the "true" model is unknown are described by functions whose complexity can be controlled by a smoothing parameter. Sometimes the model can be constructed such that population rates that must be positive will be, or they can be fit using a constrained optimization method. The smoothing parameters should be selected using the methods of Section 7 and the confidence limits on inferred quantities found using the methods of the same section, in order to incorporate model uncertainty into confidence-interval calculation.

## Literature Cited

Banks, H. T., L. W. Botsford, F. Kappel, and C. Wang. 1991. Estimation of growth and survival in size-structured cohort data: An application to larval striped bass (*Morone saxatilis*). *Journal of Mathematical Biology* 30: 125–150.

Caswell, H. 1989. *Matrix Population Models*. Sinauer, Sunderland, Mass.

Caswell, H., and S. Twombly. 1989. Estimation of stage-specific demographic parameters for zooplankton populations: Methods based on stage-classified matrix projection models. Pp. 94–107 *in* L. McDonald, B. Manly, J. Lockwood, and J. Logan, eds., *Estimation and Analysis of Insect Populations*. Springer-Verlag, Berlin.

Craven, P., and G. Wahba. 1979. Smoothing noisy data with spline functions. *Numerische Mathematik* 31: 377–403.

Gill, P. E., W. Murray, and M. H. Wright. 1981. *Practical Optimization*. Academic Press, London.

Gurney, W. S. C., R. M. Nisbet, and J. H. Lawton. 1983. The systematic formulation of tractable single species population models incorporating age structure. *Journal of Animal Ecology* 52: 479–495.

Gurney, W. S. C., R. M. Nisbet, and S. P. Blythe. 1986. The systematic

formulation of models of stage structured populations. Pp. 474–494 *in*
J. A. J. Metz and O. Diekmann, eds., *The Dynamics of Physiologically
Structured Populations.* Springer-Verlag, Berlin.

Hay, S. J., G. T. Evans, and J. C. Gamble. 1988. Birth, death and
growth rates for enclosed populations of calanoid copepods. *Journal
of Plankton Research* 10(3): 431–454.

Levenberg, K. 1944. A method for the solution of certain non-linear
problems in least squares. *Applied Mathematics* 25: 164–168.

Manly, B. F. J. 1989. A review of methods for the analysis of stage
frequency data. Pp. 1–69 *in* L. McDonald, B. Manly, J. Lockwood,
and J. Logan, eds., *Estimation and Analysis of Insect Populations.*
Springer-Verlag, Berlin.

Marquardt, D. W. 1963. An algorithm for least-squares estimation of
non-linear parameters. *Journal of the Society for Industrial & Applied
Mathematics* 11: 431–441.

Metz, J. A. J., and O. Diekmann. 1986. *The Dynamics of Physiologically Structured Populations.* Lecture Notes in Biomathematics 68.
Springer-Verlag, Berlin.

Murray, J. D. 1993. *Mathematical Biology.* 2nd ed. Springer-Verlag,
Berlin.

Nisbet, R. M., and W. S. C. Gurney. 1982. *Modelling Fluctuating Populations.* Wiley, Chichester, Engl.

———. 1983. The systematic formulation of population models for insects with dynamically varying instar durations. *Theoretical Population Biology* 23(1): 114–135.

Nisbet, R. M., S. P. Blythe, W. S. C. Gurney, and J. A. J. Metz.
1985. Stage structured models of populations with distinct growth
and development processes. *IMA Journal of Mathematics Applied in
Medicine & Biology* 2: 57–68.

Nisbet, R. M., W. S. C. Gurney, W. W. Murdoch, and E. McCauley.
1989. Structured population models: A tool for linking effects at individual and population levels. *Biological Journal of the Linnean Society*
37: 79–99.

Nychka, D. 1990. The average posterior variance of a smoothing spline
and a consistent estimate of the average squared error. *Annals of
Statistics* 18: 415–428.

Ohman, M. D., and S. N. Wood. In press. Mortality estimation for planktonic copepods: *Pseudocalanus newmani* in a temperate fjord. *Limnology & Oceanography.*

Parslow, J., N. C. Sonntag, and J. B. L. Matthews. 1979. Technique
of systems identification applied to estimating copepod population
parameters. *Journal of Plankton Research* 1: 137–151.

Press, W. H., S. A. Teukolsky, W. T. Vetterling, and B. P. Flannery.
1992. *Numerical Recipes in C.* 2nd ed. Cambridge University Press.

Villalobos, M., and G. Wahba. 1987. Inequality-constrained multivariate smoothing splines with application to the estimation of posterior probabilities. *Journal of the American Statistical Association* 82(397): 239–248.

Wahba, G. 1983. Bayesian "confidence intervals" for the cross-validated smoothing spline. *Journal of the Royal Statistical Society B* 45(1): 133–150.

———. 1990. *Spline Models of Observational Data.* SIAM, Philadelphia.

Watkins, D. S. 1991. *Fundamentals of Matrix Computations.* Wiley, New York.

Wood, S. N. 1994. Obtaining birth and mortality patterns from structured population trajectories. *Ecological Monographs* 64(1): 23–44.

Wood, S. N., and R. M. Nisbet. 1991. *Estimation of Mortality Rates in Stage-Structured Populations.* Lecture Notes in Biomathematics 90. Springer-Verlag, Berlin.

Wood, S. N., S. P. Blythe, W. S. C. Gurney, and R. M. Nisbet. 1989. Instability in mortality estimation schemes related to stage-structure population models. *IMA Journal of Mathematics Applied in Medicine & Biology* 6: 47–68.

# CHAPTER 20

# Nonlinear Models of Structured Populations: Dynamic Consequences of Stage Structure and Discrete Sampling

*John Val, Ferdinando Villa, Konstadia Lika, and Carl Boe*

The importance of linear models in the study of the dynamics of structured populations is great. Yet, probably only a few population phenomena can be assumed to be linear. Nonlinearity arises whenever interaction is incorporated into mathematical models, as for density dependence, predator-prey relationships, competition, external forcing, and spatial effects.

Even if nonlinearity can be considered the universal rule in natural populations, one should not dismiss linear models as "unrealistic." In order to describe the ecological mechanisms providing the qualitative dynamics or patterns of populations and communities, simple, admittedly unrealistic, mathematical (or otherwise conceptual) descriptions representing current ecological knowledge are of great value. Even in "applied" ecology one should not lightly dismiss linear models; see the discussion about "prediction" and "projection" by Caswell (1989).

Another reason for the major role of linear models in population ecology is that no general theory has yet been developed to account for nonlinear models. As an example, the analysis of matrix population models (Caswell 1989) is, in the linear case, a successful application of the theory of nonnegative matrices; an elegant, self-consistent theory is developed on the basis of a few fundamental theorems. Despite some recent insights (Cushing 1989), no such generality has been achieved when dealing with nonlinear matrix models, like those resulting from the incorporation of density-dependent growth.

In general, the greater ecological detail that nonlinear population models embody is achieved at a high cost in terms of mathematical tractability. It is usually impossible to find an explicit solution to nonlinear differential or difference equations. Most nonlinear analysis is done either by a study of the stability properties of the (linearized) systems in the neighborhood of a stable manifold or by numerical simulation. Both approaches have disadvantages. In linearized stability analysis only a narrow subset of the system's behavior can be revealed. Numerical simulation cannot lead to the same beautiful synthesis that analytical tools provide, since it limits analysis to a finite number of parameter combinations. With the appearance of more-powerful computers, however, the set of parameter values is greatly enlarged and a "catalogue" of system properties can be achieved.

Apart from these difficulties, analogous nonlinear models analyzed under different mathematical approaches may lead to different qualitative behavior. The richness of the possible behavior of nonlinear models can be considered both a hassle and the reason for their power. Here we consider some simple examples where richness in behavior is wanted and where richness is questionable.

## 1 Nonlinearity and Modeling Strategies

Whenever interaction among individuals takes place, nonlinear terms show up in even the simplest model describing this interaction. For example, genetic exchange can happen only if an individual meets another. Different interactions become important on different time scales. The task of model building involves choosing appropriate organizational scales, picking those phenomena that are most relevant to the dynamics of the chosen level of organization. We need nonlinear models when interactions play a basic role among the factors determining the behavior under examination. When the system description and the behavior of interest are given on different organizational levels, as in locally specified "cellular automata" models (Wolfram 1984), nonlinear (simulation) models are the only possible choice.

Biotic and abiotic phenomena causing nonlinearity in population models can be broadly classified as follows.

*Intraspecific interactions.* The rate of change in a multi-stage population depends on the simultaneous presence of individuals of the same or other stages. Examples include sex, cannibalism, and competition. Competition for space—or, more generally, for resources—among individuals is a mechanism for the self-limitation of populations.

*Interspecific interactions.* The rate of change of a multi-stage population is affected by the simultaneous presence of individuals of a different species, in a negative sense (competition, predation, parasitism) or positive sense (altruism, mutualism).

*External forcing.* Environmental factors like temperature or humidity affect the growth rates of one or more stages in a population.

*Spatial factors.* Nonlinear effects can be caused by spatial heterogeneity of the environment, particularly when local interaction between population patches takes place.

Usually, intra- or interspecific interactions cannot be treated by means of linear models, although some cases of external forcing or spatial factors can be incorporated in linear models.

Individual-based models (DeAngelis & Gross 1992), particularly *i*-state configuration models (Metz & Diekmann 1986; Caswell & John 1992), can incorporate a great deal of ecological detail, usually at the cost of being mathematically intractable. The current fascination with this and other related modeling formalisms reflects interest in capturing more-complex dynamics in time and space.

Experimental data generally determine the minimum detail to be included in a model. A modeler would like this minimum to be the maximum as well. However, while the models should be as simple as possible, they must build on coherent and biologically plausible mechanisms.

Experimental data on population dynamics generally come from discrete-time sampling, and discrete-time models are often offered to explain data obtained from the underlying continuous-time processes. If the time intervals between the samples is small enough, a discrete-time model can reveal a lot about the underlying continuous mechanistic process and even allow for a test of possible hypotheses for the underlying continuous process. In Section 2, we give an example of a situation in which data collection has been frequent and the data forced the use of a stage-structured continuous-time model. It is the well-known "Nicholson's blowfly" laboratory experiment (Nicholson 1954) and its model counterpart put forward by Gurney et al. (1980). This mechanistic model is presented in the form of a deterministic delay-differential equation. We examine the effect of some modifications on this model, one of which actually brings us closer to the data.

Coarse-grained sampling does not reveal much of an underlying mechanism, but still a simple biologically plausible model is wanted. Given that energy conservation limits the maximal size of a population, the most simple model for density-dependent regulation in a single-species model describing the data of yearly sampling is the discrete-logistic equation (see, e.g., May et al. 1974) or analogues like the Ricker function. The mechanistic idea in these models is that interacting with conspecifics affects either the individual mortality rate or the reproductive output. These models have a stable equilibrium if both $r$ and $T$ are small. In this case the discrete version is nothing more than a numerical solution of the continuous-time analogue. However, fixing $T$ and increasing $r$ opens up a region where the equilibrium becomes unstable and a chaotic pattern eventually results. But how can such dynamics explain a process with underlying continuity? In Section 3, we consider some mechanisms that might be approximated by these two discrete models.

## 2  Data Forcing Detail: The "Nicholson's Blowfly" Experiment

Sheep blowflies (*Lucilia cuprina*) have, like most insects, a multi-stage life cycle. After the egg hatches, it goes through instar stages before finally emerging as the adult fly. Nicholson (1954) closely observed a laboratory population of blowflies, grown for many generations in a single cage. The resulting population dynamics (Fig. 1a) were surprising, and there has been much discussion of the origin of the double peaks. Some argued that the intrinsic dynamics of the delay caused by egg-to-adult development, resulting in the "humped" form of the future-recruitment curve, was the cause (Gurney et al. 1980); others explained the pattern with age-specific variation in the survival and fecundity of the blowfly (Readshaw & VanGerwen 1983).

Delay-differential equations provide a conceptually simple way to describe a multi-stage population in continuous time, in which the instar stages last a fixed period of time. Using this modeling approach on the "Nicholson's blowfly" data, Gurney et al. (1980) successfully reproduced the dynamic patterns in the data. In this

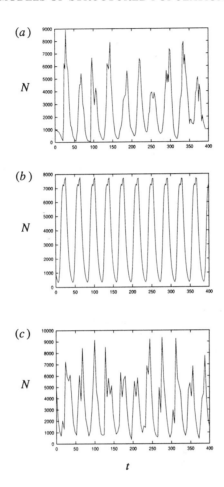

FIGURE 1. (a) *The original Nicholson (1954) data (redrawn from data published in Brillinger et al. 1980). (b) Predictions of the Nicholson's blowfly model when* $r = 6/day$, $N_0 = 850$ *adult flies,* $\delta = 0.2/day$, *and* $\tau_0 = 15$ *days. Parameters based on fitted values from Nicholson's laboratory data. (c) Results of a simulated run of model (1) with random variation in the developmental-time delay (parameter values described in the text).*

section, we examine the behavior of this model, first under the effect of random variation and then with nonlinear environmental forcing of the egg-to-adult developmental time. In the next section, the same model reappears as part of the discussion concerning the validity of discrete-time models.

Gurney et al. (1980) formulated their model as follows. The rate of change in the current population density of the adult blowflies, $dN(t)/dt$, is modeled as a function of the rate $E$ at which they were produced as eggs a developmental time $\tau$ ago. Here, $\tau_0$ represents the average time needed for an egg to produce an adult individual, under the assumption that it does not differ too much from individual to individual. The adult blowflies have a constant probability of death, and pre-adult stages are not vulnerable to death. Current egg production is density-dependent. The basic model now reads

$$\frac{dN(t)}{dt} = E(t - \tau_0) - \delta N(t) \,,$$
$$E(t) = rN(t)e^{-N(t)/N_0} \,, \tag{1}$$

where $N(t)$ is the number of adult flies at time $t$, $E(t)$ is the rate of production of vital eggs, $\tau_0$ is the reproductive delay (time from egg to adult), $\delta$ is the mortality rate of adult flies, $r$ is the fecundity per surviving adult, and $N_0$ is the "carrying capacity" for egg production by adult flies.

Depending on the region of the parameter space chosen, the coupled dynamic system (1) can exhibit stable equilibria, damped oscillations, stable limit cycles, or chaos. The original Nicholson data were adequately described by the above model, using parameters estimated from the published data (the growth rate $r$ and the death rate $\delta$) and by Nicholson himself (the developmental delay $\tau_0$). The model reproduced the period and amplitude of the oscillation well, showing the characteristic "double-peak" pattern apparent in the original data (Fig. 1).

Biological common sense suggests that most parameters in the model are subject to variation resulting from environmental influence. Particularly interesting are the effects of variation in the time needed for egg-to-adult development, $\tau_0$, which makes the model vary in population structure without losing too much of its simplicity. The next models do not allow for individual differences in developmental time; it is assumed that all the individuals in a cohort behave similarly (or that $\tau$ variance is very low). This assumption is made for the reason that the delay equations are not able to deal with individual variation. This assumption can be re-laxed if one chooses physiologically structured population models (Metz & Diekmann 1986) as the mathematical tool.

In changing developmental time, two cases are of particular bio-logical interest: random variation and periodic (seasonal or daily) variation in $\tau_0$. In fact, even Nicholson's notes seem to imply some degree of variation in the developmental delay; the measured de-velopmental delay was 15 days, but premature pupation reduced its value to 13 days in some cases (according to Mills, Nicholson's assistant, reported in Readshaw & Cuff 1980). It is highly likely that even stronger variation is common in natural populations, principally due to temperature shifts. Furthermore, it is widely known that seasonal temperature cycles may shorten and expand the developmental time in a more deterministic way, particularly in ectothermic organisms like insects. However, the seasonal patterns are made much more complex by the almost universal presence of wintering mechanisms based on stage differentiation.

Intuitively, the effects of a variable developmental time are likely to be important as well as interesting. A decrease in the develop-mental time should lead to "bursts" of adult individuals, since all juveniles having an age greater than the new developmental time would emerge. When the developmental time increases, there may be no more eggs to develop, since they would have already devel-oped in the past, and instars just before emergence would stop maturing.

Specialized numerical integration techniques (Neves & Thomp-son 1992) must be used to study delay-differential equations with state-dependent delays. Equations (1) have been studied numer-ically for different types of environmental variation in the devel-opmental delay. To perform numerical integration of the models described below, we use a fixed-step fourth-order Runge-Kutta method, in which the history is obtained via quadratic interpo-lation. In updating the system, we take into account the following biological considerations.

1. Eggs can develop into adults only once. When a variable delay makes the model refer twice to the same point in the history of the system, any egg that has developed already must be ex-cluded from the computation.

2. When the developmental delay is reduced under the effect of a forcing function, all remaining eggs that have age greater than $t - \tau$ must be converted into adult flies.

In all the simulation experiments below, the basic assumptions remain as outlined above. In natural systems, the developmental time, $\tau$, is very much dependent on the physiological status of the individuals laying the eggs; thus, $\tau$ values should be distributed rather than constant. As Murdoch et al. (1987) pointed out, the distribution of $\tau$ may alter the stability properties of a system with delay-differential equations when its variance is large enough. Here we assume a narrow distribution of $\tau$ values, with constant variance and a mean that shifts uniformly in response to an externally induced forcing factor.

## A Model with Random Variation in Developmental Delay

Here, $\tau$ varies randomly in length at random points in time. Two additional parameters are included: the mean size of the fluctuations, $\mu$; and the mean waiting time to the next change in developmental time, $\pi$. The mean $\mu$ specifies the range of a uniform distribution $U_{[-\mu/2, \, \mu/2]}$ with mean 0 from which the $\tau$ deviation is drawn; a new $\tau$ computed in this way each $d$ days, where $d$ is taken from a Poisson distribution with mean $\pi$ days. Simulations use parameter values for $\pi$ ranging from 0.5 to 5 days and for $\mu$ ranging from 0.5 to 4 days. After each waiting time, the new $\tau$ is computed as $\tau = \tau_0 + U_{[-\mu/2, \, \mu/2]}$ to approximate the range of $\tau$ found in the original experiments.

Figure 1$c$ shows the total number of adult blowflies computed in a sample run of the modified model. Parameters $r$ and $\delta$ are set equal to the ones used in Figure 1$b$ and $\tau_0 = 14$, $\pi = 2$, and $\mu = 1.5$. The resulting time series looks similar to the original Nicholson data of Figure 1$a$. Direct comparison in the time domain is not possible because the simulations are, like Nicholson's data, of stochastic nature.

A better comparison between Nicholson's data and the simulation runs can be done in the frequency domain. Figure 2 shows spectral-density plots of ($a$) the original Nicholson data, ($b$) 400 days of the simulated unforced time series of Figure 1$b$, and ($c$) 800 days of a realization of the random-$\tau$ model with $\tau_0 = 14$, $\pi = 2$, and $\mu = 2$. (Only the first 400 days of Nicholson's data are considered; after that, as explained in Brillinger et al. 1980, the pattern lost its deterministic character somewhat, probably because of parameter drift.)

The most important difference between the Nicholson data and the output of the unforced model is in the lower frequencies. The

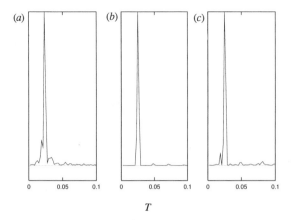

FIGURE 2. *Normalized spectral-density plots:* (a) *the original Nicholson time series;* (b) *the simulated unforced model (eqs. 1);* (c) *a simulated run of the random-$\tau$ model.*

original data show significant peaks with frequencies lower than the dominant 40-day period. This lower frequency appears in the time domain as a sine-like pattern drawn by the peak points of each 40-day period. The original model does not explain this frequency at all. The spectral analysis must be used with care, though, since the original time series is barely long enough to express two complete cycles of this lower frequency. What is clear is that the spectra of the original data and the random-$\tau$ model are more similar, and the similarity is conserved in most realizations with the same parameters. In particular, a significant peak corresponding with a frequency lower than the 40-day period appears only in the original and random-$\tau$ simulations.

The stochastic nature of the forcing function results in qualitative behavior in the population dynamics closer to the original data than to the original model. Further experimentation shows that the lowest frequency in the simulated time series is inversely related to the amplitude of the random fluctuations. This is probably because a decrease in developmental times causes more instars to emerge at once, resulting in the propagation of slower "waves" into the system. Besides having an effect on the left side of the spectrum, large fluctuations (on the order of $\pm 3.5$ days) cause shorter frequencies to become more significant in the frequency spectrum. This is regarded as a sign of shifting toward complex dynamics.

These results, while not invalidating the simple model of equations (1) in explaining the basic pattern shown by the Nicholson data, identify the random variation in developmental time as a possible cause for the appearance of the lower frequencies found in the original data. The "double peaking" apparent in the deterministic, unforced model, which has drawn so much attention from those who studied the model after Nicholson, is also more prominent in the stochastically modified model.

## A Model with Seasonal Variation in Developmental Delay

Having found closer agreement between the model and the data using random changes in developmental time, we check whether it is this random pattern of variation in $\tau$ that is important or whether any form of variation in $\tau$ serves well. The next model incorporates a periodically varying developmental delay. In the basic model, eggs laid at time $t$ survive, and each produces an adult individual at time $t+\tau$. It is biologically reasonable to assume that the developmental time, $\tau$, is influenced by seasonal factors, particularly temperature. A simple example of periodic forcing shows how important the dynamic consequences of seasonal variation can be. To mimic seasonal effects, we incorporate a sinusoidal dependence of $\tau$ on time:

$$\tau(t) = \tau_0 + A\sin(\omega t),\tag{2}$$

where $\omega$ is the angular frequency, and $A$ the amplitude of the forcing function. Note that, given the initial condition $\tau_0$, $\tau$ obeys the differential equation

$$\frac{d\tau}{dt} = A\,\omega\cos(\omega t).\tag{3}$$

The growth rate of the adult blowfly population depends on the egg production rate at time $t-\tau(t)$ modified toward the compression of the delay time. If $\tau$ expands, adults cannot emerge, since they have to wait longer as instars, and population growth is not possible. If $\tau$ shrinks, instars leave their larval state, thereby accelerating population growth. In the ideal situation, this is described by the delay-differential equation

$$\frac{dN(t)}{dt} = \left\{ E\big[t - \tau(t)\big]\left(1 - \frac{d\tau}{dt}\right)\right\} I_{-d\tau/dt} - \delta N(t),\tag{4}$$

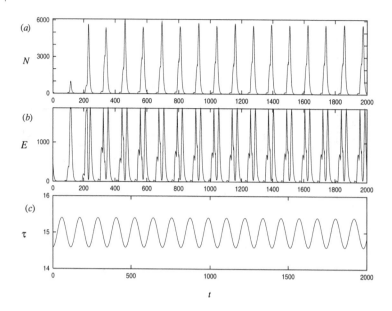

FIGURE 3. *Dynamics of Nicholson's blowfly model with periodically forced developmental delay: (a) total number of adult flies; (b) rate of egg production; (c) developmental delay $\tau$. $r = 6/day$, $N_0 = 850$ adult flies, $\delta = 0.2/day$, $A = 0.4$, $\omega = 0.054$.*

where $I_x$ is an indicator function satisfying

$$I_x = \begin{cases} 1 & x \geq 0, \\ 0 & \text{otherwise}. \end{cases} \tag{5}$$

In solving equation (4) we use a slightly modified model, which allows us to deal with the necessity of making discrete the time and the history of the system:

$$\frac{dN(t)}{dt} = \left\{ E\big[t - \tau(t)\big] + \int_{t-\tau}^{t-\tau-d\tau/dt} E(s)\,ds \right\} I_{-d\tau/dt} - \delta N(t). \tag{6}$$

Note that (6) equals (4) when $E$ is constant.

Integration of the system using the same parameters as in Figure 1b drastically changes the system dynamics (see Fig. 3). The oscillations of period 40 days have totally disappeared, replaced by peaks, with a period like the sine function, that modulate the

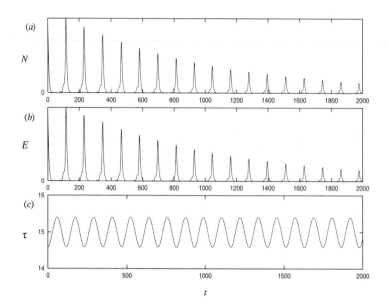

FIGURE 4. *Dynamics of Nicholson's blowfly model with periodically forced developmental delay: panel sequence as in Figure 3.* $r = 3.2/day$, $N_0 = 850$ *adult flies,* $\delta = 0.2/day$, $A = 0.4$, $\omega = 0.054$.

developmental delay. Since using other angular frequencies does not change this pattern, the effects found with the randomly varying $\tau$ are not due to simply changing $\tau$.

In playing with this model, we get a striking result when the egg production rate is lowered. Figure 4 shows that the population goes to extinction when the growth rate $r$ is reduced to a value that causes a stable limit cycle in the unforced system. The lack of emerging adult flies, resulting from the elongation of the developmental time, leads to the absence of egg production as well (Fig. 4b). During the outburst of adult flies upon the shortening of the developmental period, egg production cannot compensate for the scarcity of adult flies. In this case, the forced system undergoes immediate extinction. It is also worth noting how dramatic the effect can be: the maximum $\tau$ variation here is less than 3 percent of its mean value, and the period of nonproductivity is about 60 days, which is four times the mean developmental time. Even if the egg production rate in the unforced system is one-fifth the rate used in Figure 4, the population can be viable.

The system with periodic forcing of the developmental delay can reproduce realistic behavior for natural insect populations. Developmental delay is likely to increase during the winter, because of low temperatures. In the warm season, when temperatures rise, the developmental delay is expected to be shorter. A more realistic model would probably include a much steeper gradient in the changes in $\tau$ than the sine function used here. The "bottleneck" pattern so often observed in natural insect populations, with periods of high density alternating with periods of low density, could be explained with this simple model based on periodic variation in the egg-to-adult developmental time. Further research, including the same kind of periodic forcing of other parameters, would probably be an interesting starting point for better explanations of commonly found insect population dynamics.

## 3  Data Lacking Detail: Discrete Maps of Continuous-Time Processes

This section explores the other end of the spectrum of the detail in data acquisition. When focusing on environmental issues, most of the monitoring of ecological systems is done annually, and the data generally consist of numbers of individuals of certain key species. If fluctuations in population size are relatively small, both linear and nonlinear discrete-time models do well in forecasting population size in the next year (see, e.g., Caswell 1989; Tuljapurkar 1989). Large fluctuations may arise from either random changes in the environment or the underlying dynamics of the system. This section concentrates on the latter.

The main reason for focusing on fluctuations resulting from system dynamics is the following. Discrete sampling inclines the modeler to employ discrete-time models in explaining the data. Since deterministic fluctuations occur only in nonlinear models, these discrete-time models are nonlinear. The simplest discrete nonlinear models have arisen from the conception, based on the energy-conservation principle, that population size has to be bounded. Mechanistically, one might think of an increased mortality rate or a reduced reproductive output, both due to crowding. When these kinds of mechanisms are incorporated in discrete-time models, one can indeed find regions in the parameter space where the system shows periodic cycles or chaotic behavior. The parameter values causing the transition into these regions usually do not seem biologically unreasonable. The question is whether we are still looking at the same mechanisms that we intentionally put in the model in order to control population size. Natural events take place in continuous time, as do all individual interactions. Discrete-time models are nothing but periodic samples of the underlying continuous process, whether stochastic or not. Incorporation of the same mechanisms in ordinary differential equations never results in the bifurcation patterns found in the discrete models. This means that the discrete map must have changed its original meaning. But what meaning does it have?

Here we try to find some deterministic models in continuous time that might show, at least qualitatively, behavior similar to the single-species discrete-time logistic equation

$$N(t+1) = N(t) + rN(T)[1 - N(T)/K] \tag{7}$$

in the unstable region. In other words, we look for systems that can show periodic solutions and chaos intrinsically.

The first attempt to explain the qualitative behavior of the discrete logistic focuses on the model dealt with in the preceding section: a multi-stage single-species model with density-dependent reproduction. As shown by Gurney et al. (1980), the addition of a delay to a continuous-time logistic model for a single species allows for all the dynamics shown by the discrete logistic. Comparing the two model systems, this may not come as a complete surprise. A discrete-time model includes the obvious delay of one time interval. The difference lies in the fact that the system in the delay equation is updated continuously, whereas in the discrete version, updating occurs only once per time interval. This continuous updating causes additional dynamic features not found in the discrete system. The route to chaos in the discrete logistic can occur only via a series of period doublings: we find a multiple state attractor in which the number of states is a power of 2. Periodic sampling of the delay equation can yield a region of the parameter space in which multiple-state attractors have any number of states.

The delay models do well when the length of the nonreproductive period exceeds the life expectancy of the reproductive stage. This behavior vanishes when the situation is reversed. The population of reproductive individuals consists of a mix of many generations, and the cohort structure disappears when no other mechanisms force reproduction to be periodic. In these cases, different structuring mechanisms are required.

Other biological systems that can show cycles are predator-prey systems. Can we observe something that looks qualitatively the same as the unstable discrete logistic if we sample one single species of such a system?

### Predator-Prey Models and the Discrete Logistic

Beginning with a simple Lotka-Volterra predator-prey model, we focus on the stability analysis of mathematical models, comparing the results with the bifurcation pattern from the single-species discrete-logistic model. We then add progressively more detail: first, periodic (seasonal) forcing of parameters; then, a stage structure for the prey population; at the end, both of these factors together.

Predator-prey models are necessarily nonlinear: in order to catch a prey, a predator must meet one. The simplest model of this kind is the well-known Lotka-Volterra model:

$$\frac{dN}{dt} = N(r - AP)\,,$$
$$\frac{dP}{dt} = P(CAN - DP)\,, \tag{8}$$

where $N = N(t)$ and $P = P(t)$ are the prey and predator population densities, respectively, $r$ is the per capita growth rate of the prey, $A$ is the rate at which prey is captured by the predator, $C$ is the yield coefficient of the predator on the prey eaten, and $D$ is the per capita death rate of the predator. The equilibria for this system are

$$(N, P) = \left\{(0,0), (Dr/CA^2, r/A)\right\}\,. \tag{9}$$

An unfortunate characteristic of this model is that the nontrivial equilibrium can never be structurally stable.

Note that if the biomass of predator and prey are used, instead of numbers or densities, and if the prey is the one and only food source for the predator, the yield coefficient, $C$, must be smaller than one. This is a result of the loss of heat in any biochemical reaction. With numbers or densities, the yield coefficient can exceed one only if the predator is smaller than the prey.

From a biological viewpoint, it is not likely that predators can maintain an intake rate proportional to the density of the prey. Saturation effects clearly have a mechanistic underpinning. The

effect is caused by a handling time of predators needed to catch
and eat prey. This results in a dependence of $A$ on $N$, which generally is denoted by the functional response $f(N)$. In case of complete saturation—that is, prey density tending to infinity—$Nf(N)$
approaches a constant value (e.g., Holling Type-II functional response). This type of functional response causes the loss of the
nontrivial equilibrium, and predator and prey populations either
become extinct or tend to infinity. Thus, by introducing a bit more
biological reality, predator-prey cycles can no longer occur.

If it is assumed that the food resources for the prey are limited, its population size is bounded. For large prey densities, then,
the rate of increase of the prey must be less than or equal to the
mortality rate. Examples of models showing this type of behavior
are the logistic-growth equation and the Ricker function (see, e.g.,
Nisbet & Gurney 1982). Including a concavity like these in the
growth rate of the prey species may lead to limit cycles in models
when prey handling is taken into account (Nisbet & Gurney 1982;
Murray 1989).

Several mechanisms might affect the stability and amplitudes of
predator-prey cycles. One potentially stabilizing mechanism is an
escape in time for the prey. Suppose that the juvenile stage of the
prey is not vulnerable to predation. Examples include insects living
their larval stage in immature fruits, thereby escaping bird attacks,
and wind-dispersing plants that escape seed predation by delayed
seed setting. Another factor that might influence the amplitude of
the cycling behavior of a predator-prey system is seasonal forcing.
If the predator-prey cycle extends over periods of more than one
year, periodic forcing of prey reproduction or prey mortality might
amplify or damp the magnitude of the cycle.

In the Lotka-Volterra model, assume that density dependence
follows the logistic-growth equation and that predators handle prey
according to the Holling Type-II functional response. The resulting
system is given by

$$\frac{dN}{dt} = rN\left(1 - \frac{N}{K}\right) - \frac{ANP}{B + N},$$
$$\frac{dP}{dt} = \frac{CANP}{B + N} - DP,$$

(10)

where $r$, $A$, $C$, and $D$ are defined as before, $K$ is the carrying capacity of the prey population, and $B$ is the half-saturation constant
for predators feeding on prey.

In order to capture the essentials of the model, we transform the system into a dimensionless form. If, by definition, $x = N/B$, $y = rB^2P/KA$, and $s = rBt/K$, equations (10) become

$$\frac{dx}{ds} = x\left[k - x - \frac{y}{1+x}\right],$$
$$\frac{dy}{ds} = yd\left[\frac{x}{1+x} - a\right],$$

(11)

where $k = K/B$, $d = KAC/rB$, and $a = D/AC$. Now, replace $s$ with $t$. Before building models incorporating stage-structured prey and seasonal forcing, we present the properties of this model. The equilibria of this system are

$$\left(x^*, y^*\right) = \left\{(0,0),\ (k,0),\ \left(a/(1-a), k - a/(1-a)\right)\right\}.$$  (12)

The third pair makes biological sense only when positive. The conditions for this positive equilibrium to exist are

$$a < 1 \quad \text{and} \quad k > a/(1-a).$$  (13)

The stability of the steady states is determined by the Jacobian

$$J = \begin{pmatrix} k - 2x - \dfrac{y}{(x+1)^2} & -\dfrac{x}{x+1} \\ \dfrac{yd}{(x+1)^2} & d\left(\dfrac{x}{x+1} - a\right) \end{pmatrix}$$  (14)

evaluated at $(x^*, y^*)$.

The eigenvalues $\lambda$ can be computed from the characteristic equation

$$\lambda^2 - \text{tr}(J)\lambda + \det(J) = 0.$$  (15)

For stability of an equilibrium, all eigenvalues must have negative real parts. It is easily seen that $(0,0)$ is a saddle point for all relevant parameter values. If the internal equilibrium does not exist—if $k \leq a/(1-a)$—the equilibrium $(k,0)$ is stable. Otherwise, it is a saddle point. If an internal equilibrium exists, it might be locally stable or unstable. Necessary and sufficient conditions for local stability of the positive steady state are

$$\text{tr}(J) < 0 \Rightarrow a\left[k(a-1) + a + 1\right]/(a-1) < 0,$$
$$\det(J) > 0 \Rightarrow -ad\left[a - k(1-a)\right] > 0.$$  (16)

The latter of these conditions for stability is always satisfied. Figure 5 shows the regions of existence and stability of the internal equilibrium. In the modifications we make to this basic model, we

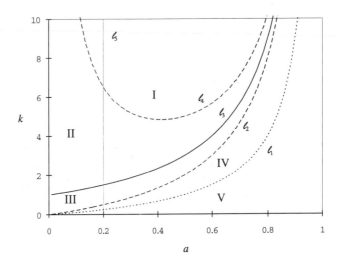

FIGURE 5. *Bifurcation diagram for the predator-prey system (eqs. 11) with respect to the parameters a and k, d = 0.25. The lines $\ell_1, \ldots, \ell_4$ are boundaries for the different stability regions of the nontrivial equilibrium. In regions I and II, the equilibrium is unstable; in III and IV, stable. In region V, a nontrivial equilibrium does not exist. Line $\ell_5$ marks the parameter space used in the modifications of this system.*

concentrate on the behavior of the system on the line $a = 0.2$ ($\ell_5$ in Fig. 5).

*Periodic fluctuations in prey characteristics.* The first modification of the system is the addition of periodic forcing. Assume that both the intrinsic growth rate, $r$, and the carrying capacity, $K$, are affected in equal proportions by periodic fluctuations in the environment. This means that we are forcing only the parameter $k$ in system (11). The resulting system is

$$\frac{dx}{dt} = x\left\{ k\left[1 + \sin(\omega t)\right] - x - \frac{y}{1+x} \right\},$$
$$\frac{dy}{dt} = yd\left[ \frac{x}{1+x} - a \right], \tag{17}$$

where $\omega$ is the angular frequency.

In forcing the system, the nontrivial equilibrium for system (11) disappears. The remaining question is whether the forcing amplifies or damps the fluctuations caused by the predator-prey interaction. Figure 6 shows the result of a constant forcing for different values

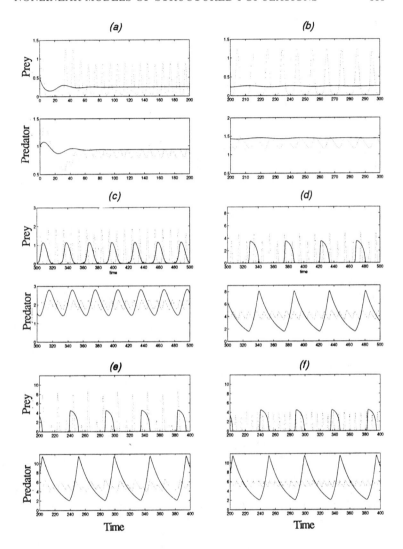

FIGURE 6. *Dynamics of system (eqs. 17) with $a = 0.2$ and $d = 0.25$. Solid lines, The unforced case ($\omega = 0$); dotted lines, the forced case ($\omega \neq 0$). (a), (b) The system is forced in the stable region; $\omega = 0.628$, $k = 1$ and $k = 1.4$, respectively. (c)–(e) The system is forced in the unstable region; $\omega = 0.628$, $k = 2$, 4, and 5, respectively. (f) $\omega = 0.864$, $k = 5$.*

of $k$ on line $\ell_5$. Forcing in the stable region (panels $a$ and $b$) causes the mean predator level to be lower than in the unforced case. In the unstable region, the frequency of the forcing takes over the predator-prey cycle (panels $6c$–$e$). In all cases, the predator amplitudes decrease while the prey amplitudes increase. Notice that the influence on the mean predator density becomes less negative and might even become positive.

From the graphs it can be seen that, as $k$ increases, the system undergoes a series of flip bifurcations (period doublings). These flip bifurcations can be reversed by increasing the angular frequency (cf. panels $6e$, $6f$). These period doublings in the forced predator-prey system are similar to bifurcations found when the reproduction rate is changed in a discrete-logistic model for a single species.

*A mechanism for the escape of prey in time.* The second modification adds an escape in time for the offspring of the prey, and the model uses a delay-differential equation comparable to the one introduced in equations (1). Assume that the juvenile stage is of constant duration and, further, that the rate of juvenile production follows the positive part of the logistic equation. The resulting system is

$$
\begin{aligned}
\frac{dx}{dt} &= \max\left\{0, x(t-\tau)\left[k - x(t-\tau)\right]\right\} - x\frac{y}{1+x}, \\
\frac{dy}{dt} &= yd\left(\frac{x}{1+x} - a\right),
\end{aligned}
\tag{18}
$$

where $\tau$ is the period (stage duration) during which the prey hides. The equilibria for this system are the same as for the basic system (11). The stability, however, now depends on the delay. The characteristic equation is obtained as follows. First, expand the system around the equilibrium, by defining $u = x - x^*$ and $v = y - y^*$. Then,

$$
\begin{aligned}
\frac{du}{dt} &= (k - 2x^*)u(t-\tau) - \frac{uy^*}{(1+x^*)^2} - \frac{vx^*}{1+x^*}, \\
\frac{dv}{dt} &= \frac{uy^*d}{1+x^*} + vd\left(\frac{x^*}{1+x^*} - a\right).
\end{aligned}
\tag{19}
$$

It is easy to see that the equilibria $(0,0)$ and $(k,0)$ have the same stability as in the nondelayed system. For the internal equilibrium, proceed by seeking a solution of the form

$$
(u, v)^{\mathrm{T}} = A^{\mathrm{T}} \exp \lambda t.
\tag{20}
$$

Substituting the solution into the linearized system yields

$$\lambda^2 + (c_1 - c_2 e^{-\lambda\tau})\lambda + c_3 = 0\,, \tag{21}$$

where $c_1 = y^*/(1 + x^*)^2$, $c_2 = k - 2x^*$, and $c_3 = adc_1$. When $\tau = 0$, this equation is equal to the characteristic equation of the basic model. Increasing $\tau$ just a little increases the number of eigenvalues from two to infinity. But increasing $\tau$ in the region where the eigenvalues for the basic system are complex leads to behavior in the delayed system comparable to that in the basic system (cf. Figs. 7a, 7b, left column). It is therefore likely that the dominant eigenvalue of the delayed system remains close to the complex pair. Further increasing $\tau$ leads to stability of the system (Fig. 7d, left). If the dominant eigenvalue is a complex pair, then this change in stability has to be a Hopf bifurcation (i.e., the eigenvalue crosses the imaginary axes).

The parameter values at which the Hopf bifurcations occur are computed as follows. Let $\lambda = \alpha + \beta i$. The characteristic equation is satisfied when

$$
\begin{aligned}
\alpha^2 - \beta^2 + c_1\alpha - c_2 e^{-\alpha\tau}\left[\alpha\cos(\beta\tau) + \beta\sin(\beta\tau)\right] + c_3 &= 0\,, \\
2\alpha\beta + c_1\beta - c_2 e^{-\alpha\tau}\left[\beta\cos(\beta\tau) - \alpha\sin(\beta\tau)\right] &= 0\,.
\end{aligned}
\tag{22}
$$

For a Hopf bifurcation to occur we must have a zero real part; that is, $\alpha = 0$. After putting $\alpha = 0$ in the preceding system, the second equation becomes

$$\beta = \left[\cos^{-1}(c_1/c_2) + 2\pi j\right]/\tau\,, \qquad j = 0, \pm1, \pm2, \ldots\,, \tag{23}$$

for $-1 \le c_1/c_2 \le 1$. Substituting (23) into (22) yields

$$
\begin{aligned}
&-\left[\cos^{-1}(c_1/c_2) + 2\pi j\tau\right]^2 \\
&- c_2\left[\cos^{-1}(c_1/c_2) + 2\pi j\right]\sin\left[\cos^{-1}(c_1/c_2)\right]/\tau + c_3 = 0\,.
\end{aligned}
\tag{24}
$$

Equation (24) allows the computation of $\tau$ values that result in integral values for $j$, thereby giving rise to Hopf bifurcations. However, the value of $\tau$ computed in this way does not match the one at which stability changes numerically. This probably indicates the presence of a subcritical Hopf bifurcation; that is, in this parameter region there exist both a locally stable equilibrium and a pair of cycles. One cycle is stable; the other is unstable. The unstable one collapses at the point of the Hopf bifurcation.

A further increase in $\tau$ leads back to situations in which stable cycles are present. The frequencies of these cycles are higher than those at low values of $\tau$. This might show that the cycles generated

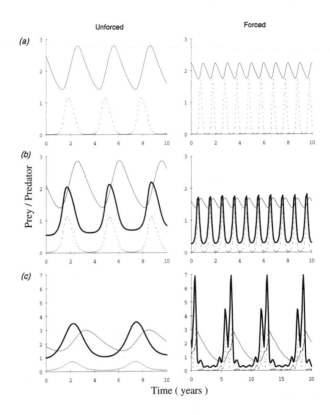

FIGURE 7. (Above and facing page) *Dynamics of the delayed system
when* $\tau = 0$, 0.5, 2, 5, 15, 30; *left columns, unforced system (eqs. 18);
right columns, forced system (eqs. 25), with* $\omega = 0.628$; $d = 0.25$,
$a = 0.2$, $k = 2$. *Thick line, Total prey; dashed line, prey vulnerable
to predation; thin line, predator.*

by the delay are starting to govern the behavior of the system.
Increasing $\tau$ even further results in an increase in subharmonic
cycles (Fig. 7f, *left*). Notice that increasing $\tau$ results in an increase
in total prey density, whereas the density of prey vulnerable to
predation and the predator density are hardly influenced.

   Figure 8 shows the effect of increasing $k$ when $\tau = 5$, the situation for which a stable equilibrium exists in the preceding case.
Increasing $k$ from 0.5 to 8 results in bifurcations from a stable
equilibrium to stable cycles of higher periodicity and chaos. Again,
total prey density increases, but since $k$ is related to the maximal
production of eggs, this is rather trivial.

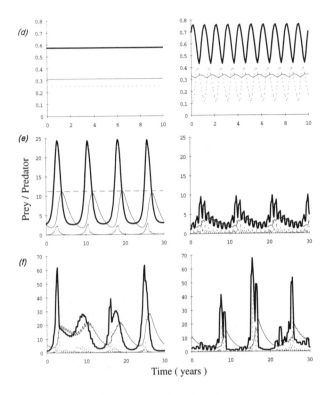

FIGURE 7. (Continued)

*Periodic fluctuations in prey characteristics **and** escape in time.*
The final modification of the basic predator-prey system combines
seasonal forcing and escape in time. The system is given by

$$\frac{dx}{dt} = \max\left(0, x(t-\tau)\{k\,[1+\sin(\omega t) - x(t-\tau)]\,\}\right) - x\frac{y}{1+x},$$
$$\frac{dy}{dt} = yd\left(\frac{x}{1+x} - a\right). \tag{25}$$

The general result here is that adding forcing to the delayed
system decreases the total number of predator and prey (Figs. 7,
8). Forcing further increases the interference of all kinds of periodic
cycles.

*Comparison of predator-prey models and the discrete logistic.* We
considered predator-prey systems, trying to reproduce the behavior of the single-species discrete logistic (SSDL). A predator-prey

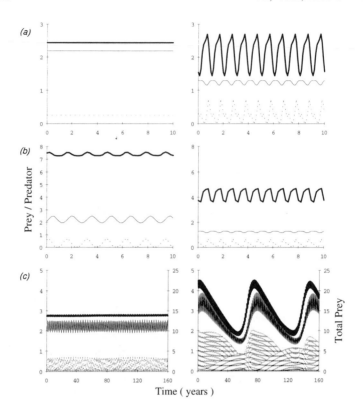

FIGURE 8. *Dynamics of the delayed system for* $k = 0.5, 4, 8$*; left column, unforced system;* right column, *forced system, with* $\omega = 0.628$*,* $d = 0.25$*,* $a = 0.2$*,* $\tau = 5$*.* Thick line, *Total prey;* dashed line, *prey vulnerable to predation;* thin line, *predator.*

system of the Lotka-Volterra type, without the introduction of seasonal forcing and stage structure, cannot show the behavior found for the SSDL. Adding seasonal fluctuation to prey production rate leads to period doublings with increases in the prey growth rate. So far, we have not found chaos with this system, but more research remains to be done. As mentioned in the beginning of this section, the use of a discrete model implies an escape in time for the prey. Increasing the growth rate leads to both period doublings and chaos. But periodicity need not be of a power of two. A combination of a delay and seasonal forcing increases the possibility of having periodicity of a different power.

## 4 Conclusions

In this chapter we consider two extremes of data collection on populations at regular time intervals. In one, Nicholson's blowfly experiment, sampling was done at short intervals, revealing astonishing but apparently explainable dynamics. In the other, a theoretical annual sampling scheme reveals relatively high fluctuations. Such data might be analyzed assuming that the fluctuations arise from stochastic events. We choose to explain such data starting with deterministic models.

In both cases we end up with nonlinear models in continuous time. With detailed sampling, the data force us to do so. Nicholson's data were analyzed by Gurney et al. (1980) with a purely deterministic multi-state model in which egg production was made density-dependent. This model does not fully explain the data. Adding some randomness in egg-to-adult developmental time leads to a better agreement with the data. Although the model's fit to the data might be due solely to changes in the developmental time by periodic forcing, this appears not to be the case. Randomness is needed. Future research, in which parameters other than egg-to-adult developmental time vary randomly, could reveal whether it is only the random variation in developmental time that can bring the model closer to the data, or whether any form of randomness provides a better explanation of the data.

The annual-sampling scheme provides a biological basis for using nonlinear discrete-time models to explain large fluctuations in population sizes. Nonlinear terms are mostly inserted into models as representatives of some mechanistic background. This mechanism falls apart if the discrete-time models are taken into the region of high fluctuations or chaos. This is because the intrinsically continuous-time dynamics resulting from a mechanistic underpinning are too fast for the sample step taken.

Supposing that large fluctuations do indeed exist in real data, we must look for continuous-time maps which, when sampled at an annual frequency, reveal a pattern comparable to the simple nonlinear discrete-time models. Two basically different mechanisms might satisfy this need. The first, equivalent to the model used in the blowfly model, is a multi-stage single-species model in which the nonreproductive period is longer than the mean lifetime of individuals in the reproductive period. Here, the species itself has a structure.

The second mechanism is a periodically forced predator-prey system. Here, all the individuals are in a reproductive state, and it is the interference of the periodicity in habitat quality with the intrinsic predator-prey cycle that generates the wanted bifurcation diagrams. In this case, the structure is not within one species but within a whole system reacting to its environment. It does not come as a surprise that mixing these two basic models also works. Note, however, that the similarity between these structured continuous models and discrete models is analyzed only qualitatively.

One common factor in both the continuous- and discrete-time models is density dependence. In the continuous-time models, density dependence is a prerequisite, but it is the structure that does the job. The discrete-time analogue does not reveal that a structure has been implicitly imposed on the model when moving into the unstable parameter region.

## Acknowledgments

We wish to thank Hal Caswell, Shripad Tuljapurkar, Roger Nisbet, André de Roos, and all the lecturers and participants in the 1993 Woods Hole Summer Course on Modeling Structured Populations in Marine, Freshwater and Terrestrial Environments, for their constant support and advice. A particular mention is due Jeff Hatfield and Quan Dong, who worked on nonlinear issues with us. We also thank James Robert Buchanan for software help, and David Brillinger for providing the original Nicholson data.

## Literature Cited

Brillinger, D. R., J. Guckenheimer, P. Guttorp, and G. Oster. 1980. Empirical modelling of population time series data: The case of age and density dependent vital rates. *Lectures on Mathematics in the Life Sciences* 13: 65–90.

Caswell, H. 1989. *Matrix Population Models.* Sinauer, Sunderland, Mass.

Caswell, H., and A. M. John. 1992. From the individual to the population in demographic models. Pp. 36–61 *in* D. L. DeAngelis and L. J. Gross, eds., *Individual-Based Models and Approaches in Ecology.* Chapman & Hall, New York.

Cushing, J. M. 1989. A strong ergodic theorem for some nonlinear matrix models for the dynamics of structured populations. *Natural Resources Modeling* 3(3): 331–357.

DeAngelis, D. L., and L. J. Gross, eds. 1992. *Individual-Based Models and Approaches in Ecology.* Chapman & Hall, New York.

Gurney, W. S. C., S. P. Blythe, and R. M. Nisbet. 1980. Nicholson's blowflies revisited. *Nature* 287: 17–21.

May, R. M., G. R. Conway, M. P. Hassell, and T. R. E. Southwood. 1974. Time delays, density-dependence and single-species oscillations. *Journal of Animal Ecology* 43: 747–770.

Metz, J. A. J., and O. Diekmann, eds. 1986. *The Dynamics of Physiologically Structured Populations.* Lecture Notes in Biomathematics 68. Springer-Verlag, Berlin.

Murdoch, W. W., R. M. Nisbet, S. P. Blythe, W. S. C. Gurney, and J. D. Reeve. 1987. An invulnerable age class and stability in delay-differential parasitoid-host models. *American Naturalist* 129: 263–282.

Murray, J. D. 1989. *Mathematical Biology.* Springer-Verlag, Berlin.

Neves, K. W., and S. Thompson. 1992. Software for the numerical solution of systems of functional differential equations with state dependent delays. *Journal of Applied Numerical Mathematics* 9: 385–401.

Nicholson, A. J. 1954. An outline of the dynamics of animal populations. *Australian Journal of Zoology* 2: 9–65.

Nisbet, R. M., and W. S. C. Gurney. 1982. *Modelling Fluctuating Populations.* Wiley, New York.

Readshaw, J. L., and W. R. Cuff. 1980. A model of Nicholson's blowfly cycles and its relevance to predation theory. *Journal of Animal Ecology* 49: 1005–1010.

Readshaw, J. L., and A. C. M. VanGerwen. 1983. Age-specific survival, fecundity and fertility of the adult blowfly, *Lucilia cuprina*, in relation to crowding, protein food, and population cycles. *Journal of Animal Ecology* 52: 879–887.

Tuljapurkar, S. D. 1989. An uncertain life: Demography in random environments. *Theoretical Population Biology* 35: 227–294.

Wolfram, S. 1984. Cellular automata as models of complexity. *Nature* 311: 419–424.

# CHAPTER 21

# Multispecies Lottery Competition: A Diffusion Analysis

## Jeff S. Hatfield and Peter L. Chesson

The lottery model is a stochastic competition model designed for space-limited communities of sedentary organisms. Examples of such communities may include coral reef fishes (Chesson & Warner 1981), aquatic sessile organisms (Fagerstrom 1988), and plant communities such as trees in a tropical forest (Leigh 1982; Hatfield et al., in press). The lottery model, and its properties and behavior, has been discussed previously (Chesson & Warner 1981; Chesson 1982, 1984, 1991, 1994; Warner & Chesson 1985; Chesson & Huntly 1988). Furthermore, explicit conditions for the coexistence of two species and the stationary distribution of the two-species model were determined (in Hatfield & Chesson 1989) using an approximation with a diffusion process (Karlin & Taylor 1981). However, a diffusion approximation for the multispecies model (for more than two species) has not been reported previously, and a stage-structured version has not been investigated. The stage-structured lottery model would be more reasonable for communities of long-lived species in which recruitment or death rates depend on the age or stage of the individuals (e.g., trees in a forest). In this chapter, we present a diffusion approximation for the multispecies lottery model and also discuss a stage-structured version of this model.

   The conditions for coexistence and the stationary distribution of the lottery model are useful because they indicate whether a given community tends toward persistence in a variable environment, and if so, they provide information about the type and magnitude of population fluctuations that the competing species experiences over time. Furthermore, if the assumptions of the lottery model apply to a given community, then the expected stationary distribution can be fit to species-abundance data collected in the field to yield information and insights about life-history parameters of the competing species (e.g., Hatfield et al., in press). The multispecies and stage-structured models are particularly important to study because these models obviously have broader application than the two-species model.

## 1 Multispecies Model

In the multispecies lottery model, the equation for the $i$th species, $i = 1, 2, \ldots, k$, is given by

$$P_i(t+1) = [1 - \delta_i(t)] P_i(t) + \left[ \sum \delta_j(t) P_j(t) \right] \frac{\beta_i(t) P_i(t)}{\sum \beta_j(t) P_j(t)}, \quad (1)$$

where $P_i(t)$ is the proportion of space occupied by species $i$ at time $t$, $\beta_i(t)$ is the per capita recruitment rate of species $i$, and $\delta_i(t)$ is the adult death rate during the time interval $(t, t+1]$. The diffusion approximation to the multispecies lottery model is determined by its infinitesimal mean, variance, and covariance coefficients (Karlin & Taylor 1981), defined as

$$\mu_i(\mathbf{p}) = \lim_{\epsilon \to 0} \frac{1}{\epsilon} \mathrm{E} \left[ P_i(t+\epsilon) - P_i(t) \right|$$

$$\mathbf{P}(t) = \mathbf{p} = p_1, p_2, \ldots, p_k], \quad (2)$$

$$\sigma_i^2(\mathbf{p}) = \lim_{\epsilon \to 0} \frac{1}{\epsilon} \mathrm{E} \left\{ \left[ P_i(t+\epsilon) - P_i(t) \right]^2 | \mathbf{P}(t) = \mathbf{p} \right\}, \quad (3)$$

$$\sigma_{ij}(\mathbf{p}) = \lim_{\epsilon \to 0} \frac{1}{\epsilon} \mathrm{E} \left\{ [P_i(t+\epsilon) - P_i(t)] \right.$$

$$\left. \times [P_j(t+\epsilon) - P_j(t)] | \mathbf{P}(t) = \mathbf{p} \right\}, \quad (4)$$

where $i, j = 1, 2, \ldots, k$ and $i \neq j$.

To approximate the multispecies lottery model with a diffusion process, the model must first be rescaled for continuous time. A description is given elsewhere (Hatfield 1986; Hatfield & Chesson 1989) and involves replacing $P_i(t + 1)$ by $P_i(t + \epsilon)$ in the left-hand side of equation (1) and letting $\epsilon$ approach 0. Furthermore, the amount of change occurring per time unit must be decreased, which is accomplished by defining $X_i(t) = \ln [\beta_i(t)/\delta_i(t)]$ and assuming that $\mathrm{E}[X_i(t)] = \epsilon \mu_i$ and $\mathrm{var}[X_i(t)] = \epsilon \sigma_i^2$. In addition, let $\mathrm{cov}[X_i(t), X_j(t)] = \epsilon \theta_{ij}$, $i \neq j$, $\delta_u(t) = d_u \exp[Y_u(t)]$, $\mathrm{E}[Y_u(t)] = 0$, and cov $[Y_u(t), X_i(t) - X_j(t)] = \epsilon\theta_{uij}$. Higher-order moments are assumed to be terms of order $o(\epsilon)$.

Employing power-series expansions of $e^{\pm X_i(t)}$ and $e^{\pm Y_i(t)}$, substituting the definitions and assumptions given above into equations (2), (3), and (4), and letting $\epsilon$ approach 0 yields the diffusion coefficients for the multispecies lottery model:

$$\mu_i(\mathbf{p}) = \frac{d_i p_i}{\sum_n d_n p_n}\left\{\sum_u d_u p_u\left[(\mu_i - \mu_u)\right.\right.$$

$$+\left(1 - \frac{d_u p_u}{\sum_n d_n p_n}\right)\theta_{uiu}$$

$$+\left(\frac{d_u p_u}{\sum_n d_n p_n}\right)\theta_{iiu}$$

$$+\left(\frac{d_u p_u}{\sum_n d_n p_n} - \frac{1}{2}\right)$$

$$\left.\times\left(\sigma_i^2 + \sigma_u^2 - 2\theta_{iu}\right)\right]$$

$$+\frac{1}{\sum_n d_n p_n}\sum_u\sum_m d_u d_m p_u p_m$$

$$\times\left(\sigma_i^2 - \theta_{iu} - \theta_{im} + \theta_{um}\right.$$

$$\left.\left.+ \theta_{iim} - \theta_{uim}\right)\right\}, \tag{5}$$

$$\sigma_i^2(\mathbf{p}) = \frac{d_i^2 p_i^2}{\left(\sum_n d_n p_n\right)^2}\sum_u\sum_m d_u d_m p_u p_m$$

$$\times\left(\sigma_i^2 - \theta_{iu} - \theta_{im} + \theta_{um}\right), \tag{6}$$

$$\sigma_{ij}(\mathbf{p}) = \frac{d_i d_j p_i p_j}{\left(\sum_n d_n p_n\right)^2}\sum_u\sum_m d_u d_m p_u p_m$$

$$\times\left(\theta_{ij} - \theta_{ju} - \theta_{im} + \theta_{um}\right). \tag{7}$$

The stationary distribution, $\psi(\mathbf{p})$, of the diffusion process represented by equations (5), (6), and (7) must satisfy the Kolmogorov forward equation (Keilson 1965),

$$0 = \sum_{i=1}^{k-1}\frac{\partial}{\partial p_i}\left[\sum_{j=1}^{k-1}\frac{\partial\sigma_{ij}(\mathbf{p})\psi(\mathbf{p})}{2\partial p_j} - \mu_i(\mathbf{p})\psi(\mathbf{p})\right], \tag{8}$$

where $\sigma_{ii}(\mathbf{p}) = \sigma_i^2(\mathbf{p})$. Solving equation (8) requires the following simplifying assumptions: $\sigma_i^2 = \sigma^2$; $\theta_{ij} = \rho\sigma^2$, $i \neq j$, $0 \leq |\rho| \leq 1$; $\theta_{uiu} = \tau\sigma^2$, $\theta_{iiu} = -\tau\sigma^2$, $u \neq i$; and $\theta_{uim} = 0$, $u \neq m$, $u \neq i$, $m \neq i$. The assumption concerning the variances implies that each species experiences a similar amount of variation (i.e., environmental fluctuations) in the $X_i$. The assumption concerning the $\theta_{ij}$ means that the correlation between all pairs $X_i$ and $X_j$ is identical. The assumptions concerning $\theta_{uiu}$, $\theta_{iiu}$, and $\theta_{uim}$ would result from a dependence between $Y_u$ and $X_u$, in addition to the independence of $Y_u$ and $X_i$ ($m \neq i$). Defining $\varphi = \sigma^2(1-\rho)$ and $\varphi' = \sigma^2(1-\rho-\tau)$, these assumptions yield

$$\mu_i(\mathbf{p}) = d_i p_i \left\{ \mu_i - \sum d_n \mu_n p_n \sum d_n p_n \right.$$

$$\left. + \varphi' \left[ \frac{\sum d_n^2 p_n^2}{\left(\sum d_n p_n\right)^2} - \frac{d_i p_i}{\sum d_n p_n} \right] \right\}, \qquad (9)$$

$$\sigma_i^2(\mathbf{p}) = d_i^2 \varphi p_i^2 \left[ 1 + \sum d_n^2 p_n^2 \left(\sum d_n p_n\right)^2 - \frac{2 d_i p_i}{\sum d_n p_n} \right], \qquad (10)$$

$$\sigma_{ij}(\mathbf{p}) = d_i d_j \varphi p_i p_j \left[ \frac{\sum d_n^2 p_n^2}{\left(\sum d_n p_n\right)^2} - \frac{d_i p_i}{\sum d_n p_n} \right.$$

$$\left. - d_j p_j \sum d_n p_n \right]. \qquad (11)$$

Equation (8) is still too difficult to solve for coefficients (9), (10), and (11). Keilson (1965) proved that if the process is irrotational—that is, $(\partial/\partial p_j)\,\partial\psi/\partial p_i = (\partial/\partial p_i)\,\partial\psi/\partial p_j$—then a unique solution exists to the inner expression of equation (8):

$$0 = \sum_{j=1}^{k-1} \frac{\partial \sigma_{ij}(\mathbf{p})\psi(\mathbf{p})}{2\partial p_j} - \mu_i(\mathbf{p})\psi(\mathbf{p}), \quad i = 1, 2, \ldots, k. \qquad (12)$$

For the process determined by (9), (10), and (11), extensive algebra shows that the process is irrotational if $d_1 = d_2 = \ldots = d_k = d$. Adding this assumption to (9), (10), and (11) yields the coefficients

$$\mu_i(\mathbf{p}) = d p_i \left[ \mu_i - \sum \mu_n p_n + \varphi' \left( \sum p_n^2 - p_i \right) \right], \qquad (13)$$

$$\sigma_i^2(\mathbf{p}) = d^2 \varphi p_i^2 \left( 1 + \sum p_n^2 - 2 p_i \right), \qquad (14)$$

$$\sigma_{ij}(\mathbf{p}) = d^2 \varphi p_i p_j \left( \sum p_n^2 - p_i - p_j \right). \qquad (15)$$

In this case, these equations are identical to those for a diffusion process from the SAS-CFF model in population genetics (Gillespie 1980, 1991) and thus the solution, $\psi(\mathbf{p})$, to (12) is a Dirichlet distribution:

$$\psi(\mathbf{p}) = c \prod_{i=1}^{k} p_i^{v_i - 1}, \tag{16}$$

where $v_i = 2 \left[ k \left( \mu_i - \bar{\mu} \right) + \varphi' - d\varphi \right] / d\varphi k$, $\bar{\mu} = \sum \mu_n / k$, and $c$ is the constant of integration, which allows $\psi(\mathbf{p})$ to be a probability density.

The condition for the existence of (16) as a probability density is that $v_i > 0$, which implies that

$$d < \left[ \varphi' - k \max \left( \bar{\mu} - \mu_i \right) \right] / \varphi \tag{17}$$

and

$$\sigma^2 > k \max \left( \bar{\mu} - \mu_i \right) / \left[ (1 - d)(1 - \rho) - \tau \right]. \tag{18}$$

Thus, equations (17) and (18) are the conditions for the coexistence of $k$ species in this community. This does not prove that (17) and (18) are the conditions for the existence of a stationary distribution; it proves only that (17) and (18) guarantee that (16) is a probability distribution. However, Seno and Shiga (1984) showed that $v_i > 0$ is in fact the condition for the existence of the stationary distribution and thus the condition for the coexistence of the $k$ species in this model.

Given that the $k$ competing species are able to coexist, the stationary distribution provides information about the year-to-year population fluctuations that these species experience over time. This statistical distribution can also be used to generate expected means and variances and to evaluate the stability of the coexistence by looking at the shape of the distribution. (For a discussion of stability in population fluctuations for a community of two species, see Hatfield & Chesson 1989.)

## 2 Stage-Structured Model

For a stage-structured version of the lottery model, let species $i$ have $A_i$ stages (or ages), and let $\gamma_{ij}(t)$ be the probability that stage $j$ of species $i$ proceeds to stage $j+1$ during the time interval $(t, t+1]$, $i = 1, 2, \ldots, k$, $j = 1, 2, \ldots, A_i$. (Note that $\gamma_{ij} = 1$ for all $i$ and $j$ yields the age-structured model.) Let $\beta_{ij}(t)$ be the per capita recruitment rate, and let $\delta_{ij}(t)$ be the death rate for stage $j$ of species $i$ during the time interval $(t, t+1]$. Then, the equations for the stage-structured lottery model are

$$P_{i1}(t+1) = [1 - \gamma_{i1}(t)]\,[1 - \delta_{i1}(t)]\,P_{i1}(t)$$
$$+ \sum_i \sum_j \delta_{ij}(t) P_{ij}(t)$$
$$\times \frac{\sum_j \beta_{ij}(t) P_{ij}(t)}{\sum_i \sum_j \beta_{ij}(t) P_{ij}(t)} \tag{19}$$
$$P_{i2}(t+1) = [1 - \gamma_{i2}(t)]\,[1 - \delta_{i2}(t)]\,P_{i2}(t)$$
$$+ \gamma_{i1}(t)\,[1 - \delta_{i1}(t)]\,P_{i1}(t), \tag{20}$$

$$\vdots$$

$$P_{iA_i}(t+1) = [1 - \gamma_{iA_i}(t)]\,[1 - \delta_{iA_i}(t)]\,P_{iA_i}(t)$$
$$+ \gamma_{iA_{i-1}}(t)\,\big[1 - \delta_{iA_{i-1}}(t)\big]\,P_{iA_{i-1}}(t)\,,$$

$i = 1, 2, \ldots, k$. Unfortunately, the stage-structured version of the lottery model cannot be investigated with a diffusion approximation because the rescaled continuous-time model does not appear to converge to a diffusion process. In this case, computer simulations could be used to provide insight into the behavior of this model.

The lottery model provides a simple example of species coexistence by a mechanism known as the storage effect (Chesson 1984; Warner & Chesson 1985). It is one of two general coexistence mechanisms that depend on temporal environmental fluctuations (Chesson 1994). The storage effect is most likely to occur when the various individuals in a population differ in their sensitivities to environmental factors and competition. Such differences in sensitivity occur, for example, as a consequence of the distinction between juveniles and adults in the lottery model and in general scenarios involving iteroparous organisms (Chesson 1984; Chesson & Huntly 1988). Study of the lottery model has provided important general information about the operation of the storage effect. Its stage-structured version may well provide important new understanding of this general coexistence mechanism.

## Acknowledgments

Most of this work applies to J.S.H.'s Ph.D. dissertation at The Ohio State University. J.S.H. is also grateful to have interacted with and learned from Shripad Tuljapurkar, Hal Caswell, the many other fine instructors, and fellow students during the summer course at Cornell University.

## Literature Cited

Chesson, P. L. 1982. The stabilizing effect of a random environment. *Journal of Mathematical Biology* 15: 1–36.

———. 1984. The storage effect in stochastic population models. Pp. 76–89 *in* S. A. Levin and T. Hallam, eds., *Mathematical Ecology*. Lecture Notes in Biomathematics 54. Springer-Verlag, New York.

———. 1991. A need for niches? *Trends in Ecology and Evolution* 6: 26–28.

———. 1994. Multispecies competition in variable environments. *Theoretical Population Biology* 45: 227–276.

Chesson, P. L., and N. Huntly. 1988. Community consequences of life-history traits in a variable environment. *Annales Zoologici Fennici* 25: 5–16.

Chesson, P. L., and R. R. Warner. 1981. Environmental variability promotes coexistence in lottery competitive systems. *American Naturalist* 117: 923–943.

Fagerstrom, T. 1988. Lotteries in communities of sessile organisms. *Trends in Ecology and Evolution* 3: 303–306.

Gillespie, J. H. 1980. The stationary distribution of an asymmetrical model of selection in a random environment. *Theoretical Population Biology* 17: 129–140.

———. 1991. *The Causes of Molecular Evolution*. Oxford University Press.

Hatfield, J. S. 1986. Diffusion analysis and stationary distribution of the lottery competition model. Ph.D. diss. Ohio State University, Columbus.

Hatfield, J. S., and P. L. Chesson. 1989. Diffusion analysis and stationary distribution of the two-species lottery competition model. *Theoretical Population Biology* 36: 251–266.

Hatfield, J. S., W. A. Link, D. K. Dawson, and E. L. Lindquist. In press. Coexistence and community structure of tropical trees in a Hawaiian montane rain forest. *Biotropica*.

Karlin, S., and H. M. Taylor. 1981. *A Second Course in Stochastic Processes*. Academic Press, New York.

Keilson, J. 1965. A review of transient behavior in regular diffusion and birth-death processes. Part II. *Journal of Applied Probability* 2: 405–428.

Leigh, E. G., Jr. 1982. Introduction: Why are there so many kinds of tropical trees? Pp. 63–66 *in* E. G. Leigh, A. S. Rand, and D. W. Windsor, eds., *The Ecology of a Tropical Forest: Seasonal Rhythms and Long-Term Changes.* Smithsonian Institution Press, Washington, D.C.

Seno, S., and T. Shiga. 1984. Diffusion models of temporally varying selection in population genetics. *Advances in Applied Probability* 16: 260–280.

Warner, R. R., and P. L. Chesson. 1985. Coexistence mediated by recruitment fluctuations: A field guide to the storage effect. *American Naturalist* 125: 769-787.

# About the Authors

Shripad TULJAPURKAR is an ecologist and demographer. He previously held faculty appointments at Stanford, Berkeley, and Portland State. His research focuses on dynamics and uncertainty in population analysis. Mountain View Research, 2251 Grant Road, A, Los Altos, CA 94024. tulja@bose.stanford.edu.

Hal CASWELL is a senior scientist at the Woods Hole Oceanographic Institution. He earned his doctorate in zoology from Michigan State University in 1974. As a mathematical population ecologist, he is the author of *Matrix Population Models: Construction, Analysis and Interpretation* (Sinauer, 1989). Biology Department, Woods Hole Oceanographic Institution, Woods Hole, MA 02543. (508) 548-1400 ×2751. hcaswell@whoi.edu.

Put O. ANG earned his doctorate from the Unversity of British Columbia. He did his postdoctoral research with the Canadian Federal Department of Fisheries and Ocean, Halifax Fisheries Research Laboratory, working on the population dynamics and harvesting strategies for marine macroalgae. His areas of research are the population and community ecology of marine algae, coral-reef ecology, and environmental biology. Department of Biology, The Chinese University of Hong Kong, Shatin, New Territories, Hong Kong. put-ang@cuhk.hk.

Carl BOE earned his doctorate in 1994 in demography from the University of California, Berkeley. He has worked on feedback models of fertility in humans, crabs, and fish. He is currently with the Entomology Department at the University of California, Davis, using insect models of aging to investigate old-age mortality consequences of reproductive events occurring earlier in the life cycle. boe@bose.stanford.edu.

Louis W. BOTSFORD's research interest is the application of age, size, and spatially structured models to practical population problems in the fields of fisheries and conservation biology. His research involves modeling and field studies of physical oceanographic influences on spatially distributed invertebrate populations, studies of endangered salmonids and other fish in rivers and estuaries in the western United States, and studies of the influence of environment and density on California quail. Department of Wildlife, Fish, and Conservation Biology, University of California, Davis, CA 95616. (916)752-6169. loo@megalopa.ucdavis.edu.

Mark BOYCE holds an endowed professorship and is a Wisconsin Distinguished Professor. His current research interests focus on ecosystem management, natural-resource modeling, and population-viability analysis. Studies are under way on grizzly bears in the Rocky Mountains, Karner blue butterflies in central Wisconsin, and jackpine budworm on pine barrens in northwestern Wisconsin. College of Natural Resources, University of Wisconsin–Stevens Point, Stevens Point, WI 54481-3897. mboyce@uwspmail.uwsp.edu.

Bing CAI is interested in individual-based and large-scale ecosystem modeling, as well as object-oriented analysis, design, and programming. Department of Zoology, Systems Analysis, Miami University, Oxford OH 45056. (513) 529-1679. bingcai@miamiu.muohio.edu.

Carlos CASTILLO-CHAVEZ joined the Biometrics Unit in 1988. He was promoted to the associate professor in 1991 and named chair of the Unit in 1995. He received a Presidential Faculty Fellowship in 1992. Biometrics Unit, 332 Warren Hall, Cornell University, Ithaca, NY 14853-7801. (607) 255-5488; FAX (607) 255-4698. cc32@cornell.edu.

Peter L. CHESSON's primary field of research is community ecology. Research School of Biological Sciences, Australian National University, Canberra, A.C.T. 0200, Australia. 61 (6) 249-0606; FAX 61 (6) 249-5095. chesson@rsbs-central.anu.edu.au.

Kathleen CROWE's 1991 doctorate in applied mathematics is from the University of Arizona. Her research areas are structured-population dynamics, the evolution of traits, and dynamical systems. She is a competitive swimmer, runner, and rower. Department of Mathematics, Humboldt State University, Arcata, CA 95521. (707) 826-4951; FAX (707) 826-3140. kmc4@axe.humboldt.edu.

J. M. CUSHING is the author of over a hundred scientific articles and serves on the editorial board of several leading journals. He has been a member of the Interdisciplinary Program on Applied Mathematics since its inception at the University of Arizona in 1978. Department of Mathematics, University of Arizona, Tucson, AZ 85721. (602) 621-6863. cushing@math.arizona.edu.

André DE ROOS received his doctorate from the University of Leiden, The Netherlands, where he developed the numerical techniques to study physiologically structured population models under the supervision of Hans Metz and Odo Diekmann. He held postdoctoral positions at the University of Calgary, Canada, and the University of StrathClyde, Scotland. In recent years his work has been extended to include the dynamics of spatially structured populations, in addition to his work on the size-dependent dynamics of waterfleas and fishes. Population Biology Section, University of Amsterdam, Kruislaan 320, 1098 SM Amsterdam, The Netherlands. 31 (20) 525 7747; FAX 31 (20) 525 7754. andre@shogun.bio.uva.nl.

Robert DESHARNAIS earned his doctoral degree in zoology in 1982 at the University of Rhode Island. His research interests include nonlinear population dynamics, population genetics, and theoretical population biology. His research combines mathematical modeling with laboratory experiments using flour beetles of the genus *Tribolium*. He is coauthor, with Robert Costantino, of the book *Population Dynamics and the 'Tribolium' Model: Genetics and Demography* (Springer-Verlag, 1991). Biology Department, California State University, 5151 State University Drive, Los Angeles, CA 90032. (213) 343-2056. bob@biol1next.calstatela.edu.

Philip DIXON, currently an ecological statistician, is interested in the development of better methods for drawing conclusions from ecological data, the analysis of spatial relationships, and the conservation of rare plants. When not in the office, he can be found tramping, playing bassoon, folk dancing, or helping raise two young boys. Savannah River Ecology Laboratory, University of Georgia, Drawer E, Aiken, SC 29802. dixon@srel.edu.

Janet M. FISCHER's dissertation focuses on zooplankton community dynamics in freshwater lakes. She is using an experimental approach to assess the role of functional compensation (the maintenance of community function despite changes in species composition) in community responses to stress. Center for Limnology, University of Wisconsin, 680 N. Park Strett, Madison, WI 53706. jfischer@macc.wisc.edu.

Nancy FRIDAY is interested in the population dynamics and abundance estimation of cetacean populations. Her doctoral dissertation is an analysis of heterogeneity and capture-recapture estimates of the abundance of North Atlantic humpback whale. When not working on her dissertation, she is enjoying the company of her partner, their two cats and one dog, making jewelry, or escaping to the Berkshire Mountains of Massachusetts. University of Rhode Island, Graduate School of Oceanography, Narragansett Bay Campus, South Ferry Road, Narragansett, RI 02882. nfriday@gsosun1.gso.uri.edu.

Oscar E. GAGGIOTTI is a native of Argentina, where he obtained his undergraduate degree in biological oceanography; his master's and doctoral degrees are from Rutgers University. His research is aimed at bridging the gap between population genetics and population ecology by extending existing population-genetics theory to include realistic ecological scenarios. In this regard, he has developed ecological models for the maintenance of sex and models to explore the effect that different stochastic migration patterns have on the genetic structure of metapopulations. NOAA–Southwest Fisheries Science Center, P.O. Box 271, La Jolla, CA 92038-0271. (619) 546-7086; FAX (619) 546-7003. ogaggiot@sgilj.ucsd.edu.

Brian GRANTHAM's dissertation research into marine community ecology focuses on biophysical interactions influencing larval transport and recruitment, determinants of intertidal community structure, and zooplankton dynamics. Department of Biological Sciences, Stanford University, Stanford, CA 94305-5020. brian@ecology.stanford.edu.

Jeff S. HATFIELD's primary field of research is quantitative ecology. National Biological Service, Patuxent Wildlife Research Center, 11510 American Holly Drive, Laurel, MD 20708. (301) 497-5633; FAX (301) 497-5666. Jeff_Hatfield@nbs.gov.

Selina S. HEPPELL anticipates completing the requirements for a doctoral degree by the end of 1997. Her field of research is population ecology and life-history evolution, with an emphasis on how life-history attributes should affect management policies for endangered species. Duke University Marine Laboratory, 135 Duke Marine Lab Road, Beaufort, NC 28516-9721. (919) 504-7567, 7636; FAX (919) 504-7648. ssh4@acpub.duke.edu.

Kevin HIGGINS' work focuses on structured-population models and nonlinear time-series analysis. Division of Environmental Studies and Institute of Theoretical Dynamics, University of California, Davis, CA 95616. khiggins@ucdavis.edu.

Eileen HOFMANN earned her master's and doctoral degrees in marine science and engineering from North Carolina State University. Her research interests are mathematical modeling of circulation, biological interactions in marine ecosystems, and descriptive physical oceanography. Department of Oceanography, Center for Coastal Physical Oceanography, Old Dominion University, Norfolk, VA 23529. (804) 683-5334. hofmann@ccpo.odu.edu.

Carol C. HORVITZ received her doctorate in biology in 1980 from Northwestern University. Her research, which links empirical studies with mathematical models of the population dynamics of understory plants in tropical forests, focuses on environmental variation caused by natural disturbances and plant-animal interactions. Her work takes her to lowland tropical rain forests in Mexico and Costa Rica and subtropical hardwood hammock forests in southern Florida. Biology Department, University of Miami, Coral Gables, FL 33124. 0004028316@mcimail.com.

Shu-Fang HSU SCHMITZ received her doctorate from the Biometrics Unit at Cornell University and is now a postdoctoral researcher with interests in mathematical modeling in biology, demography, and epidemiology. She is also involved in biostatistical research for clinical trials on cancer patients. Institute of Mathematical Statistics, University of Bern, Sidlerstrasse 5, 3012 Bern, Switzerland. 41 (31) 631 8805; FAX 41 (31) 631 3870. sfshsu@math-stat.unibe.ch.

Mrigesh KSHATRIYA is currently at the University of Miami. His field of research is modeling population and community dynamics of terrestrial ecosystems for conservation purposes. He is currently working on modeling the community dynamics of African ungulates that inhabit East African savannas. This work focuses on modeling feeding constraints of ungulate species and on modeling the population dynamics given these constraints. This work is for his doctoral dissertation and is partly funded by Dr. Western, Director of Kenya Wildlife Service. Department of Biology, University of Miami, Coral Gables, Florida 33124. ukfpyx5m@impala.ir.miami.edu.

Jochen KUMM received his doctoral degree in biological sciences from Stanford University. He is currently a postdoctoral research associate at Stanford studying population-genetics models allowing for interactions of genes and cultural practices in human evolution. His long-term interests involve statistical models of disease expression in complex environmental and genetic backgrounds. Department of Biological Sciences, Stanford University, Stanford, CA 94305-5020. iochen@charles.stanford.edu.

Carol E. LEE earned her bachelor's and master's degrees at Stanford University and is currently a doctoral candidate. She is examining species invasion as an evolutionary process in aquatic habits. Specifically, she is studying the evolution of freshwater adaptation in a cosmopolitan copepod, *Eurytemora affinis*. Her approach combines methods from evolutionary genetics and physiological ecology to determine whether freshwater colonization of a primarily marine and estuarine species has occurred multiple independent times or is rarely followed by dispersal among similar freshwater habitats. Her research interests include phylogenetics, evolutionary genetics, physiological ecology, and zooplankton ecology. Marine Molecular Biotechnology Laboratory, University of Washington, School of Oceanography, Box 357940, Seattle, WA 98195-7940. carolee@ocean.washington.edu.

Dina LIKA is a doctoral candidate in mathematical ecology. She received her bachelor's in mathematics from the University of Crete, Greece, in 1987 and her master's in mathematics from the University of Tennessee, Knoxville, in 1992. She is working on resource-directed movement for structured-population models. Department of Mathematics, 121 Ayres Hall, University of Tennessee, Knoxville, TN 37996-1300. (615) 974-2461; FAX (615) 974-6576. lika@math.utk.edu.

Vicki L. MEDLAND is a doctoral candidate in ecology. Her research interests include aquatic ecology, ecology of ephemeral ponds, life-history evolution, and ecological modeling. Her dissertation focuses on modeling the effects of environmental variability on plankton community structure. Savannah River Ecology Laboratory, University of Georgia, Drawer E, Aiken, SC 29802. (803) 725-7455; FAX (803) 725-3309. medland@srel.edu.

Bruce MONGER's affiliation is with the Center for the Environment, Cornell University, Ithaca, New York. His research topics include the physical mechanisms of prey capture by marine protists; the mediation of material and energy flow in the open ocean by protistan grazers; and satellite remote sensing in connection with modeling biological and physical interactions on an ocean-basin scale. NASA Goddard Space Flight Center, Greenbelt, MD 20771. bmonger@neptune.gsfc.nasa.gov.

Sido MYLIUS. Department of Theoretical Biology, Institute of Evolutionary and Ecological Sciences, Leiden University, P.O. Box 9516, 2300 RA Leiden, The Netherlands. (31) 71 274907; FAX 31 (20) 6223922. sbqhsm@rulsfb.leidenuniv.nl.

Chris NATIONS obtained a master's degree in zoology under the guidance of Mark Boyce at the University of Wyoming and is now working toward a doctorate in statistics. His interests include biometrics, particularly survival estimation from mark-recapture data and viable-population analysis. Department of Statistics, University of Wyoming, Laramie, WY 82071. nations@uwyo.edu.

Roger NISBET's doctorate was in statistical physics. His main research interest has been ecological modeling, and recent work has emphasized individual-based and structured-population models. In particular, he has focused on methods for formulating tractable models of stage-structured populations. These (and other) methods have been applied to the study of fluctuations in zooplankton populations, to investigations of stability and fluctuations in host-parasitoid systems, and to an investigation of the effects of toxicants on the dynamics of mussel populations. Department of Biological Sciences, University of California, Santa Barbara, CA 93016. (805) 893-7115. nisbet@islay.lscf.ucsb.edu.

Steven Hecht ORZACK received his doctorate in biology from Harvard University. His areas of interest include sex-ratio evolution, life-history evolution, population dynamics, and the philosophy of biology. He also plays basketball and jazz guitar whenever possible. Fresh Pond Research Institute, 64 Fairfield Street, Cambridge, MA 02140. (617) 864-4307. Department of Ecology and Evolution, University of Chicago, 1101 E. 57th Street, Chicago, IL 60637. orza@midway.uchicago.edu.

Daniel PROMISLOW completed his graduate work in zoology at Oxford University, followed by postdoctoral studies in Paris, Queen's University in Canada, and the University of Minnesota. His previous studies have included comparative work on life-history strategies, costs of sexual selection, and the evolution of aging. His current work focuses on sexual selection and life-history evolution, using quantitative genetic analyses of demographic traits in *Drosophila*, comparative studies of vertebrates, and computer models. Department of Genetics, University of Georgia, Athens, GA 30602-7223. promislow@bscr.uga.edu.

Douglas SCHEMSKE received his doctorate from the University of Illinois in 1977. His research interests include plant-animal interactions, speciation, adaptation, plant mating systems, demography, and conservation biology. Current projects focus on the genetics of reproductive isolation and the evolution of flower color polymorphisms. Department of Botany, Box 355325, University of Washington, Seattle, WA 98195-5325. (206) 543-9450; FAX (206) 685-1728. schem@u.washington.edu.

John VAL's doctorate is in mathematical biology, which allows him the possibility of remaining a generalist in biology. His postdoctoral research is on models for the cell-division cycle of eukaryotic cells. Other interests include landscape ecology, biotechnology, and evolutionary ecology. Virginia Technical Institute and State University. john@rulbii.leidenuniv.nl. `http://leibniz.biol.vt.edu/jval/index.html`.

Ferdinando VILLA earned his doctorate in 1993 in ecology from the University of Parma, Italy. He has worked on theoretical community ecology, island biogeography, and landscape ecology. He is currently studying the island biogeography of small, tourism-disturbed islands with individual-oriented simulation models. Institute of Ecology, University of Parma, Viale delle Scienze, 43100 Parma, Italy. 39 (521) 905 615; FAX 39 (521) 905 402. villa@eagle.bio.unipr.it.

Glenda M. WARDLE obtained her undergraduate and master's degrees in botany from the University of Auckland, New Zealand. At the University of Chicago, she studied in the Committee of Evolutionary Biology, and her doctoral research combined demographic models and estimates of selection in natural populations to explain the maintenance of life-history diversity in the plant *Campanula americana* (Campanulaceae). Her research interests include evolutionary ecology, population dynamics, and plant life-history evolution. School of Biological Sciences A08, University of Sydney, NSW 2006, Australia. gwardle@bio.usyd.edu.

Simon N. WOOD's current research includes providing theoretical models of biological control, modeling competition in changing climates, and developing methods of inference for structured populations. Statistical Ecology Group, Mathematical Institute, University of St. Andrews KY16 9SS, United Kingdom. 44 (1334) 463799; FAX 44 (1334) 463748. snw@st-and.ac.uk.

# Index

weak resonance, 47
invariant loop, 230
Naimark-Sacker, 46, 230
of positive equilibria, 218
parameter, 213
period-doubling, 54, 228, 318,
    378, 606
pitchfork, 46, 228
point, 213
predator-prey model and, 601
saddle-node, 46, 227
spectrum, 218, 219
stability, 241
stable, 223, 225
subcritical, 51, 223, 225, 314
supercritical, 51, 223, 225
to the left, 223, 225
to the right, 223
transcritical, 45, 48, 227
types of, 45
unstable, 223, 225
vertical, 218
biotic-abiotic coupling, 435
birth-death model, 462
blowfly, 99, 589–599, 611
*Boophilus*, *see* ticks, cattle
bottleneck, 599
boundary condition, 141–142,
    146, 155, 160, 161, 164, 165,
    167–170, 174, 175, 177, 192,
    193, 200
breakpoints, 464

*Calathea ovandensis*, 249, 256
California red scale, 108
*Cancer magister*, 375
cannibalism, 226, 232, 304–306,
    388–394
in Dungeness crab, 378
recruitment and, 388
*Tribolium*, 392

carbon cycling, 415
carrying capacity, 305
chaos, 54, 230, 318, 320, 398,
    463, 592, 600
effect of adult survival, 337
characteristic equation, 27, 155,
    157, 158, 181–184, 198–202
predator-prey model, 603
circulation, 421
circulation pattern, *see* marine ecosystems
class-transition matrix, 207
cline, 355–368
cohort mixing
equilibrium population size,
    339
rate of genetic change, 339
stage-structured model with
    genetic variability, 334
comparative statics, 457
compensation, 40
competition
interspecific, 303, 520, 615
intraspecific, 237
size structure and, 523
symmetric and asymmetric,
    520, 523
symmetric, 515
conservation, 451, 471–508
corn oil sensitive (*cos*) mutant, *see* *T. castaneum*
coupling
abiotic-biotic, 435
physical-biological, 434
courtship period
battle of the sexes, 347
cross-validation, 580
cycles
2-cycle, 46, 49, 324, 373
4-cycle, 49
amplitude, 602
decaying, 378